M&F, Second Edition
David Knox

Executive Editor: Mark Kerr

Acquisitions Editor: Seth Dobrin

Senior Developmental Editor: Robert Jucha

Assistant Editor: Mallory Ortberg

Editorial Assistant: Nicole Bator

Media Editor: John Chell

Senior Brand Manager: Liz Rhoden

Senior Market Development Manager: Michelle Williams

Product Development Manager, 4LTR Press: Steven E. Joos

Senior Content Project Manager: Cheri Palmer

Senior Art Director: Caryl Gorska

Manufacturing Planner: Judy Inouye

Rights Acquisitions Specialist: Roberta Broyer

Production Service and Composition: Lachina Publishing Services

Photo Researcher: Q2a/Bill Smith

Text Researcher: PreMediaGlobal

Illustrator: Lachina Publishing Services

Text and Cover Designer: Riezebos Holzbaur

Cover Image: © Steve Hix/Somos Images/ Corbis

Inside Front Cover: © iStockphoto.com/ sdominick; © iStockphoto.com/alexsl; © iStockphoto.com/A-Digit

Page i: © iStockphoto.com/CostinT; © iStockphoto.com/photovideostock; © iStockphoto.com/Leontura

Back Cover: © iStockphoto.com/René Mansi

For product information and technology assistance, contact us at
Cengage Learning Customer & Sales Support, 1-800-354-9706

For permission to use material from this text or product, submit all requests online at **cengage.com/permissions** Further permissions questions can be emailed to **permissionrequest@cengage.com**

Library of Congress Control Number: 2012943409

ISBN-13: 978-1-133-58791-0

ISBN-10: 1-133-58791-7

Wadsworth
20 Davis Drive
Belmont, CA 94002-3098
USA

Cengage Learning is a leading provider of customized learning solutions with office locations around the globe, including Singapore, the United Kingdom, Australia, Mexico, Brazil, and Japan. Locate your local office at **www.cengage.com/global.**

Cengage Learning products are represented in Canada by Nelson Education, Ltd.

To learn more about Wadsworth, visit **www.cengage.com/wadsworth.**

Purchase any of our products at your local college store or at our preferred online store **www.CengageBrain.com.**

Printed in the United States of America
2 3 4 5 6 7 16 15 14 13

Brief Contents

1 Marriage and Family: An Introduction **2**

2 Gender **30**

3 Communication **50**

4 Singlehood, Hanging Out, Hooking Up, and Cohabitation **68**

5 Love and Selecting a Partner **82**

6 Sexuality in Relationships **112**

7 Marriage Relationships **130**

8 Same-Sex Couples and Families **148**

9 Work, Marriage, and Family **168**

10 Planning for Children **182**

11 Parenting **200**

12 Stress and Crisis in Relationships **218**

13 Abuse in Relationships **236**

14 Divorce and Remarriage **252**

15 Relationships in the Later Years **284**

References **302**

Name Index **325**

Subject Index **332**

4LTR Press solutions are designed for today's learners through the continuous feedback of students like you. Tell us what you think about **M&F** and help us improve the learning experience for future students.

YOUR FEEDBACK MATTERS.

Complete the Speak Up
survey in CourseMate at
www.cengagebrain.com

 Follow us at
www.facebook.com/4ltrpress

Contents

1 Marriage and Family: An Introduction 2

1-1 Marriage 2
- 1-1a Elements of Marriage 4
- 1-1b Types of Marriage 5

1-2 Family 6
- 1-2a Definitions of Family 6
- 1-2b Types of Families 7

1-3 Changes in Marriage and the Family 9
- 1-3a The Industrial Revolution and Family Change 9
- 1-3b Changes in the Last 60 Years 10

1-4 Theoretical Frameworks for Viewing Marriage and the Family 12
- 1-4a Social Exchange Framework 12
- 1-4b Family Life Course Development Framework 12
- 1-4c Structure-Function Framework 13
- 1-4d Conflict Framework 14
- 1-4e Symbolic Interaction Framework 15
- 1-4f Family Systems Framework 16
- 1-4g Feminist Framework 16

1-5 Choices in Relationships— View of the Text 17
- 1-5a Facts About Choices in Relationships 17
- 1-5b Global, Structural/Cultural, and Media Influences on Choices 20
- 1-5c Other Influences on Relationship Choices 24

1-6 Research—Process and Evaluation 25
- 1-6a Steps in the Research Process 25
- 1-6b Evaluating Research in Marriage and the Family 26

© aceshot1/Shutterstock.com

2 Gender 30

2-1 Terminology of Gender Roles 31
- 2-1a Sex 31
- 2-1b Gender 32
- 2-1c Gender Identity 34
- 2-1d Transgenderism 34
- 2-1e Gender Roles 35
- 2-1f Gender Role Ideology 35

2-2 Theories of Gender Role Development 36
- 2-2a Biosocial 36
- 2-2b Social Learning 36
- 2-2c Identification 37
- 2-2d Cognitive-Developmental 37

2-3 Agents of Socialization 37
- 2-3a Family 38
- 2-3b Race/Ethnicity 38
- 2-3c Peers 38
- 2-3d Religion 39
- 2-3e Education 39

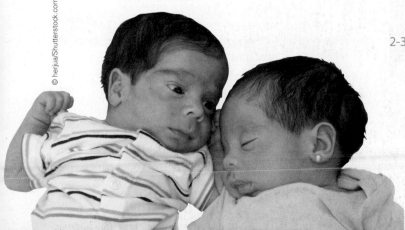

© herjua/Shutterstock.com

2-3f   Economy 39
2-3g   Mass Media 39

2-4   Consequences of Traditional Gender Role Socialization 40

2-4a   Consequences of Traditional Female Role Socialization 40

2-4b   Consequences of Traditional Male Role Socialization 43

2-5   Changing Gender Roles 46

2-5a   Androgyny 46
2-5b   Gender Role Transcendence 47
2-5c   Gender Postmodernism 48

3 Communication 50

3-1   The Nature of Interpersonal Communication 51

3-2   Conflicts in Relationships 52

3-2a   Inevitability of Conflict 52
3-2b   Benefits of Conflict 52
3-2c   Sources of Conflict 53
3-2d   Styles of Conflict 54

3-3   Principles and Techniques of Effective Communication 55

3-4   Self-Disclosure, Lying, Secrets, and Cheating 58

3-4a   Self-Disclosure 58
3-4b   Lying 59
3-4c   Secrets 59
3-4d   Cheating 60

3-5   Gender Differences in Communication 60

3-6   Theories Applied to Relationship Communication 61

3-6a   Symbolic Interactionism 61
3-6b   Social Exchange 61

3-7   Fighting Fair: Seven Steps in Conflict Resolution 62

3-7a   Address Recurring, Disturbing Issues 62

Sean Justice/Jupiterimages

3-7b   Identify New Desired Behaviors 63
3-7c   Identify Perceptions to Change 63
3-7d   Summarize Your Partner's Perspective 63
3-7e   Generate Alternative Win-Win Solutions 63
3-7f   Forgive 65
3-7g   Be Alert to Defense Mechanisms 66
3-7h   When Silence is Golden 67

4 Singlehood, Hanging Out, Hooking Up, and Cohabitation 68

© iStockphoto.com/craftvision

4-1   Singlehood 68

4-1a   Individuals Are Delaying Marriage Longer 69
4-1b   Social Movements and the Acceptance of Singlehood 71
4-1c   Alternatives to Marriage Project 72
4-1d   Legal Blurring of the Married and Unmarried 72

4-2   Categories of Singles 72

4-2a   Never-Married Singles 72
4-2b   Divorced Singles 73
4-2c   Widowed Singles 74

4-3 Ways of Finding a Partner 74

4-3a *Hanging Out* 74
4-3b *Hooking Up* 75
4-3c *The Internet—Meeting Online: Upside and Downside* 76
4-3d *Video Chatting* 78
4-3e *Speed-Dating: The Eight-Minute Date* 78

4-3f *High-End Matchmaking* 79
4-3g *International Dating* 79

4-4 Cohabitation 79

4-4a *Definitions of Cohabitation* 80
4-4b *Eight Types of Cohabitation Relationships* 80

5 Love and Selecting a Partner 82

5-1 Ways of Conceptualizing Love 84

5-1a *Love Styles* 84
5-1b *Romantic Versus Realistic Love* 85
5-1c *Triangular View of Love* 86

5-2 Love in Social and Historical Context 87

5-2a *Social Control of Love* 87
5-2b *Love in Medieval Europe—From Economics to Romance* 88
5-2c *Love in Colonial America* 89

5-3 How Love Develops in a New Relationship 89

5-3a *Social Conditions for Love* 89
5-3b *Body Type Condition for Love* 89
5-3c *Psychological Conditions for Love* 90
5-3d *Physiological and Cognitive Conditions for Love* 91
5-3e *Other Factors Associated With Falling in Love* 92

5-4 Jealousy in Relationships 92

5-4a *Types of Jealousy* 92
5-4b *Causes of Jealousy* 92
5-4c *Consequences of Jealousy* 94

5-5 Cultural Restrictions on Whom an Individual can Love or Marry 95

5-5a *Endogamy* 96
5-5b *Exogamy* 96

5-6 Sociological Factors Operative in Partnering 96

5-6a *Homogamy—Twelve Factors* 96

5-7 Psychological Factors Operative in Partnering 100

5-7a *Complementary-Needs Theory* 100
5-7b *Exchange Theory* 101
5-7c *Parental Characteristics* 102
5-7d *Personality Types* 102

5-8 **Sociobiological Factors Operative in Partnering 104**

    5-8a *Definition of Sociobiology* 104

    5-8b *Criticisms of the Sociobiological Perspective* 104

5-9 **Engagement 104**

    5-9a *Asking Specific Questions* 105

    5-9b *Visiting Your Partner's Parents* 105

    5-9c *Participating in Premarital Education Programs* 105

5-10 **Marrying for the Wrong Reasons 106**

    5-10a *Rebound* 106

    5-10b *Escape* 106

    5-10c *Unplanned Pregnancy* 106

    5-10d *Psychological Blackmail* 107

    5-10e *Insurance Benefits* 107

    5-10f *Pity* 107

    5-10g *Filling a Void* 107

5-11 **Consider Calling Off the Wedding If . . . 108**

    5-11a *Age 18 or Younger* 108

    5-11b *Known Partner Less Than Two Years* 108

    5-11c *Abusive Relationship* 109

    5-11d *Numerous Significant Differences* 109

    5-11e *On-and-Off Relationship* 109

    5-11f *Parental Disapproval* 110

    5-11g *Low Sexual Satisfaction* 110

6 Sexuality in Relationships 112

6-1 **Sexual Values 112**

6-2 **Alternative Sexual Values 114**

    6-2a *Absolutism* 114

    6-2b *Relativism* 116

    6-2c *Friends With Benefits* 117

    6.2d *Hedonism* 117

6-3 **Sexual Double Standard 117**

6-4 **Sources of Sexual Values 118**

6-5 **Gender Differences in Sexuality 119**

    6-5a *Gender Differences in Sexual Beliefs* 119

    6-5b *Gender Differences in Sexual Behavior* 119

6-6 **Pheromones and Sexual Behavior 121**

6-7 **Sexuality in Relationships 122**

    6-7a *Sexual Relationships Among Never-Married Individuals* 122

    6-7b *Sexual Relationships Among Married Individuals* 122

    6-7c *Sexual Relationships Among Divorced Individuals* 122

6-8 **Safe Sex: Avoiding Sexually Transmitted Infections 123**

    6-8a *Transmission of HIV and High-Risk Behaviors* 123

    6-8b *STI Transmission—The Illusion of Safety in a "Monogamous" Relationship* 124

    6-8c *Prevention of HIV and STI Transmission* 125

6-9 **Sexual Fulfillment: Some Prerequisites 126**

    6-9a *Self-Knowledge, Self-Esteem, and Health* 126

    6-9b *A Good Relationship, Positive Motives* 126

    6-9c *An Equal Relationship* 127

    6-9d *Open Sexual Communication, Sexual Self-Disclosure, and Feedback* 127

    6-9e *Having Realistic Expectations* 127

    6-9f *Avoiding Spectatoring* 128

    6-9g *Debunking Sexual Myths* 129

7 Marriage Relationships 130

7-1 **Individual Motivations for Marriage 131**

    7-1a *Love* 131

    7-1b *Personal Fulfillment* 132

    7-1c *Companionship* 132

    7-1d *Parenthood* 132

    7-1e *Economic Security* 132

© Serg Zastavkin/Shutterstock.com

7-2 **Societal Functions of Marriage 133**

7-3 **Marriage as a Commitment 133**

    7-3a Person-to-Person Commitment 134
    7-3b Family-to-Family Commitment 134
    7-3c Couple-to-State Commitment 134

7-4 **Marriage as a Rite of Passage 134**

    7-4a Weddings 135
    7-4b Honeymoons 136

7-5 **Changes After Marriage 136**

    7-5a Legal Changes 136
    7-5b Personal Changes 136
    7-5c Friendship Changes 136
    7-5d Relational Changes 137
    7-5e Parents and In-Law Changes 138
    7-5f Financial Changes 139

7-6 **Diversity in Marriage 139**

    7-6a Age-Discrepant Relationships and
    Marriages 139
    7-6b Interracial Marriages 140
    7-6c Interreligious Marriages 141

privilege /Shutterstock.com

    7-6d Cross-National Marriages 141
    7-6d Military Marriages 142

7-7 **Marital Success 145**

    7-7a Marital Happiness Across Time 147

8 Same-Sex Couples and Families   148

8-1 **Prevalence of Homosexuality, Bisexuality, and Same-Sex Couples 151**

    8-1a Problems Associated With Identifying and Classifying Sexual Orientation 151
    8-1b Prevalence of Homosexuality, Heterosexuality, and Bisexuality 152
    8-1c Prevalence of Same-Sex Couple Households 152

8-2 **Origins of Sexual-Orientation Diversity 153**

    8-2a Beliefs About What "Causes" Homosexuality 153

    8-2b Can Homosexuals Change Their Sexual Orientation? 153

8-3 **Heterosexism, Homonegativity, Homophobia, and Biphobia 154**

    8-3a Homonegativity and Homophobia 155
    8-3b Biphobia 155

8-4 **Gay, Lesbian, Bisexual, and Mixed-Orientation Relationships 156**

    8-4a Relationship Satisfaction 156
    8-4b Conflict Resolution and Breakup Aftermath 157
    8-4c Monogamy and Sexuality 157
    8-4d Relationships of Bisexuals 158
    8-4e Mixed-Orientation Relationships 158

8-5 **Legal Recognition and Support of Same-Sex Couples and Families 159**

    8-5a Decriminalization of Sodomy 159
    8-5b Registered Partnerships, Civil Unions, and Domestic Partnerships 159
    8-5c Same-Sex Marriage 160

8-6 **LGBT Parenting Issues 163**

    8-6a Development and Well-Being of Children With Gay or Lesbian Parents 163
    8-6b Discrimination in Child Custody, Visitation, Adoption, and Foster Care 164

8-7 **Effects of Antigay Bias and Discrimination on Heterosexuals 165**

© Chelsea Curry

9 Work, Marriage, and Family   168

9-1   **Effects of Employment on Spouses 168**

9-1a   *Money as Power in a Couple's Relationship* 168
9-1b   *Working Wives* 170
9-1c   *Wives Who "Opt Out"* 171
9-1d   *Types of Dual-Career Marriages* 171

9-2   **Effects of Employment on Children 172**

9-2a   *Quality Time* 172
9-2b   *Day-Care Considerations* 173

9-3   **Balancing Work and Family 174**

9-3a   *Superperson Strategy* 174
9-3b   *Cognitive Restructuring* 174
9-3c   *Delegation of Responsibility and Limiting Commitments* 174
9-3d   *Time Management* 176
9-3e   *Role Compartmentalization* 176

9-4   **Debt 176**

9-4a   *Income Distribution and Poverty* 177
9-4b   *Effects of Poverty on Marriages and Families* 178
9-4c   *Global Inequality* 178
9-4d   *Being Wise About Credit* 179
9-4e   *Credit Rating and Identity Theft* 179
9-4f   *Discussing Debt and Money: Do It Now* 180

© Ryan McVay/Jupiter Images

10 Planning for Children   182

10-1   **Do You Want to Have Children? 184**

10-1a   *Social Influences Motivating Individuals to Have Children* 184
10-1b   *Individual Motivations for Having Children* 185
10-1c   *Lifestyle Changes and Economic Costs of Parenthood* 185

10-2   **How Many Children Do You Want? 186**

10-2a   *Child-free Marriage* 186
10-2b   *One Child* 186
10-2c   *Two Children* 186
10-2d   *Three Children* 186
10-2e   *Four Children—New Standard for the Affluent?* 187

10-3   **Teenage Motherhood 187**

10-3a   *Problems Associated With Teenage Motherhood* 188

10-4   **Infertility 188**

10-4a   *Causes of Infertility* 189
10-4b   *Assisted Reproductive Technology* 189

10-5   **Planning for Adoption 192**

10-5a   *Demographic Characteristics of People Seeking to Adopt a Child* 192
10-5b   *Characteristics of Children Available for Adoption* 192

IT'S A BOY

© rSnapshotPhotos/Shutterstock.com

10-5c Costs of Adoption 193
10-5d Transracial Adoption 193
10-5e Open Versus Closed Adoptions 193

10-6 **Foster Parenting 194**

10-7 **Abortion 195**

10-7a Incidence of Abortion 195

10-7b Reasons for an Abortion 195
10-7c Pro-Life and Pro-Choice Abortion
         Positions 197
10-7d Physical Effects of Abortion 197
10-7e Psychological Effects of Abortion 197
10-7f Postabortion Attitudes of Men 198

11 Parenting **200**

11-1 **Roles Involved in Parenting 200**

11-2 **Choices Perspective of Parenting 205**

11-2a Nature of Parenting Choices 205
11-2b Five Basic Parenting Choices 205

11-3 **Transition to Parenthood 205**

11-3a Transition to Motherhood 205
11-3b Transition to Fatherhood 207
11-3c Transition from a Couple to a Family 208

11-4 **Parenthood: Some Facts 208**

11-4a Views of Children Differ Historically 209
11-4b Each Child Is Unique 209
11-4c Birth Order Effects on Personality 209
11-4d Parents Are Only One Influence in a Child's
         Development 210
11-4e Parenting Styles Differ 211

11-5 **Principles of Effective Parenting 211**

11-5a Give Time, Love, Praise, and
         Encouragement 212
11-5b Be Realistic 212
11-5c Avoid Overindulgence 212
11-5d Monitor Activities and Drug Use 212
11-5e Set Limits and Discipline Children for
         Inappropriate Behavior 213
11-5f Provide Security 213
11-5g Encourage Responsibility 213

11-5h Teach Emotional Competence 214
11-5i Provide Sex Education and Teach
         Nonviolence 214
11-5j Establish Norm of Forgiveness 214
11-5k Keep Children Connected With Nature 214
11-5l Respond to the Teen Years Creatively 215

11-6 **Single-Parenting Issues 215**

11-6a Single Mothers by Choice 215
11-6b Challenges Faced by Single Parents 216

12 Stress and Crisis in Relationships **218**

12-1 **Personal Stress and Crisis Events 218**

12-1a Definitions of Stress and Crisis
         Events 218
12-1b Resilient Families 220

12-2 **Positive Stress-
       Management
       Strategies 220**

12-2a Changing Basic Values
         and Perspective 220
12-2b Exercise 221
12-2c Friends and
         Relatives 221
12-2d Love 221

12-2e Religion and Spirituality 222
12-2f Humor 222
12-2g Sleep 222
12-2h Biofeedback 222
12-2i Deep Muscle Relaxation 223
12-2j Education 223
12-2k Pets 223

12-3 **Harmful Stress-Management
       Strategies 223**

12-4 **Family Crisis Examples 224**

12-4a Physical Illness and Disability 224
12-4b Mental Illness 225

12-4c Middle-Age Crazy (Midlife Crisis) 226
12-4d Extramarital Affair (and Successful Recovery) 226
12-4e Unemployment 231

12-4f Substance Abuse 232
12-4g Death of a Family Member 234
12-4h Suicide of a Family Member 234

13 Abuse in Relationships 236

13-1 **Nature of Relationship Abuse 236**

13-1a Violence 237
13-1b Emotional Abuse 238
13-1c Female Abuse of Partner 239
13-1d Stalking in Person 239
13-1e Stalking Online—Cybervictimization 240

13-2 **Explanations for Violence and Abuse in Relationships 240**

13-2a Cultural Factors 240
13-2b Community Factors 242
13-2c Individual Factors 242
13-2d Family Factors 243

13-3 **Sexual Abuse in Undergraduate Relationships 244**

13-3a Acquaintance and Date Rape 244
13-3b Rophypnol—The Date Rape Drug 246

13-4 **Abuse in Marriage Relationships 246**

13-4a General Abuse in Marriage 246
13-4b Rape in Marriage 246

13-5 **Effects of Abuse 246**

13-5a Effects of Partner Abuse on Victims 248
13-5b Effects of Partner Abuse on Children 248

© Nonstock/Jupiterimages

13-6 **The Cycle of Abuse 248**

13-6a Why People Stay in Abusive Relationships 249
13-6b How to Leave an Abusive Relationship 249
13-6c Strategies to Prevent Domestic Abuse 250
13-6d Treatment of Partner Abusers 251

14 Divorce and Remarriage 252

14-1 **Macro Factors Contributing to Divorce 252**

14-1a Increased Economic Independence of Women 253
14-1b Changing Family Functions and Structure 254
14-1c Liberal Divorce Laws 254
14-1d Prenuptial Agreements and the Internet. 254
14-1e Fewer Moral and Religious Sanctions 254
14-1f More Divorce Models 254
14-1g Mobility and Anonymity 255
14-1h Social Class, Ethnicity, and Culture 255

14-2 **Micro Factors Contributing to Divorce 255**

14-2a Falling Out of Love 255
14-2b Spending Limited Time Together 256
14-2c Decreasing Positive Behaviors 256

14-2d Having an Affair 256
14-2e Having Poor Conflict Resolution Skills 256
14-2f Changing Values 256
14-2g Onset of Satiation 256
14-2h Having the Perception That One Would Be Happier If Divorced 258

14-3 **Ending an Unsatisfactory Relationship 258**

14-3a Remaining Unhappily Married 259

14-4 **Gender Differences in Filing for Divorce 259**

14-5 **Consequences for Spouses Who Divorce 260**

14-5a Psychological Consequences of Divorce 260
14-5b Recovering From a Broken Heart 260
14-5c Financial Consequences 262
14-5d Fathers' Separation From Children 262

14-5e   Shared Parenting
         Dysfunction 263
14-5f   Parental Alienation 263

14-6 **Effects of Divorce on Children 264**

14-6a   Who Gets the
         Children? 265
14-6b   Minimizing Negative
         Effects of Divorce on
         Children 266

14-7 **Prerequisites for Having a "Successful" Divorce 268**

14-8 **Divorce Prevention 270**

14-9 **Remarriage 271**

14-9a   Remarriage for Divorced
         Individuals 271
14-9b   Preparation for
         Remarriage 272
14-9c   Issues of
         Remarriage 272
14-9d   Remarriage for the
         Widowed 273
14-9e   Stages of Involvement
         With a New
         Partner 274

14-10 **Stepfamilies 274**

14-10a   Definition and Types of
         Stepfamilies 275
14-10b   Myths of
         Stepfamilies 275
14-10c   Unique Aspects of
         Stepfamilies 275
14-10d   Stepfamilies in
         Theoretical Perspective 278
14-10e   Stages in Becoming a Stepfamily 279

14-11 **Strengths of Stepfamilies 279**

14-11a   Exposure to a Variety of Behavior
         Patterns 279
14-11b   Happier Parents 280
14-11c   Opportunity for a New Relationship With
         Stepsiblings 280
14-11d   More Objective Stepparents 280

14-12 **Developmental Tasks for Stepfamilies 280**

14-12a   Acknowledge Losses and Changes 280
14-12b   Nurture the New Marriage
         Relationship 281
14-12c   Integrate the Stepfather Into
         the Child's Life 281
14-12d   Allow Time for the Relationship Between
         the New Partner and One's Children to
         Develop 281
14-12e   Have Realistic Expectations 282
14-12f   Accept Your Stepchildren 282
14-12g   Establish Your Own Family Rituals 282
14-12h   Decide About Money 282
14-12i   Give Parental Authority to Your Spouse/
         Coparent 282
14-12j   Support the Children's Relationship With
         Their Absent Parent 283
14-12k   Cooperate With the Children's Biological
         Parent and Coparent 283
14-12l   Support the Children's Relationship With
         Grandparents 283
14-12m   Participate in a Stepfamily Education
         Web-based Program 283

AISPIX by Image Source/Shutterstock.com

15 Relationships in the Later Years 284

15-1 **Age and Ageism 285**
  15-1a The Concept of Age 285
  15-1b Ageism 287
  15-1c Theories of Aging 289

15-2 **Caring for the Frail Elderly—the "Sandwich Generation" 289**

15-3 **Issues Confronting the Elderly 290**
  15-3a Income 290
  15-3b Housing 291
  15-3c Physical Health 291
  15-3d Mental Health 292
  15-3e Retirement 292
  15-3f Sexuality 294

15-4 **Successful Aging 295**

15-5 **Relationships and the Elderly 296**
  15-5a Relationships Between Elderly Spouses 296
  15-5b Renewing an Old Love Relationship 296
  15-5c Use of Technology to Maintain Relationships 298
  15-5d Relationships With Siblings at Age 85 and Beyond 298
  15-5e Relationships With Children at Age 85 and Beyond 298

15-6 **The End of Life 298**
  15-6a Death of a Spouse 298
  15-6b Involvement With New Partners at Age 80 and Beyond 299
  15-6c Preparing for Death 300

References 302

Name Index 325

Subject Index 332

Marriage and Family: An Introduction

"Call it a **clan,** call it a **network,** call it a **tribe,** call it a **family.** Whatever you call it, whoever you are, **you need one.**"

—JANE HOWARD, JOURNALIST

SECTIONS

1-1 Marriage

1-2 Family

1-3 Changes in Marriage and the Family

1-4 Theoretical Frameworks for Viewing Marriage and the Family

1-5 Choices in Relationships—View of the Text

1-6 Research—Process and Evaluation

An aging professor of marriage and family revealed a pattern of always giving the same tests each year. Colleagues thought the teacher both lazy and unfair because the old tests would get out and be used by new students. But the teacher justified giving the same exams every year on the premise that although the questions were the same, the answers kept changing. For example, in 1950, 9% of all households were single households. More recently, the percentage has changed to 28%. The rush to marry soon after high school has been replaced by great caution and delay. Sociologist Klinenberg (2012) noted that "living alone helps to pursue sacred modern values such as individual freedom, personal control and self realization." In this chapter we review the meaning of marriage and family, how these institutions have changed, a choices framework (as well as other theoretical views of the family), and how to be cautious about reading research studies in marriage and family.

1-1 Marriage

All of us were born into a family and will end up in a family (however one defines this concept) of our own. "Raising a family" remains one of the top values in life for undergraduates. In a nationwide study of 203,967 undergraduates in 270 colleges

© aceshot1/Shutterstock.com

and universities, 73% identified raising a family as an essential objective (80% chose financial success as their top goal; Pryor et al., 2011). Although individuals today are delaying marriage, the overwhelming majority (in the past) ended up getting married—96% of U.S. adult women and men age 75 and older report having been married at least once (*Statistical Abstract of the United States*, 2012, Table 57).

Traditionally, **marriage** in the U.S. has been viewed as a legal relationship that binds a couple together for the reproduction, physical care, and socialization of children. In the United States, marriage is a legal contract between a heterosexual couple (although eight states now recognize same-sex marriages) and the state in which they reside that specifies the economic relationship between the couple (they become joint owners of their income and debt) and encourages sexual fidelity. The fine print of what marriage involves includes the following elements.

marriage a legal contract between a couple and the state in which they reside that regulates their economic and sexual relationship.

1-1a Elements of Marriage

Several elements comprise the meaning of marriage in the United States.

Legal Contract Marriage in our society is a legal contract into which two people of different sexes and legal age (usually 18 or older) may enter when they are not already married to someone else. The marriage license certifies that a legally empowered representative of the state married the individuals, often with two witnesses present. The marriage contract actually gives more power to the government and its control over the couple (Aulette, 2010). The government will dictate not only who may marry (e.g., heterosexuals in most states, age 18 or above, not already married) but also the conditions of divorce (e.g., alimony and child support).

Under the laws of the state, the license means that spouses will jointly own all future property acquired and that each will share in the estate of the other. In most states, whatever the deceased spouse owns is legally transferred to the surviving spouse at the time of death. In the event of divorce and unless the couple had a prenuptial agreement, the property is usually divided equally regardless of the contribution of each partner. The license also implies the expectation of sexual fidelity in the marriage. Though less frequent because of no-fault divorce, in some states, infidelity is a legal ground for both divorce and alimony.

The marriage license is also an economic license that entitles a spouse to receive payment from a health insurance company for medical bills if the partner is insured, to collect Social Security benefits at the death of the other spouse, and to inherit from the estate of the deceased. One of the goals of gay rights advocates who seek the legalization of marriage between homosexuals is that the couple will have the same rights and benefits as heterosexuals. Though the courts are reconsidering the definition of what constitutes a "family," the law is currently designed to protect spouses, not lovers or cohabitants. An exception is **common-law marriage,** in which a heterosexual couple who presents themselves as married will be regarded as legally married in those states (currently 14) that recognize such marriages.

Emotional Relationship Ninety-three percent of adults point to love as the top reason to get married. Other reasons include making a lifelong commitment (87%), having companionship (81%), and having children (59%; Pew Research Center, 2010). Love is also the top reason to stay married.

Forty percent of 2,922 undergraduates in one study reported that they would divorce if they no longer loved their spouse (Knox and Hall, 2010). This emphasis on love is not shared throughout the world, however. Individuals in other cultures (for example, India and Iran) do not require feelings of love to marry—love is expected to follow, not precede, marriage. In these countries, parental approval and similarity of religion, culture, and education are considered more important criteria for marriage than love.

Sexual Monogamy Marital partners expect sexual fidelity. Over half (54%) of 2,922 undergraduates agreed with the statement "I would divorce a spouse who had an affair," and over two thirds (67%) agreed that they would end a relationship with a partner who cheated on them (Knox and Hall, 2010).

Legal Responsibility for Children Although individuals marry for love and companionship, one of the most important reasons for the existence of marriage from the viewpoint of society is to legally bind a male and a female for the nurture and support of any children they may have. In our society, child rearing is the primary responsibility of the family, not the state.

Marriage is a relatively stable relationship that helps to ensure that children will have adequate care

© Tamera Rees/iStockphoto.com/© Michal Rozanski/iStockphoto.com

common-law marriage a marriage by mutual agreement between cohabitants without a marriage license or ceremony (recognized in some, but not all, states).

Table 1.1

Benefits of Marriage and Liabilities of Singlehood

	Benefits of Marriage	Liabilities of Singlehood
Health	Spouses have fewer hospital admissions, see a physician more regularly, and are "sick" less often.	Single people are hospitalized more often, have fewer medical checkups, and are "sick" more often.
Longevity	Spouses live longer than single people.	Single people die sooner than married people.
Happiness	Spouses report being happier than single people.	Single people report less happiness than married people.
Sexual satisfaction	Spouses report being more satisfied with their sex lives, both physically and emotionally.	Single people report being less satisfied with their sex lives, both physically and emotionally.
Money	Spouses have more economic resources than single people.	Single people have fewer economic resources than married people.
Lower expenses	Two can live more cheaply together than separately.	Cost is greater for two singles than one couple.
Drug use	Spouses have lower rates of drug use and abuse.	Single people have higher rates of drug use and abuse.
Connectedness	Spouses are connected to more individuals who provide a support system—partner, in-laws, etc.	Single people have fewer individuals upon whom they can rely for help.
Children	Rates of high school dropouts, teen pregnancies, and poverty are lower among children reared in two-parent homes.	Rates of high school dropouts, teen pregnancies, and poverty are higher among children reared by single parents.
History	Spouses develop a shared history across time with significant others.	Single people may lack continuity and commitment across time with significant others.
Crime	Spouses are less likely to be involved in crime.	Single people are more likely to be involved in crime.
Loneliness	Spouses are less likely to report loneliness.	Single people are more likely to report being lonely.

© Cengage Learning 2013

and protection, will be socialized for productive roles in society, and will not become the burden of those who did not conceive them. Even at divorce, the legal obligation of the noncustodial parent to the child is maintained through child-support payments.

Announcement/Ceremony The legal binding of a couple is often preceded by an announcement in the local newspaper and then followed by a formal ceremony in a church or synagogue. The presence of parents, siblings, and friends at the wedding helps to verify the commitment of the partners to each other and helps to marshal social and economic support to launch the couple into married life. Most people in our society decide to marry. They not only benefit from being married but also avoid the liabilities of being single (see Table 1.1).

1-1b Types of Marriage

Although we think of marriage in the United States as involving one man and one woman, other societies view marriage differently. **Polygamy** is a form of marriage involving more than two spouses. Polygamy occurs "throughout the world . . . and is found on all continents and among adherents of all world religions" (Zeitzen, 2008). There are three forms of polygamy: polygyny, polyandry, and pantagamy.

Polygyny **Polygyny** involves one husband and two or more wives and is practiced illegally in the United States by some religious fundamentalist groups. These groups are primarily in Arizona, New Mexico, and Utah (as well as Canada) and have splintered off from the Church of Jesus Christ of Latter-day Saints (commonly known as the Mormon Church). To be clear, the Mormon Church does not practice or condone polygyny. Those that split off from the Mormon Church represent only about 5% of Mormons in Utah. The largest offshoot is called the Fundamentalist Church of Jesus Christ of Latter-day Saints (FLDS). Members of the group feel that the practice of polygyny is God's will. Joe Jessop, age 88, is an elder of the FLDS. He has 5 wives, 46 children, and 239 grandchildren. Although the practice is illegal, polygynous individuals are rarely prosecuted because a husband will have only one legal wife and the

polygamy a generic term referring to a marriage involving more than two spouses.

polygyny a form of polygamy in which one husband has two or more wives.

© Stephan Gladieu/Getty Images

others will just live with and be a part of the larger family. Women are socialized to bear as many children as possible to build up the "celestial family" that will remain together for eternity (Anderson, 2010).

Polyandry Tibetan Buddhists foster yet another brand of polygamy, referred to as **polyandry**, in which one wife has two or more (up to five) husbands. These husbands, who may be brothers, pool their resources to support one wife. Polyandry is a much less common form of polygamy than polygyny. The major reason for polyandry is economic. A family that cannot afford wives or marriages for each of its sons may find a wife for the eldest son only. Polyandry allows the younger brothers to also have sexual access to the one wife that the family is able to afford.

Polamory **Polyamory** means "multiple loves" (poly means "many"; amorous means "love") and is a lifestyle in which lovers embrace the idea of having multiple emotional and sexual partners. During the mid-1800s, the Oneida Community of Oneida, New York, embraced a form of

polyandry a form of polygamy in which one wife has two or more husbands.

polyamory literally, multiple loves (poly means "many"; amorous means "love"); a lifestyle in which lovers embrace the idea of having multiple emotional and sexual partners.

pantagamy a group marriage in which each member of the group is "married" to the others.

family a group of two or more people related by blood, marriage, or adoption.

polyamory (called *complex marriage* in which every man was married to every woman). Today in Louisa, Virginia, 25% of the 100 members of Twin Oaks Intentional Community are polyamorous in that each partner may have several emotional/physical relationships with others at the same time. Although not legally married, these adults view themselves as emotionally bonded to each other and may even rear children together. Polyamory is not swinging, as polyamorous lovers are concerned about enduring, intimate relationships that include sex.

Pantagamy **Pantagamy** describes a group marriage in which each member of the group is "married" to the others. Pantagamy is a formal arrangement that was practiced in communes (for example, Oneida) of the 19th and 20th centuries. Pantagamy is, of course, illegal in the United States.

1-2 Family

Most people who marry choose to have children and become a family. However, the definition of what constitutes a family is sometimes unclear. This section examines how families are defined, their numerous types, and how marriages and families have changed in the past 60 years.

1-2a Definitions of Family

The U.S. Census Bureau defines **family** as a group of two or more people related by blood, marriage, or adoption. This definition has been challenged because it does not include foster families or long-term couples (heterosexual or homosexual) that live together. The answer to the question "Who is family?" is important because access to resources such as health care, Social Security, and retirement benefits is involved. In some cases, families are being defined by function rather than by structure—what is the level of emotional and financial commitment and interdependence? How long have they lived together? Do the partners view themselves as a family?

Sociologically, a family is defined as a kinship system of all relatives living together or recognized as a social unit, including adopted people. The family is regarded as the basic social institution because of its important functions of procreation and socialization, and because it is found in some form in all societies.

Same-sex couples (e.g. Ellen DeGeneres and her partner, Portia de Rossi) define themselves as family. Eight states (and the number continues to increase) recognize marriages between same-sex individuals. Short of marriage, other states (the number of which keeps changing) recognize committed gay relationships as **civil unions** (pair-bonded relationships given legal significance in terms of rights and privileges). In France, civil unions have become more popular than marriage—there are now two civil unions for every three marriages. In France a civil union can be dissolved with a registered letter (Sayare & De La Baume, 2010).

Although other states typically do not recognize same-sex marriages or civil unions (and, thus, people moving from these states to another state lose the privileges associated with marriage), about 25 (the number keeps changing) cities and countries (including Canada) recognize some form of domestic partnership. **Domestic partnerships** are relationships in which cohabitating individuals are given official recognition by a city or corporation so as to receive partner benefits (for example, health insurance). The Walt Disney Company recognizes domestic partnerships.

Some individuals view their pets as part of their family. Examples of treating pets like children include

hanging a stocking and/or buying presents for the pet at Christmas, buying "clothes" for one's pet, and leaving money in one's will for the care of a pet. Some pet owners buy accident insurance—Progressive© car insurance covers pets. Pets are also included in divorce agreements—the "parents" are granted custody and visitation rights (Gregory, 2010).

1-2b Types of Families

There are various types of families.

Family of Origin Also referred to as the **family of orientation**, this is the family into which you were born or the family in which you were reared. It involves you, your parents, and your siblings. When you go to your parents' home for the holidays, you return to your **family of origin**. Your experiences in your family of origin have an impact on your own relationships. Individuals whose parents had a high-quality relationship are likely to have a similar high-quality marriage and to have a familistic orientation (Chiu and Busby, 2010).

Pets are routinely included in family photos.

civil union
a pair-bond, or monogamous relationship, given legal significance in terms of rights and privileges.

domestic partnership
a relationship in which individuals who live together are emotionally and financially interdependent and are given some kind of official recognition by a city or corporation so as to receive partner benefits.

family of orientation the family of origin into which a person is born.

family of origin the family into which an individual is born or reared, usually including a mother, father, and children.

Sisters are the most enduring of all relationships.

Siblings represent an important part of one's family of origin. Indeed, the relationship with one's siblings, particularly the sister–sister relationship, represents the most enduring relationship in a person's lifetime. And sisters who live near one another and who do not have children report the greatest amount of intimacy and contact (Meinhold et al., 2006).

Family of Procreation

The **family of procreation** represents the family that you will begin should you marry and have children. Of U.S. citizens living in the United States, 95.83% marry and establish their own family of procreation (*Statistical Abstract*, 2012, Table 57). Across the life cycle, individuals move from the family of orientation to the family of procreation.

Nuclear Family

The **nuclear family** refers to either a family of origin or a family of procreation. In practice, your nuclear family consists either of you, your parents, and your siblings, or of you, your spouse, and your children. Generally, one-parent households are not referred to as nuclear families. They are binuclear families if both parents are involved in the child's life or single-parent families if one parent is involved in the child's life and the other parent is totally out of the picture.

The nuclear family is a "universal social grouping" found in all of the 250 societies studied by sociologist George Peter Murdock (1949). The nuclear family channels not only sexual energy between two adult partners who reproduce but also cooperation in the care of offspring and their socialization to be productive members of society. "This universal social structure, produced through cultural evolution in every human society, as presumably the only feasible adjustment to a series of basic needs, forms a crucial part of the environment in which every individual grows to maturity" (p. 11).

Traditional, Modern, and Postmodern Families Sociologists have identified three central concepts of the family (Silverstein & Auerbach, 2005). The **traditional family** is the two-parent nuclear family, with the husband as breadwinner and wife as homemaker. The **modern family** is the dual-earner family, in which both spouses work outside the home. **Postmodern families** represent a departure from these models and include lesbian or gay couples and mothers who are single by choice, which emphasizes that a healthy family need not be heterosexual or include two parents.

Binuclear Family A **binuclear family** is a family in which the members live in two separate households. This family type is created when the parents of the children divorce and live separately, setting up two separate units, with the children remaining a part of each unit. Each of these units may also change again when the parents remarry and bring additional children into the respective units (**blended family**). Hence, the children may go from a nuclear family with both parents to a binuclear unit with parents living in separate homes to a blended family when parents remarry and bring additional children into the respective units.

Extended Family The **extended family** includes not only the nuclear family (or parts of it) but other relatives as well. These relatives include grandparents, aunts, uncles, and cousins. An example of an extended family living together would be a husband and wife, their children, and the husband's parents (the children's grandparents).

family of procreation the family a person begins by getting married and having children.

nuclear family family consisting of an individual, his or her spouse, and his or her children, or of an individual and his or her parents and siblings.

traditional family the two-parent nuclear family with the husband as breadwinner and wife as homemaker.

modern family the dual-earner family, in which both spouses work outside the home.

Table 1.2

Differences Between Marriage and the Family in the United States

Marriage	Family
Usually initiated by a formal ceremony.	Formal ceremony not essential.
Involves two people.	Usually involves more than two people.
Ages of the individuals tend to be similar.	Individuals represent more than one generation.
Individuals usually choose each other.	Members are born or adopted into the family.
Ends when one spouse dies or the couple is divorced.	Continues beyond the life of the individual.
Sex between spouses is expected and approved.	Sex between near kin is neither expected nor approved.
Requires a license.	No license is needed to become a parent.
Procreation expected.	Consequence of procreation.
Spouses are focused on each other.	Focus changes with addition of children.
Spouses can voluntarily withdraw from marriage.	Parents cannot remove themselves from their obliation to children via divorce.
Money in unit is spent on the couple.	Money is used for the needs of the children.
Recreation revolves around adults.	Recreation revolves around children.

Reprinted by permission of Dr. Lee Axelson.

Asians are more likely than European Americans to live with their extended families. Among Asians, the status of the elderly in the extended family derives from religion. Confucian philosophy, for example, prescribes that all relationships are of the superordinate–subordinate type—that is, husband–wife, parent–child, and teacher–pupil. For traditional Asians to abandon their elderly rather than include them in larger family units would be unthinkable. However, commitment to the elderly may be changing as a result of the westernization of Asian countries such as China, Japan, and Korea.

1-3 Changes in Marriage and the Family

Whatever family we experience today was different previously and will change yet again. A look back at some changes in marriage and the family follows.

1-3a The Industrial Revolution and Family Change

The Industrial Revolution refers to the social and economic changes that occurred when machines and factories, rather than human labor, became the dominant mode for the production of goods. Industrialization occurred in the United States during the early and mid-1800s and represents one of the most profound influences on the family.

Before industrialization, the family functioned as an economic unit that produced goods and services for its own consumption. Parents and children worked together in or near the home to meet the survival needs of the family. As the United States became industrialized, more men and women left the home to sell their labor for wages. The family was no longer a self-sufficient unit that determined its work hours. Rather, employers determined where and when family members would work. Whereas children in preindustrialized America worked on farms and contributed to the economic survival of the family, children in industrialized America became economic liabilities rather than assets. Child labor laws and

postmodern family nontraditional family emphasizing that a healthy family need not be heterosexual or have two parents.

binuclear family a family in which the members live in two separate households.

blended family (stepfamily) a family created when two individuals marry and at least one of them has a child or children from a previous relationship or marriage.

extended family the nuclear family or parts of it plus other relatives.

mandatory education removed children from the labor force and lengthened their dependence on parental support. Eventually, both parents had to work away from the home to support their children. The dual-income family had begun.

During the Industrial Revolution, urbanization occurred as cities were built around factories and families moved to the city to work in the factories. Living space in cities was crowded and expensive, which contributed to a decline in the birthrate and thus smaller families. The development of transportation systems during the Industrial Revolution made it possible for family members to travel to work sites away from the home and to move away from extended kin. With increased mobility, many extended families became separated into smaller nuclear family units consisting of parents and their children. As a result of parents leaving the home to earn wages and the absence of extended kin in or near the family household, children had less adult supervision and moral guidance. Unsupervised children roamed the streets, increasing the potential for crime and delinquency.

Industrialization also affected the role of the father in the family. Employment outside the home removed men from playing a primary role in child care and in other domestic activities. The contribution men made to the household became primarily economic.

Finally, the advent of industrialization, urbanization, and mobility is associated with the demise of familism and the rise of individualism. When family members functioned together as an economic unit, they were dependent on one another for survival and were concerned about what was good for the family. This familistic focus on the needs of the family has since shifted to a focus on self-fulfillment—individualism. Families from familistic cultures such as China who immigrate to the United States soon discover that their norms, roles, and values begin to alter to accommodate the industrialized, urbanized, individualistic behavior patterns and thinking. Individualism and the quest for personal fulfillment are thought to have contributed to high divorce rates, absent fathers, and parents spending less time with their children.

Hence, although the family is sometimes blamed for juvenile delinquency, violence, and

marriage-resilience perspective the view that changes in the institution of marriage are not indicative of a decline and do not have negative effects.

> With increased mobility, many extended families became separated into smaller nuclear family units consisting of parents and their children.

divorce, it is more accurate to emphasize changing social norms and conditions of which the family is a part. When industrialization takes parents out of the home so that they can no longer be constant nurturers and supervisors, the likelihood of aberrant acts by their children increases. One explanation for school violence is that absent, career-focused parents have failed to provide close supervision for their children.

1-3b Changes in the Last 60 Years

Enormous changes have occurred in marriage and the family in the last 60 years. Divorce has replaced death as the endpoint for the majority of marriages, marriage and intimate relations have become legitimate objects of scientific study, feminism and changes in gender roles in marriage have risen, and remarriages have declined (Amato et al., 2007). Other changes include the use of technology (e.g. text messaging) in establishing and maintaining relationships, a delay in age at marriage, and increased acceptance of singlehood, cohabitation, and child-free marriages. Even the definition of what constitutes a family is being revised, with some emphasizing that a durable emotional bond between individuals is the core of "family" and others insisting on a more legalistic view of connections by blood, marriage, or adoption. Table 1.3 contrasts marriages and families in the 1950s with marriages and families in 2015.

In spite of the persistent and dramatic changes in marriage and the family, marriage and the family continue to be resilient. From this **marriage-resilience perspective**, changes in the institution of marriage are neither negative nor indicative that marriage is in a state of decline. Indeed, these changes are thought to have "few negative consequences for adults, children, or the wider society" (Amato et al., 2007, p. 6).

Table 1.3

Changes in Marriages and Families From 1950 to 2015

	1950	2015
Family Relationship Values	Strong values for marriage and the family. Individuals who want to remain single or childfree are considered deviant, even pathological. Husband and wife should not be separated by jobs or careers.	Individuals who remain single or childfree experience social understanding and sometimes encouragement. Single and child-free people are no longer considered deviant or pathological but are seen as self-actuating individuals with strong job or career commitments. Husbands and wives can be separated for reasons of job or career and live in a commuter marriage. Married women in large numbers have left the role of full-time mother and housewife to join the labor market.
Gender Roles	Rigid gender roles, with men dominant and earning income while wives stay home, taking care of children.	Egalitarian gender roles with both spouses earning income. Greater involvement of men in fatherhood.
Sexual Values	Marriage is regarded as the only appropriate context for intercourse in middle-class America. Living together is unacceptable, and a child born out of wedlock is stigmatized. Virginity is sometimes exchanged for marital commitment.	For many, concerns about safer sex have taken precedence over the marital context for sex. Virginity is rarely exchanged for anything. Living together is regarded as not only acceptable but sometimes preferable to marriage. For some, unmarried, single parenthood is regarded as a lifestyle option. Hooking-up is the new courtship norm.
Homogamous Mating	Strong social pressure exists to date and marry within one's own racial, ethnic, religious, and social class group. Emotional and legal attachments are heavily influenced by obligation to parents and kin.	Dating and mating have become more heterogamous, with more freedom to select a partner outside one's own racial, ethnic, religious, and social class group. Attachments are more often by choice.
Cultural Silence on Intimate Relationships	Intimate relationships are not an appropriate subject for the media.	Individuals on talk shows, interviews, and magazine surveys are open about sexuality and relationships behind closed doors.
Divorce	Society strongly disapproves of divorce. Familistic values encourage spouses to stay married for the children. Strong legal constraints keep couples together. Marriage is forever.	Divorce has replaced death as the endpoint for a majority of marriages. Less stigma is associated with divorce. Individualistic values lead spouses to seek personal happiness. No-fault divorce allows for easy severance. Marriage is tenuous. Increasing numbers of children are being reared in single-parent households.
Familism versus Individualism	Families are focused on the needs of children. Mothers stay home to ensure that the needs of their children are met. Adult concerns are less important.	Adult agenda of work and recreation has taken on increased importance, with less attention being given to children. Children are viewed as more sophisticated and capable of thinking as adults, which frees adults to pursue their own interests. Day care is used regularly.
Homosexuality	Same-sex emotional and sexual relationships are a culturally hidden phenomenon. Gay relationships are not socially recognized.	Gay relationships are increasingly a culturally open phenomenon. Some definitions of the family include same-sex partners. Domestic partnerships are increasingly given legal status in some states. Same-sex marriage is a hot social and political issue. More states are legalizing same-sex marriage. "Don't ask, don't tell" has been repealed.
Scientific Scrutiny	Aside from Kinsey's, few studies are conducted on intimate relationships.	Acceptance of scientific study of marriage and intimate relationships.
Family Housing	Husbands and wives live in same house.	Husbands and wives may "live apart together" (LAT), which means that, although they are emotionally and economically connected, they (by choice) maintain two households, houses, condos, or apartments.
Technology	Nonexistent except phone.	Cell phones, texting, sexting.

Theoretical Frameworks for Viewing Marriage and the Family

Theoretical frameworks are a set of interrelated principles designed to explain a particular phenomenon and provide a point of view. The more common frameworks follow:

1-4a Social Exchange Framework

theoretical framework a set of interrelated principles designed to explain a particular phenomenon and to provide a point of view.

social exchange framework marital perspective in which spouses exchange resources and decisions are made on the basis of perceived profit and loss.

utilitarianism the doctrine holding that individuals rationally weigh the rewards and costs associated with behavioral choices.

family life course development the stages and process of how families change over time.

family life cycle stages which identify the various challenges faced by members of a family across time.

The **social exchange framework** is one of the most commonly used theoretical perspectives in marriage and the family. The framework views interactions/choices in terms of cost/profit and operates from a premise of **utilitarianism**—the theory that individuals rationally weigh the rewards and costs associated with behavioral choices. Each interaction between spouses, parents, and children can be understood in terms of each individual seeking the most benefits at the least cost so as to have the highest "profit" and avoid a "loss" (White & Klein, 2002). Both men and women marry because they perceive more benefits than costs for doing so. The person they marry also offers the highest profit and the lowest cost of all alternatives available.

A social exchange view of marital roles emphasizes that spouses negotiate the division of labor on the basis of exchange. For example, a man participates in child care in exchange for his wife's earning an income, which relieves him of the total financial responsibility. Social exchange theorists also emphasize that power in relationships is the ability to influence, and avoid being influenced by, the partner.

The various bases of power, such as money, the need for a partner, and brute force, may be expressed in various ways, including withholding resources, decreasing investment in the relationship, and violence.

1-4b Family Life Course Development Framework

The **family life course development** framework emphasizes the important role transitions of individuals that occur in different periods of life and in different social contexts. For example, young unmarried lovers may become cohabitants, then parents, grandparents, retirees, and widows. The family life cycle should be viewed as a basic set of stages through which not all individuals pass (e.g., the child-free) and in which there is great diversity, particularly in regard to race and education (e.g., African Americans are less likely to marry; the highly educated are less likely to divorce; Cherlin, 2010).

The family life course developmental framework has a basis in sociology (for example, role transitions), whereas the **family life cycle** has a basis in psychology, with its emphasis on stages that identify the various developmental tasks family members face across time (e.g., establishing the marital dyad, having chil-

> "Getting **married** is a way to **show** family and friends that you have a **successful personal life.** It's the ultimate merit badge."
>
> —ANDREW CHERLIN, SOCIOLOGIST

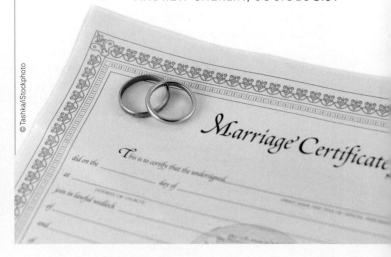

© Tashka/iStockphoto

Gay relationships are becoming more culturally accepted.

© David Knox

dren, etc.). If developmental tasks at one stage are not accomplished, functioning in subsequent stages will be impaired. For example, one of the developmental tasks of early marriage is to emotionally and financially separate from one's family of origin. If such separation does not take place, independence as individuals and as a couple is impaired.

1-4c Structure-Function Framework

The **structure-function framework** emphasizes how marriage and family contribute to society. Just as the human body is made up of different parts that work together for the good of the individual, society is made up of different institutions (family, education, economics, and so on) that work together for the good of society. **Functionalists** view the family as an institution with values, norms, and activities meant to provide stability for the larger society. Such stability depends on families performing various functions for society.

First, families replenish society with socialized members. Because our society cannot continue to exist without new members, we must have some way of ensuring a continuing supply. However, just having new members is not enough. We need socialized members—those who can speak our language and know the norms and roles of our society. Societies "rely on families to produce upright people who make for conscientious, law-abiding citizens lessening the demand for governmental policing and social services" (Girgis et al., 2011, p. 245).

Genie, a young girl who was discovered in the 1970s, had been kept in isolation in one room in her

California home for 12 years by her abusive father (James, 2008). She could barely walk and could not talk. Although provided intensive therapy at UCLA, and the object of thousands of dollars of funded research, Genie progressed only slightly. Today, she is in her mid-50s, institutionalized, and speechless. Her story illustrates the need for socialization; the legal bond of marriage and the obligation to nurture and socialize offspring help to ensure that this socialization will occur.

Second, marriage and the family promote the emotional stability of the respective spouses. Society cannot provide enough counselors to help us whenever we have problems. Marriage ideally provides in-residence counselors who are loving and caring partners with whom people share their most difficult experiences.

Children also need people to love them and to give them a sense of belonging. This need can be fulfilled in a variety of family contexts (two-parent families, single-parent families, extended families). The affective function of the family is one of its major offerings. No other institution focuses so completely on meeting the emotional needs of its members as marriage and the family.

Third, families provide economic support for their members. Although modern families are no longer self-sufficient economic units, they provide food, shelter, and clothing for their members. One need only consider the homeless in our society to be reminded of this important function of the family.

In addition to the primary functions of replacement, emotional stability, and economic support, other functions of the family include the following:

- **Physical care**—Families provide the primary care for their infants, children, and aging parents. Other agencies (neonatal units, day care centers, assisted-living residences) may help, but the family remains the primary and recurring caretaker. Spouses also support the physical health of one another by encouraging each other to take medications and to see the doctor.

structure-function framework
theoretical perspective that emphasizes how marriage and family contribute to the larger society.

functionalists
theorists who view the family as an institution with values, norms, and activities meant to provide stability for the larger society.

- **Regulation of sexual behavior**—Spouses are expected to confine their sexual behavior to each other, which reduces the risk of having children who do not have socially and legally bonded parents and of contracting or spreading sexually transmitted infections.

- **Status placement**—Being born in a family provides social placement of the individual in society. One's family of origin largely determines one's social class, religious affiliation, and future occupation. Prince William, the son of Prince Charles, was automatically in the upper class and destined to be in politics by virtue of being born into a political family.

- **Social control**—Spouses in high-quality, durable marriages provide social control for each other that results in less criminal behavior. Parole boards often note that the best guarantee that a person released from prison will not return to prison is a spouse who expects the partner to get a job and avoid criminal behavior and who reinforces these goals.

1-4d Conflict Framework

The **conflict framework** views individuals in relationships as competing for valuable resources. Conflict theorists recognize that family members have different goals and values that result in conflict. For example, a wife and mother who has devoted her life to staying home and taking care of her husband and children may decide to return to school or to seek full-time employment. This decision might be good for her personally, but her husband and children might not like it. Similarly, divorce may have a positive outcome for spouses in turmoil but a negative outcome for children, whose standard of living and access to the noncustodial parent are likely to decrease.

Conflict theorists also view conflict not as good or bad but as a natural and normal part of relationships. They regard conflict as necessary for change and growth of individuals, marriages, and families. Cohabitation relationships, marriages, and families all have the potential for conflict. Cohabitants are in conflict about commitment to marry, spouses are in conflict about the division of labor, and parents are in conflict with their children over rules such as curfew, chores, and homework. These three units may also be in conflict with other systems. For example, cohabitants are in conflict with the economic institution for health benefits for their partners. Similarly, employed parents are in conflict with their employers for flexible work hours, maternity or paternity benefits, and day care facilities.

Conflict theorists emphasize conflict over power in relationships. Premarital partners, spouses, parents, and teenagers use power to control one another. The reluctance of some courtship partners to make a marital commitment is an expression of wanting to maintain their autonomy because marriage implies a relinquishment of power from each partner to the other. Spouse abuse is sometimes the expression of one partner trying to control the other through fear, intimidation, or force. Divorce may also illustrate control. The person who executes the divorce is often the person with the least interest in the relationship. Having the least interest gives that person power in the relationship.

conflict framework view that individuals in relationships compete for valuable resources.

> The person who executes the divorce is often the person with the least interest in the relationship.

Parents and adolescents are also in a continuous struggle over power. Parents attempt to use privileges and resources as power tactics to produce compliance in their adolescent. However, adolescents may use the threat of suicide as their ultimate power ploy to bring their parents under control.

1-4e Symbolic Interaction Framework

The **symbolic interaction framework** views marriages and families as symbolic worlds in which the various members give meaning to each other's behavior. In one study African Americans of different ages identified the two most common meanings they attributed to marriage—commitment and love (Curran et al., 2010). Concepts inherent in the symbolic interaction framework include the definition of the situation, the looking-glass self, and the self-fulfilling prophecy (Blumer, 1969).

Definition of the Situation Two people who have just spotted each other at a party are constantly defining the situation and responding to those definitions. Is the glance from the other person (1) an invitation to approach, (2) an approach, or (3) a misinterpretation—he or she was looking at someone else? The definition a person arrives at will affect subsequent interaction.

Marriage also has different definitions/meanings. For "marriage naturalists" it is an event that is a natural progression of a relationship (often begun in high school) and is expected of oneself, one's partner, and the families. Persons in rural areas more often have this view. In contrast, "marriage planners" are more metropolitan and view marriage as an event one "gets ready for" by completing one's college or graduate school education, establishing oneself in a job/career, and maturing emotionally and psychologically. These individuals may cohabit and have children before they decide to marry (Kefalas et al., 2011).

Looking-Glass Self The image people have of themselves is a reflection of what other people tell them about themselves (Cooley, 1964). People may develop an idea of who they are by the way others act toward them. If no one looks at or speaks to them, they will begin to feel unsettled, according to Cooley. Similarly, family members constantly hold up social mirrors for one another into which the respective members look for definitions of self.

© @erics/Shutterstock

The importance of the family (and other caregivers) as an influence on the development and maintenance of a positive self-concept cannot be overemphasized (Brown et al., 2009). Orson Welles, known especially for his film *Citizen Kane,* once said that he was taught that he was wonderful and that everything he did was perfect. He never suffered from a negative self-concept. Cole Porter, known for creating such memorable songs as "I Get a Kick Out of You," had a mother who held up social mirrors offering nothing but praise and adoration. Because children spend their formative years surrounded by their family, the self-concept they develop in that setting is important to their feelings about themselves and their positive interaction with others.

People are not passive sponges but evaluate the perceived appraisals of others, accepting some opinions and not others (Mead, 1934). Although some parents teach their children that they are worthless, those children may eventually overcome the definition.

Self-Fulfilling Prophecy Once people define situations and the behaviors in which they are expected to engage, they are able to behave toward one another in predictable ways. Such predictability of behavior also tends to exert influence on subsequent behavior. If you feel that your partner expects you to be faithful, your behavior is likely to conform to these expectations. The expectations thus create a self-fulfilling prophecy.

symbolic interaction framework
theoretical perspective that views marriage and families as symbolic worlds in which the various members give meaning to each other's behavior.

Symbolic interactionism as a theoretical framework helps to explain various choices in relationships. Individuals who decide to marry have defined their situation as a committed, reciprocal love relationship. This choice is supported by the belief that the partners will view each other positively (looking-glass self) and be faithful spouses and cooperative parents (self-fulfilling prophecies).

1-4f Family Systems Framework

The **family systems framework** views each member of the family as part of a system and the family as a unit that develops norms of interacting, which may be explicit (for example, parents specify chores for the children) or implicit (for example, spouses expect fidelity from each other). These rules serve various functions, such as allocating the resources (money for vacation), specifying the division of power (who decides how money is spent), and defining closeness and distance between systems (seeing or avoiding parents or grandparents).

Rules are most efficient if they are flexible. For example, they should be adjusted over time in response to children's growing competence. A rule about not leaving the yard when playing may be appropriate for a 4-year-old but inappropriate for a 15-year-old.

Family members also develop boundaries that define the individual and the group and separate one system or subsystem from another. A boundary is a "border between the system and its environment that affects the flow of information and energy between the environment and the system" (White & Klein, 2002, p. 124). A boundary may be physical, such as a closed bedroom door, or social, such as expectations that family problems will not be aired in public. Boundaries may also be emotional, such as communication, which maintains closeness or distance in a relationship. Some family systems are cold and abusive; others are warm and nurturing.

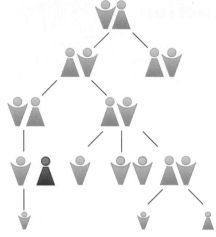

© LongQuattro/Shutterstock

In addition to rules and boundaries, family systems have roles (leader, follower, scapegoat) for the respective family members. These roles may be shared by more than one person or may shift from person to person during an interaction or across time. In healthy families, individuals are allowed to alternate roles rather than being locked into one role. In problem families, one family member is often allocated the role of scapegoat, or the cause of all the family's problems (for example, an alcoholic spouse).

Family systems may be open, in that they are open to information and interaction with the outside world, or closed, in that they feel threatened by such contact. The Amish have closed family systems and minimize contact with the outside world. Some communes also encourage minimal outside exposure. Twin Oaks Intentional Community of Louisa, Virginia, does not permit any of its almost 100 members to own a television or keep one in their rooms. Exposure to the negative drumbeat of the evening news is seen as harmful.

Family systems theory also emphasizes how one system affects another. For example, in a recent study, spouses who were happy with each other were also happy with their children (Malinen et al., 2010). In effect, the marital system had a positive effect on the family system.

1-4g Feminist Framework

Although a **feminist framework** views marriage and family as contexts of inequality and oppression for women, feminists seek equality in their relationships with their partners. There are 11 feminist perspectives, including lesbian feminism (emphasizing oppressive heterosexuality and men's domination of social spaces), psychoanalytic feminism (focusing on cultural domination of men's phallic-oriented ideas and repressed emotions), and standpoint feminism (stressing the neglect of women's perspectives and experiences in the production of knowledge; Lorber, 1998). Regardless of which feminist framework is being discussed, all feminist frameworks have the themes of inequality and oppression. According to feminist theory, gender

family systems framework theoretical perspective that views each member of the family as part of a system and the family as a unit that develops norms of interaction.

feminist framework perspective that views marriage and the family as contexts for inequality and oppression.

structures our experiences (for example, women and men will experience life differently because there are different expectations for the respective genders).

Although the previous theoretical frameworks are useful in understanding marriage and the family, in this text we encourage a proactive approach of taking charge of your life and making wise relationship choices. Making the right choices in your relationships, including marriage and family relationships, is critical to your health, happiness, and sense of well-being. Your times of greatest elation and sadness are in reference to your love relationships.

Among the most important of your relationship choices are whether to marry, whom to marry, when to marry, whether to have children, whether to remain emotionally and sexually faithful to your partner, and whether to use a condom. Though structural and cultural influences are operative, a choices framework emphasizes that you have some control over your relationship destiny by making deliberate choices to initiate, respond to, nurture, or terminate intimate relationships.

1-5a Facts About Choices in Relationships

The facts to keep in mind when making relationship choices include the following.

Not to Decide Is to Decide Not making a decision is a decision by default. If you are sexually active and decide not to use a condom, you have made a decision to increase your risk for contracting a sexually transmissible infection, including HIV. If you don't make a deliberate choice to end a relationship that is unfulfilling, abusive, or going nowhere, you have made a choice to continue in that relationship and to have little chance of getting into a more positive and satisfying relationship. If you don't

make a decision to ignore a text message from an ex-partner, you will respond.

Some Choices Require Correction Some of our choices, although appearing correct at the time that we make them, turn out to be disasters. Once we realize that a choice is having consistently negative consequences, we need to stop defending the old choice, reverse the position, make new choices, and move forward. Otherwise, we remain consistently locked into continued negative outcomes of "bad" choices. For example, choosing a partner who was loving and kind but who becomes abusive and dangerous requires correcting that choice. To stay in the abusive relationship will have predictable disastrous consequences. Making the decision to disengage and to move on opens the opportunity for a loving relationship with another partner. In the meantime, living alone may be a better alternative than living in a relationship in which you are abused and may end up dead. Other examples of making corrections include ending dead or loveless relationships (perhaps after investing time and effort to improve the relationship or love feelings), changing jobs or careers, and changing friends.

Choices Involve Trade-Offs By making one choice, you relinquish others. Every relationship choice you make will have a downside and an upside. If you decide to stay in a relationship that

Some couples in long-distance relationships decide to be faithful.

becomes a long-distance relationship, you are continuing involvement in a relationship that is obviously important to you. However, you may spend a lot of time alone and wonder if you made the right decision in continuing the relationship. If you decide to marry, you will give up your freedom to pursue other emotional or sexual relationships, or both, and you will also give up some of your control over how you spend your money—but you may also get a wonderful companion with whom to share life.

Choices Include Selecting a Positive or Negative View

As Thomas Edison progressed toward inventing the lightbulb, he said, "I have not failed. I have found ten thousand ways that won't work." In spite of an unfortunate event in your life, you can choose to see the bright side. Regardless of your circumstances, you can choose to view a situation in positive terms. A breakup with a partner you have loved can be viewed as the end of your happiness or an opportunity to become involved in a new, more fulfilling relationship. The discovery of your partner cheating on you can be viewed as the end of the relationship or an opportunity to examine your relationship, to open up communication channels with your partner, and to develop a stronger relationship. Finally, discovering that you are infertile can be viewed as a catastrophe or as a challenge to face adversity with your partner. Our point of view does make a difference—it is the one thing we have control over.

Choices Involve Different Decision-Making Styles

One study of 148 college students revealed four patterns of decision making (Allen et al., 2008). These patterns and the percentage using each pattern included (1) "I am in control" (45%), (2) "I am experimenting and learning" (33%), (3) "I am struggling but growing" (14%), and (4) "I have been irresponsible" (3%). Of those who reported that they were in control, about a third (34%) said that they were "taking it slow," and about 11% reported that they were "waiting it out." Men were more likely to report that they were "in control." Hence, these college students could conceptualize their decision-making style; they

sequential ambivalence the individual experiences one wish and then another.

simultaneous ambivalence the person experiences two conflicting wishes at the same time.

knew what they were doing. Of note, only 3% labeled themselves as being irresponsible.

Choices Produce Ambivalence

Choosing among options and trade-offs often creates ambivalence—conflicting feelings that produce uncertainty or indecisiveness as to a course of action to take. There are two forms of ambivalence: sequential and simultaneous. In **sequential ambivalence**, the individual experiences one wish and then another. For example, a person may vacillate between wanting to stay in a less-than-fulfilling relationship and wanting to end it. In **simultaneous ambivalence**, the person experiences two conflicting wishes at the same time. For example, the individual may feel both the desire to stay with the partner and the desire to break up at the same time. The latter dilemma is reflected in the saying, "You can't live with them, and you can't live without them." Some anxiety about choices is normal and should be embraced.

Most Choices Are Revocable; Some Are Not

Most choices can be changed. For example, a person who has chosen to be sexually active with multiple partners can later decide to be monogamous or to abstain from sexual relations. Individuals who have in the past chosen to emphasize career, money, or advancement over marriage and family can choose to prioritize relationships over economic and career-climbing behaviors. People who have been unfaithful in the past can elect to be emotionally and sexually committed to a new partner.

Other choices are less revocable. For example, backing out of the role of spouse is much easier than backing out of the role of parent. Whereas the law permits disengagement from the role of spouse (formal divorce decree), the law ties parents to dependent offspring (e.g., through child support). Hence, the decision to have a child is usually irrevocable. Choosing to have unprotected sex can also result in a lifetime of coping with sexually transmitted infections.

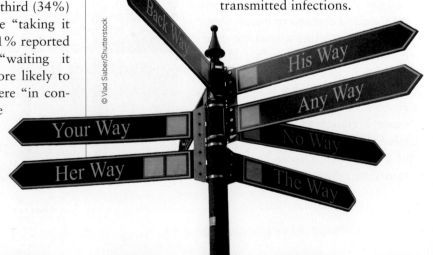

Choices of Generation Y Those in **Generation Y** (typically born between 1979 and 1984) are the children of the baby boomers. Currently numbering about 40 million, these Generation Yers (also known as the Millennial or Internet Generation) have been the focus of their parents' attention. They have been nurtured, coddled, and scheduled into day care to help them get ahead. The result is a generation of high self-esteem, self-absorbed individuals who believe they "are the best." Unlike their parents, who believe in paying one's dues, getting credentials, and sacrificing through hard work to achieve economic stability, Generation Yers focus on fun, enjoyment, and flexibility. They might choose a summer job at the beach if it buys a burger and a room with six friends over an internship at IBM that smacks of the corporate America sellout. Generation Yers, well aware that some college graduates are working at McDonald's, may wonder, "Why bother attending college?" and instead seek innovative ways to drift through life; they may continue to live with their parents, live communally, or get food by "dumpster diving." In effect, they are the generation of immediate gratification; they focus only on the here and now. The need for Social Security is too far off, and they feel they can get "free health care at the local hospital emergency room."

Generation Yers are also relaxed about relationship choices. Rather than forming a pair-bond, they hang out, hook up, and live together. They are in no hurry to find "the one," to marry, or to begin a family. To be sure, not all youth fit this characterization. Some have internalized their parents' values and are focused on education, credentials, a stable job, a retirement plan, and health care. They may view education as the ticket to a good job and expect their college to provide a credential they can market.

© Andresr/Shutterstock

Choices Are Influenced by the Stage in the Family Life Cycle The choices a person makes tend to be individualistic or familistic, depending on the stage of the family life cycle that the person is in. Before marriage, individualism characterizes the thinking and choices of most individuals. Individuals need only be concerned with their own needs. Most people delay marriage in favor of completing school, becoming established in a career, and enjoying the freedom of singlehood.

Once married, and particularly after having children, the person's familistic values and choices ensue as the needs of a spouse and children begin to influence behavior. Evidence of familistic choices is reflected in the fact that spouses with children are less likely to divorce than spouses without children.

Making Wise Choices Is Facilitated by Learning Decision-Making Skills Choices occur at the individual, couple, and family levels. Deciding to transfer to another school

> **Generation Y**
> children of the baby boomers, typically born between 1979 and 1984. Also known as the Millennial or Internet Generation.

"This is the generation that won't commit to going to a party on Saturday night because something better might come along— someone better might come along."

—SHANNON FOX, PSYCHOTHERAPIST

Table 1.4

"Best" and "Worst" Choices Identified by University Students

Best Choice	Worst Choice
Waiting to have sex until I was older and involved.	Cheating on my partner.
Ending a relationship with someone I did not love.	Getting involved with someone on the rebound.
Insisting on using a condom with a new partner.	Making decisions about sex when drunk.
Ending a relationship with an abusive partner.	Staying in an abusive relationship.
Forgiving my partner and getting over cheating.	Changing schools to be near my partner.
Getting out of a relationship with an alcoholic.	Not going after someone I really wanted.

© Cengage Learning 2011

or take a job out of state may involve all three levels, whereas the decision to lose weight is more likely to be an individual decision. Regardless of the level, the steps in decision making include setting aside enough time to evaluate the issues involved in making a choice, identifying alternative courses of action, carefully weighing the consequences for each choice, and being attentive to your own inner voice ("Listen to your senses"). The goal of most people is to make relationship choices that result in the most positive and least negative consequences. Students in my marriage and family class identified their "best" and "worst" relationship choices (see Table 1.4).

1-5b Global, Structural/Cultural, and Media Influences on Choices

Choices in relationships are influenced by global, structural/cultural, and media factors. Although a major theme of this book is the importance of taking active control of your life in making relationship choices, it is important to be aware that the social world in which you live restricts and channels such choices. For example, enormous social disapproval for marrying someone of another race is part of the reason that more than 90% of individuals in the United States marry someone of the same race.

institutions
established and enduring patterns of social relationships.

Globalization Families exist in the context of world globalization. Economic, political, and religious happenings throughout the world affect what happens in your marriage and family in the United States. When the price of oil per barrel increases in the Middle East, gasoline costs more, leaving fewer dollars to spend on other items. When the stock market in Hong Kong drops 500 points, Wall Street reacts, and U.S. stocks drop. The politics of the Middle East (for example, terrorist or nuclear threats from Iran) impact Homeland Security measures so that getting through airport security to board a plane may take longer.

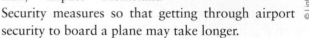

© Lightspring/Shutterstock

The country in which you live also affects your happiness and well-being. For example, in a study, citizens of 13 countries were asked to indicate their level of life satisfaction on a scale from 1 (dissatisfied) to 10 (satisfied): Citizens in Switzerland averaged 8.3, those in Zimbabwe averaged 3.3, and those in the United States averaged 7.4 (Veenhoven, 2007). The Internet, CNN, and mass communications provide global awareness so that families are no longer isolated units.

Social Structure The social structure of a society consists of institutions, social groups, statuses, and roles.

1. **Institutions.** The largest elements of society are social **institutions**, which may be defined as established and enduring patterns of social relationships. The institution of the family in the United States is definitely biased toward heterosexism. For example, there is no federal recognition of gay marriage. Indeed, in the face of traditional terms in marriage such as *husband* and *wife*, gay and lesbian individuals often have difficulty deciding which term to use (*lover, partner, significant other, companion, spouse, life mate,* and *wife* or *husband*).

    In addition to the family, major institutions of society include the economy, education, and religion. Institutions affect individual decision making. For example, you live in a capitalistic society where economic security is paramount—it is the number one value held by college students (Pryor et al., 2011). In effect, the more time you spend focused on obtaining money, the less time you have for relationships. You are now involved in the educational institution that will impact your choice of a mate (college-educated people tend to select and marry one another). Religion also affects sexual and relationship choices (for example, religion may result in delaying first intercourse, not using a condom, or marrying someone of the same faith). The family is a universal institution. Spouses who "believe in the institution of the family" are highly committed to maintaining their marriage and do not regard divorce as an option. The government is an institution that has an enormous impact on your choices. A dramatic example of this impact is the Australian government's policy in regard to Aboriginal children (see the box feature, "When Choices Are Taken Away by the Government").

When Choices Are Taken Away by the Government: The Australian Aboriginal Example

Bob Randall and his partner are companions in Australia to help Aborigines attain basic human rights.

In Australia, between 1885 and 1969, between 50,000 and 100,000 half-caste (one White parent) Aboriginal children were taken by force from their parents by the Australian police. The White society wanted to convert these children to Christianity and to destroy their Aboriginal culture, which was viewed as primitive and without value. The children were forced to walk or were taken by camel (depending on the territory in Australia where they lived) hundreds of miles away from their parents to church missions (see the *Rabbit-Proof Fence* DVD).

Bob Randall (2008) is one of the children who was taken by force from his parents at the age of 7, never to see them again. Of his experience, he wrote,

Instead of the wide open spaces of my desert home, we were housed in corrugated iron dormitories with rows and rows of bunk beds. After dinner we were bathed by the older women, put in clothing they called pajamas, and then tucked into one of the iron beds between the sheets. This was a horrible experience for me. I couldn't stand the feel of the cloth touching my skin. (p. 35)

The Australian government subsequently apologized for the laws and policies of successive parliaments and governments that inflicted profound grief, suffering, and loss on the Aborigines. However, Randall notes that the Aborigines continue to be marginalized and that nothing has been done to compensate for the horror of taking children from their families.

2. **Social groups.** Institutions are made up of social groups, defined as two or more people who share a common identity, interact, and form a social relationship. Most individuals spend their day going between social groups. You may awaken in the context (social group) of a roommate, partner, or spouse. From there you go to class with other students, have lunch with friends, work with your boss, and talk on the phone to your parents. So, within 24 hours you have participated in at least five social groups. These social groups have varying influences on your choices. Your roommate influences what other people you can have in your room for how long, your friends may want to eat at a particular place, your boss will assign you certain duties, and your parents may want you to come home for the weekend.

Your interpersonal choices are influenced mostly by your partner and peers (for example, your sexual values, use of condoms, and the amount of alcohol you consume). Thus, selecting a partner and peers carefully is important. As Falstaff, one of Shakespeare's characters, said, "Company, villainous company, hath been the spoil of me" (*Henry IV, Pt. 1*, 1.3).

The age of the partner you select to become romantically involved with is influenced by social context. If you are a woman, your parents and peers will probably approve of you dating and marrying "someone a little older." Likewise, if you are a man, your parents and peers will probably approve of you dating and marrying someone "a little younger." This pattern of women dating older men and men dating younger women is referred to as the **mating gradient**, which suggests that women marry "up" in age and education and men marry "down" in age and education.

Social groups may be categorized as primary or secondary. **Primary groups**, which tend to involve small numbers of individuals, are character-

ized by interaction that is intimate and informal. Our family is an example of our primary group. Persons in our primary groups are those who love us and have lifetime relationships with us. In contrast to primary groups, **secondary groups**, which may be small or large, are characterized by interaction that is impersonal and formal. Your fellow students and teachers or your coworkers are examples of secondary groups. Members of our secondary groups do not have an emotional connection with us and are more transient.

Meryl Streep alluded to both secondary and primary groups when she said:

I have won a long list of awards, and while I'm very proud of the work, I can assure you that awards have very little bearing on my personal happiness and sense of purpose and well-being. That comes from . . . involvement in the lives of those I love. (Couric, 2011, p. 188)

3. **Statuses.** Just as institutions consist of social groups, social groups consist of statuses. A **status** is a position a person occupies within a social group. The statuses we occupy largely define our social identity. The statuses in a family may consist of mother, father, child, sibling, stepparent, and so on. In discussing family issues, we refer to statuses such as teenager, cohabitant, and spouse. Statuses are relevant to choices in that many choices can significantly change one's status. Making decisions that change one's status from single person to spouse to divorced person can influence how people feel about themselves and how others treat them.

4. **Roles.** Every status is associated with many **roles**, or sets of rights, obligations, and expectations associated with a status. Our social statuses identify who we are; our roles identify what we

> ## Statuses are relevant to choices in that many choices can significantly change one's status.

mating gradient the tendency for husbands to marry wives who are younger and have less education and less occupational success.

primary group small, intimate, informal group.

secondary group large or small group characterized by impersonal and formal interaction.

status a social position a person occupies within a social group.

roles the behaviors in which individuals in certain status positions are expected to engage.

are expected to do. Roles guide our behavior and allow us to predict the behavior of others. Spouses adopt a set of obligations and expectations associated with their status. By doing so, they are better able to influence and predict each other's behavior.

Because individuals occupy a number of statuses and roles simultaneously, they may experience role conflict. For example, the role of the parent may conflict with the role of the spouse, employee, or student. If your child needs to be driven to the math tutor, your spouse needs to be picked up at the airport, your employer wants you to work late, and you have a final exam all at the same time, you are experiencing role conflict.

© olly/Shutterstock

Culture Just as social structure refers to the parts of society, culture refers to the meanings and ways of living that characterize people in a society. Two central elements of culture are beliefs and values.

1. **Beliefs. Beliefs** refer to definitions and explanations about what is true. The beliefs of an individual or a couple influence the choices they make. Couples who believe that young children flourish best with a full-time parent in the home make different job and child-care decisions than do couples who believe that day care offers opportunities for enrichment.

2. **Values. Values** are standards regarding what is good and bad, right and wrong, desirable and undesirable. Values influence choices. Valuing **individualism** leads to making decisions that serve the individual's interests rather than the family's interests (**familism**). "What makes me happy," not "what makes my family happy," is the focus of the individualist. European Americans are characteristically individualistic, whereas Hispanic Americans are characteristically familistic. There is a paradox in American values—Americans value marriage and abhor divorce, but they are also individualistic, which ensures a high divorce rate (Cherlin, 2009). Forty percent of

2,922 undergraduates reported that they would divorce their spouse if they fell out of love (Knox & Hall, 2010). Allowing one's personal love feelings to dictate the stability of a marriage is a highly individualistic value.

These elements of social structure and culture play a central role in making interpersonal choices and decisions. One of the goals of this text is to emphasize the influence of social structure and culture on your interpersonal decisions. Sociologists refer to this awareness as the **sociological imagination** (or *sociological mindfulness*). For example, though most people in the United States assume that they are free to select their own sex partner, this choice (or lack of it) is in fact heavily influenced by structural and cultural factors. For example, most people date, have sex with, and marry a person of the same racial background.

Media A Kaiser Family Foundation (2010) study of 8- to 18-year-olds revealed that daily consumption of watching television averages 270 minutes, listening to music averages 151 minutes, and playing video games averages 73 minutes. Media in all of its forms (television, music, video games, print) influences how we think about and make our relationship choices. For example, media exposure transmits the "acceptability" of various sexual values and norms. At the 2011 Academy Awards, unmarried and pregnant Natalie Portman accepted her award for best actress. Other events given exposure by the media that impact us in regard to relationships, marriage, and family include the sexting scandal of Representative Anthony Weiner (while his

beliefs definitions and explanations about what is thought to be true.

values standards regarding what is good and bad, right and wrong, desirable and undesirable.

individualism philosophy in which decisions are made on the basis of what is best for the individual.

familism philosophy in which decisions are made in reference to what is best for the family as a collective unit.

sociological imagination the perspective of how powerful social structure and culture are in influencing personal decision making.

Open display of female sexuality (at the 2012 Academy Awards) is an American cultural value. Such a display in countries such as Jordan, Afghanistan, and Iran would not occur.

© Jason Merritt/Getty Images

wife was pregnant), the love child conceived by Arnold Schwarzenegger and his maid, and the divorce of Tiger Woods (on the heels of revelations of his 13 mistresses).

1-5c Other Influences on Relationship Choices

Aside from structural and cultural influences on relationship choices, other influences include family of origin (the family in which you were reared), unconscious motivations, habit patterns, individual personality, and previous experiences.

1. **Family of origin.** Your family of origin is a major influence on your subsequent choices and relationships. For example, Fish et al. (2011) examined data on 353 respondents and concluded: "Individuals who reported that they were aware of their parents' affair were significantly more likely to report having been unfaithful themselves: physically, emotionally, and composite. This makes sense considering the tendency of individuals to use family patterns through generations."

2. **Unconscious motivations.** Unconscious processes are operative in our choices. A person reared in a lower-class home without adequate food and shelter may become overly concerned about the accumulation of money and may make all decisions in reference to obtaining, holding, or hoarding economic resources.

3. **Habits.** Habit patterns also influence choices. People who are accustomed to and enjoy spending a great deal of time alone may be reluctant to make a commitment to live with people who make demands on their time. A person who has workaholic tendencies is unlikely to allocate enough time to a relationship to make it flourish. Alcohol abuse is associated with having a higher number of sexual partners and with not using condoms to avoid pregnancy and contracting sexually transmitted infections.

4. **Personality.** One's personality (e.g., introverted, extroverted; passive, assertive) also influences choices. For example, people who are assertive are more likely than those who are passive to initiate a conversation with someone they are attracted to at a party. People who are very quiet and withdrawn may never choose to initiate a conversation even though they are attracted to someone. People with a bipolar disorder who are manic one part of the semester (or relationship) and depressed the other part are likely to make different choices when each process is operative.

5. **Relationships and life experiences.** Current and past relationship experiences also influence one's perceptions and choices. Individuals currently in a relationship are more likely to hold relativistic sexual values (e.g., choose intercourse over abstinence). Similarly, people who have been cheated on in a previous relationship are vulnerable to not trusting a new partner. Pop singer

> **"The heart** has its reasons that
> **reason knows nothing of."**
>
> —BLAISE PASCAL, *PENSÉES*, 1670

Rihanna noted that the physical attack on her by Chris Brown shocked her—she was in love and had given her heart to him. But confronting him about texting another woman led to his beating her up (photos are widely available on the Internet), which changed her view of love, relationships, and men.

Having emphasized that making choices in relationships is the theme of this text and having reviewed the mechanisms operative in those choices, we now look at how research is conducted. And, why we should be cautious in our view of research.

1-6 Research—Process and Evaluation

Research in marriage and family is valuable in that it replaces guesswork with facts. But the research process must be systematic and be sensitive to various caveats. Following a list of steps in the research process, we identify some specific issues to keep in mind when evaluating research.

1-6a Steps in the Research Process

Several steps are followed in conducting research.

1. **Identify the topic or focus of your research.** Select a subject about which you are passionate. For example, are you interested in studying cohabitation of college students? Give your projected study a title in the form of a question—"Do People Who Cohabit Before Marriage Have Happier Marriages Than Those Who Do Not?"

2. **Review the literature.** Go online to the various databases of your college or university and read research that has already been published on cohabitation. Not only will this step prevent you

from "reinventing the wheel" (you might find that a research study has already been conducted on exactly what you want to study), but it will give you ideas for your study.

3. **Develop hypotheses.** A **hypothesis** is a suggested explanation for a phenomenon. For example, you might suggest that cohabitation results in greater marital happiness and less divorce because the partners have a chance to "test-drive" each other.

4. **Decide on a method of data collection.** To test your hypothesis, do you want to interview college students, give them a questionnaire, or ask them to complete an online questionnaire? Of course, you will also need to develop a list of questions for your interview or survey questionnaire.

5. **Get IRB approval.** To ensure the protection of people who agree to be interviewed or who complete questionnaires, researchers must obtain **IRB approval** by submitting a summary of their proposed research to the Institutional Review Board (IRB) of their institution. The IRB reviews the research plan to ensure that the project is consistent with research ethics and poses no undue harm to participants. When collecting data from individuals, it is important that they are told that their participation is completely voluntary, that the study maintains their anonymity, and that the results are confidential. Respondents under age 18 need the consent of their parents. Community members were aghast when some parents of students at Memorial Middle School in Massachusetts reported that they never received a consent form for their seventh grader to take a survey

hypothesis a suggested explanation for a phenomenon.

IRB (Institutional Review Board) approval the endorsement by one's college, university, or institution that the proposed research is consistent with research ethics standards and poses no undo harm to participants.

on youth risk behavior that included items on oral sex and drug use (Starnes, 2011).

6. **Collect and analyze data.** Various statistical packages are available to analyze data to discover if your hypotheses are true or false.

7. **Write up and publish results.** Writing up and submitting your findings for publication are important so that your study becomes part of the literature in cohabitation.

College students are overrepresented in research on relationships.

© lightpoet/Shutterstock.com

1-6b Evaluating Research in Marriage and the Family

"New Research Study" is a frequent headline in popular magazines such as *Cosmopolitan* promising accurate information about "hooking up," "sexting," "what women/men want," or other sexual, communication, and gender issues. As you read such articles, as well as the research in such texts as this, be alert to their potential flaws.

Sample Some of the research on marriage and the family is based on random samples. In a **random sample**, each individual in the population has an equal chance of being included in the study. Random sampling involves randomly selecting individuals from an identified population. That population often does not include the homeless or people living on military bases. Studies that use random samples are based on the assumption that the individuals studied are similar to, and therefore representative of, the population that the researcher is interested in.

Because of the trouble and expense of obtaining random samples, most researchers study subjects to whom they have convenient access. This often means students in the researchers' classes. The result is an overabundance of research on "convenience" samples consisting of White, Protestant, middle-class college students. Because college students cannot be assumed to be similar in their attitudes, feelings, and behaviors to their noncollege peers or older adults, research based on college students cannot be generalized beyond the base population.

In addition to having a random sample, having a large sample is important. The American Council on Education and the University of California collected a national sample of 203,967 first-semester undergraduates at 270 colleges and universities throughout the United States; the sample was designed to reflect the responses of all full-time students entering four-year colleges and universities (Pryor et al., 2011). If only 50 college students had been in the sample, the results would have been less valuable in terms of generalizing beyond that sample.

Be alert to the sample size of the research you read. Most studies are based on small samples. In addition, considerable research has been conducted on college students, leaving us to wonder about the attitudes, values, and behaviors of noncollege students.

Control Groups In an example of a study that concludes that an abortion (or any independent variable) is associated with negative outcomes (or any dependent variable), the study must necessarily include two groups: (1) women who have had an

random sample research sample in which each person in the population being studied has an equal chance of being included in the study.

abortion, and (2) women who have not had an abortion. The latter would serve as a **control group**—the group not exposed to the independent variable you are studying (**experimental group**). Hence, if you find that women in both groups in your study develop negative attitudes toward sex, you know that abortion cannot be the cause. Be alert to the existence of a control group, which is usually not included in research studies.

Age and Cohort Effects

In some research designs, different cohorts or age groups are observed or tested or both at one point in time. One problem that plagues such research is the difficulty—even impossibility—of discerning whether observed differences between the subjects studied are due to the research variable of interest, to cohort differences, or to some variable associated with the passage of time (e.g., biological aging).

A good illustration of this problem is found in research on changes in marital satisfaction over the course of the family life

© 3DDock/Shutterstock

cycle. In such studies, researchers may compare the levels of marital happiness reported by couples that have been married for different lengths of time. For example, a researcher may compare the marital happiness of two groups of people—those who have been married for 50 years and those who have been married for 5 years. However, differences between these two groups may be due to (1) differences in age (age effect), (2) differences in the historical time period that the two groups have lived through (cohort effect), or (3) differences in the lengths of time the couples have been married (research variable).

Terminology

In addition to being alert to potential shortcomings in sampling and control groups, you should consider how the phenomenon being researched is defined. For example, if you are conducting research on living together, how would you define this term? How many people, of what sex, spending what amount of time, in what place, engaging in what behaviors will constitute your definition? Indeed, researchers have used more than 20 definitions of what constitutes living together.

What about other terms? Considerable research has been conducted on marital success, but how is the term to be defined? What is meant by marital satisfaction, commitment, interpersonal violence, and sexual fulfillment? Before reading too far in a research study, be alert to the definitions of the terms being used. Exactly what is the researcher trying to measure?

Researcher Bias

Although one of the goals of scientific studies is to gather data objectively, it may be impossible for researchers to be totally objective. Researchers are human and have values, attitudes, and beliefs that may influence their research methods and findings. It may be important to know what the researcher's bias is in order to evaluate the findings. For example, a researcher who does not support abortion rights may conduct research that focuses only on the negative effects of abortion. Similarly, researchers funded by corporations that have a vested interest in having their products endorsed are also suspect.

Time Lag

Typically, a two-year lag exists between the time a study is completed and the study's appearance in a professional journal. Because book production takes even longer than getting an article printed in a professional journal, textbooks do not always present the most current research, especially on topics in flux. In addition, even though a study may have been published recently, the data on which the study was based may be old. Be aware that the research you read in this or any other text may not reflect the most current, cutting-edge research.

Distortion and Deception

In some cases there is outright deception by professionals. In order to continue receiving funding for his study, Dr. Anil Potti at Duke University changed data on a report and provided fraudulent research results (Darton, 2012). As a result, patients took ineffective medications for treatment and were given false hope.

control group group used to compare with the experimental group that is not exposed to the independent variable being studied.

experimental group the group exposed to the independent variable.

Distortion and deception, deliberate or not, also exist in marriage and family research. Marriage is a very private relationship that happens behind closed doors; individual interviewees and respondents to questionnaires have been socialized not to reveal the intimate details of their lives to strangers. Hence, they are prone to distort, omit, or exaggerate information, perhaps unconsciously, to cover up what they may feel is no one else's business. Thus, researchers sometimes obtain inaccurate information.

Marriage and family researchers know more about what people say they do than about what they actually do. An unintentional and probably more common form of distortion is inaccurate recall. Sometimes researchers ask respondents to recall details of their relationships that occurred years ago. Time tends to blur some memories, and respondents may relate not what actually happened but, rather, what they remember to have happened or, worse, what they wish had happened.

Other Research Problems Nonresponse on surveys and the discrepancy between attitudes and

> Distortion and deception, deliberate or not, also exist in marriage and family research.

behaviors are other research problems. With regard to nonresponse, not all individuals who complete questionnaires or agree to participate in an interview are willing to provide information about such personal issues as date rape and partner abuse. Such individuals leave the questionnaire blank or tell the interviewer they would rather not respond. Others respond but give only socially desirable answers. The implication for research is that data gatherers do not know the extent to which something may be a problem because people are reluctant to provide accurate information.

Diversity in Relationships

The "one-size-fits-all" model of relationships and marriage is nonexistent. Individuals may be described as existing on a continuum from heterosexuality to homosexuality, from rural to urban dwellers, and from being single and living alone to being married and living in communes. Emotional relationships range from being close and loving to being distant and violent. Family diversity includes two-parent (other- or same-sex) families, single-parent families, blended families, families with adopted children, multigenerational families, extended families, and families representing different racial, religious, and ethnic backgrounds. *Diverse* is the term that accurately describes marriage and family relationships today.

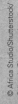

© Africa Studio/Shutterstock/
© Iiaszlo/Shutterstock

It is sometimes assumed that if people have a certain attitude (e.g., a belief that extramarital sex is wrong), then their behavior will be consistent with that attitude (avoidance of extramarital sex). However, this assumption is not always accurate. People do indeed say one thing and do another. This potential discrepancy should be kept in mind when reading research on various attitudes.

Finally, most research reflects information that volunteers provide. However, volunteers may not represent nonvolunteers when they are completing surveys. In view of the research cautions identified here, you might ask, "Why bother to report the findings?" The quality of some family science research is excellent. For example, articles published in the *Journal of Marriage and Family* (among other journals) reflect a high level of sound methodology. There are even some less sophisticated journals that provide useful information on marital, family, and other relationship data. Table 1.5 summarizes potential inadequacies of any research study.

Table 1.5
Potential Research Problems in Marriage and Family

Weakness	Consequences	Example
Sample not random	Cannot generalize findings	Opinions of college students do not reflect opinions of other adults.
No control group	Inaccurate conclusions	Study on the effect of divorce on children needs control group of children whose parents are still together.
Age differences between groups of respondents	Inaccurate conclusions	Effect may be due to passage of time or to cohort differences.
Unclear terminology	Inability to measure what is not clearly defined	How are living together, marital happiness, sexual fulfillment, good communication, and quality time defined?
Researcher bias	Slanted conclusions	A researcher studying the value of a product (e.g., The Atkins Diet) should not be funded by the organization being studied.
Time lag	Outdated conclusions	Often-quoted Kinsey sex research is over 65 years old.
Distortion	Invalid conclusions	Research subjects exaggerate, omit information, and/or recall facts or events inaccurately. Respondents may remember what they wish had happened.

STUDY TOOLS ➲ 1

Ready to study? In this book, you can:

- ➲ Rip out the Chapter Review card in the back of the book to study for exams
- ➲ Take the Self Assessment for this chapter (card in the back of the book) and see where you stand on the vital issues raised in the chapter

Or you can go online to CourseMate at www.cengagebrain.com for these resources:

- ➲ Complete Practice Quizzes to prepare for tests
- ➲ Review Key Terms Flash Cards (online or print)
- ➲ Read about Marriage and Family in the news
- ➲ Play "Beat the Clock" to master concepts
- ➲ Check out Personal Applications

Gender

"Newsflash: You're the boy. I'm the girl. You **text me first** or we **don't talk today."**

—UNKNOWN, UNIVERSITY NEWSPAPER

SECTIONS

2-1 Terminology of Gender Roles

2-2 Theories of Gender Role Development

2-3 Agents of Socialization

2-4 Consequences of Traditional Gender Role Socialization

2-5 Changing Gender Roles

Caster Semenya won gold in the women's 800-m race at the World Athletics Championship. But some colleagues questioned whether the 18-year-old South African was, in fact, a woman. Tests showed that she had three times the level of testosterone of a typical woman and that she was intersexed, which blurred her gender and gave her an unfair advantage. Her right to keep her gold medal has been controversial, and her case emphasizes the blurred boundary of gender—what is a male and a female? What factors determine whether one is identified as a man or a woman? What happens when one of these factors is altered—does the gender identity of the person automatically change?

Although the case of Caster Semenya reflects that gender is not always a clear-cut issue, sociologists note that one of the defining moments in an individual's life is when the sex of a fetus (in the case of an ultrasound) or infant (in the case of a birth) is announced. The phrase "It's a boy" or "It's a girl" immediately summons an onslaught of cultural programming affecting the color of the nursery (blue for a boy and pink for a girl), the name of the baby (there are few gender-free names such as Chris), and occupational choices (construction or nursing). The social script for women and men is so radically different that being identified as either is to put one's life on a path quite different than if the person had been reared as the other gender. In this chapter, we examine the ways in which gender is defined, factors associated with gender, the theories and influences of gender role development, and the consequences of being a woman or a man. We begin by looking at the terms used to discuss gender issues.

© herjua/Shutterstock.com

2-1 Terminology of Gender Roles

In common usage, the terms *sex* and *gender* are often interchangeable, but sociologists, family or consumer science educators, human development specialists, and health educators do not find these terms synonymous. After clarifying the distinction between *sex* and *gender,* we discuss other relevant terminology, including *gender identity, gender role,* and *gender role ideology.*

2-1a Sex

Sex refers to the biological distinction between females and males. Hence, to be assigned as female or male, several factors are used to determine the biological sex of an individual:

sex the biological distinction between being female and being male.

- *Chromosomes:* XX for females; XY for males

- *Gonads:* Ovaries for females; testes for males

- *Hormones:* Greater proportion of estrogen and progesterone than testosterone in females; greater proportion of testosterone than estrogen and progesterone in males

- *Internal sex organs:* Fallopian tubes, uterus, and vagina for females; epididymis, vas deferens, and seminal vesicles for males

- *External genitals:* Vulva for females; penis and scrotum for males

Even though we commonly think of biological sex as consisting of two dichotomous categories (female and male), biological sex exists on a continuum. Sometimes not all of the items just identified are found neatly in one person (who would be labeled as a female or a male). Rather, items typically associated with females or males might be found together in one person, resulting in mixed or ambiguous genitals; such persons are called **intersexed individuals** (see the opening paragraph about Caster Semenya). Indeed, the genitals in these intersexed (or middlesexed) individuals (about 2% of all births) are not clearly male or female (Crawley et al., 2008). **Intersex development** refers to congenital variations in the reproductive system, sometimes resulting in ambiguous genitals. The self-concept of these individuals is variable. Some may view themselves as one sex or the other or as a mix. Because our culture does not know how to relate to mixed-sex individuals, the individual is typically reared as a woman or as a man. However, an intersexed person who is reared as and presents as a woman may have no ovaries and will not be able to have children.

Even if one's chromosomal makeup is XX or XY, too much or too little of the wrong kind of hormone during gestation can also cause variations in sex development. Genetics, hormones, and brain mechanisms might all underlie neuroanatomic changes. Increasingly, professionals are becoming more informed about variations in gender—lesbian, gay, bisexual, transgendered, and intersexed. These individuals often feel a sense of exclusion, isolation, and fear due to negative social attitudes (Dysart-Gale, 2010).

2-1b Gender

Gender refers to the social and psychological characteristics associated with being female or male. For example, women see themselves (and men agree) as moody and easily embarrassed; men see themselves (and women agree) as competitive, sarcastic, and sexual (Knox et al., 2004).

In popular usage, gender is dichotomized as an either-or concept (feminine or masculine). Each gender has some characteristics of the other. However, gender may also be viewed as existing along a continuum of femininity and masculinity. Third- and fourth-grade children value their own sex's personality traits, but boys with many same-sex friends tend to denigrate female traits. Boys who act like girls are often criticized (Robnett and Susskind, 2010).

There is an ongoing controversy about whether gender differences are innate as opposed to learned or socially determined. Just as sexual orientation may be best explained as an interaction of biological and social or psychological variables, gender differences may also be seen as a consequence of both biological and social or psychological influences. For example, Irvolino and colleagues (2005) studied the genetic and environmental effects on the sex-typed behavior of 3,999 sibling pairs of 3- to 4-year-old twins and non-twins and concluded that their gender role behavior was a function of both genetic inheritance (e.g., chromosomes and hormones) and social factors (e.g., male/female models such as parents, siblings, and peers).

Whereas some researchers emphasize an interaction of the biological and social, others emphasize a biological imperative as the basis of gender role behavior. As evidence for the latter, the late John Money, psychologist and former director of the now-defunct Gender Identity Clinic at Johns Hopkins University School of Medicine, encouraged the parents of a boy (Bruce) to rear him as a girl (Brenda) because of a botched circumcision that left the infant without a penis. Money argued that social mirrors dictate one's

intersexed individuals people with mixed or ambiguous genitals.

intersex development refers to congenital variations in the reproductive system, sometimes resulting in ambiguous genitals.

gender the social and psychological behaviors associated with being female or male.

© dencg/Shutterstock

Feminine

Masculine

© Hasan Kursad Ergan/iStockphoto.com

© Margo Harrison/Shutterstock.com

gender identity, and thus, if the parents treated the child as a girl (e.g., name, dress, toys), the child would adopt the role of a girl and later that of a woman. The child was castrated and sex reassignment began.

However, the experiment failed miserably; as an adult, David Reimer (the child's real name) reported that he never felt comfortable in the role of a girl and had always viewed himself as a boy. He later married and adopted his wife's two children. For the book *As Nature Made Him: The Boy Who Was Raised as a Girl* (Colapinto, 2000), David worked with the writer to tell his story. His courageous decision to make his poignant personal story public has shed light on scientific debate on the "nature/nurture" question. In the past, David's situation was used as a textbook example of how "nurture" is the more important influence in gender identity, if a reassignment is done early enough. Today, his case makes the point that one's biological wiring dictates gender outcome (Colapinto, 2000). Indeed, David Reimer noted in a television interview, "I was scammed," referring to the absurdity of trying to rear him as a girl. Distraught with the ordeal of his upbringing and beset with financial difficulties, he committed suicide in May 2004 via a gunshot to the head.

The story of David Reimer emphasizes the power of biology in determining gender identity. Other research supports the critical role of biology. Cohen-Kettenis (2005) emphasized that biological influences in the form of androgens in the prenatal brain are very much involved in creating one's gender identity.

Nevertheless, **socialization** (the process through which we learn attitudes, values, beliefs, and behaviors appropriate to the social positions we occupy) does impact gender role behaviors, and social scientists tend to emphasize the role of social influences in gender differences. Although her research is controversial, Margaret Mead (1935) focused on the role of social learning in the development of gender roles in her study of three cultures.

She visited three New Guinea tribes in the early 1930s and observed that the Arapesh socialized both men and women to be feminine, by Western standards. The Arapesh people were taught to be cooperative and responsive to the needs of others. In contrast, the Tchambuli were known for dominant women and submissive men—just the opposite of our society. Both of these societies were unlike the Mundugumor, which socialized only ruthless, aggressive, "masculine" personalities. The inescapable conclusion of this cross-cultural study is that human beings are products of their social and cultural environments and that gender roles are learned. As Peoples observed, "cultures construct gender in different ways" (2001, p. 18). Indeed, the very terms we use to describe various body parts of women and men carry notions of power and use. A penis is a "probe" (active) whereas a vagina is a "hole" (to be filled; Crawley et al., 2008).

© Laura Neal/iStockphoto.com

socialization the process through which we learn attitudes, values, beliefs, and behaviors appropriate to the social positions we occupy.

© David Knox

2-1c Gender Identity

Gender identity is the psychological state of viewing oneself as a girl or a boy and, later, as a woman or a man. Such identity is largely learned and is a reflection of society's conceptions of femininity and masculinity.

Some individuals experience **gender dysphoria**, a condition in which one's gender identity does not match one's biological sex. An example of gender dysphoria is transsexualism (discussed in the next section).

2-1d Transgenderism

The word **transgender** is a generic term for a person of one biological sex who displays characteristics of the other sex. **Cross-dresser** is a broad term for individuals who may dress or present themselves in the gender of the other sex. Some cross-dressers are heterosexual adult males who enjoy dressing and presenting themselves as women. Cross-dressers may also be women who dress as men and present themselves as men. Cross-dressers may be heterosexual, homosexual, or bisexual.

Transsexuals are individuals with the biological and anatomical sex of one gender (e.g., male) but the self-concept of the opposite sex (that is, female). "I am a woman trapped in a man's body" reflects the feelings of the male-to-female transsexual (MtF), who may take hormones to develop breasts and reduce facial hair and may have surgery to artificially construct a vagina. Such a person lives full time as a woman. The female-to-male transsexual (FtM) is one who is a biological and anatomical female but feels "I am a man trapped in a woman's body." This person may take male hormones to grow facial hair and deepen her voice and may have surgery to create an artificial penis. This person lives full time as a man. Thomas Beatie, born a biological woman, viewed himself as a man and transitioned from living as a woman to living as a man. His wife, Nancy, had two adult children before having a hysterectomy (which rendered her unable to bear more children) so her husband agreed to be artificially inseminated (because he had ovaries), became pregnant, and delivered a child. Because he was legally a male, the media referred to him as "The Pregnant Man"; they asked him: if he gave birth to the child, could he be the father? Technically, Oregon law defines birth as an expulsion or extraction from the mother, so Tom is the technical mother. However, the new parents could petition the courts and have Tom declared the father and his wife the mother (Heller, 2008). Chaz Bono, the child of Cher, is also a female-to-male transsexual (Bono, 2011).

© s_bukley/Shutterstock.com

The openness or Chaz Bono on various talk shows reflects an emerging cultural norm of gender diversity in the U.S.

gender identity the psychological state of viewing oneself as a girl or a boy and, later, as a woman or a man.

gender dysphoria the condition in which one's gender identity does not match one's biological sex.

transgender a generic term for a person of one biological sex who displays characteristics of the opposite sex.

cross-dresser a generic term for individuals who may dress or present themselves in the gender of the opposite sex.

transsexual an individual who has the anatomical and genetic characteristics of one sex but the self-concept of the other.

Individuals need not take hormones or have surgery to be regarded as transsexuals. The distinguishing variable is living full time in the role of the gender opposite one's biological sex. A man or woman who presents full time as the opposite gender is a transsexual by definition.

2-1e Gender Roles

Gender roles are the social norms that dictate what is socially regarded as appropriate female and male behavior. All societies have expectations of how boys and girls, men and women "should" behave. Gender roles influence women and men in virtually every sphere of life, including family and occupation. One prevalent norm in family life is that women end up devoting more time to child rearing and child care. Lareau and Weininger (2008) studied the division of labor of parents transporting their children to various leisure activities and found that traditional gender roles were the norm and that mothers "are the ones who must satisfy these demands" (p. 450).

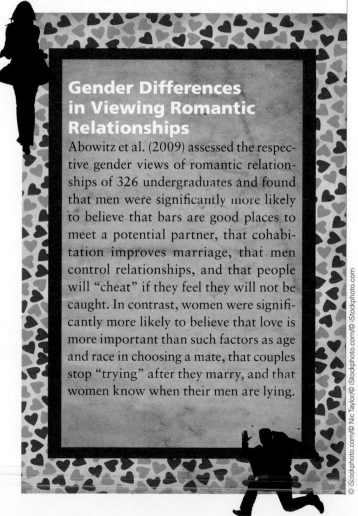

Gender Differences in Viewing Romantic Relationships

Abowitz et al. (2009) assessed the respective gender views of romantic relationships of 326 undergraduates and found that men were significantly more likely to believe that bars are good places to meet a potential partner, that cohabitation improves marriage, that men control relationships, and that people will "cheat" if they feel they will not be caught. In contrast, women were significantly more likely to believe that love is more important than such factors as age and race in choosing a mate, that couples stop "trying" after they marry, and that women know when their men are lying.

Bloch and Taylor (2012) discussed hour mismatches (the actual number of hours a person works that is different from the number of hours one wants to work—over or under) of women and men. The researchers found that hours spent on childcare affect men's hour mismatches, while hours spent on housework affect women's. Their findings illustrate the way that gender structures perceptions of ideal work hours and the ability to achieve them.

Walzer (2008) noted that marriage is a place where men and women "do gender" in the sense that roles tend to be identified as breadwinning, housework, parenting, and emotional expression and are gender-differentiated. She also noted that divorce generates "redoing" gender in the sense that it changes the expectations for masculine and feminine behavior in families (e.g., women become breadwinners, men become single parents, and so on).

The term **sex roles** is often confused with and used interchangeably with the term *gender roles*. However, whereas gender roles are socially defined and can be enacted by either women or men, sex roles are defined by biological constraints and can be enacted by members of one biological sex only—for example, wet nurse, sperm donor, childbearer.

2-1f Gender Role Ideology

Gender role ideology refers to beliefs about the proper role relationships between women and men in any given society. Where there is gender equality, there is enhanced relationship satisfaction (Walker & Luszcz, 2009). Egalitarian wives are also most happy with their marriages if their husbands share both the work and the emotions of managing the home and caring for the children (Wilcox & Nock, 2006).

In spite of the rhetoric regarding the entrenchment of egalitarian interaction between women and men in the United States, there is evidence of traditional gender roles in mate selection, with men in the role of initiating relationships. When 692 undergraduate females were asked if they had ever asked a new guy to go out, 60.1% responded no (Ross et al., 2008). In another study, 30% of the female respondents reported a preference for marrying a traditional man (one

gender roles behaviors assigned to women and men in a society.

sex roles behaviors defined by biological constraints.

gender role ideology the proper role relationships between women and men in a society.

who saw his role as provider and who was supportive of his wife staying home to rear children; Abowitz et al., 2011). Of 335 undergraduate men, 31% preferred a traditional wife (Knox & Zusman, 2007).

Traditional American gender role ideology has perpetuated and reflected patriarchal male dominance and male bias in almost every sphere of life. Even our language reflects this male bias. For example, the words *man* and *mankind* have traditionally been used to refer to all humans. There has been a growing trend away from using male-biased language. Dictionaries have begun to replace *chairman* with *chairperson* and *mankind* with *humankind*.

2-2 Theories of Gender Role Development

Various theories attempt to explain why women and men exhibit different characteristics and behaviors.

2-2a Biosocial

In the discussion of gender at the beginning of the chapter, we noted the profound influence of biology on one's gender. **Biosocial theory** (also referred to as *sociobiology*) emphasizes that social behaviors (e.g., gender roles) are biologically based and have an evolutionary survival function. For example, women tend to select and mate with men whom they deem will provide the maximum parental investment in their offspring. The term **parental investment** refers to any investment by a parent that increases the offspring's chance of surviving and thus increases reproductive success. Parental investments require time and energy. Women have a great deal of parental investment in their offspring (including nine months of gestation), and they tend to mate with men who have high status, economic resources, and a willingness to share those economic resources.

biosocial theory (sociobiology) emphasizes the interaction of one's biological or genetic inheritance with one's social environment to explain and predict human behavior.

parental investment any investment by a parent that increases the chance that the offspring will survive and thrive.

The biosocial explanation for mate selection is extremely controversial. Critics argue that women may show concern for the earning capacity of a potential mate because they have been systematically denied access to similar economic resources, and selecting a mate with these resources is one of their remaining options. In addition, it is argued that both women and men, when selecting a mate, think more about their partners as companions than as future parents of their offspring.

2-2b Social Learning

Derived from the school of behavioral psychology, the social learning theory emphasizes the roles of reward and punishment in explaining how a child learns gender role behavior. This is in contrast to the biological explanation for gender roles. For example, consider two young brothers who enjoy playing "lady"; each of them puts on a dress, wears high-heeled shoes, and carries a pocketbook. Their father comes home early one day and angrily demands, "Take those clothes off and never put them on again. Those things are for women." The boys are punished for "playing lady" but rewarded with their father's approval for boxing or playing football (both of which involve hurting others).

Reward and punishment alone are not sufficient to account for the way in which children learn gender roles. Children also learn gender roles when parents or peers offer direct instruction (e.g., "girls wear dresses" or "a man stands up and shakes hands"). In addition, many of society's gender rules are learned through modeling. In modeling, children observe and imitate another's behavior. Gender role models include parents, peers, siblings, and characters portrayed in the media.

The impact of modeling on the development of gender role behavior is controversial. For example, a

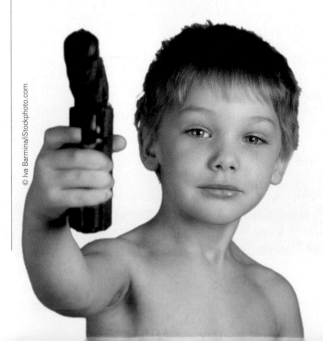

© Iva Barmina/iStockphoto.com

© Felix Mizioznikov/Shutterstock

modeling perspective implies that children will tend to imitate the parent of the same sex, but children in all cultures are usually reared mainly by women. Yet this persistent female model does not seem to interfere with the male's development of the behavior that is considered appropriate for his gender. One explanation suggests that boys learn early that our society generally grants boys and men more status and privileges than it does to girls and women. Therefore, boys devalue the feminine and emphasize the masculine aspects of themselves.

2-2c Identification

Freud was one of the first researchers to study gender role development. He suggested that children acquire the characteristics and behaviors of their same-sex parent through a process of identification. Girls identify with their mothers; boys identify with their fathers. For example, girls are more likely to become involved in taking care of children because they see women as the primary caregivers of young children. In effect, they identify with their mothers and will see their own primary identity and role as those of a mother. Likewise, boys will observe their fathers and engage in similar behaviors to lock in their own gender identity. The classic example is the son who observes his father shaving and wants to do likewise (be a man, too).

2-2d Cognitive-Developmental

The cognitive-developmental theory of gender role development reflects a blend of the biological and social learning views. According to this theory, the biological readiness of the child, in terms of cognitive development, influences how the child responds to gender cues in the environment (Kohlberg, 1966). For example, gender discrimination (the ability to identify social and psychological characteristics associated with being female or male) begins at about age 30 months. However, at this age, children do not view gender as a permanent characteristic. Thus, even though young children may define people who wear long hair as girls and those who never wear dresses as boys, they also believe they can change their gender by altering their hair or changing clothes.

Not until age 6 or 7 do children view gender as permanent (Kohlberg, 1966, 1969). In Kohlberg's view, this cognitive understanding involves the development of a specific mental ability to grasp the idea that certain basic characteristics of people do not change. Once children learn the concept of gender permanence, they seek to become competent and proper members of their gender group. For example, a child standing on the edge of a school playground may observe one group of children jumping rope while another group is playing football. That child's gender identity as either a girl or a boy connects with the observed gender-typed behavior, and the child joins one of the two groups. Once in the group, the child seeks to develop behaviors that are socially defined as gender-appropriate.

2-3 Agents of Socialization

Three of the four theories discussed in the preceding section emphasize that gender roles are learned through interaction with the environment. Indeed, though biology may provide a basis for one's gender identity, cultural influences in the form of various socialization agents (parents, peers, religion, and the media) shape the individual toward various gender roles. These powerful influences in large part dictate what people think, feel, and do in their roles as man or woman. In the next section, we look at the different sources influencing gender socialization.

2-3a Family

The family is a gendered institution with female and male roles highly structured by gender. The names parents assign their children, the clothes they dress them in, and the toys they buy them all reflect gender. Parents may also be stricter on female children—determining the age they are allowed to leave the house at night on a date, texting "mama when [they] get to the party," and being assigned more chores (Mandara et al., 2010).

The importance of the father in the family was noted in Pollack's (2001) study of adolescent boys. "America's boys are crying out for a new gender revolution that does for them what the last forty years of feminism has tried to do for girls and women," he stated (p. 18). This new revolution will depend on fathers who teach their sons that feelings and relationships are important. How equipped do you feel today's fathers are to provide these new models for their sons?

Siblings also influence gender role learning. As noted in Chapter 1, the relationship with one's sibling (particularly in sister–sister relationships) is likely to be the most enduring of all relationships (Meinhold et al., 2006). Also, growing up in a family of all sisters or all brothers intensifies social learning experiences toward femininity or masculinity. A male reared with five sisters and a single-parent mother is likely to reflect more feminine characteristics than a male reared in a home with six brothers and a stay-at-home dad.

2-3b Race/Ethnicity

The race and ethnicity of one's family also influence gender roles. Although African American families are often stereotyped as being matriarchal, the more common pattern of authority in these families is egalitarian (Taylor, 2002). Both President Barack Obama and First Lady Michelle Obama have law degrees from Harvard, and their relationship appears to be very egalitarian.

The fact that African American women have increased economic independence provides a powerful role model for young African American women. A similar situation exists among Hispanics, who represent the fastest-growing segment of the U.S. population. Mexican American marriages have great variability, but where the Hispanic woman works outside the home, her power may increase inside the home. However, because Hispanic men are much more likely to be in the labor force than Hispanic women, traditional role relationships in the family are more likely to be the norm.

2-3c Peers

Though parents are usually the first socializing agents that influence a child's gender role development, peers become increasingly important during the adolescent/school years. In regard to the degree to which peers influence the use of alcohol, 371 adolescents ages 11 to 13 years old (55.5% female, 83.0% Caucasian) participated in a study (Trucco et al., 2011). The researchers found that having peers who used alcohol and who approved of alcohol use by the participant predicted initiation of alcohol use. In another study, though an adolescent's best friend was an important influence for deviant behavior (e.g., getting drunk, smoking, fighting), the number of friends an adolescent had and

"Women always worth about the things that men forget; men always worry about the things women remember."

—AUTHOR UNKNOWN

those friends' behavior was also influential (Rees and Pogarsky, 2011). Having only one close friend who engaged in a high frequency of deviant behavior (e.g., smoking, drinking, driving recklessly) resulted in a greater influence than having several close friends (only one of whom engaged in deviant behavior).

Peers also influence gender roles throughout the family life cycle. In Chapter 1, we discussed the family life cycle and noted the various developmental tasks throughout the cycle. With each new stage, role changes are made, and one's peers influence those role changes. For example, when a couple moves from being child-free to being parents, peers who are also parents quickly socialize them into the role of parent and the attendant responsibilities.

2-3d Religion

Of Americans age 18 and older, 56% say that religion is "very important" in their lives; 48% of those ages 18 to 29 report this level of importance for religion (Pew Research Center, 2010). Exposure to religion includes a traditional framing of gender roles. Male dominance is indisputable in the hierarchy of religious organizations. Mormons, particularly, adhere to traditional roles in marriage where men are regarded as the undisputed head of the household. Other research confirms that the stronger the religiosity for men, the more traditional and sexist their view of women (Maltby et al., 2010). The book *Raising a Modern-Day Knight* (Lewis, 2007) is a religious parenting guide exhorting fathers to encourage their sons to look forward to marriage and to be the "leader/head" of the family.

Other examples of religion and gender roles include the Roman Catholic Church, which does not have female clergy. Men dominate the 19 top positions in the U.S. dioceses. In addition, the Mormon Church is dominated by men and does not provide positions of leadership for women.

2-3e Education

The educational institution serves as an additional socialization agent for gender role ideology. Schools are basic cultures of transmission in that they make deliberate efforts to reproduce the culture from one generation to the next. In the 1950s, "Dick and Jane" books read by legions of school-age children conveyed traditional gender roles and relationships. Today, the message is more egalitarian. In a study of gender role attitudes in 24 highly developed countries, higher educational attainment was associated with approving egalitarian gender roles (Boehnke, 2011).

2-3f Economy

The economy is a very gendered institution. **Occupational sex segregation** is the concentration of men and women in different occupations which has grown out of traditional gender roles. Men dominate as vascular surgeons, airline pilots, architects, and auto mechanics; women dominate as elementary school teachers, florists, and hair stylists (Weisgram et al., 2010). Female-dominated occupations tend to require less education and to have lower status. The gains of women have been uneven because women have had a strong incentive to enter male jobs, but men have had little incentive to take on female activities or jobs (England, 2010). The salaries in female occupations have remained relatively low across time. Women are aware that there are gender inequities in pay, and they expect to be paid less than men (Hogue et al., 2010).

2-3g Mass Media

Media images (e.g., movies, television, books, magazines, newspapers, computer games, music) of women and men typically conform to traditional gender stereotypes, and media portrayals depicting the exploitation, victimization, and sexual objectification of women are common. Observe the array of magazine covers at any newsstand. Compare the number of provocative females with males. Similarly, notice the traditional gender role scripting in popular television programs. Kim and colleagues (2007) identified the gender or sexual scripting of 25 prime-time television shows and found evidence of the traditional scripting (e.g., men are dominant and "need sex"; women are passive and valued for their bodies).

In regard to the influence of music on gender roles, a preference for hip-hop music is associated with gender stereotypes (e.g., men are sex driven and tough, women are sex objects; Ter Gogt et al., 2010). As for the influence of music television videos,

occupational sex segregation the concentration of women in certain occupations and men in other occupations.

a content analysis of 34 music videos revealed stereotypical notions of women as sexual objects, females as subordinate, and males as aggressive (Wallis, 2011). Rivadeneyraa and Lebob (2008) studied ninth-grade students and found that watching "romantic" television (e.g., soaps, Lifetime movies) was associated with having more traditional gender role attitudes in dating situations.

The cumulative effects of family, peers, religion, education, the economy, and mass media perpetuate gender stereotypes. Each agent of socialization reinforces gender roles that are learned from other agents of socialization, thereby creating a gender role system that is deeply embedded in our culture.

2-4 Consequences of Traditional Gender Role Socialization

This section discusses different consequences, both negative and positive, for women and men, of traditional female and male socialization in the United States.

2-4a Consequences of Traditional Female Role Socialization

Table 2.1 summarizes some of the negative and positive consequences of being socialized as a woman in U.S. society. Each consequence may or may not be true for a specific woman. For example, although women in general have less education and income, a particular woman may have more education and a higher income than a particular man.

Table 2.1
Consequences of Traditional Female Role Socialization

Negative Consequences	Positive Consequences
Less income (more dependent)	Longer life
Feminization of poverty	Stronger relationship focus
Higher STD/HIV infection risk	Keep relationships on track
Negative body image	Bonding with children
Less marital satisfaction	Identity not tied to job

© Cengage Learning 2013

Negative Consequences of Traditional Female Role Socialization There are several negative consequences of being socialized as a woman in our society.

1. **Less Income.** Although women are increasingly more likely than men to complete a four-year college degree (Wells et al., 2011), their incomes have not caught up. Women still earn about two thirds of what men earn, even when the level of educational achievement is identical (see Table 2.2).

© COlga Bogatyrenko/Shutterstock

Table 2.2
Women's and Men's Median Income with Similar Education

	Bachelor's Degree	Master's Degree	Doctoral Degree
Men	$54,091	$69,825	$89,845
Women	$35,972	$50,576	$65,587

Source: *Statistical Abstract of the United States* (2012, 131th ed.). Washington, DC: U.S. Census Bureau, Table 702.

That women with the same education earn less than men reflects two realities. One, sexism continues to exist as revealed in the occupational sex segregation discussed earlier. If a woman occupies the position, the salary is lower. The assumption is sometimes made that the woman is married, that her husband has a higher income to supplement her income, and that she will work for less.

In addition, women are paid less because they continue to prioritize children and family over income and career advancement. Women often seek jobs that dovetail with their child's school hours (e.g., elementary school teacher) or jobs that are flexible so as to allow them to be with their children when they need to. Increasingly, men are being socialized to become actively involved coparents, but such involvement may come at a price. Men who tell their employers that they must have flexible hours to help with their children are not placed in positions which earn the most income or provide for career advancement, or both. In one family we know, the parents have similar educations. But his income is four times hers, which will take care of their four children. If he "cuts back," her income will not make up the slack, so they are forced into traditional gender roles by economics.

2. **Feminization of poverty.** Another reason many women are relegated to a lower income status is the **feminization of poverty**. This term refers to the disproportionate percentage of poverty experienced by women living alone or with their children. Single mothers are particularly associated with poverty.

When head-of-household women are compared with married-couple households, the median income is $32,597 versus $71,830 (*Statistical Abstract of the United States,* 2012 Table 692). The process is cyclical—poverty contributes to teenage pregnancy—because teens have limited supervision and few alternatives to parenthood.

Such early childbearing interferes with educational advancement and restricts women's earning capacity, which keeps them in poverty. Their offspring are born into poverty, and the cycle begins anew.

Even if they get a job, women tend to be employed fewer hours than men, and they earn less money, even when they work full time. Not only is discrimination in the labor force operating against women, but women also usually make their families a priority over their employment, which translates into less income. Such prioritization is based on the patriarchal family, which ensures that women stay economically dependent on men and are relegated to domestic roles. Such dependence limits the choices of many women.

Low pay for women is also related to the fact that they tend to work in occupations that pay relatively low incomes. Indeed, women's lack of economic power stems from the relative dispensability of women's labor (it is easy to replace) and how work is organized (men control positions of power). Women also live longer than men, and poverty is associated with being elderly.

© Zak Waters/Alamy

When women move into certain occupations, such as teaching, there is a tendency in the marketplace to segregate these occupations from men's, and the result is a concentration of women in lower-paid occupations. The salaries of women in these occupational roles increase at slower rates. For example, salaries in the elementary and secondary teaching profession, which is predominantly female, have not kept pace with inflation.

Conflict theorists assert that men are in more powerful roles than women and use this power to dictate incomes and salaries of women and "female professions." Functionalists also note that keeping salaries low for women keeps women dependent and in childcare roles so as to keep equilibrium in the family. Hence, for both conflict and structural reasons, poverty is primarily a feminine issue. One of the consequences of being a woman is to have an increased chance of feeling economic strain throughout life.

feminization of poverty the idea that women disproportionately experience poverty.

3. **Higher risk for sexually transmitted infections.** Gender roles influence a woman's vulnerability to STIs and HIV not only because women receive more bodily fluids from men but also because heterosexual women have sex with men who have a greater number of partners. In addition, some women feel limited power to influence their partners to wear condoms (East et al., 2011).

4. **Negative body image.** Just as young girls tend to have less positive self-concepts than do boys (Yu and Zie, 2010), they also feel more negatively about their bodies due to a Western cultural emphasis on being thin and trim (Grogan, 2010). Gender differences in body dissatisfaction do not exist throughout the world. In a study of body satisfaction/dissatisfaction of 513 Malay, Indian, and Chinese adolescent boys and girls, the researchers found no overall body dissatisfaction and no differences between the genders (Mellor et al., 2010).

American women also live in a society that devalues them. **Sexism** is an attitude, action, or institutional structure that subordinates or discriminates against individuals or groups because of their sex. Sexism against women reflects the tradition of male dominance and presumed male superiority in American society.

Benevolent sexism (reviewed by Maltby et al., 2010) is a related term and reflects the belief that women are innocent creatures who should be protected and supported by men. Although such a view has many positive aspects, it assumes that women are best suited for domestic roles and need to be taken care of by a man (e.g., fathers and husbands) because they are not capable of taking care of themselves.

sexism an attitude, action, or institutional structure that subordinates or discriminates against an individual or group because of their sex.

5. **Less marital satisfaction.** In a study of 2,511 married couples, the wives had poorer mental health than their husbands (Read & Grundy, 2011). Women also tend to be more angry and more depressed than men. Their anger is related to the sense of powerlessness that women feel in America—inequitable division of labor, lower status/lower wage jobs, less power in relationships, and the like. (Simon & Lively, 2010). Analysis of data from the General Social Survey over a 30-year period (1972–2002) found that women reported less marital satisfaction than men (Corra et al., 2009). The same finding was revealed by Bulanda (2011), who suggested that the lower marital satisfaction of wives is attributed to power differentials in the marriage. Traditional husbands expect to be dominant, which translates into their earning an income and having the expectation that the wife will not only earn an income but also take care of the house and children.

> Sexism against women reflects the tradition of male dominance and presumed male superiority in American society.

Positive Consequences of Traditional Female Role Socialization We have discussed the negative consequences of being born and socialized as a woman. However, there are also decided benefits.

1. **Longer life expectancy.** Females born in 2015 can expect to live to the age of 81.4; men can expect to live to the age of 76.4 (*Statistical Abstract of the United States*, 2012, Table 104).

2. **Stronger relationship focus.** Women prioritize family over work and do more child care than men (Craig & Mullan, 2010). Mothers provide more "emotion work," helping children with whatever they are struggling with, and so on. (Minnottea et al., 2010).

© Kristian Dowling/Getty Images

3. **Keeping relationships on track.** Because women show more concern for relationships, they are more likely to move the couple to seek help (Eubanks Fleming and Córdova, 2012) and to initiate conversation when there is a problem. In regard to the latter, when undergraduates report that their partner has cheated, women are significantly more likely to confront the partner about the infidelity (Barnes et al., 2012). Women are also more likely to bring up a discussion of the future of the relationship and doing so tends to have a positive outcome for the relationship (Nelms et al., 2012).

4. **Bonding with children.** Another advantage of being socialized as a woman is the potential to have a closer bond with children. In general, women tend to be more emotionally bonded with their children than are men. Although the new cultural image of the father is of one who engages emotionally with his children, many fathers continue to be content for their wives to take care of their children, with the result that mothers, not fathers, become more emotionally bonded with their children.

2-4b Consequences of Traditional Male Role Socialization

Male socialization in American society is associated with its own set of consequences. Both the negative and positive consequences are summarized in Table 2.3. As with women, each consequence may or may not be true for a specific man.

Negative Consequences of Traditional Male Role Socialization There are several negative consequences associated with being socialized as a man in U.S. society.

1. **Identity synonymous with occupation.** Ask men who they are, and many will tell you

Table 2.3
Consequences of Traditional Male Role Socialization

Negative Consequences	Positive Consequences
Identity tied to work role	Higher income and occupational status
Limited emotionality	More positive self-concept
Fear of intimacy; more lonely	Less job discrimination
Disadvantaged in getting custody	Freedom of movement; more partners to select from; more normative to initiate relationships
Shorter life	Happier marriage

© Cengage Learning 2013

© Radius Images/Jupiterimages

what they do. Society tends to equate a man's identity with his occupational role. Male socialization toward greater dedication to their job is evident in governmental statistics.

Maume (2006) analyzed national data on taking vacation time and found that men were much less likely to do so. They cited fear that doing so would affect their job or career performance evaluation. Women, on the other hand, were much more likely to use all of their vacation time. However, the "work equals identity" equation for men may be changing. Increasingly, as women are more present in the labor force and become co-providers, men become co-nurturers and co-homemakers. In addition, more stay-at-home dads and fathers are seeking full custody in divorce litigation. These changes challenge cultural notions of masculinity.

That men work more and play less may translate into fewer friendships and relationships. In a study of 377 university students, 25.9% of the men compared to 16.7% of the women reported feeling a "deep sense of loneliness" (Vail-Smith et al., 2007). Similarly, Grief (2006) reported that a quarter of 386 adult men reported that they did not have enough friends. Grief also suggested some possible reasons for men having few friends—homophobia, lack of role models, fear of being vulnerable, and competition between men. McPherson and colleagues (2006) also found that men reported fewer confidants than women.

2. **Limited expression of emotions.** Some men feel caught between society's expectations that they be competitive, aggressive, and unemotional and their own desire to be more cooperative, passive, and emotional. Congressman John Boehner's being

© Luis Louro/Shutterstock

tearful and emotional has been commented on in the media—a congresswoman being emotional would go without notice or comment. Most men not only cry less (Barnes et al., 2012) but are pressured to disavow any expression that could be interpreted as feminine (e.g., emotional). As evidence, 55% of the soldiers serving in Iraq or Afghanistan reported that they feared they would appear "weak" if they expressed feelings of fear or symptoms of posttraumatic stress disorder (Thompson, 2008). In addition, Cordova and colleagues (2005) studied a sample of husbands and wives and confirmed that the men were less able to express their emotions than the women. Notice that men are repeatedly told to "prove their manhood" (which implies not being emotional), whereas women in our culture have no dictum "to 'prove their womanhood'—the phrase itself sounds ridiculous" (Kimmel, 2001, p. 33). Men who are emotional are forced to hide their tears, speak loudly with authority, and engage in "manly behavior."

3. **Fear of intimacy.** Lease and colleagues (2010) confirmed that men are socialized to be less emotional, less supportive, and more competitive

in their relationships with others. The emotional detachment of men stems from the provider role, which requires them to stay in control. Being emotional is seen as weakness (Garfield, 2010).

4. **Custody disadvantages.** Courts are sometimes biased against divorced men who want custody of their children. Because divorced fathers are typically regarded as career-focused and uninvolved in child care, some are relegated to seeing their children on a limited basis, such as every other weekend or four evenings a month.

5. **Shorter life expectancy.** Men typically die five years sooner (at age 76.4) than women (*Statistical Abstract*, 2012, Table 104). One explanation is that the traditional male role emphasizes achievement, competition, and suppression of feelings, all of which may produce stress. Not only is stress itself harmful to physical health, but it may lead to compensatory behaviors such as smoking, alcohol or other drug abuse, and dangerous, risk-taking behavior (all of which are higher in males).

In sum, the traditional male gender role is hazardous to men's physical health. However, as women have begun to experience many of the same stresses and behaviors as men, their susceptibility to stress-related diseases has increased. For example, since the 1950s, male smoking has declined whereas female smoking has increased, resulting in an increased incidence of lung cancer in women.

Benefits of Traditional Male Socialization As a result of higher status and power in society, men tend to have a more positive self-concept and greater confidence in themselves. In a sample of 288 undergraduates and graduates, 48% of men, in contrast to 30% of undergraduate women, agreed that "we determine whatever happens to us, and nothing is predestined" (Dotson-Blake et al., 2008). Men also enjoy higher incomes and an easier climb up the good-old-boy corporate ladder; they are rarely stalked or are targets of sexual harassment. Other benefits are the following:

1. **Freedom of movement.** Men typically have no fear of going anywhere, anytime. Their freedom of movement is unlimited. Unlike women, who are taught to fear rape and to be aware of their surroundings, walk in well-lit places, and not walk alone after dark, men are oblivious to these fears and perceptions. They can be alone in public and be anxiety-free about something ominous happening to them.

2. **Greater available pool of potential partners.** Because of the mating gradient (men marry "down" in age and education whereas women marry "up"), men tend to marry younger women so that a 35-year-old man may view women from 20 to 40 years of age as possible mates. However, a woman of age 35 is more likely to view men her same age or older as potential mates. As she ages, fewer men are available; less so for men.

3. **Norm of initiating a relationship.** Men are advantaged because traditional norms allow men to be aggressive in initiating relationships with women. In contrast, women are less often aggressive in initiating a relationship. In a study of 1,027 undergraduates, 61.1% of the female respondents reported that they had not "asked a guy to go out" (Ross et al., 2008).

We have been discussing the respective ways in which traditional gender role socialization affects women and men. Table 2.4 summarizes 12 implications that traditional gender role socialization has for the relationships of women and men.

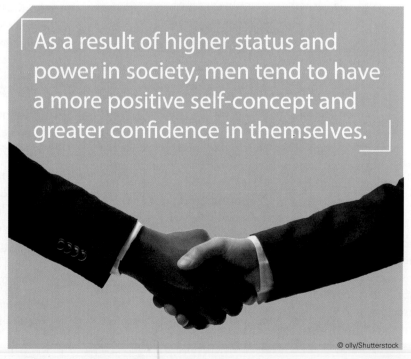

As a result of higher status and power in society, men tend to have a more positive self-concept and greater confidence in themselves.

© olly/Shutterstock

Table 2.4
Effects of Gender Role Socialization on Relationship Choices

Women

1. Women who are socialized to invest in relationships, to be in love, and to be nurturing may find it challenging to leave an emotionally or physically abusive relationship.
2. Women who are socialized to play a passive role and not initiate relationships are limiting interactions that could develop into valued relationships.
3. Women who are socialized to accept that they are less valuable and important than men are less likely to seek or achieve egalitarian relationships with men.
4. Women who internalize society's standards of beauty and view their worth in terms of their age and appearance are likely to feel bad about themselves as they age. Their negative self-concept, more than their age or appearance, may interfere with their relationships.
5. Women who are socialized to accept that they are solely responsible for taking care of their parents, children, and husband are likely to experience role overload. Potentially, this could result in feelings of resentment in their relationships.
6. Women who are socialized to emphasize the importance of relationships in their lives will continue to seek relationships that are emotionally satisfying.

Men

1. Men who are socialized to define themselves more in terms of their occupational success and income and less in terms of positive individual qualities leave their self-esteem and masculinity vulnerable should they become unemployed or work in a low-status job.
2. Men who are socialized to restrict their experience and expression of emotions are denied the opportunity to discover the rewards of emotional interpersonal sharing.
3. Men who are socialized to believe it is not their role to participate in domestic activities (child rearing, food preparation, house cleaning) will not develop competencies in these life skills. Potential partners often view domestic skills as desirable qualities.
4. Heterosexual men who focus on cultural definitions of female beauty overlook potential partners who might not fit the cultural beauty ideal but who would nevertheless be good life companions.
5. Men who are socialized to view women who initiate relationships in negative ways are restricted in their relationship opportunities.
6. Men who are socialized to be in control of relationship encounters may alienate their partners, who may desire equal influence in relationships.

© Cengage Learning 2011

2-5 Changing Gender Roles

Imagine a society in which women and men each develop characteristics, lifestyles, and values that are independent of gender role stereotypes. Characteristics such as strength, independence, logical thinking, and aggressiveness are no longer associated with maleness, just as passivity, dependence, emotions, intuitiveness, and nurturance are no longer associated with femaleness. Both sexes are considered equal, and women and men may pursue the same occupational, political, and domestic roles. Some gender scholars have suggested that people in such a society would be neither feminine nor masculine but would be described as androgynous. The next sections discuss androgyny, gender role transcendence, and gender postmodernism.

androgyny a blend of traits that are stereotypically associated with masculinity and femininity.

2-5a Androgyny

Androgyny typically refers to being neither male nor female but a blend of both traits. Two forms of androgyny are described here:

1. *Physiological androgyny* refers to intersexed individuals, discussed earlier in the chapter. The genitals are neither clearly male nor female, and there is a mixing of "female" and "male" chromosomes and hormones.

2. *Behavioral androgyny* refers to the blending or reversal of traditional

© HBSS/Corbis

male and female behavior, so that a biological male may be very passive, gentle, and nurturing and a biological female may be very assertive, rough, and selfish. Congresswoman Nancy Pelosi has both masculine and feminine qualities (Dabbous & Ladley, 2010). She has a "heart of gold" as mother and grandmother but a "spine of steel" when negotiating in Congress.

Androgyny may also imply flexibility of traits; for example, an androgynous individual may be emotional in one situation, logical in another, assertive in another, and so forth. Gender role identity (androgyny, masculinity, femininity) was assessed in a sample of Korean and American college students with androgyny emerging as the largest proportion in the American sample and femininity in the Korean sample (Shin et al., 2010).

Cheng (2005) found that androgynous individuals have a broad coping repertoire and are much more able to cope with stress. As evidence, Moore et al. (2005) found that androgynous individuals with Parkinson's disease not only were better able to cope with their disease but also reported having a better quality of life than those with the same disease

© evemilla/iStock

© Carolina K. Smith, M.D./Shutterstock

"When I look in the mirror, I don't see a female. I see a soldier."

—TERESA KING, FIRST ARMY DRILL SERGEANT

who expressed the characteristics of one gender only. Similarly, androgynous individuals reported much less likelihood of having an eating disorder (Hepp et al., 2005).

Woodhill and Samuels (2003) emphasized the need to differentiate between positive and negative androgyny. **Positive androgyny** is devoid of the negative traits associated with masculinity (aggression, hard-heartedness, indifference, selfishness, showing off, and vindictiveness). Antisocial behavior has also been associated with masculinity (Ma, 2005). Negative aspects of femininity include being passive, submissive, temperamental, and fragile. The researchers also found that positive androgyny is associated with psychological health and well-being.

2-5b Gender Role Transcendence

Beyond the concept of androgyny is that of gender role transcendence. We associate many aspects of our world, including colors, foods, social or occupational roles, and personality traits, with either masculinity or femininity. The concept of **gender role transcendence** involves abandoning gender schema (for example, becoming "gender aschematic"; Bem, 1983) so that personality traits, social or occupational roles, and other aspects of our lives become divorced from gender categories. However, such transcendence is not equal for women and men. Although females are becoming more masculine, in part because our society values whatever is masculine, men are not becoming more feminine. Indeed, adolescent boys may be described as very gender-entrenched. Beyond gender role transcendence is gender postmodernism.

positive androgyny a view of androgyny that is devoid of the negative traits associated with masculinity and femininity.

gender role transcendence abandoning gender frameworks and looking at phenomena independent of traditional gender categories.

> Beyond gender role transcendence is gender postmodernism.

2-5c Gender Postmodernism

Mirchandani (2005) emphasized that empirical post-modernism can help us see into the future. Such a view would abandon the notion that the genders are natural and focus on the social construction of individuals in a gender-fluid society. Monro (2000) previously noted that people would no longer be categorized as male or female but be recognized as capable of many identities—"a third sex" (p. 37). A new conceptualization of "trans" people calls for new social structures, "based on the principles of equality, diversity and the right to self-determination" (p. 42). No longer would our society telegraph transphobia but embrace pluralization "as an indication of social evolution, allowing greater choice and means of self-expression concerning gender" (p. 42).

© s_bukley/Shutterstock.com

Lady Gaga dressing as a man at the 2011 Annual MTV Awards reflects US society's openness to a fluidity of gender roles.

STUDY TOOLS 2

Ready to study? In this book, you can:

- ⮌ Rip out the Chapter Review card in the back of the book to study for exams
- ⮌ Take the Self Assessment for this chapter (card in the back of the book) and see where you stand on the vital issues raised in the chapter

Or you can go online to CourseMate at www.cengagebrain.com for these resources:

- ⮌ Complete Practice Quizzes to prepare for tests
- ⮌ Review Key Terms Flash Cards (online or print)
- ⮌ Read about Marriage and Family in the news
- ⮌ Play "Beat the Clock" to master concepts
- ⮌ Check out Personal Applications

USE THE TOOLS.

- Rip out the Review Cards in the back of your book to study.
Or Visit CourseMate to:
- Read, search, highlight, and take notes in the Interactive eBook
- Review Flashcards (Print or Online) to master key terms
- Test yourself with Auto-Graded Quizzes
- Bring concepts to life with Videos!

Go to CourseMate for **M&F** to begin using these tools.
Access at **www.cengagebrain.com**

Complete the Speak Up
survey in CourseMate at
www.cengagebrain.com

f Follow us at
www.facebook.com/4ltrpress

Communication

"I didn't say that I **didn't say it.** I said that **I didn't say** that **I said it.** I want to make that **very clear."**

—GEORGE ROMNEY, FORMER MICHIGAN GOVERNOR

SECTIONS

3-1 The Nature of Interpersonal Communication

3-2 Conflicts in Relationships

3-3 Principles and Techniques of Effective Communication

3-4 Self-Disclosure, Lying, Secrets, and Cheating

3-5 Gender Differences in Communication

3-6 Theories Applied to Relationship Communication

3-7 Fighting Fair: Seven Steps in Conflict Resolution

Famed magician Harry Houdini and his new bride, Bess, had a fight which ended in his putting her on a train back to her home in Bridgeport, Connecticut. He had told her if she "disobeyed" him he would send her home. "I always keep my word. Good bye, Mrs. Houdini," he said mockingly as he watched the train depart. In Bridgeport, Bess was distraught, but at 2:00 a.m. the doorbell rang. She rushed to the door where Houdini stood. "See, darling, I told you I would send you away if you disobeyed, but I didn't say I wouldn't fly after you and bring you back" (Kalush & Sloman 2006, p. 37). The Houdini spat and resolution reflect communication and conflict in relationships—the subject of this chapter.

"Good communication" is regarded as the primary factor responsible for a good relationship. Individuals report that communication confirms the quality of their relationship ("We can talk all night about anything and everything") or condemns their relationship ("We have nothing to say to each other; we are getting a divorce"). Positive marital communication also has a positive effect on mental health (Braun et al., 2010).

Communication between romantic partners today is influenced by a technological explosion including text messaging, Facebook, sexting, Skype, and more. Lovers now "stay connected all day" and may send sexual photos via their cell phone to each other out of attachment anxiety (Weisskirch and Delevi, 2011). This chapter acknowledges the effect of technology on relationships today.

3-1 The Nature of Interpersonal Communication

Communication is both verbal and nonverbal. Although most communication is focused on verbal content, most interpersonal communication (estimated to be as high as 80%) is nonverbal. **Nonverbal communication** is the "message about the message," the gestures, eye contact, body posture, tone, volume, and rapidity of speech. Even though a person says, "I love you and am faithful to you," crossed arms and lack of eye contact will convey a very different (and more important) meaning than the same words accompanied by a tender embrace and sustained eye-to-eye contact.

nonverbal communication
the "message about the message," using gestures, eye contact, body posture, tone, volume, and rapidity of speech.

© Petr Vaclavek/Shutterstock

We tend to assign more importance to nonverbal cues than to verbal cues. In effect, we like to hear sweet words, but we feel more confident when we see behavior that supports the words. "Your behavior is so loud I can't hear what you are saying" is a phrase that reflects a partner's focus on behavior rather than words.

Flirting is a good example of both nonverbal and verbal behavior. One researcher defined flirting as showing another person romantic interest without serious intent (Moore, 2010). Examples of how interest is shown include preening, such as stroking one's hair or adjusting one's clothing, and positional cues, such as leaning toward or away from the target person. Individuals go through a series of steps on their way to sexual intercourse, and the female is responsible for signaling the male that she is interested so that he moves the interaction forward. At each stage, she must signal readiness to move to the next stage. For example, if the male holds the hand of the female, she must squeeze his hand before he can entwine their fingers. She will control the speed of the interaction.

Communication via texting has become commonplace in the development/maintenance (Looi et al. 2010) and ending (Faircloth et al., 2012; Gershon, 2010) of romantic relationships (as well as coparenting after divorce; Ganong et al. 2011). Individuals use their cell phones to send text messages more than any other technology. Coyne et al. (2011) found that the most common reasons for texting are to express affection (75%), discuss serious issues (25%) and apologize (12%). Women text more than men. In a Nielsen State of the Media Survey, women spent 818 minutes texting 640 messages; men, 716 minutes texting 555 messages (Carey and Salazar, 2011).

conflict the interaction that occurs when the behavior or desires of one person interfere with the behavior or desires of another.

Texting in romantic relationships (and also used in families; Coyne et al. 2011b) occurs at a very high rate with an average of 2,272 messages a month (Bauerlein, 2010). Perry and Werner-Wilson (2011) assessed the use of technology—referred to as computer-mediated communication, or CMC)—in conflict resolution in romantic relationships, and found that CMC allowed for de-escalation of conflict and provided time for couples to construct ideas and think about what they were going to say.

3-2 Conflicts in Relationships

Conflict can be defined as the process of interaction that results when the behavior of one person interferes with the behavior of another. A professor in a marriage and family class said, "If you haven't had a conflict with your partner, you haven't been going together long enough." This section explores the inevitability, desirability, sources, and styles of conflict in relationships.

3-2a Inevitability of Conflict

If you are alone this Saturday evening from 6:00 until midnight, you are assured of six conflict-free hours. However, if you plan to be with your partner or roommate during that time, the potential for conflict exists. Whether you eat out, where you eat, where you go after dinner, and how long you stay must be negotiated. Although it may be relatively easy for you and your companion to agree on the evening's agenda, marriage involves the meshing of desires on an array of issues for potentially 60 years or more. Conflict is inevitable in any intimate relationship.

3-2b Benefits of Conflict

Conflict can be healthy and productive for a couple's relationship. Ignoring an issue may result in the partners becoming increasingly

© wavebreakmedia ltd/Shutterstock

resentful and dissatisfied with the relationship. Couples in trouble are not those who disagree but those who never discuss their disagreements. However, sustained conflict over chronic problems contributes to poor mental and physical health of the partners in the relationship (Wickrama et al., 2010).

3-2c Sources of Conflict

Conflict has numerous sources, some of which are easily recognized, whereas others are hidden inside the web of interaction.

1. **Behavior.** Your partner's behavior affects how you feel. Drinking too much, lying to you, and spending money are all behaviors which may result in discord and conflict. Think of what your partner does that upsets you, and you have an example of how behavior can be a source of conflict. Relationship happiness can be thought of as an exchange of behavior of a kind and at a rate that is mutually satisfying to the respective partners. When your partner engages in the kind of behavior that you like (compliments, expressions of love, affection, sex) and does so at a frequency that you enjoy (usually a high frequency), *and* avoids doing things you do not like (criticize you, cheat, be late) at the desired rate (zero), you will define yourself as happy with your partner.

2. **Cognitions and perceptions.** Aside from your partner's actual behavior, your cognitions and perceptions of a behavior can be a source of satisfaction or dissatisfaction. One husband complained that his wife "had boxes of coupons everywhere and always kept the house a wreck." The wife made the husband aware that she saved $100 on their grocery bill every week and asked him to view the boxes and the mess as "saving money." He changed his view, and the clutter ceased to be a problem.

3. **Value differences.** Because you and your partner have had different socialization experiences, you may also have different values—about religion (one feels religion is a central part of life; the other does not), money (one feels uncomfortable being in debt; the other has the buy-now-pay-later philosophy), in-laws (one feels responsible for parents when they are old; the other does not), and children (number, timing, discipline). The effect of value differences depends less on the degree of the difference than on the degree of rigidity with which each partner holds his or her values. Dogmatic and rigid thinkers, feeling threatened by value disagreement, may try to squash alternative views and thus produce more conflict. Partners who recognize the inevitability of difference may consider the positives of an alternative view and move toward acceptance.

When both partners do this, the relationship takes priority, and the value differences suddenly become less important.

4. **Inconsistent rules.** Partners in all relationships develop a set of rules to help them function smoothly. These unwritten but mutually understood rules include what time you are supposed to be home after work, whether you should call if you are going to be late, how often you can see friends alone, and when and how to make love. Conflict results when the partners disagree on the rules or when inconsistent rules develop in the relationship. For example, one wife expected her husband to take a second job so they could afford a new car, but she also expected him to spend more time at home with the family.

5. **Leadership ambiguity.** Unless a couple has an understanding about which partner will make decisions in which area (for example, the wife may make decisions about money management, and the husband may make decisions about rearing the children), unnecessary conflict may result. Whereas some couples may want to discuss certain issues, others may want to develop a clear specification of roles.

3-2d Styles of Conflict

Spouses develop various styles of conflict. If you were watching a DVD of various spouses disagreeing over the same issue, you would notice at least six styles of conflict. These styles have been described by Greeff and De Bruyne (2000) as the following:

Competing Style The partners are both assertive and uncooperative. Both try to force their way on the other so that there is a winner and a loser. A couple arguing over whether to discipline a child with a spanking or a time-out would resolve the argument with the dominant partner forcing a decision.

Collaborating Style The respective partners are both assertive and cooperative. Both partners express their views and cooperate to find a solution. A spanking, a time-out, or just a talk with the child might resolve the previous issue, but both partners would be satisfied with the resolution.

competing style of conflict conflict style in which partners are both assertive and uncooperative. Each tries to force a way on the other so that there is a winner and a loser.

Compromising Style Here there is an intermediate solution: Both partners find a middle ground they can live with—perhaps spanking the child for serious infractions such as playing with matches in the house and imposing a time-out for talking back.

Avoiding Style The partners are neither assertive nor cooperative. They avoid a confrontation and let either parent control the disciplining of the child. Thus the child might be both spanked and put in time-out. The avoidance style is a way of keeping conflict low in a relationship. The downside is that partners may become resentful.

Accommodating Style The respective partners are not assertive in their positions, but each accommodates the other's point of view. Each attempts to soothe the other and to avoid conflict. Although the goal of this style is to rise above the conflict and keep harmony in the relationship, fundamental feelings about the "rightness" of one's own approach may be maintained.

Parallel Style Both partners deny, ignore, and retreat from addressing a problem issue. "Don't talk about it, and it will go away" is the theme of this conflict style. Gaps begin to develop in the relationship; neither partner feels free to talk, and both believe that they are misunderstood. They eventually become involved in separate activities rather than spending time together.

Greeff and De Bruyne (2000) studied 57 couples who had been married at least 10 years and found that the collaborating style was associated with the highest level of marital and spousal satisfaction. The competitive style, used by either partner, was associated with the lowest level of marital satisfaction. Regardless of the style of conflict, partners who say positive things to each other at a ratio of 5:1 (positives to negatives) seem to stay together (Gottman, 1994).

> Conflict results when the partners disagree on the rules or when inconsistent rules develop in the relationship.

3-3 Principles and Techniques of Effective Communication

People who want effective communication in their relationship follow various principles and techniques, including the following:

1. **Make communication a priority.** Communicating effectively implies making communication an important priority in a couple's relationship. When communication is a priority, partners make time for it to occur in a setting without interruptions: They are alone, they do not answer the phone/text, and they turn the television off. Making communication a priority results in the exchange of more information between partners, which increases the knowledge each partner has about the other.

2. **Establish and maintain eye contact.** Partners who look at each other when they are talking not only communicate an interest in each other but also are able to gain information about the partner's feelings and responses to what is being said. Not looking at your partner may be interpreted as lack of interest and prevents you from observing nonverbal cues.

3. **Ask open-ended questions.** When your goal is to find out your partner's thoughts and feelings about an issue, using **open-ended questions** is best. Such questions (for example, "How do you feel about me?") encourage your partner to give an answer that contains a

> *How was your day?*

lot of information. **Closed-ended questions** (for example, "Do you love me?"), which elicit a one-word answer such as yes or no, do not provide the opportunity for the partner to express a range of thoughts and feelings.

4. **Use reflective listening.** Effective communication requires being a good listener. One of the skills of a good listener is the ability to use the technique of **reflective listening**, which involves paraphrasing or restating what the person has said to you while being sensitive to what the partner is feeling. For example, suppose you ask your partner, "How was your day?" and your partner responds, "I felt exploited today at work because I went in early and stayed late and a memo from my new boss said that future bonuses would be eliminated because of a company takeover." Listening to what your partner is both saying and feeling, you might respond, "You feel frustrated because you really worked hard and felt unappreciated . . . and it's going to get worse."

Reflective listening serves the following functions:
   a. It creates the feeling for speakers that they are being listened to and are being understood
   b. It increases the accuracy of the listener's understanding of what the speaker is saying.

If a reflective statement does not accurately reflect what a speaker thinks and feels, the speaker can correct the inaccuracy by restating her or his thoughts and feelings.

An important quality of reflective statements is that they are nonjudgmental. For example, suppose two lovers are arguing about spending time with their respective friends and one says, "I'd like to spend one night each week with my friends and not feel guilty about it." The partner may respond by making a statement that is judgmental (critical or evaluative), such as those in Table 3.1. Judgmental

open-ended question question that encourages answers that contain a great deal of information.

closed-ended question question that allows for a one-word answer and does not elicit much information.

reflective listening paraphrasing or restating what a person has said to indicate that the listener understands.

© Fotosearch/Jupiterimages; Bubble: SoleilC/Shutterstock.com

responses punish or criticize people for what they think, feel, or want and often result in frustration and resentment. Table 3.1 also provides several examples of nonjudgmental reflective statements.

5. **Use "I" statements.** "I" statements focus on the feelings and thoughts of the communicator without making a judgment on others. Because "I" statements are a clear and nonthreatening way of expressing what you want and how you feel, they are likely to result in a positive change in the listener's behavior.

In contrast, **"you" statements** blame or criticize the listener and often result in increasing negative feelings and behavior in the relationship. For example, suppose you are angry with your partner for being late. Rather than say, "You are always late and irresponsible" (which is a "you" statement), you might say, "I get upset when you are late and will feel better if you call and let me know." The latter focuses on your feelings and a desirable future behavior rather than blaming the partner for being late.

6. **Touch.** Hertenstein and colleagues (2007) identified the various meanings of touch, such as conveying emotion, attachment, bonding, compliance, power, and intimacy. The researchers also emphasized the importance of using touch as a mechanism of nonverbal communication to emphasize one's point or meaning.

7. **Use "soft" emotions.** Sanford (2007) identified "hard" emotions (e.g., angry or aggravated) and "soft" emotions (e.g., sad or hurt) displayed during conflict. The use of hard emotions resulted in an escalation of negative communication, whereas the display of soft emotions resulted in more benign

"I" statements statements that focus on the feelings and thoughts of the communicator without making a judgment on others.

"you" statements statements that blame or criticize the listener and often result in increasing negative feelings and behavior in the relationship.

Table 3.1

Judgmental and Nonjudgmental Responses to a Partner's Saying, "I'd Like to Spend One Evening a Week With My Friends"

Nonjudgmental, Reflective Statements	Judgmental Statements
You value your friends and want to maintain good relationships with them.	You only think about what you want.
You think it is healthy for us to be with our friends some of the time.	Your friends are more important to you than I am.
You really enjoy your friends and want to spend some time with them.	You just want a night out so that you can meet someone new.
You think it is important that we not abandon our friends just because we are involved.	You just want to get away so you can drink.
You think that our being apart one night each week will make us even closer.	You are selfish.

© Cengage Learning 2013

communication and more positive feelings about the importance of resolving interpersonal conflict.

8. **Avoid negative and hurtful statements to your partner.** It is important to avoid brutal statements to your partner. Indeed, be very careful how you give negative feedback or communicate disapproval to your partner. Couples in marriage counseling often report "it was a bad week" based on one negative comment made by the partner (Markman et al., 2010).

9. **Make positive statements.** We noted earlier that research has confirmed that partners who say positive things to each other at a ratio of 5:1 (positives to negatives) are more likely to report a satisfactory relationship and more likely to stay together (Gottman, 1994).

People like to hear others say positive things about them. These positive statements may be in the form of a compliment ("You look terrific!"), appreciation ("Thanks for putting gas in the car"), acknowledgment of kindness ("I appreciate your taking care of my mother"), and awareness of motivation to nurture a positive relationship ("You really love me and our relationship").

© Ariwasabi/Shutterstock

I wish you'd tell me when you buy something . . .

Photo: © Brand X Pictures/Jupiterimages; Bubble: SoleilC/Shutterstock.com

will do in similar circumstances in the future is important. For example, if going to a party together results in one partner's drinking too much and drifting off with someone else, what needs to be done in the future to ensure an enjoyable evening together? In this example, a specific resolution would be to decide how many drinks the partner will have within a given time period.

13. Give congruent messages. Congruent messages are those in which the verbal and nonverbal behaviors match. A person who says, "Okay, you're right" and smiles while embracing the partner is communicating a congruent message. In contrast, the same words accompanied by leaving the room and slamming the door communicate a very different message.

14. Share power. Power is the ability to impose one's will on the partner and to avoid being influenced by the partner. In a study of 227 relationships, most participants viewed their relationships to be ones of equal power (Oyamot ct al., 2010). In another study, over half (53%) of 2,922 undergraduates from two universities reported that they had the same amount of power in the relationship as their partner. Thirteen percent felt that they had less power, and 17%, more power (Knox & Hall, 2010). Power may take the form of love and sex. The person who loves less and who needs sex less has enormous power over the partner who is very much in love and who is dependent on the partner for sex. This pattern reflects the principle of least interest whereby the partner who has the least interest in the relationship controls the relationship.

10. Tell your partner what you want. Focus on what you want rather than on what you don't want. Rather than say, "You always leave the bathroom a wreck," an alternative might be "Please hang up your towel after you take a shower." Rather than say, "You never call me when you are going to be late," say "Please call me when you are going to be late."

> Focus on what you want rather than on what you don't want.

11. Stay focused on the issue. Branching refers to going out on different limbs of an issue rather than staying focused on the issue. If you are discussing the overdrawn checkbook, stay focused on the checkbook. To remind your partner that he or she is equally irresponsible when it comes to getting things repaired or doing housework is to get off the issue of the checkbook. Stay focused.

12. Make specific resolutions to disagreements. To prevent the same issues or problems from recurring, agreeing on what each partner

branching in communication, going out on different limbs of an issue rather than staying focused on the issue.

congruent message one in which verbal and nonverbal behaviors match.

power the ability to impose one's will on one's partner and to avoid being influenced by the partner

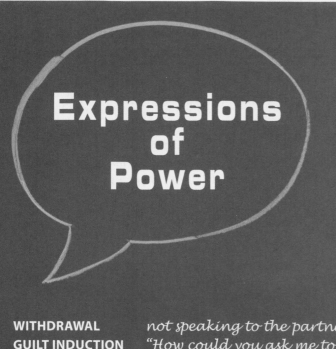

Expressions of Power

WITHDRAWAL	*not speaking to the partner*
GUILT INDUCTION	*"How could you ask me to do this?"*
BEING PLEASANT	*"Kiss me and help me move the sofa."*
NEGOTIATION	*"We can go to the movie if we study for a couple of hours before we go."*
DECEPTION	*Running up credit card debts of which the partner is unaware*
BLACKMAIL	*"I'll find someone else if you won't have sex with me."*
PHYSICAL ABUSE OR VERBAL THREATS	*"I'll kill you if you leave."*
CRITICISM	*"I can't think of anything good about you."*

15. **Keep the process of communication going.** Communication includes both content (verbal and nonverbal information) and process (interaction). It is important not to allow difficult content to shut down the communication process. To ensure that the process continues, the partners should focus on the fact that sharing information is essential and reinforce each other for keeping the process alive. For example, if your partner tells you

something that you do that bothers him or her, it is important to thank your partner for telling you rather than becoming defensive. In this way, your partner's feelings about you stay out in the open rather than being hidden behind a wall of resentment. If you punish such disclosure because you don't like the content, disclosure will stop.

Although effective communication skills can be learned, Robbins (2005) noted that physiological capacities may enhance or impede the acquisition of these skills. She noted that people with attention deficit/hyperactivity disorder (ADHD) might have deficiencies in basic communication and social skills. Being able to communicate effectively is valuable. Individuals and couples who do so feel connected and able to negotiate conflict. Those who do not feel adrift, alone, and helpless.

3-4 Self-Disclosure, Lying, Secrets, and Cheating

Shakespeare noted in *Macbeth* that "the false face must hide what the false heart doth know," suggesting that dishonesty may affect the way one feels about oneself and one's relationships with others. In this section we explore how lying, secrets, and cheating impact romantic relationships.

3-4a Self-Disclosure

Relationships become more stable when individuals disclose themselves—their formative years, previous relationships (positive and negative), experiences of elation and sadness or depression, and goals (achieved and thwarted). As we will note in the discussion of love in Chapter 5, self-disclosure is a psychological condition necessary for the development of love. To the degree that you disclose yourself to another, you invest yourself in and feel closer to that person. People who disclose nothing are investing nothing and remain aloof. One way to encourage disclosure in one's partner is to make disclosures about one's own life and then ask about the partner's life.

Individuals in the early phases of a relationship are typically judicious in what they disclose. Too much openness, too quickly, may create anxiety in the

© SW Productions/Getty Images

recipient ("Why is my partner telling me all of this?"). But no disclosure may create frustration ("My partner never tells me about anything personal—I wonder if they are hiding something?").

3-4b Lying

Lying is pervasive in American society. There is no shortage of examples. Arnold Schwarzenegger fathered a child with the family's maid while married to his wife, Maria. Presidential candidate John Edwards fathered a child with his mistress, and Tiger Woods had 13 alleged mistresses (all three marriages ended over these lies).

Politicians routinely lie to citizens ("Lobbyists can't buy my vote"), and citizens lie to the government (by cheating on their taxes). Teachers lie to students ("The test will be easy"), and students lie to teachers ("I studied all night"). Parents lie to their children ("It won't hurt"), and children lie to their parents about where they have been, whom they were with, and what they did. Dating partners lie to each other ("I've had a couple of previous sex partners"), women lie to men ("I had an orgasm"), and men lie to women ("I'll call"). When 2,922 undergraduates were asked if they had lied to a partner they were involved with, 57% reported that they had done so (Knox & Hall, 2010). The price of lying is high—distrust and alienation are the cost. One student wrote:

At this moment in my life I do not have any love relationship. I find college dating to be very hard. The guys here lie to you about anything and you wouldn't know the truth. I find it's mostly about sex here and having a good time before you really have to get serious. That is fine, but that is just not what I am all about.

Information posted on social networking sites may also be a lie. Only 31% of a large sample of social network site users reported that they were "totally honest" about what they put on Facebook, Twitter, and other sites (Carey & Trap, 2010).

3-4c Secrets

In addition to telling an outright lie, people may conceal the truth, pretend, or withhold information. Regarding the latter, in virtually every relationship, partners may not share things with each other about themselves or their past. They keep secrets.

In a study of 431 undergraduates, Easterling and colleagues (2012) found the following:

1. **Most keep secrets.** Over 60% of respondents reported ever having kept a secret from a romantic partner, and over one quarter of respondents reported currently doing so.

2. **Females keep more secrets.** Sensitivity to the partner's reaction, desire to avoid hurting the partner, and desire to avoid damaging the relationship may be the primary reasons why females were more likely than males to keep a secret from a romantic partner.

3. **Spouses keep more secrets.** Spouses have a great deal to lose if there is an indiscretion or if one partner does something the other will disapprove of (e.g., spending money on a purchase). Partners who are dating or "seeing each other" have less to lose and are less likely to keep secrets.

Lying is pervasive in American society. There is no shortage of examples.

4. **Blacks keep more secrets.** Blacks are a minority who are still victimized by the White majority. One way to avoid such victimization is to keep one's thoughts to oneself—to keep a secret. This skill of deception may generalize to one's romantic relationships.

5. **Homosexuals keep more secrets.** Indeed, the phrase "in the closet" means "keeping a secret." Transgendered individuals in Europe are required to reveal their secret before marriage (Sharpe, 2012).

Respondents were asked why they kept a personal secret from a romantic partner. "To avoid hurting the partner" was the top reason reported by 38.9% of the respondents. "It would alter our relationship" and "I feel so ashamed for what I did" were reported by 17.7% and 10.7% of the respondents, respectively.

3-4d Cheating

Even in "monogamous" relationships, there is considerable cheating. In a study of 1,341 undergraduates, 27.2% of the males and 19.8% of the females reported having oral, vaginal, or anal sex outside a relationship that their partner considered monogamous (Vail-Smith et al., 2010). People most likely to cheat in these "monogamous" relationships were men over the age of 20, those who were binge drinkers, members of a fraternity, male NCAA athletes, and those who reported that they were "nonreligious."

Cheating (infidelity) may be conceptualized as both sexual and nonsexual (Strickler & Hans, 2010). Sexual cheating involves intercourse, oral sex, and kissing. Nonsexual cheating can be interpersonal (secret time together, flirting), electronic (text messaging, e-mailing), or solitary (sexual fantasies with others, pornography, masturbation). The researchers found that of 400 undergraduates, 74% of the males and 67% of the females in a committed relationship reported that they had cheated according to their own criteria. Hence, in the survey, they identified a specific behavior as cheating and later reported that they had engaged in that behavior.

© Maciej Oleksy/Shutterstock

Gender Differences in Communication

Women and men differ in their approach to and patterns of communication. Women are more communicative about relationship issues, view a situation emotionally, and initiate discussions about relationship problems. Deborah Tannen (1990, 2006) is a specialist in communication. She observed that, to women, conversations are negotiations for closeness in which they try "to seek and give confirmations and support, and to reach consensus" (1990, p. 25). A woman's goal is to preserve intimacy and avoid isolation. To men, conversations are about winning and achieving the upper hand.

Women also tend to approach a situation emotionally. A husband might react to a child's being seriously ill by putting pressure on the wife to be mature about the situation (e.g., to stop crying) and by encouraging stoicism (e.g., asking her not to feel sorry for herself). The wife, on the other hand, wants her husband to be more emotional (e.g., to cry to show that he really cares that the child is ill). Mothers and fathers also speak differently to their children. Shinn and O'Brien (2008) observed the interactions between parents and their third-grade children and found that mothers used more affiliative (relationship) speech than fathers and that fathers used more assertive speech than mothers. No sex differences in children's speech were found, suggesting that these differences do not emerge until later.

Women disclose more in their relationships than men do (Gallmeier et al., 1997). In this study of 360 undergraduates, women were more likely to disclose information about previous love relationships, previous sexual relationships, their love feelings for the partner, and what they wanted for the future of the relationship. They also wanted their partners to reciprocate their (the women's) disclosure, but such disclosure was not forthcoming. Punyanunt-Carter (2006) confirmed that

With regard to resolving a conflict over how to spend the semester break (e.g., vacation alone or go to see parents), the respective partners must negotiate their definitions of the situation (is it about their time together as a couple or their loyalty to their parents?). The looking-glass self involves looking at each other and seeing the reflected image of someone who is loved and cared for and someone with whom a productive resolution is sought. Taking the role of the other involves each partner's understanding the other's logic and feelings about how to spend the break.

3-6b Social Exchange

Exchange theorists suggest that the partners' communication can be described as a ratio of rewards to costs. Rewards are positive exchanges, such as compliments, compromises, and agreements. Costs refer to negative exchanges, such as critical remarks, complaints, and attacks. When the rewards are high and the costs are low, the outcome is likely to be positive for both partners (profit). When the costs are high and the rewards low, neither may be satisfied with the outcome (loss).

When discussing how to spend the semester break, the partners are continually in the process of exchange—not only in the words they use but also in the way they use them. If the communication is to continue, both partners need to feel acknowledged for their points of view and to feel a sense of legitimacy and respect. Communication in abusive relationships is characterized by the parties criticizing and denigrating each other, which usually results in a shutdown of the communication process.

female college students are more likely to disclose than are male college students.

Behringer (2005) found that in spite of the fact that women and men may have different communication foci, they both value openness, honesty, respect, humor, and resolution as principal components of good communication. They also each endeavor to create a common reality. Hence, although spouses may be on different pages, they are reading the same book.

3-6 Theories Applied to Relationship Communication

Symbolic interactionism and social exchange are theories that help to explain the communication process.

3-6a Symbolic Interactionism

Interactionists examine the process of communication between two actors in terms of the meanings each attaches to the actions of the other. Definition of the situation, the looking-glass self, and taking the role of the other (discussed in Chapter 1) are all relevant to understanding how partners communicate.

3-7 Fighting Fair: Seven Steps in Conflict Resolution

When a disagreement occurs, it is important to establish rules for fighting that will leave the partners and their relationship undamaged after the disagreement. Such guidelines for fair fighting include not calling each other names, not bringing up past misdeeds, not attacking each other, and not beginning a heated discussion late at night. In some cases, a good night's sleep has a way of altering how a situation is viewed and may even result in the problem no longer being an issue.

Fighting fairly also involves keeping the interaction focused, being respectful, and moving toward a win-win outcome. If recurring issues are not discussed and resolved, conflict may create tension and distance in the relationship, with the result that the partners stop talking, stop spending time together, and stop being intimate. A conflictual, unsatisfactory marriage creates misery in much the same way that divorce diminishes the psychological, social, and physical well-being of the partners. Developing and using skills for fair fighting and conflict resolution are critical for the maintenance of a good relationship. Resolving issues via communication is not easy.

It is important to resolve conflict in a way that will leave the partners and their relationship undamaged. A study of 464 newlyweds over a four-year period emphasized that relationships are very precarious. Even couples who reported considerable satisfaction at one time ended up getting divorced (Lavner & Bradbury, 2010). The researchers recommended that couples "impose and regularly maintain ground rules for safe and nonthreatening communication." Such guidelines include not calling each other names, not bringing up past misdeeds, and not attacking each other.

Other factors are discussed in the following sections. Markman et al. (2010) found that communication skills that reflect the ability to handle conflict, which they call "constructive arguing," are the single biggest predictor of marital success over time.

3-7a Address Recurring, Disturbing Issues

Addressing issues in a relationship is important. As noted earlier, couples who stack resentments rather than discuss conflictual issues do no service to their relationship. Indeed, the healthiest response to feeling upset about a partner's behavior is to engage the partner in a discussion about the behavior. Not to do so is to let the negative feelings fester, which will result in emotional and physical withdrawal from the relationship. For example, Pam is jealous that Mark spends more time with other people at parties than with her. "When we go someplace together," she blurts out, "he drops me to disappear with someone else for two hours." Her jealousy is spreading to other areas of their relationship. "When we are walking down the street and he turns his head to look at another woman, I get furious." If Pam and Mark don't discuss her feelings about Mark's behavior, their relationship may deteriorate as a result of a negative response cycle: He looks at another woman and she gets angry; he gets angry at her getting angry and finds that he is even more attracted to other women; she gets angrier because he escalates his looking at other women, and so on.

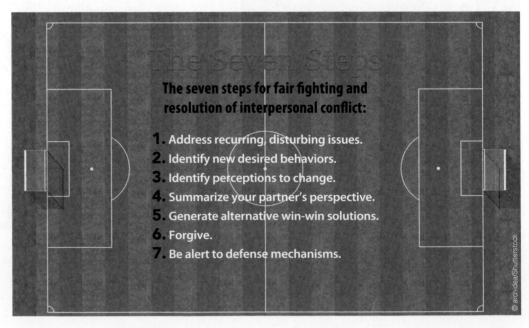

The Seven Steps

The seven steps for fair fighting and resolution of interpersonal conflict:

1. Address recurring, disturbing issues.
2. Identify new desired behaviors.
3. Identify perceptions to change.
4. Summarize your partner's perspective.
5. Generate alternative win-win solutions.
6. Forgive.
7. Be alert to defense mechanisms.

© archidea/Shutterstock

© Edw/Shutterstock

To bring the matter up, Pam might say, "I feel jealous when you spend more time with other women at parties than with me. I need some help in dealing with these feelings." By expressing her concern in this way, she has identified the problem from her perspective and asked her partner's cooperation in handling it.

When discussing difficult relationship issues, it is important to avoid attacking, blaming, or being negative. Such reactions reduce the motivation of the partner to talk about an issue and thus reduce the probability of a positive outcome.

Using good timing in discussing difficult issues with your partner is also important. In general, it is best to discuss issues or conflicts when:

1. You and your partner are in private.

2. You and your partner have ample time to talk.

3. You and your partner are rested and feeling generally good (avoid discussing conflict issues when one of you is tired, upset, or under unusual stress).

3-7b Identify New Desired Behaviors

Dealing with conflict is more likely to result in resolution if the partners focus on what they want rather than on what they don't want. For example, rather than tell Mark she doesn't want him to spend so much

time with other women at parties, Pam might tell him that she wants him to spend more time with her at parties.

3-7c Identify Perceptions to Change

Rather than change behavior, changing one's perception of a behavior may be easier and quicker. Rather than expect one's partner to always be "on time," it may be easier to drop the expectation that one's partner be on time and to stop being mad about something that doesn't matter. Pam might also decide that it does not matter that Mark looks at and talks to other women. If she feels secure in his love for her, the behavior is inconsequential.

3-7d Summarize Your Partner's Perspective

We often assume that we know what our partner thinks and why he or she does things. Sometimes we are wrong. Rather than assume how our partner thinks and feels about a particular issue, we might ask open-ended questions in an effort to learn our partner's thoughts and feelings about a particular situation.

Pam's words to Mark might be, "What is it like for you when we go to parties?" and "How do you feel about my jealousy?" Once your partner has shared thoughts about an issue with you, summarizing your partner's perspective in a nonjudgmental way is important. After Mark has told Pam how he feels about their being at parties together, she can summarize his perspective by saying, "You feel that I cling to you more than I should, and you would like me to let you wander around without feeling like you're making me angry." (She may not agree with his view, but she knows exactly what it is—and Mark knows that she knows.) In addition, Mark should summarize Pam's view—"You enjoy our being together and prefer that we hang relatively close to each other when we go to parties. You do not want me off in a corner talking to another girl or dancing."

3-7e Generate Alternative Win-Win Solutions

Looking for win-win solutions to conflicts is imperative. Solutions in which one person wins means that one person is not getting his or her needs met. As a result, the person who loses may develop feelings of resentment, anger, hurt, and hostility toward the

© Erik Snyder

brainstorming
suggesting as many alternatives as possible without evaluating them.

win-win relationship a relationship in which conflict is resolved so that each partner derives benefits from the resolution.

win-lose solution a solution to a conflict in which one partner benefits at the expense of the other.

lose-lose solution a solution to a conflict in which neither partner benefits.

winner and may even look for ways to get even. In this way, the winner is also a loser. In intimate relationships, one winner really means two losers.

Generating win-win solutions to interpersonal conflict often requires **brainstorming**. The technique of brainstorming involves suggesting as many alternatives as possible without evaluating them. The movie *Moneyball* with Brad Pitt emphasized the value of finding unique ways to solve a problem. Pitt played the role of real life baseball Oakland A's general manager Billy Beane, who brainstormed his way into having a competitive team with relatively little revenue.

Knox and colleagues (1995) conducted research on the degree to which 200 college students who were involved in ongoing relationships reported that they were in win-win, win-lose, and lose-lose relationships. Descriptions of the various relationships follow:

Win-win relationships are those in which conflict is resolved so that each partner derives benefits from the resolution. For example, suppose a couple have a limited amount of money and disagree on whether to spend it on eating out or on seeing a current movie. One possible win-win solution might be for the couple to eat a relatively inexpensive dinner and rent a movie. An example of a **win-lose solution** would be for one of the partners to get what he or she wanted (eat out or go to a movie), with the other partner getting nothing of what he or she wanted. Caughlin and Ramey (2005) studied demand-and-withdraw patterns in parent–adolescent dyads and found that the demand on the part of one of them was usually met by withdrawal on the part of the other partner. Such demand may reflect a win-lose interaction.

A **lose-lose solution** is one in which both partners get nothing that they want—in the preceding scenario, the partners would neither go out to eat nor see a movie and would be mad at each other.

In the study mentioned earlier (Knox et al., 1995), more than three quarters (77.1%) of the students reported being involved in a win-win relationship, with men and women reporting similar percentages. Of the respondents, 20% were involved in win-lose relationships. Only 2% reported that they were involved in lose-lose relationships. Of the students in win-win relationships, 85% reported that they expected to continue their relationship, in contrast to only 15% of students in win-lose relationships. No student in a lose-lose relationship expected the relationship to last.

After a number of solutions are generated, each solution should be evaluated and the best one selected. In evaluating solutions to conflicts, it may be helpful to ask the following questions:

1. *Does the solution satisfy both individuals?* Is it a win-win solution?

2. *Is the solution specific?* Does it specify exactly who is to do what, how, and when?

3. *Is the solution realistic?* Can both parties realistically follow through with what they have agreed to do?

4. *Does the solution prevent the problem from recurring?*

5. *Does the solution specify what is to happen if the problem recurs?*

Kurdek (1995) emphasized that conflict-resolution styles that stress agreement, compromise, and humor are associated with marital satisfaction, whereas conflict engagement, withdrawal, and defensiveness styles are associated with lower marital satisfaction. In Kurdek's study of 155 married couples, the style in which the wife engaged the husband in conflict and the husband withdrew was particularly associated with low marital satisfaction for both spouses.

Communicating effectively and creating a win-win context in one's relationship contributes to a high-quality marital relationship—which research indicates is good for one's health.

3-7f Forgive

Forgiveness is a quality which can transform a conflict from a hopeless deadlock to a satisfying resolution. Offender remorse positively predicts forgiveness, and such forgiveness is associated with helping to repair the damage to a relationship (Merolla & Zhang, 2011). It is also less helpful to try to "will" oneself to forgive the transgressions of another than to engage in a process of self-reflection—to understand that one also has made mistakes, has hurt others, and is guilty and to accept that we are all fallible and need forgiveness—in short, to empathize with the transgressor (Hill, 2010).

A heated discussion is often cooled when both partners decide that they are going to forgive each other for whatever and move forward.

David Knox

Forgiveness ultimately means "letting go" of one's anger, resentment, and hurt, and its power comes from its symbolic expression of love for the person who has betrayed one. Forgiveness also has a personal benefit—it releases one from feeling hurt and reduces hypertension and feelings of stress. To forgive is to restore the relationship—to pump life back into it. Of course, forgiveness given too quickly may be foolish. A person who has deliberately hurt his or her partner without remorse may not deserve forgiveness.

3-7g Be Alert to Defense Mechanisms

defense mechanisms unconscious techniques that function to protect individuals from anxiety and minimize emotional hurt.

escapism the simultaneous denial of and withdrawal from a problem.

rationalization the cognitive justification for one's own behavior that unconsciously conceals one's true motives.

projection attributing one's own feelings, attitudes, or desires to one's partner while avoiding recognition that these are one's own thoughts, feelings, and desires.

displacement shifting one's feelings, thoughts, or behaviors from the person who evokes them onto someone else.

Effective conflict resolution is sometimes blocked by **defense mechanisms**—unconscious techniques that function to protect individuals from anxiety and to minimize emotional hurt. The following paragraphs discuss some common defense mechanisms.

Escapism is the simultaneous denial of and withdrawal from a problem. The usual form of escape is avoidance. The spouse becomes "busy" and "doesn't have time" to think about or deal with the problem, or the partner may escape into recreation, sleep, alcohol, marijuana, or work. Denying and withdrawing from problems in relationships offer no possibility for confronting and resolving those problems.

Rationalization is the cognitive justification for one's own behavior that unconsciously conceals one's true motives. For example, one wife complained that her husband spent too much time at the health club in the evenings. The underlying reason for the husband's going to the health club was to escape an unsatisfying home life. However, the idea that he was in a dead marriage was too painful and difficult for the husband to face, so he rationalized to himself and his wife that he spent so much time at the health club because he made a lot of important business contacts there. Thus, the husband concealed his own true motives from himself (and his wife).

Projection occurs when one spouse unconsciously attributes individual feelings, attitudes, or desires to the partner. For example, the wife who desires to have an affair may accuse her husband of being unfaithful to her. Projection may be seen in such statements as "You spend too much money" (projection for "I spend too much money") and "You want to break up" (projection for "I want to break up"). Projection interferes with conflict resolution by creating a mood of hostility and defensiveness in both partners. The issues to be resolved in the relationship remain unchanged and become more difficult to discuss.

Displacement involves shifting your feelings, thoughts, or behaviors from the person who evokes them onto someone else. The wife who is turned down for a promotion and the husband who is driven to exhaustion by his boss may direct their hostilities (displace them) onto each other rather than toward their respective employers. Similarly, spouses who are angry at each other may displace this anger onto someone else, such as the children.

> To forgive is to restore the relationship—to pump life back into it. Of course, forgiveness given too quickly may be foolish. A person who has deliberately hurt his or her partner without remorse may not deserve forgiveness.

By knowing about defense mechanisms and their negative impact on resolving conflict, you can be alert to them in your own relationships.

3-7h When Silence is Golden

Even in the midst of a heated quarrel, some words should never be spoken. To do so is to destroy the relationship forever. William Berle, adopted son of comedian Milton Berle, recalled being in an argument with his dad who lashed out at him, saying "Oh yeah? Well, I did make one mistake and that was twenty-seven years ago when we adopted you . . . how do you like that you little prick!" (Berle, 1999, p. 188). The son recalled being devastated and walking into the next room to take out a pistol to kill himself. Luckily, a knock on the door interrupted his plan.

Truth is better than friction.
—*Charles Herguth*

© Polka Dot Images/Jupiterimages

3

Ready to study? In this book, you can:

- ➲ Rip out the Chapter Review card in the back of the book to study for exams

- ➲ Take the Self Assessment for this chapter (card in the back of the book) and see where you stand on the vital issues raised in the chapter

Or you can go online to CourseMate at www.cengagebrain.com for these resources:

- ➲ Complete Practice Quizzes to prepare for tests
- ➲ Review Key Terms Flash Cards (online or print)
- ➲ Read about Marriage and Family in the news
- ➲ Play "Beat the Clock" to master concepts
- ➲ Check out Personal Applications

M&F CHAPTER 4

Singlehood, Hanging Out, Hooking Up, and Cohabitation

"I never married because I have three pets at home that serve the **same purpose** as a **husband.** I have a **dog** that growls every **morning,** a **parrot** that sleeps all **afternoon,** and a **cat** that comes home **late at night."**

—MARIE CORELLI, BRITISH NOVELIST

SECTIONS

4-1 Singlehood

4-2 Categories of Singles

4-3 Ways of Finding a Partner

4-4 Cohabitation

© iStockphoto.com/craftvision

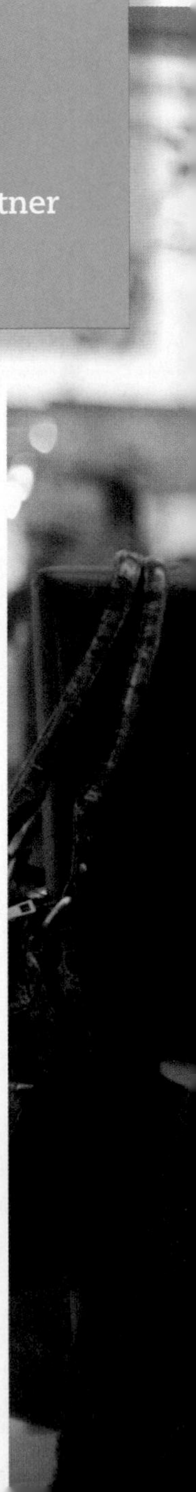

Susan Boyle is the single British singer who, at age 50, captured audiences with her beautiful voice and best-selling album *I Dreamed a Dream*. But her manager and siblings report that after her performances worldwide she returns alone to her hotel room. "Susan would be the first to admit that sometimes she feels lonely," says her manager, Jerry. Susan is not alone in her sentiments. Although all of us are lonely from time to time, we are socialized to seek and enjoy being in a relationship.

Though most end up married, youth today are in no hurry. They enjoy the freedom of singlehood, and most put off marriage until their late 20s. In the meantime, they hang out (undergraduates less often use the term "dating" but say they are "seeing someone"), hook up (the new term for "one-night stand"), and pair off in exclusive relationships, which may include cohabitation.

4-1 Singlehood

In this section, we discuss how social movements have increased the acceptance of singlehood, the various categories of single people, the choice to be permanently unmarried, the human immunodeficiency virus (HIV) infection risk associated with this choice, and the fact that more people are delaying marriage.

4-1a Individuals Are Delaying Marriage Longer

Although most Americans will eventually marry, there is a trend toward delaying when one gets married. As evidence, in 1960, two thirds (68%) of all 20-somethings were married. More recently, just over a fourth (26%) in this age category was married (Pew Research Center, 2010; see Figure 4.1). The median age at marriage for men was 28.7; for women, 26.5 (Pew Research Center, 2011). Young Americans are not alone in their delay of marriage—individuals in France, Germany, and Italy are engaging in a similar pattern.

Table 4.1 lists the standard reasons people give for remaining single. The primary advantages of remaining single are freedom and control over one's life. Others do not set out to be single but drift into singlehood longer than they anticipated, discover that they like it, and remain single.

Still other individuals delay getting married because they are afraid. In interviews with single women, some reported they were afraid to get married (Manning et al., 2010).

Figure 4.1

Median Age at First Marriage in America in Selected Years, by Sex

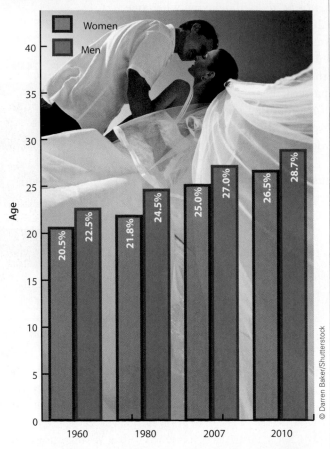

Source: U.S. Census Bureau (2006), *America's Families and Living Arrangements: 2006*, Table MS-2, Estimated Median Age at First Marriage, by Sex (updated for 2010), retrieved from http://www.census.gov; Pew Research Center, 2011.

One respondent, Tania, a 24-year-old cohabiting mother of three, reported

> I'm scared to get married . . . and being committed to somebody for so long . . . cause I know some people do get married and there are a lot of arguments in the house. It really depends on the relationship and on the person. And I know there can be a lot of arguing . . . And I think that's what scares me.

Some fear divorce.

> It's like, I mean I do [want to get married] and then I don't . . .

Because . . . here every time I turn around somebody's marriage is ending, and people are getting divorced. And that just bothers me. (Manning et al., 2010)

There is also a scarcity of men. In her article "The End of Men," Rosin (2010) noted that (due to the recession) women now hold the majority of the nation's jobs. The result is that men are being marginalized and bring less to the table economically. Rosin also states that modern industrial society no longer values men's size and strength and that "social intelligence, open communication, the ability to sit still and focus are, at a minimum, not predominantly male"— yet these are precisely the qualities now demanded in the global economy. There is more singlehood among Blacks (Chambers & Kravitz, 2011). Black women, particularly, complain that there are few Black men who have the education and job stability they prefer in a mate.

Regardless of the reason, will those who delay getting married eventually marry? Eighty-five percent of 2,922 undergraduates at two universities agreed that "someday, I want to marry" (Knox & Hall, 2010). And the older a person, the more likely the person is to have married. By age 75, 96% of Americans have married at least once (*Statistical Abstract of the United States*, 2012). In the meantime, the enjoyment of singlehood is facilitated by three social movements.

Table 4.1

Reasons to Remain Single

Benefits of Singlehood	Limitations of Marriage
Freedom to do as one wishes	Restricted by spouse or children
Variety of lovers	One sexual partner
Spontaneous lifestyle	Routine, predictable lifestyle
Close friends of both sexes	Pressure to avoid close other-sex friendships
Responsible for one person only	Responsible for spouse and children
Spend money as one wishes	Expenditures influenced by needs of spouse and children
Freedom to move as career dictates	Restrictions on career mobility
Avoid being controlled by spouse	Potential to be controlled by spouse
Avoid emotional and financial stress of divorce	Possibility of divorce

© Cengage Learning 2011

4-1b Social Movements and the Acceptance of Singlehood

The acceptance of singlehood as a lifestyle can be attributed to social movements—the sexual revolution, the women's movement, and the gay liberation movement. The sexual revolution involved openness about sexuality and permitted intercourse outside the context of marriage. No longer did people feel compelled to wait until marriage for involvement in a sexual relationship. Hence, the sequence changed from dating, love, maybe intercourse with a future spouse, and then marriage and parenthood to "hanging out," "hooking up" with numerous partners, maybe living together (in one or more relationships), marriage, and children. (For some, living together is a required relationship stage.)

The women's movement emphasized equality in education, employment, and income for women. As a result, rather than get married and depend on a husband for income, women earned higher degrees, sought career opportunities, and earned their own income. This economic independence brought with it independence of choice. Women could afford to remain single or to leave an unfulfilling or abusive relationship. Commanding respect and an egalitarian relationship have become normative to today's relationships.

The gay liberation movement, with its desire for recognition of same-sex marriage (now legal in several states), has increased the visibility of gay people and relationships. The repeal of "Don't ask, don't tell" and the increasing number of openly gay politicians (e.g., Barney Frank) and celebrities (e.g., Ellen DeGeneres) infuse new norms into our society. Though some gay people still marry heterosexuals to provide a traditional social front, the gay liberation movement has

© Creatas/Photolibrary

Marriage is only one of many acceptable family forms.

provided support for a lifestyle consistent with one's sexual orientation. This includes rejecting traditional heterosexual marriage. Today, some gay pair-bonded couples regard themselves as married even though they are not legally wed. Some gay couples have formal wedding ceremonies in which they exchange rings and vows of love and commitment.

In effect, there is a new wave of youth who feel that their commitment is to themselves in early adulthood and to marriage in their late 20s and 30s, if at all. The increased acceptance of singlehood translates into staying in school or getting a job, establishing oneself in a career, and becoming economically and emotionally independent from one's parents. The old pattern was to leap from high school into marriage. The new pattern of these Generation Yers (discussed in Chapter 1) is to wait until after college, become established in a career, and enjoy themselves. A few (less than 10% of American adults) opt for remaining single forever.

"**Marriage itself** has become increasingly optional as a context for intimate partnerships and parenthood."

—MEGAN SWEENEY, DEMOGRAPHER

The value of singlehood may vary by race. Brewster (2006) interviewed 40 African American men 18 to 25 years of age in New York City. These men dated numerous women, and some maintained serious relationships with multiple women simultaneously. These men "redefined family for themselves, and marriage is not included in the definition" (p. 32).

4-1c Alternatives to Marriage Project

According to the website of the Alternatives to Marriage Project (AtMP; *http://www.unmarried.org/aboutus.php*), the organization's mission is to advocate "for equality and fairness for unmarried people, including people who are single, choose not to marry, cannot marry, or live together before marriage." The nonprofit organization is not against marriage but provides support and information for the unmarried and "fights discrimination on the basis of marital status. . . . We believe that marriage is only one of many acceptable family forms, and that society should recognize and support healthy relationships in all their diversity."

The Alternatives to Marriage Project is open to everyone, "including singles, couples, married people, individuals in relationships with more than two people, and people of all genders and sexual orientations. We welcome our married supporters, who are among the many friends, relatives, and allies of unmarried people."

4-1d Legal Blurring of the Married and Unmarried

The legal distinction between married and unmarried couples is blurring. Whether it is called the deregulation of marriage or the deinstitutionalization of marriage, the result is the same—more of the privileges previously reserved for the married are now available to unmarried or same-sex couples, or both. As noted earlier, domestic partnership conveys rights and privileges (e.g., health benefits for a partner) previously available only to married people.

singlehood state of being unmarried.

4-2 Categories of Singles

The term **singlehood** is most often associated with young, unmarried individuals. However, there are three categories of single people: the never-married, the divorced, and the widowed. Figure 4.2 shows the distribution of the American adult population by relationship status.

4-2a Never-Married Singles

It is rare for adults to remain unmarried their entire lives. Men who never marry are more likely to be less educated and have lower incomes (Murray, 2012). Women who never marry may have lower incomes, health issues, drug abuse problems, or have children with multiple partners (Manning et al. 2010). Being obese is also a liability on the marriage market for both women and men (Sobal and Hansen, 2011).

Are people who have never married less happy? Wienke and Hill (2009) compared single people with married people and cohabitants (both heterosexual and homosexual) and found that single people were less

Figure 4.2

U.S. Adult Population by Relationship Status N = 229,100,000

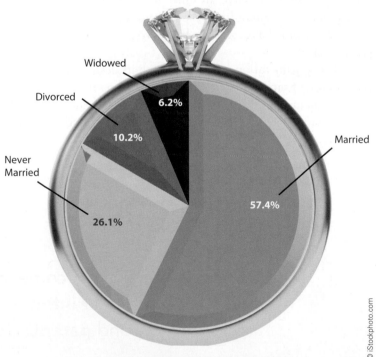

Widowed 6.2%

Divorced 10.2%

Never Married 26.1%

Married 57.4%

Source: *Statistical Abstract of the United States* (2012, 131th ed.), Table 56 (Washington, DC).

© iStockphoto.com

© Andrest/Shutterstock

happy regardless of their sexual orientation. In regard to feeling more lonely, with **loneliness** defined as the subjective evaluation that the number of relationships a person has is smaller than the person desires or that intimacy the person wants has not been realized, Fokkema et al. (2012) noted that loneliness has been associated being single.

In spite of the viability of singlehood as a lifestyle, stereotypes remain, as never-married people are sometimes viewed as desperate or as swingers. Indeed, a cultural norm still disapproves of singlehood as a lifelong lifestyle.

Jessica Donn (2005) emphasized that because mature "adulthood" implies that one is married, single people are left to negotiate a positive identity outside of marriage. She studied the subjective well-being of 171 self-identified heterosexual, never-married singles (40 men and 131 women), ages 35 to 45 years, who were not currently living with a romantic partner. Women participants reported higher life satisfaction and positive affect than did men. Donn hypothesized that having social connections and close friendships (greater frequency of social contact, more close friends, someone to turn to in times of distress, and greater reciprocity with a confidant) was the variable associated with higher subjective well-being for both women and men (but women evidenced greater connectedness). Other findings revealed that relationships, career, older age, and avoiding thoughts of old age contributed to a sense of well-being for never-married men. For women, financial security, relationships, achievement, and control over their environment contributed to a sense of well-being.

Sharp and Ganong (2007) interviewed 32 White, never-married, college-educated women ages 28 to 34 who revealed a sense of uncertainty about their lives.

> In spite of the viability of singlehood as a lifestyle, stereotypes remain, as never-married people are sometimes viewed as desperate or as swingers.

Although some were despondent that they would ever meet a man and have children, others (particularly when they became older) reminded themselves of the advantages of being single: freedom, financial independence, ability to travel, and so on. However, the overriding consensus of these respondents was that they were "running out of time to marry and to have children." The researchers emphasized the enormous cultural expectation to follow age-graded life transitions and to stay on time and on course. People outside the norm struggle with managing their differences, with varying degrees of success.

4-2b Divorced Singles

Divorced people are also regarded as single. There are 13.7 million divorced females and 9.9 million divorced males in the United States (*Statistical Abstract of the United States*, 2012, Table 57). Their return to singlehood is not easy. Divorcing spouses sometimes report that they should have been less eager to begin divorce proceedings and recommend to other spouses contemplating divorce to "work it out" and to "see a counselor" (Knox & Corte, 2007).

The divorced have a higher suicide risk. One researcher examined the living arrangements of over 800,000 adults and found that not being

loneliness
the subjective evaluation that the number of relationships a person has is smaller than the person desires or that intimacy the person wants has not been realized.

> "No, I don't have a **boyfriend.** No, **I don't need** a boyfriend. **I am enough.** I am complete just the way I am. **I choose to be single,** just like **I choose to not listen** to people who make marriage seem like the only **possible pinnacle** a life can have."
>
> —LAUREN ROHRER

married or living with children increased one's risk of suicide (Denney, 2010). Not only is the suicide rate of singles higher but they die sooner (have a shorter life span) than spouses. One explanation for greater longevity among spouses is the protective aspect of marriage. "The protection against diseases and mortality that marriage provides may take the form of easier access to social support, social control, and integration, which leads to risk avoidance, healthier lifestyles, and reduced vulnerability" (Denney, 2010, p. 375).

Married people also look out for the health of the other. Spouses often prod each other to "go to the doctor," "have that rash on your skin looked at," and "remember to take your medication." Single people often have no one in their life to nudge them toward regular health maintenance. Regarding the insulation against suicide, married people are more likely to be "connected" to intimates, and it is this connection between individuals that seems to insulate a person from suicide.

4-2c Widowed Singles

The widowed are forced into coping with the death of a spouse. There are 11.4 million widowed females and 2.9 million widowed males in the United States (*Statistical Abstract*, 2012, Table 57). These figures approximate a 10:1 ratio of females to males. The stereotype (often true) of the widow and widower is one of utter loneliness. But for some widows, there are compensations. Some widows were trapped in an unhappy marriage. Not only does the death of the spouse provide an escape from a less than fulfilling marriage but the widow

hanging out going out in groups where the agenda is to meet others and have fun.

© sprinter81/Shutterstock

may also receive the Social Security benefits of the deceased. Similarly, widowers may have been in an unhappy marriage and may receive Social Security benefits from the deceased wife. When widows were compared with spouses, the former were less likely to have a confidant but reported greater support from children, friends, and relatives (Ha, 2008). Hence, the widowed may have a broader array of relationships than those who are married. Nevertheless, widowhood represents one of the most difficult crisis events an individual confronts.

4-3 Ways of Finding a Partner

It is not just married people who have partners. Single people either have them or want them. Being connected to someone seems to be a goal for most singles. In a sample of 2,922 undergraduates at two universities, 60% reported that they were involved with someone; 30% reported no involvement at the time of the survey (Knox & Hall, 2010). In this section we review the various ways of finding a partner.

4-3a Hanging Out

Hanging out, also referred to as *getting together*, refers to going out in groups where the agenda is to meet others and have fun. The individuals go to a bar or party or eat out, or both. Of 2,922 undergraduates, 85% reported that "hanging out for me is basically about meeting people and having fun" (Knox & Hall, 2010). There is usually no agenda beyond meeting and having fun. Of the 2,922 respondents, only 4% said that hanging out was about beginning a relationship that may lead to marriage (Knox & Hall, 2010). Hanging out is basically about screening and interviewing a number of potential partners in one setting. At a party, one can drift over to a potential partner who "looks good" and start talking. If there is chemistry in the banter and perceived interest from the person, the interaction will continue and can include the exchange of phone numbers for subsequent texting. If there is no chemistry, the individual can move on to the next person without having invested any significant time. Both partners are in the process of assessing the other.

4-3b Hooking Up

Hooking up is a sexual encounter that occurs between individuals who have no relationship commitment. There is generally no expectation of seeing one another again, and alcohol usually facilitates the interaction. Data on the frequency of college students having experienced a hook-up varies. Reporting on more than 17,000 students at 20 colleges and universities (from Stanford University's Social Life Survey), sociologist Paula England revealed that 72% of both women and men reported having hooked up (40% intercourse; 35% kiss, touch; 12% hand/genital; 12% oral sex). Cunnilingus was less likely than fellatio to occur in hook-ups (Blackstrom et al., 2012). Men had 10 hook-ups; women, 7 (Jayson, 2011). In another study of 832 undergraduates, 60% of Caucasian Americans and 35% of African Americans reported having "hooked up." More alcohol use and more favorable attitudes toward hooking up were associated with a higher likelihood of having hooked up (Owen et al., 2010). Finally, in a survey of 118 first-semester female college students on their hook-up experiences, 60% had done so. Most hook-ups

involved friends (47%) or acquaintances (23%) rather than strangers (Fieldera & Careya, 2010).

These data reflect that hooking up is becoming the norm on college and university campuses. Not only do female students outnumber male students (60 to 40) on today's college and university campuses (which means women are less able to bargain sex for commitment because men have a lot of women from which to choose), but also individuals want to remain free for summer internships and study abroad (Uecker & Regnerus, 2011).

Undergraduate Interest in Finding a Partner

Many college students are not involved in an emotional relationship but are looking for a partner. In a sample of 2,922 university students, 38% reported that they were not dating and were not involved with anyone (Knox & Hall, 2010). In the same sample of undergraduates, 54% of the women and 50% of the men reported that finding a partner with whom to have a happy marriage was a top value (Knox & Hall, 2010). Indeed, all societies have a way of moving women and men from same-sex groups into pair-bonded, legal relationships for the reproduction and socialization of children.

© Yellow Dog Productions/Getty Images

hooking up having a one-time sexual encounter in which there are generally no expectations of seeing one another again.

Although hooking up may be an exciting sexual adventure, it is fraught with feelings of regret.

Women and men experience hooking up differently. Women tend to view the experience as the beginning of a relationship, the possibility that "he will call tomorrow and we will start seeing each other." Men are more hedonistic and view the hook-up as a night of free sex. They have little intention of calling the next day but will say anything during the hook-up to create the context for sex ("You are beautiful" and "Where have you been all of my life?").

One study revealed that men benefited more since they were able to have casual sex with a willing partner and no commitment. Women were more at risk for feeling guilty, becoming depressed, and defining the experience negatively. Both experienced the hook-up in a context of deception (neither being open about their relationship goals) and may have exposed themselves to an STI (Bradshaw et al., 2010).

Bogle (2008) interviewed 57 college students and alumni about their experiences with dating and sex. She found that hooking up had become the *primary* means for heterosexuals to get together on campus. About half (47%) of hook-ups start at a party and involve alcohol, with men averaging five drinks and women three drinks (England & Thomas, 2006).

Bogle (2002, 2008) noted three outcomes of hooking up. In the first, previously described, nothing results from the first-night sexual encounter. In the second, the college students will repeatedly hook up with each other on subsequent occasions of "hanging out." However, a low level of commitment characterizes this type of relationship, in that each person is still open to hooking up with someone else. A third outcome of hooking up, and the least likely, is that the two people begin going out or spending time together in an exclusive relationship. Hence, hooking up is most often a sexual adventure that rarely results in the development of a relationship. Stepp (2007) interviewed two groups of high school and one group of college students and discovered the same outcome in most relationships involving hooking up—they rarely end with the couple becoming a monogamous pair. However, there are exceptions. When 2,922 undergraduates were presented with the statement, "People who 'hook up' and have sex the first night don't end up in a stable relationship," 14% disagreed (Knox & Hall, 2010).

4-3c The Internet— Meeting Online: Upside and Downside

"We could never let anyone know how we really met," remarked a couple who had met on the Internet. "People who meet online are stigmatized as desperate losers." As more individuals use the Internet, such stigmatization is changing. Over 16 million U.S. adults report that they have looked online for a partner. Their mission is clear—to go online to shop for potential romantic partners and to "sell" themselves in hopes of creating a successful romantic relationship (Heino et al., 2010). The profiles that individuals construct or provide for others to view reflect impression management, presenting an image that is perceived to be what the target audience wants. In this regard, men tend to emphasize their status characteristics (e.g., income, education, career), whereas women tend to emphasize their youth, trim body, and beauty (Spitzberg & Cupach, 2007).

Internet Partners: The Upside The advantages of meeting someone online include its efficiency and focus on the nonphysical. In regard to efficiency, it takes time and effort to meet someone at a coffee shop for an hour, only to discover that the person has habits (e.g., does or does not smoke) or values (too religious or agnostic) that would eliminate him or her as a potential partner. Using the Internet, one can spend a short period of time and literally scan hundreds of profiles and potential partners without leaving the house. For noncollege people who are busy in their job or career, the Internet offers the chance to meet someone outside their immediate social circle.

Right Mate at Heartchoice.com is one of many dating sites that offers not only a way to meet others (through Match.com) but also provides various articles to help in the assessment of a partner. Some websites target specific demographic groups, such as Black singles (BlackPlanet.com), Jewish singles (Jdate.com), and gay people (Gay.com). In one study of online dating, women received an average of 55 replies compared to men, who reported receiving 39 replies. Younger women (average age of 35), attractive women, and those who wrote longer profiles were more successful in generating replies (Whitty, 2007).

The Internet may also be used to find out information about a partner. Argali.com can be used to find out where the Internet mystery person lives, Zabasearch .com to learn how long the person has lived there, and Zoominfo.com to locate where the person works. The person's birth date can be found at Birthdatabase.com. Women might want to see if any red flags have been posted on the Internet at Dontdatehimgirl.com.

For individuals who learn about each other online, what is it like to finally meet? First, they tend to meet each other relatively quickly; often, after three e-mail exchanges, they will move to a phone call and set up a time to meet during that phone call (McKenna, 2007). When they do meet, Baker (2007) is clear: "If they have presented themselves accurately and honestly online, they encounter few or minor surprises at the first meeting offline or later on in further encounters" (p. 108). About 7% end up marrying someone they met online (Albright, 2007).

Internet Partners: The Downside There are also downsides to meeting on the Internet. Lying is a particular problem. Seven categories of misrepresentation used by 5,020 individuals who posted profiles in search of an Internet date include: personal assets ("I own a house at the beach"), relationship goals ("I want to get married"), personal interests ("I love to exercise"), personal attributes ("I am religious"), past relationships ("I have only been married once"), weight, and age. Men were most likely to misrepresent personal assets, relationship goals, personal interests, and personal attributes, whereas women were more likely to misrepresent weight (J. A. Hall et al., 2010). In a recent survey, 34 online daters revealed that when using the Internet to find a partner, they assume there is exaggeration, and

Pros:
- Highly efficient
- Develop a relationship without visual distraction
- Avoid crowded, loud, uncomfortable locations, like bars
- The opportunity to try on new identities

Online Meeting

Cons:
- Ease of deception
- Relationship involvement escalates too quickly
- Inability to assess "chemistry" through the computer
- A lot of competition

© Iveta Angelova/Shutterstock

they compensate for such exaggeration. For example, female respondents noted that men exaggerate how tall they are, so the women automatically reduce a man's height. If a man says he is 5'11", the woman assumes he is 5'9" (Heino et al., 2010).

Physical appearance is another topic of deception. There is a relationship between the online daters' level of physical attractiveness and the use of deception. The less attractive the person, the more likely the person is to enhance her or his photographs or lie about physical descriptors (Toma & Hancock, 2010).

Online users may also lie about being single. "Saleh" was married yet maintained 50 simultaneous online relationships with other women. He allegedly wrote intoxicating love letters, many of which were cut-and-paste jobs, to various women. He made marriage proposals to several, and some bought wedding gowns in anticipation of the wedding (Albright, 2007). Although Saleh is an "Internet guy," it is important to keep in mind that people not on the Internet may also be very deceptive and cunning. To suggest that the Internet is the only place where deceivers lurk is to turn a blind eye to those people who meet through traditional channels.

It is important to be cautious of meeting someone online. There are horror stories of some people who met online. Although the Internet is a good place to meet new people, it allows someone you rejected or an old lover to monitor your online behavior. Most sites note when you have been online last, so if you reject someone online by saying, "I'm really not ready for a relationship," that same person can log on and see that you are still looking. Some individuals become obsessed with a person they meet online and turn into cyberstalkers when rejected.

Other disadvantages of online meeting include the potential to fall in love too quickly as a result of intense mutual disclosure; not being able to assess "chemistry" or to observe nonverbal cues and gestures or how a person interacts with your friends or family; and the tendency to move too quickly (from texting to phone to meeting to first date to marriage) without spending much time to get to know each other.

Another disadvantage of using the Internet to find a partner is that having an unlimited number of options sometimes results in not looking carefully at the options one has. Two researchers studied undergraduates looking for romantic partners on the Internet who had 30, 60, and 90 people to review and found that the more options the person had, the less time the undergraduate spent carefully considering each profile (Wu & Chiou, 2009). They suggested that it was better to examine a small number of potential online partners carefully than to be distracted by a large pool of applicants.

4-3d Video Chatting

As noted on its website, iSpQ Video Chat ("I speak"; *http://www.ispq.com/*) is a "free webcam program which connects you with people from around the world in a live chat community. Connecting is easy with public directories, video rooms, and a live contact list. Breakthrough video and voice technology enable a real and natural experience that shrinks any distance between two or more people." Much like Skype, iSpQ Video Chat allows one to see, hear, and interact with people on one's video screen. "Meeting people and having fun" is a primary motivation for video chatting (Hodge, 2003). The iSpQ software is free for Windows and Mac users, and versions in Dutch, German, Italian, French, and Spanish are available.

4-3e Speed-Dating: The Eight-Minute Date

© rangizzz/Shutterstock

Dating innovations that highlight the concept of speed include the eight-minute date. The website 8minutedating.com identifies "Eight-Minute Dating Events" throughout the United States at which a person has eight one-on-one dates that last eight minutes each. If both parties are interested in seeing each other again, the organizer provides contact information so that the individuals can set up another date.

Ivan places an ad in a local newspaper for a young (ages 20 to 35), trim (less than 130 pounds), single woman "interested in meeting an American male for romance and eventual marriage" and waits in a hotel for the phone to ring. Ivan and his client then meet and interview "candidates" in the hotel lobby. The contract is fulfilled when the man finds a woman he likes.

© sven hagolani/Corbis

Speed-dating is cost-effective because it allows daters to meet face-to-face without burning up a whole evening. Adams and colleagues (2008) interviewed participants who had experienced speed-dating to assess how they conceptualized the event. The researchers found that women were more likely to view speed-dating as an investment of time and energy to find someone (58% of women versus 25% of men), whereas men were more likely to see the event as one of exploration (e.g., to see how flexible a person was; 75% of men versus 17% of women). Wilson and colleagues (2006) collected data on 19 young men who had three-minute social exchanges with 19 young women and found that those partners who wanted to see each other again had more in common than those who did not want to see each other again. Common interests were assessed using the compatibility quotient (CQ).

4-3f High-End Matchmaking

Wealthy, busy clients looking for marriage partners pay Selective Search (www.selectivesearch.com) $20,000 to find them a mate. The Chicago-based service personally interviews the client and then searches the database for a partner (there are 140,000 women in the database). The women may be personally interviewed to ensure a match. Eighty-eight percent of the company's clients meet their eventual spouse within the first nine months (Stein, 2011).

4-3g International Dating

Ivan Thompson specializes in finding Mexican women for his American clients. As documented in the movie *Cupid Cowboy* (*Cowboy del Amor*, Ohayon, 2005), Ivan takes males (one at a time) to Mexico (Torreón is his favorite place). For $3,000,

4-4 Cohabitation

Cohabitation, also known as *living together*, is increasing in the United States. There are 6.5 million unmarried-couple households in the United States (*Statistical Abstract*, 2012, Table 63). Of 2,922 undergraduates, over two thirds (68%) reported that they would live with a partner they were not married to, and 16% were currently cohabiting or had already done so (Knox & Hall, 2010). Living together is typically seen, particularly by single women, as movement toward marriage (Lichter et al., 2010). Almost 60% of women in the United States have lived with a partner before age 24. In a study of 1,624 mothers cohabiting at the time of a nonmarital birth, two thirds ended their relationship within five years (with young, Black, multipartnered mothers more likely to do so; Kamp Dush, 2011). Most cohabitation relationships are short-lived, with 20% resulting in marriage (Schoen et al., 2007).

Cohabitation is on the increase. Reasons include career or educational commitments; increased tolerance by society, parents, and peers; improved birth control technology; desire for a stable emotional and sexual relationship without legal ties; and avoidance of loneliness (Kasearu, 2010).

Whether a person decides to cohabit is related to the society in which the person lives. Virtually all marriages in Sweden are preceded by the couple cohabiting (Thomson & Bernhardt, 2010); however, only 12% of first marriages in Italy are preceded by cohabitation (Kiernan,

cohabitation (living together) two unrelated adults (by blood or by law) involved in an emotional and sexual relationship who sleep in the same residence at least four nights a week.

2000). Italy is primarily Catholic, which helps to account for the low cohabitation rate.

4-4a Definitions of Cohabitation

Researchers have used more than 20 definitions of cohabitation. These various definitions involve variables such as the duration of the relationship, frequency of overnight visits, emotional or sexual nature of the relationship, and sex of the partners. Most research on cohabitation has been conducted on heterosexual live-in couples. Even partners in relationships may view the meaning of their cohabitation differently, with women viewing it more as a sign of a committed relationship moving toward marriage and men viewing it as an alternative to marriage or as a test to see whether future commitment is something to pursue. We define cohabitation as two unrelated adults involved in an emotional and sexual relationship who sleep overnight in the same residence at least four nights a week and have done so for at least three months. The terms used to describe live-in couples include *cohabitants* and **POSSLQs** (people of the opposite sex sharing living quarters), the latter term used by the U.S. Census Bureau.

4-4b Eight Types of Cohabitation Relationships

There are various types of cohabitation:

1. **Here and now.** These new partners agree that they enjoy spending time together and do not want to be apart. They orchestrate their day so that they stay connected via texting and return to the same abode at night. Being together in the "here and now" is what is important to them. There is also a gender difference with women focusing on companionship and movement toward marriage and men focusing on sex (Huang et al., 2011).

2. **Testers.** An example of "testers" follows:

POSSLQ an acronym used by the U.S. Census Bureau that stands for "people of the opposite sex sharing living quarters."

© Chelsea Curry

"*Cohabitation without children or marriage needs to be viewed not only as a legitimate end-state in itself but also as a legitimate form of premarriage.*"
—William Pinsof, family psychologist

We're trying to see how it is going to work. So if our relationship's going to continue to grow and prosper, this would be one way for us to kind of gauge, you know, maybe we will get married. And if we can't live together, we're not going to get married. So this will help us make future decisions and stuff like that. (Sassler & Miller, 2011)

3. **Engaged.** These couples are in love and are planning to marry. Chelsea Clinton and her partner were living together and engaged as they planned their wedding. Among those who report having lived together, about two thirds (64%) say they thought of their living arrangement as a step toward marriage (Pew Research Center, 2010). For some this step is formal in that they have told their parents of their intent and have a wedding date. Others may view themselves as "engaged" in the sense that they do plan to marry but have not worked out all the details. In the meantime, they are sure that they want a life together and are expressing this commitment by living together. Consistent with previous research, these couples are typically very happy—they are young, in a new relationship, committed, without children, and focused on a life together.

4. **Money savers.** These couples live together primarily out of economic convenience. Working-class individuals tend to transition more quickly than middle-class individuals to cohabitation out of economic necessity or to meet a housing need

(Sassler & Miller, 2011). Financial disagreements are also a primary reason that cohabiting couples break up (Dew, 2011).

5. **Pension partners.** These individuals are older, have been married before, still derive benefits from their previous relationships, and are living with someone new. Getting married would mean giving up their pension benefits from the previous marriage. An example is a military widow who was given military benefits due to a spouse's death. If remarried, the widow forfeits both health and pension benefits; cohabitation involves no economic losses.

6. **Security blanket cohabiters.** Some of the individuals in these cohabitation relationships are drawn to each other out of a need for security rather than mutual attraction. They may have an on-and-off relationship and live with each other intermittently (Cross-Barnet et al., 2011).

7. **Rebellious cohabiters.** Some couples use cohabitation as a way of making a statement to their parents that they are independent and can make their own choices. Their cohabitation is more about rebelling from parents than being drawn to each other.

8. **Marriage never (cohabitants forever).** Nineteen percent of 1,336 adults in cohabitation relationships in Australia planned never to marry (Buchler et al., 2009). These cohabitants, also known as "marriage renouncing cohabiters," are typically older and may have children or do not want more children. Compared to couples who are married, cohabitants who plan never to marry tend to report lower life and relationship satisfaction. An example of a "marriage never" cohabitation couple are the celebrities Goldie Hawn and Kurt Russell.

The median length of time that a never-married woman cohabits with a partner in the United States is 14 months. In contrast, in France it is 51 months; Canada, 40 months; and Germany, 27 months. In European countries never-married women often have children during cohabitation, which is less common in the United States (Cherlin, 2010).

There are various motivations for living together as a permanent alternative to marriage. Some may fear divorce. Two thirds of 122 cohabitants reported concerns about divorce (Miller et al., 2011). Others feel that the real bond between two people is (or should be) emotional. They contend that many couples stay together because of the legal contract. Some couples who view their living together as "permanent" seek to have it defined as a **domestic partnership**. Such partnerships involve individuals who live together in an emotional or sexual relationship and are given some kind of official recognition by the city in which they live or corporation where they work so as to receive partner benefits.

domestic partnership a relationship in which individuals who live together are emotionally and financially interdependent and are given some kind of official recognition by a city or corporation so as to receive partner benefits.

Ready to study? In this book, you can:

⮕ Rip out the Chapter Review card in the back of the book to study for exams

⮕ Take the Self Assessment for this chapter (card in the back of the book) and see where you stand on the vital issues raised in the chapter

Or you can go online to CourseMate at www.cengagebrain.com for these resources:

⮕ Complete Practice Quizzes to prepare for tests

⮕ Review Key Terms Flash Cards (online or print)

⮕ Read about Marriage and Family in the news

⮕ Play "Beat the Clock" to master concepts

⮕ Check out Personal Applications

Love and Selecting a Partner

"The minute **you settle** for **less** than you **deserve,** you **get even less** than you **settled for."**

—MAUREEN DOWD, JOURNALIST

SECTIONS

5-1 Ways of Conceptualizing Love

5-2 Love in Social and Historical Context

5-3 How Love Develops in a New Relationship

5-4 Jealousy in Relationships

5-5 Cultural Restrictions on Whom an Individual can Love or Marry

5-6 Sociological Factors Operative in Partnering

5-7 Psychological Factors Operative in Partnering

5-8 Sociobiological Factors Operative in Partnering

5-9 Engagement

5-10 Marrying for the Wrong Reasons

5-11 Consider Calling Off the Wedding If . . .

Anyone who has been in love knows how mad and bizarre it can be. The story about love that takes the cake for being bizarre is about Burt and Linda. Burt Pugach, 31, was married when he fell in love with 21-year-old Linda Riss. This New York love story of the 1950s featured Linda finding out that Burt was married, breaking off the relationship, and getting involved with someone new (Larry Schwartz). Burt asked her to end the relationship with Larry and threatened that if she did not, he would "fix" it so no one else would have her. After Burt found out that she was engaged to Larry, he hired three men to throw lye in her face, which blinded her in one eye (she eventually lost sight in the other eye). Burt was sent to prison for 14 years and wrote to Linda throughout his incarceration.

Meanwhile, Linda's fiancé left her, and other men found it difficult to cope with her blindness and eye disfigurement. At the time of Burt's release from prison, she was alone, unemployed, and financially destitute. She redefined Burt's insane, abusive act as intense love for her and thought it okay for him to support her, which he was willing to do. They married shortly after his release from prison in 1974 and were still married in 2012. *Crazy Love* is a documentary on this couple's unusual love relationship and is available on DVD.

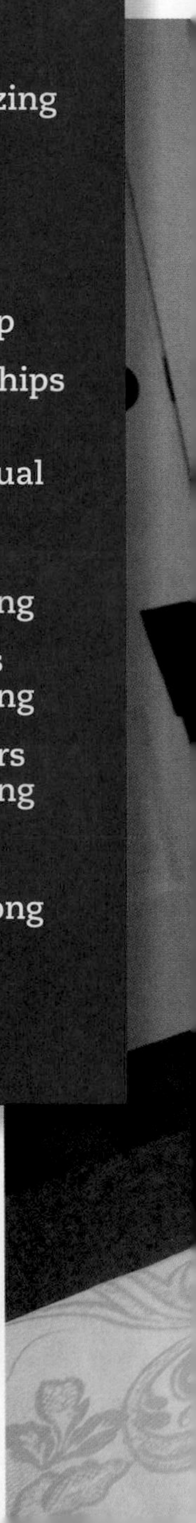

© David Knox

Love, and its romantic version of passionate love, is a universal emotion and motivation (Hatfield et al., 2012). It is also entrenched in American culture. A team of researchers surveyed 14,121 adolescents and found that the romantic ideal continues regardless of gender or sexual identity—an overwhelming proportion of the respondents rated love, faithfulness, and lifelong commitment as extremely important for marriage or long-term relationships (Meier et al., 2009). For many, being in love is associated with relationship happiness (Graham, 2011) and fewer mental health problems (Braithwaithe et al., 2010). The latter finding is based on a survey of 1,621 undergraduates in committed relationships.

© windu/Shutterstock

5-1 Ways of Conceptualizing Love

Love is often confused with lust and infatuation. Love is about deep, abiding feelings for the well-being of another with a focus on the long-term relationship (Foster, 2010); **lust** is about sexual desire and focuses on the present; **infatuation** is about emotional feelings based on little actual exposure to the love object. In the following sections, we look at the various ways of conceptualizing love.

5-1a Love Styles

Various love styles have been identified that describe the way lovers view love and relate to each other (Lee, 1973, 1988). Keep in mind that the same individual may view love in more than one way at a time or may view love in different ways at different times. These love styles are also independent of one's sexual orientation—no one love style is characteristic of heterosexuals or homosexuals.

lust sexual desire.

infatuation emotional feelings based on little actual exposure to the love object.

ludic love style love style that views love as a game in which the love interest is one of several partners, is never seen too often, and is kept at an emotional distance.

pragma love style love style that is logical and rational; the love partner is evaluated in terms of assets and liabilities.

eros love style love style characterized by passion and romance.

1. **Ludic.** The ludic lover views love as a game, refuses to become dependent on any one person and does not encourage another's intimacy. Two essential skills of the ludic lover are to juggle several partners at the same time and to manage each relationship so that no one partner is seen too often. These strategies help to ensure that the relationship does not deepen into an all-consuming love. Don Juan represented the classic ludic lover. "Love 'em and leave 'em" is the motto of the ludic lover. Men are more likely than women to be ludic lovers (indeed, they are called "players"). And, ludic love is the least frequent love style among college students (Tzeng et al., 2003).

   Ludic lovers in television series are pervasive. Don Draper in *Mad Men* (winner of the 2011 Emmy Award for Outstanding Drama Series) is the quintessential ludic lover. He feigns love and desire for the new secretary or flight attendant or conventioneer. But he is soon off to the new woman. In one episode, he was enamored with the elementary school teacher of his daughter. They flirted a time or two at the child's school and on the playground until one night he showed up at her house. She let him in and asked, "What are you doing here?" He said, "I'd just like to talk." The woman hesitated as though deciding whether to let him kiss her and said, "I know how this will end." Her prediction was true (they lasted a month).

   The **ludic love style** is sometimes characterized as manipulative and uncaring. However, ludic lovers may also be compassionate and very protective of another's feelings. For example, some uninvolved, soon-to-graduate seniors avoid involvement with anyone new and become ludic lovers so as not to encourage anyone.

2. **Pragma.** The **pragma love style** is the love of the pragmatic—that which is logical and rational. Pragma lovers assess their partners on the basis of assets and liabilities. Economic security may be a guiding consideration. One single-parent mother of four married a neurosurgeon who paid for the expensive college educations of her children at Ivy League schools. The day the last child graduated, she divorced her physician husband, who may have been a little suspicious that her love for him was all about the money.

3. **Eros.** Just the opposite of the pragmatic love style, the **eros love style**, also known as romantic love, is imbued with passion and sexual desire. After her divorce from Marc Anthony, Jennifer Lopez defined herself as "a hopeless romantic and passionate person when it comes to love." Eros is the most common love style of college women and men (Tzeng et al., 2003). They are looking

for an intense, romantic lover with whom to share their life and future. The idea of a soul mate is part of American culture. In a study of over 1,000 adults, 66% reported "belief in the idea of a soul mate;" 34% disagreed (Carey & Gellers, 2010).

4. **Mania.** The **mania love style** is the out-of-control love whereby the person "must have" the love object. Obsessive jealousy and controlling behavior are symptoms of manic love. The lover is possessive, dependent, and "must have" the beloved. People who are extremely jealous and controlling reflect manic love. "If I can't have you, no one else will" is sometimes the mantra of the manic lover. The vignette which began this chapter detailed the manic love of Burt Pugach for Linda.

5. **Storge.** The **storge love style** is a calm, soothing, nonsexual love devoid of intense passion. Respect, friendship, commitment, and familiarity are characteristics that help to define the relationship. The partners care deeply about each other but not in a romantic or lustful sense. Their love is also more likely to endure than fleeting romance. One's grandparents who have been married 50 or 60 years and who still love and enjoy each other are likely to have a storge type of love.

6. **Agape.** One of the forms of love identified by the ancient Greeks, the **agape love style** involves selflessness and giving, without expecting anything in return. These nurturing and caring partners are concerned only about the welfare and growth of each other. The love parents have for their children is often described as agape love.

© iStockphoto.com

5-1b Romantic Versus Realistic Love

Love may also be described as being on a continuum from romanticism to realism. For some people, love is romantic; for others, it is realistic. **Romantic love**, said to have appeared in all human groups at all times in human history (Berscheid, 2010), is an intense love characterized by such beliefs as "love at first sight," "one true love," and "love conquers all." In a sample of 2,922 undergraduates, 21% reported that they had experienced love at first sight (Knox & Hall, 2010). Men were more likely than women to report having experienced love at first sight (38.3% versus 32.7%). One explanation is that men must be visually attracted to young, healthy females to inseminate them. This biologically based reproductive attraction is interpreted as a love attraction so that the male feels immediately drawn to the female, but he may actually see an egg which he is driven to fertilize. Further evidence that males are more romantic than females is from Dotson-Blake and colleagues (2008), who found that men were significantly more likely than women (85% versus 73%) to believe that they could solve any relationship problem as long as they were in love. Hence, men are drawn to women and believe love can solve all problems.

In regard to love at first sight, Barelds and Barelds-Dijkstra (2007) studied the relationships of 137 married couples or cohabitants (together for an average of 25 years) and found that the relationship quality of those who fell in love at first sight was similar to that of couples who came to know each other more gradually. Huston and colleagues (2001) found that, after two years of marriage, the couples that had fallen in love more slowly were just as happy as couples that fell in love at first sight.

Romantic love may also create the context for risk taking. Elliott and colleagues (2012)

mania love style an out-of-control love whereby the person "must have" the love object; obsessive jealousy and controlling behavior are symptoms of manic love.

storge love style a love consisting of friendship that is calm and nonsexual.

agape love style love style characterized by a focus on the well-being of the love object, with little regard for reciprocation; the love of parents for their children is agape love.

romantic love an intense love whereby the lover believes in love at first sight, only one true love, and love conquers all.

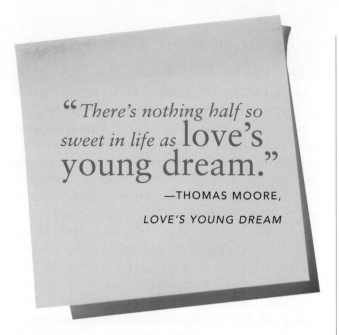

> *" There's nothing half so sweet in life as* love's young dream."
>
> —THOMAS MOORE,
>
> *LOVE'S YOUNG DREAM*

analyzed data from 381 undergraduates, 60% of whom reported that they were vulnerable to taking chances when they were in love (66% said that alcohol induced the same tendency). Engaging in unprotected sex, being involved in a "friends with benefits" relationship, breaking up with a partner to explore alternatives, and having sex before being ready were the most frequent risk-taking behaviors.

Infatuation is sometimes regarded as synonymous with romantic love. The word infatuation comes from the same root word as fatuous, meaning "silly" or "foolish," and refers to a state of passion or attraction that is not based on reason. Infatuation is characterized by the tendency to idealize the love partner. People who are infatuated magnify their lovers' positive qualities ("My partner is always happy") and overlook or minimize their negative qualities ("My partner doesn't have a problem with alcohol; he just likes to have a good time").

In contrast to romantic love is realistic love. Realistic love is also known as conjugal love. **Conjugal (married) love** is less emotional, passionate, and exciting than romantic love and is characterized by companionship, calmness, comfort, and security. The Love Attitudes Scale provides a way for you to assess the degree to which you tend to be romantic or realistic (conjugal) in your view of love. When you determine your score from the Love Attitudes Scale, be aware that your tendency to be a romantic or a realist is neither good nor

conjugal (married) love the love between married people characterized by companionship, calmness, comfort, and security.

bad. Both romantics and realists can be happy individuals and successful relationship partners.

5-1c Triangular View of Love

Sternberg (1986) developed the "triangular" view of love, which consists of three basic elements: intimacy, passion, and commitment. The presence or absence of these three elements creates various types of love experienced between individuals, regardless of their sexual orientation. These various types include the following:

1. **Nonlove**—the absence of intimacy, passion, and commitment. Two strangers looking at each other from afar have a nonlove.

2. **Liking**—intimacy without passion or commitment. A new friendship may be described in these terms.

3. **Infatuation**—passion without intimacy or commitment. Two people flirting with each other in a bar may be infatuated with each other.

4. **Romantic love**—intimacy and passion without commitment. Love at first sight reflects this type of love.

5. **Conjugal love (also known as companionate love)**—intimacy and commitment without passion. A couple that has been married for 50 years is said to illustrate conjugal love.

6. **Fatuous love**—passion and commitment without intimacy. Couples who are passionately wild about each other and talk of the future but do not have an intimate connection with each other have a fatuous love.

7. **Empty love**—commitment without passion or intimacy. A couple who stay together for social and legal reasons but who have no spark or emotional sharing between them have an empty love.

8. **Consummate love**—combination of intimacy, passion, and commitment; Sternberg's view of the ultimate, all-consuming love.

Individuals bring different combinations of the elements of intimacy, passion, and commitment (the triangle) to the table of love. One lover may bring a predominance of passion, with some intimacy but no commitment (romantic love), whereas the other person brings commitment but no passion or intimacy (empty love). The triangular theory of love allows lovers to see the degree to which they are matched in terms of passion, intimacy, and commitment in their relationship.

A common class exercise among professors who teach about marriage and the family is to randomly ask class members to identify one word they most closely associate with love. Invariably, students identify different words (commitment, feeling, trust, altruism, and so on), suggesting great variability in the way we think about love. Indeed, just the words "I love you" have different meanings, depending on whether they are said by a man or a woman. In a study of 147 undergraduates (72% female, 28% male), men (more than women) reported that saying "I love you" was a ploy to get a partner to have sex, whereas women (more than men) reported that saying "I love you" was a reflection of their feelings, independent of a specific motive (Brantley et al., 2002).

Intimacy

Commitment

Passion

India, and Indonesia (three countries representing 40% of the world's population). The parents select the mate for their child to prevent a love relationship from forming with the "wrong" person and to ensure that the child marries the "right" person. Such a person must belong to the desired social class and have the economic resources that the parents want for their daughter. In most Eastern countries, marriage is regarded as the linking of two families; the love feelings of the respective partners are irrelevant. Love is expected to follow marriage, not precede it.

Selecting a spouse for a daughter may begin early in the child's life. In some countries (e.g., Nepal and Afghanistan), child marriage occurs whereby young females (ages 8 to 12) are required to marry an older man identified by their parents. Suicide is the only alternative "choice" for these children.

More often, parents may know a family who has a son or daughter whom they would regard as a suitable partner for their offspring. If not, they may put an advertisement in the "matchmaker" section of the newspaper identifying the qualities they are seeking. The prospective mate is then interviewed by the family to confirm his or her suitability. Or a third person—a matchmaker—may be hired to do the screening and introducing.

Once the families agree that the prospective son-in-law or daughter-in-law has the desired qualities, the individuals are allowed to meet in the presence of the parents. Being alone is allowed but only for brief periods during which touching and kissing are not acceptable. Indeed, sex (not a priority) rarely occurs before marriage in these unions.

Because romantic love is such a powerful emotion and

arranged marriage mate selection pattern whereby parents select the spouse of their offspring.

5-2 Love in Social and Historical Context

Though we think of love as an individual experience, the society in which we live exercises considerable control over our love object or choice and conceptualizes that choice in various ways.

5-2a Social Control of Love

The ultimate social control of love is **arranged marriage**. Parents arrange 80% of marriages in China,

"Romantic love provides an even stronger motivation than the sex drive."

—HELEN FISHER, ANTHROPOLOGIST

from a study of 1,335 respondents on Match. com. Being from the South and identifying with religion (Catholic, Protestant, Jewish) were also associated with less openness toward interracial relationships. The greatest resistance to interracial relationships of Whites was to African Americans, then Hispanics, then Asians. Gays also tended to select partners of the same racial and ethnic background.

Another example of the social control of love is that individuals attracted to someone of the same sex quickly feel the social and cultural disapproval of this attraction (only eight states currently approve of same-sex marriage). However, regardless of the law, same-sex love is common. Individuals are biologically wired for, and capable of, falling in love and establishing intense emotional bonds with members of their own or the opposite sex (Diamond, 2003).

marriage such an important relationship, mate selection in the United States is also not left to chance when bringing an outsider into an existing family. Parents influence who their children meet by moving to certain neighborhoods, joining certain churches, and enrolling their children in certain schools. Doing so increases the chance that their offspring will "hang out" with, fall in love with, and marry someone who is similar in race, religion, education, and social class. The fact that you are a college student increases the chance that you will meet, fall in love with, and marry a college student.

America is a country that prides itself on the value of individualism. We proclaim that we are free to make our own choices. Not so fast. Love may be blind, but it knows what color a person's skin is. The data are clear—potential spouses seem to see and select people of similar color, as about 90% of people marry someone of their own racial background (less than 1% of the 60 million married couples in the United States consist of a Black person and a White person; *Statistical Abstract of the United States*, 2012. Table 60). Hence, parents and peers may approve of their offspring's and friends' love choice when the partner is of the same race and disapprove of the selection when the partner is not. These approval and disapproval mechanisms illustrate the social control of love.

Blacks, younger individuals, and politically liberal and educated Whites are more willing to cross racial lines for dating and for marriage than are Whites, older individuals, conservatives, and educated Blacks (Tsunokai & McGrath, 2011). These findings emerged

5-2b Love in Medieval Europe—From Economics to Romance

Love in the 1100s was a concept influenced by economic, political, and family structure. In medieval Europe, land and wealth were owned by kings, who controlled geographical regions—kingdoms. When so much wealth and power were at stake, love was not to be trusted as the mechanism for choosing spouses for royal offspring. Rather, marriages of the sons and daughters of the aristocracy were arranged with the heirs of other states with whom an alliance was sought. Love was not tied to marriage but was conceptualized as an adoration of physical beauty (often between a knight and his beloved) and as spiritual and romantic, even between people not married or of the same sex. Hence, romantic love had its origin in extramarital love and was not expected between spouses (Trachman & Bluestone, 2005).

The extent of kingdoms and estates and the patrimonial households declined with the English revolutions of 1642 and 1688 and the French Revolution of 1789. No longer did aristocratic families hold power; the power was transferred to individuals through parliaments or other national bodies. Even today, English monarchs are figureheads, with parliament handling the real business of international diplomacy. Because wealth and power were no longer in the hands of individual aristocrats, the need to control mate selection decreased, and the role of love changed. Marriage became less a political and business arrangement and more a mutually desired emotional union. Just as bureaucratic structure held together partners in medieval society, a new mechanism—love—would now provide the emotional and social bonding.

Hence, love changed from a feeling irrelevant to marriage in medieval times—because individuals (representing aristocratic families) were to marry even though they were not in love—to a feeling that bonded a woman and a man together for marriage.

> Marriage became less a political and business arrangement and more a mutually desired emotional union.

5-2c Love in Colonial America

Love in colonial America was similar to that in medieval times. Marriage was regarded as a business arrangement between the fathers of the respective families. An interested suitor would approach the father of a girl to express his desire to court the daughter. The fathers would generally confer on the amount of the **dowry** (also known as the **trousseau**), which included the money or valuables, or both, that the girl's father would pay the boy's father. Because unmarried women were stigmatized, marrying them off was desirable; thus, the dowry was an added inducement for a boy to marry a girl. Fathers could deny their daughters a dowry if the daughters were unwilling to marry the man their father chose. Love was not totally absent, however; sometimes a girl could persuade her father to tell the suitor she was not interested.

5-3 How Love Develops in a New Relationship

Various social, physical, psychological, physiological, and cognitive conditions affect the development of love relationships.

5-3a Social Conditions for Love

Love is a social label given to an internal feeling. Our society promotes love through popular music, movies, and novels. These media convey the message that love is an experience to pursue, enjoy, and maintain. People who fall out of love are encouraged to try again: "Love is lovelier the second time you fall." Unlike people reared in Eastern cultures, Americans grow up in a context of encouragement to turn on their radar for love.

5-3b Body Type Condition for Love

The probability of being involved in a love relationship is influenced by approximating the cultural ideal of physical appearance. Halpern and colleagues (2005) analyzed data on a nationally representative sample of 5,487 African American, Caucasian, and Hispanic adolescent females and found that, for each 1-point increase in body mass index (BMI), the probability of involvement in a romantic relationship dropped by 6%. Hence, to the degree that a woman approximates the cultural ideal of being trim and "not being fat," she increases the chance of attracting

dowry (trousseau) the amount of money or valuables a woman's father pays a man's father for the man to marry his daughter. It functioned to entice the man to marry the woman, because an unmarried daughter stigmatized the family of the woman's father.

a partner and becoming involved in a romantic love relationship. In another study of weight and marital status, Sobal and Hanson (2011) found that never-married women were more likely to be obese than married women. The researchers suggested that weight for a woman was associated with not being selected as a marriage partner. Married men were heavier than men who were separated or divorced.

Ambwani and Strauss (2007) found that body image has an effect on sexual relations and that relationships affect self-image. Hence, women who felt positively about their body were more likely to report having sexual relations with a partner. The fact that they were in a relationship was associated with positive feelings about themselves.

5-3c Psychological Conditions for Love

Various psychological conditions are associated with falling in love.

Perception of Reciprocal Liking Researchers have found that one of the most important psychological factors associated with falling in love is the perception of reciprocal liking (Riela et al., 2010). When one perceives that he or she is desired by someone else, this perception has the effect of increasing the attraction toward that person. Such an increase is particularly strong if the person feels strong physical attraction (Greitemeyer, 2010).

Personality Qualities The personality of the love object has an important effect on falling in love (Riela et al., 2010). Viewing the partner as intelligent or as having a sense of humor are examples of qualities associated with the lover wanting to be with the beloved.

Self-Esteem High self-esteem is also important for falling in love because it enables individuals to feel worthy of being loved. Feeling good about yourself allows you to believe that others are capable of loving you. Individuals with low self-esteem doubt that someone else can love and accept them.

In contrast, low self-esteem has devastating consequences for individuals and the relationships in which they become involved. Not feeling loved as a

Benefits of Self-Esteem

1. It allows one to be open and honest with others about both strengths and weaknesses.
2. It allows one to feel generally equal to others.
3. It allows one to take responsibility for one's own feelings, ideas, mistakes, and failings.
4. It allows for the acceptance of both strengths and weaknesses in oneself and others.
5. It allows one to validate oneself and not to expect the partner to do this.
6. It permits one to feel empathy—a very important skill in relationships.
7. It allows separateness and interdependence, as opposed to fusion and dependence.

child and, worse, feeling rejected and abandoned creates the context for the development of a negative self-concept and mistrust of others. People who have never felt loved and wanted may require constant affirmation from a partner as to their worth and may cling desperately to that person out of fear of being abandoned. Such dependence (the modern term is codependency) may also encourage staying in unhealthy relationships (e.g., abusive and alcoholic relationships), because the person may feel "this is all I deserve."

One characteristic of individuals with low self-esteem is that they may be addicted to unhealthy love relationships. In a study of 52 women who reported that they were involved in unhealthy love relationships in which they had selected men with problems (such as alcohol or other drug addiction), which they attempted

to solve at the expense of neglecting themselves, the researchers noted, "Their preoccupation with correcting the problems of others may be an attempt to achieve self-esteem" (Petrie et al., 1992). In effect the motif of these women was "I know I can help this man and I am someone of worth because I can."

Although having positive feelings about oneself when entering into a love relationship is helpful, sometimes such feelings develop after one becomes involved in the relationship. "I've always felt like an ugly duckling," said one woman. "But once I fell in love with him and him with me, I felt very different. I felt very good about myself then because I knew that I was somebody that someone else loved." High self-esteem, then, is not necessarily a prerequisite for falling in love. People who have low self-esteem may fall in love with someone else as a result of feeling deficient. The love they perceive the other person has for them may compensate for the perceived deficiency and improve their self-esteem. This phenomenon can happen with two individuals with low self-esteem—love can elevate the self-concepts of both individuals.

Self-Disclosure Disclosing oneself is necessary if one is to love—to feel invested in another (Radmacher & Azmitia, 2006). Disclosed feelings about the partner include "how much I like the partner," "my feelings about our sexual relationship," "how much I trust my partner," "things I dislike about my partner," and "my thoughts about the future of our relationship"—all of which are associated with relationship satisfaction. Of interest in Ross's (2006) findings is that disclosing one's tastes and interests was negatively associated with relationship satisfaction. By telling a partner too much detail about what one likes, partners may discover something that turns them off and lowers relationship satisfaction.

It is not easy for some people to let others know who they are, what they feel, or what they think. *Alexithymia* is a personality trait which describes a person with little affect. The term means "lack of words for emotions," which suggests that the person does not experience or convey emotion. Persons with alexithymia are not capable of experiencing psychological intimacy. Hesse and Floyd (2011) found that persons who seek emotional relationships are not attracted to alexithymics.

Trust is the condition under which people are most willing to disclose themselves. When people trust someone, they tend to feel that whatever feelings or information they share will not be judged and will be

Eight Dimensions of Self-Disclosure

1. Background and history
2. Feelings toward the partner
3. Feelings toward self
4. Feelings about one's body
5. Attitudes toward social issues
6. Tastes and interests
7. Money and work
8. Feelings about friends

© i3alda/Shutterstock

kept safe with that person. If trust is betrayed, people may become bitterly resentful and vow never to disclose themselves again. One woman said, "After I told my partner that I had had an abortion, he told me that I was a murderer and he never wanted to see me again. I was devastated and felt I had made a mistake telling him about my past. You can bet I'll be careful before I disclose myself to someone else" (personal communication).

Women are significantly more likely than men to disclose information about themselves. Specific areas of disclosure include previous love relationships and the desire for the future of the relationship (Gallmeier et al., 1997).

Gratitude Feelings of gratitude are associated with love. Partners who feel a sense of gratitude to each other report increased relationship satisfaction. Gratitude acts like a booster shot of love for the partners and for their relationship (Algoe et al., 2010).

5-3d Physiological and Cognitive Conditions for Love

Arousal (strong physiological reactions when in the presence of the other) is associated with falling in love (Riela et al., 2010). This physical chemistry is powerful. Partners who feel strong chemistry escalate their relationship; those who do not, end it.

Another physiological change is feeling excited or anxious and interpreting this stirred-up state as love (Walster & Walster, 1978). But cognitive function must be operative or love feelings will not develop. Individuals with brain cancer who have had the front part of their brain (between the eyebrows) removed are incapable of love. Indeed, emotions are not present in them at all (Ackerman, 1994).

The social, physical, psychological, physiological, and cognitive conditions are not the only factors important for the development of love feelings. The timing must be right. There are only certain times in life (e.g., when educational and career goals are met or within sight) when people seek a love relationship. When those times occur, a person is likely to fall in love with another person who is there and who is also seeking a love relationship. Hence, many love pairings exist because each of the individuals is available to the other at the right time—not because they are particularly suited for each other.

5-3e Other Factors Associated With Falling in Love

In addition to those already identified, other factors are associated with falling in love. These include appearance, common interests, and similar friends (Riela et al., 2010). How one acts on a date is also important, with undergraduates identifying authenticity ("be yourself"), chivalry ("open the door"), communication ("ask questions and listen to the answers"), and planning ("the guy should not ask a girl what she wants to do but give her choices") as priorities (Johnson et al., 2011).

How fast a person falls in love has also been studied by researchers. In a study of 239 undergraduates from both America and China who wrote about their experience of falling in love, 14% reported that they fell in love very fast, 42% fell in love fast, 36% fell in love slowly, and 8% fell in love very slowly (Riela et al., 2010).

jealousy an emotional response to a perceived or real threat to an important or valued relationship.

reactive jealousy feelings that the partner may be straying.

anxious jealousy obsessive rumination about the partner's alleged infidelity.

possessive jealousy attacking the partner who is perceived as being unfaithful.

5-4 Jealousy in Relationships

Jealousy is an emotional response to a perceived or real threat to an important or valued relationship. When one partner views the other texting a previous partner, feelings of jealousy may emerge. These feelings include fear of being abandoned by the partner and being replaced by someone else. "My greatest fear," reported one of our students, "was that my boyfriend's ex-girlfriend was going to talk him into giving her another chance. And that is exactly what happened. Just after I caught him texting her, he told me he needed some 'space.' He ended up going back to her. I was enraged that he dumped me to be with her again. Seeing them on campus is unbearable."

Although jealousy does not occur in all cultures (polyandrous societies value cooperation, not sexual exclusivity; Cassidy & Lee, 1989), it does occur in U.S. society and among both heterosexuals and homosexuals. Of 2,922 university students, 41% agreed with the statement "I am a jealous person" (Knox & Hall, 2010).

Jealously is more likely to occur early in a couple's relationship. Consistent with this theme, Gatzeva and Paik (2011) compared noncohabiting, cohabiting, and married couples in regard to satisfaction as related to jealous conflict. They found that married couples were less likely to report jealous conflict and that satisfaction was higher in the absence of such conflict.

5-4a Types of Jealousy

Three types of jealousy have been identified: reactive jealousy, anxious jealousy, and possessive jealousy (Barelds-Dijkstra & Barelds, 2007). **Reactive jealousy** is a reaction to something the partner is doing (e.g., texting a former lover) that may signal infidelity. **Anxious jealousy** is obsessive rumination about the partner's alleged behavior (e.g., infidelity). **Possessive jealousy** involves attacking the partner or the person to whom the partner is showing attention. Sometimes the attack is "competitor derogation" whereby the person talks negatively about the alleged suitor (M. Fisher & Cox, 2011).

5-4b Causes of Jealousy

Jealousy can be triggered by external or internal factors.

Love as a Context for Problems

Though love may bring great joy, it also creates a context for problems. Five such problems are the destruction of existing relationships, simultaneous loves, unrequited or unfulfilling loves, making risky or dangerous choices, involvement in an abusive relationship, and the emergence of stalking.

Destruction of Existing Relationships

Sometimes the development of one love relationship is at the expense of another. A student in our classes noted that when she was 16, she fell in love with a person at work who was 25. Her parents were adamant in their disapproval and threatened to terminate the relationship with their daughter if she continued to see this man. The student noted that she initially continued to see her lover and to keep their relationship hidden. However, eventually she decided that giving up her family was not worth the relationship, so she stopped seeing him permanently. Others in the same situation would end the relationship with their parents and continue the relationship with their beloved. Choosing to end a relationship with one's parents is a downside of love.

Simultaneous Loves

For all the wonder of love, awareness that one's partner is in love with or having sex with someone else can create heartbreak. Multiple involvements are not a problem for some individuals or couples. Some feel **compersion**—the opposite of jealousy whereby a person feels joy in having the person one loves and has sex with enjoying others both emotionally and sexually. However, most people are not comfortable knowing their partner has other emotional or sexual relationships. Only 2.6% of 2,922 undergraduates agreed with the statement "I can feel good about my partner having an emotional/sexual relationship with someone else" (Knox & Hall, 2010). Hence, for most undergraduates, simultaneous lovers are viewed

as a problem. "I tried to be accepting of my partner having other lovers but it did not work for me," said one of our students. "I'm just too jealous."

Unrequited or Unfulfilling Love Relationships

Sometimes love is one-sided. One person loves deeply but the other is indifferent or turned off. Such a love is referred to as *unrequited love*. In the 16th century such love was termed *lovesickness*, and today it is called *erotomania* or *erotic melancholy* (Kem, 2010). An interesting question is whether being the person who loves more in a relationship is better than being the person who loves less. The person who loves more may suffer more anguish but experience more joy in loving.

Love as a Context for Risky, Dangerous, or Questionable Choices

Plato said that "love is a grave mental illness," and some research suggests that individuals in love make risky, dangerous, or questionable decisions. In a study on "what I did for love," college students reported that "driving drunk," "dropping out of school to be with my partner," and "having sex without protection" were among the more dubious choices they had made while they were under the spell of love (Knox et al., 1998). Similarly, nonsmokers who become romantically involved with smokers are more likely to begin smoking (Kennedy et al., 2011).

Abusive or Stalking Relationships

Thirty percent of 2,922 undergraduates reported that they had been involved in an emotionally abusive relationship with a partner. As for physical abuse, 7.5% reported such previous involvement. Most stay in these abusive relationships because they are "in love." In addition, rejected lovers may stalk (obsessively pursue and harass or threaten) a partner because of anger and jealousy and try to win the partner back (Knox & Hall, 2010).

> **compersion**
> the opposite of jealousy, whereby a person feels joy in having the person one loves and has sex with enjoying others both emotionally and sexually.

External Causes External factors are behaviors a partner engages in that are interpreted as (1) an emotional and/or sexual interest in someone (or something) else, or (2) a lack of emotional and/or sexual interest in the primary partner. Respondents (185) identified talking to (34%) or about (19%) a previous partner as the most common sources of jealousy. Men were more likely to report feeling jealous when their partner talked to a previous partner, whereas women were more likely to report feeling jealous when their partner danced with someone else. What makes a woman jealous is the physical attractiveness of a rival; what makes a man jealous is the rival's physical dominance (Buunk et al., 2010).

Internal Causes Jealousy may also exist even when no external behavior indicates the partner is involved or interested in an **extradyadic relationship**—an emotional or sexual involvement between a member of a pair and someone other than the partner. Internal causes of jealousy reflect characteristics of individuals that predispose them to jealous feelings, independent of their partner's behavior, such as:

1. **Mistrust.** If an individual has been cheated on in a previous relationship, that individual may learn to be mistrustful in subsequent relationships. Such mistrust may manifest itself in jealousy. Mistrust and jealousy may be intertwined. Indeed, one must be careful if cheated on in a previous relationship not to transfer those feelings to a new partner. Disregarding the past is not easy but constantly reminding oneself of reality ("My new partner has given me zero reason to be distrustful") will help.

2. **Low self-esteem.** Individuals who have low self-esteem tend to be jealous because they lack a sense of self-worth and, hence, find it difficult to believe anyone can value and love them. Feelings of worthlessness may contribute to suspicions that someone else is valued more. It is devastating to a person with low self-esteem to discover that a partner has, indeed, selected someone else.

3. **Lack of perceived alternatives.** Individuals who have no alternative person or who feel inadequate in attracting others may be particularly vulnerable to jealousy. They feel that, if they do not keep the person they have, they will be alone.

4. **Insecurity.** Individuals who feel insecure in a relationship with their partner often experience higher levels of jealousy. They feel at any moment their partner could find someone more attractive and end the relationship with them.

5-4c Consequences of Jealousy

Jealousy can have both desirable and undesirable consequences.

Desirable Outcomes Reactive jealousy (reacting to the partner's behavior) is associated with a positive effect on the relationship (Barelds-Dijkstra & Barelds, 2007). Such a reaction may not only signify that the partner is cared for (the implied message is "I love you and don't want to lose you to someone else") but also communicate to the partner that the development of other romantic and sexual relationships is unacceptable ("You can't have me and the other person too").

The researchers noted that making the partner jealous may also have the positive function of assessing the partner's commitment and of alerting the partner that one could leave for greener pastures. Hence, one partner may deliberately evoke jealousy to solidify commitment and ward off being taken for granted. In addition, sexual passion may be reignited if one partner perceives that another would take the love object away. "The best sex I ever had was when my partner thought I was interested in someone else," said one of our students.

© Don Carstens/© Brand X Pictures/Jupiterimages

extradyadic relationship emotional or sexual involvement between a member of a couple and someone other than the partner.

Gender Differences in Coping With Jealousy

1. **Food.** Women were significantly more likely than men to report that they turned to food when they felt jealous: 30.3% of women, in contrast to 22% of men, said that they "always, often, or sometimes" looked to food when they felt jealous.

2. **Alcohol.** Men were significantly more likely than women to report that they drank alcohol or used drugs when they felt jealous: 46.9% of men, in contrast to 27.1% of women, said that they "always, often, or sometimes" drank or used drugs to make the pain of jealousy go away.

3. **Friends.** Women were significantly more likely than men to report that they turned to friends when they felt jealous: 37.9% of women, in contrast to 13.5% of men, said that they "always" turned to friends for support when feeling jealous.

4. **Nonbelief that "jealousy shows love."** Women were significantly more likely than men to disagree or to strongly disagree that "jealousy shows how much your partner loves you": 63.2% of women, in contrast to 42.6% of men, disagreed with the statement.

Source: Data taken from a survey of 291 undergraduate students (Knox et al., 2007).

© Simon Potter/Image Source

Undesirable Outcomes Shakespeare referred to jealousy as the "green-eyed monster," suggesting that it sometimes leads to undesirable outcomes for relationships. Anxious jealousy with its obsessive ruminations about the partner's alleged infidelity can make individuals miserable. They are constantly thinking about the partner being with the new person, which they interpret as confirmation of their own inadequacy. And, if the anxious jealousy results in repeated unwarranted accusations, a partner can tire of such attacks and seek to end the relationship with the accusing partner.

In its extreme form, jealousy may have devastating consequences. Possessive jealousy involves an attack on a partner or an alleged person to whom the partner is showing attention. In the name of love, people have stalked or killed their beloved and then killed themselves in reaction to rejected love. An example is the 2010 murder of Virginia Tech student Yeardley Love by George Hugely. Hugely's jealousy over Love breaking up with him and seeing someone else were contributing motives to the brutal murder—he beat her to death.

5-5 Cultural Restrictions on Whom an Individual can Love or Marry

University students routinely assert, "I can marry whomever I want!" Hardly. Rather, their culture and society radically restrict and influence their choice. One example is that homosexuals are not free to marry whomever they choose. Indeed, although several states now recognize same-sex marriage, federal law does not, reflecting wide societal disapproval. In addition, up until 1967 when the Supreme Court ruled that mixed marriages (e.g. between Whites and Blacks) were legal, such marriages in most states were a felony and individuals were put in jail.

Endogamy and exogamy are two forms of cultural pressure operative in mate selection.

endogamy the cultural expectation to select a marriage partner within one's social group.

endogamous pressures cultural attitudes reflecting approval for selecting a partner within one's social group and disapproval for selecting a partner outside one's social group.

exogamy the cultural expectation that one will marry outside the family group.

exogamous pressures cultural attitudes reflecting approval for selecting a partner outside one's family group.

pool of eligibles the population from which a person selects an appropriate mate.

homogamy tendency to select someone with similar characteristics.

homogamy theory of mate selection theory that individuals tend to be attracted to and become involved with those who have similar characteristics.

racial homogamy tendency for individuals to marry someone of the same race.

5-5a Endogamy

Endogamy is the cultural expectation to select a marriage partner within one's own social group, such as in the same race, religion, and social class. **Endogamous pressures** involve social approval and encouragement for selecting a partner within your own group (e.g., someone of your own race and religion) and disapproval for selecting someone outside your own group. The pressure toward an endogamous mate choice is especially strong when race is involved. Love may be blind, but it knows the color of one's partner. The overwhelming majority of individuals end up selecting someone of the same race to marry.

5-5b Exogamy

In addition to the cultural pressure to marry within one's social group, there is also the cultural expectation that one will marry outside the family group. This expectation is known as **exogamy**. **Exogamous pressures** involve social approval and encouragement for selecting a partner outside one's own group (e.g., someone outside your own family).

Incest taboos are universal; in addition, children are not permitted to marry the parent of the other sex in any society. In the United States, siblings and (in some states) first cousins are also prohibited from marrying each other. The reason for such restrictions is fear of genetic defects in children whose parents are too closely related.

Once cultural factors have determined the general **pool of eligibles** (the population from which a person selects an appropriate mate), individual mate choice becomes more operative. However, even when individuals feel that they are making their own choices, social influences are still operative.

5-6 Sociological Factors Operative in Partnering

Numerous sociological factors are at work in bringing two people together who eventually marry.

5-6a Homogamy— Twelve Factors

Whereas endogamy refers to cultural pressure, **homogamy** refers to individual initiative toward sameness, or "likes attract." The **homogamy theory of mate selection** states that we tend to be attracted to and become involved with those who are similar to ourselves in such characteristics as age, race, religion, and social class. In general, the more couples have in common, the higher the reported relationship satisfaction and the more durable the relationship (Clarkwest, 2007; Amato et al., 2007).

Race Race is a socially constructed concept and refers to physical characteristics (e.g., skin color, hair) that are given social significance. **Racial homogamy** operates strongly in selecting a live-in or marital partner (with greater homogamy for marital partners). As one moves from dating to living together to marriage, the willingness to marry interracially decreases. In a sample of 2,922 undergraduates, 35% of females and 30% of males reported that they had dated interra-

> **"Before you run** in double harness, **look well** to the other **horse."**
>
> —OVID, POET

cially (Knox & Hall, 2010). But only a small percentage will marry interracially. Black/White marriages, like that of Grace Hightower and Robert De Niro are unusual.

Although homogamy states that similar individuals are more comfortable with each other, racism is also operative. In their book *Two-Faced Racism*, Picca and Feagin (2007) found that three fourths of the 9,000 journal entries of college students included reports of racist remarks on campus. The researchers pointed out that, although college students commonly say "I'm not prejudiced" in a public context, racist comments are sometimes made when they are alone (backstage) with their friends. Sometimes racist comments are unintentionally hurtful, as one student reported in this journal entry:

On Thursday . . . my friend Megan (a White female) went on her first date with Steve. As their conversation began they discussed typical first date topics like family, friends, home, etc. Somehow Megan began talking about how squirrels in the Bronx are black and so people call them "squiggers." When Steve did not laugh Megan wondered why he did not think it was funny. As the conversation progressed and Steve began to talk about his family, he revealed to Megan that his dad is Black and mom is White. Right then, Megan realized why Steve did not find her joke to be so funny. Megan felt horrible. Without realizing it, Megan hurt someone that appeared to be just as White as she was. (p. 195)

Source: Excerpt from *Two-Faced Racism*, by Picca and Feagin. Routledge, 2007.

© Marc Fisher/iStockphoto.com

Ozay and colleagues (2012) analyzed data on 251 parents of undergraduates and found that that over a third (35%) disapproved of their child's involvement in an interracial relationship. Parental disapproval (which did not vary by gender of parent) increased with the seriousness of relationship involvement (e.g., dating or marriage), and parents were more disapproving of their daughters' than of their sons' interracial involvement.

In regard to parental approval of their son's or daughter's interracial involvement, Black mothers and White fathers have different roles in their respective communities in terms of setting the norms of interracial relationships. Hence, the Black mother who approves of her son's or daughter's interracial relationship may be less likely to be overruled than the White mother (the White husband may be more disapproving than the Black husband). Race may also affect one's perceptions. In a study of racial perceptions among college students, DeCuzzi and colleagues (2006) found that, although both races tended to view women and men of their own and the other race positively, there was a pronounced tendency to view women and men of their own race more positively and members of the other race more negatively.

Hohmann-Marriott and Amato (2008) found that individuals (both men and women) in interethnic marriages and cohabitation relationships (including interracial relationships such as Hispanic-Caucasian, African American–Hispanic, and African American–Caucasian) have lower quality relationships than those in same-ethnicity relationships. Lower relationship quality was defined in terms of reporting less satisfaction, more problems, higher conflict, and lower commitment to the relationship.

Similarly, Bratter and King (2008) analyzed national data and found higher divorce rates among interracial couples (compared to same-race couples). They also found race and gender variation. Compared to White couples, White female/Black male and Caucasian female/Asian male marriages were more prone to divorce, whereas those involving non-White females and White males and Hispanic and non-Hispanic people had similar or lower risks of divorce.

Age Most individuals select someone who is relatively close to themselves in age. Men tend to select women three to five years younger than themselves. The result is the "**marriage squeeze**," which is the imbalance of the ratio of marriageable-age men to marriageable-age women. In effect, women have fewer partners to select from because men choose from not only their same age group but also those younger than themselves. One 40-year-old, recently divorced woman said, "What chance do I have with all these guys looking at all these younger women?"

Education Educational **homogamy** is the tendency to select a cohabitant or marital partner with similar education. Being a college student ensures that you are surrounded by

marriage squeeze the imbalance of the ratio of marriageable-age men to marriageable-age women.

educational homogamy the tendency to select a cohabitant or marital partner with similar education.

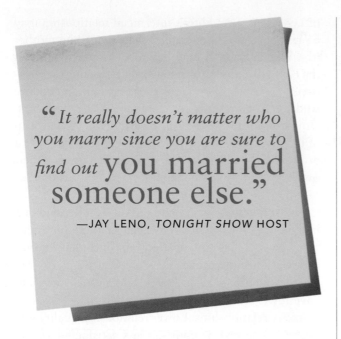

> *"It really doesn't matter who you marry since you are sure to find out* you married someone else.*"*
>
> —JAY LENO, *TONIGHT SHOW* HOST

other college students who become your pool of eligibles to meet, date, live with, and marry. Indeed, as a college graduate you are not likely to marry a person who has no college exposure. Not only does being in college predict that you will be exposed to other college students, it also increases the chance that only a college-educated partner becomes acceptable as a potential cohabitant or spouse. Indeed, the very pursuit of education becomes a value that you seek in another as a partner. "Having someone who can spell and use correct grammar become prerequisites in selecting an Internet partner," said one student.

Open-Mindedness People vary in the degree to which they are **open-minded** (receptive to understanding alternative points of view, values, and behaviors). Homogamous pairings in regard to open-mindedness are those that reflect partners who are relatively open or closed to new points of view, behaviors, and experiences. For example, an evangelical may not be open to people of alternative religions. (See the Open-Minded Scale at the end of this text.)

open-minded being open to understanding alternative points of view, values, and behaviors.

mating gradient the tendency for husbands to be more advanced than their wives with regard to age, education, and occupational success.

religion specific fundamental set of beliefs and practices generally agreed upon by a number of people or sects.

Social Class You have been reared in a particular social class that reflects your parents' occupations, incomes, and educations as well as your residence, language, and values. If you were brought up in a home in which both parents were physicians, you probably lived in a large house in a nice residential area—summer vacations and a college education were givens. Alternatively, if your parents dropped out of high school and worked "blue-collar" jobs, your home would be smaller and in a less expensive part of town, and your opportunities would be more limited (e.g., education). Social class affects one's comfort in interacting with others—we tend to feel more comfortable with others from our same social class.

The **mating gradient** refers to the tendency for husbands to be more advanced than their wives with regard to age, education, and occupational success. Indeed, husbands are typically older than their wives, have more advanced education, and earn higher incomes (*Statistical Abstract of the United States*, 2012).

Physical Appearance Homogamy is operative in regard to physical appearance in that people tend to become involved with those who are similar in degree of physical attractiveness. However, a partner's attractiveness may be a more important consideration for men than for women. In a study of homogamous preferences in mate selection, men and women rated physical appearance an average of 7.7 and 6.8 (out of 10) in importance, respectively (Knox et al., 1997).

Career Individuals tend to meet and select others who are in their same careers. In 2012, married country western singers Blake Shelton and Miranda Lambert won the Male and Female Vocalist of the Year awards, respectively. Michelle and Barack Obama are both attorneys; Angelina Jolie and Brad Pitt also share the same profession.

Marital Status Never-married people tend to select other never-married people as marriage partners, divorced people tend to select other divorced people, and widowed people tend to select other widowed people. Similar marital status may be more important to women than to men. In the study of homogamous preferences in mate selection, women and men rated similarity of marital status an average of 7.2 and 6.3 (out of 10) in importance, respectively (Knox et al., 1997).

Religion/Spirituality/Politics Religion may be defined as a set of beliefs in reference to a supreme being which involve practices or rituals (e.g., communion) generally agreed upon by a group of people.

Most of today's youth claim a religious affiliation. In a study of millennials (born after 1980 and who came of age in 2000) ages 18 to 39, only 25% reported that they have no religious affiliation (Pew Research Center, 2010). Similarly, some individuals view themselves as "not religious" but "spiritual," with spirituality defined as belief in the spirit as the seat of the moral or religious nature that guides one's decisions and behavior.

Religious homogamy is operative in that people of similar religion or spiritual philosophy tend to seek out each other. Over a third (33.6%) of 2,922 undergraduates agreed with the statement, "It is important that I marry someone of my same religion." Forty-seven percent reported that they had dated someone of another religion (Knox & Hall, 2010).

Religiosity is positively associated with delaying first intercourse, having a good relationship with one's parents, and holding negative attitudes toward divorce and cohabitation—all factors predictive of marital quality and stability (MacArthur, 2010).

The phrase "the couple that prays together, stays together" is more than just a cliché. However, it should also be pointed out that religion sometimes serves as a divisive force. For example, when one partner becomes "born again" or "saved," the relationship can be dramatically altered and eventually terminated unless the other partner shares the experience.

A rock-and-roller, hard-drinking wife (who had married a similar rock-and-roller) noted that, when her husband "got saved," it was the end of their marriage. "He gave our money to the church as the 'tithe' when we couldn't even pay the light bill," she said. "And when he told me I could no longer wear pants or lipstick or have a beer, I left."

Alford and colleagues (2011) emphasized that homogamy is operative in regard to political thinking. An analysis of national data on thousands of spousal pairs in the United States revealed that homogamous political attitudes were the strongest of all social, physical, and personality traits. Further, this similarity derived from initial mate choice "rather than persuasion and accommodation over the life of the relationship."

Attachment Individuals who report similar levels of attachment to each other report high levels of relationship satisfaction. This is the conclusion of Luo

The secure attachment of a child to a parent is the foundation of future love relationships.

© iStockphoto.com/Goldmund Lukic

and Klohnen (2005), who studied 291 newlyweds to assess the degree to which similarity affected marital quality. Indeed, similarity of attachment was *the* variable most predictive of relationship quality.

Personality In a study of 417 eHarmony married couples, the partners who were similar to each other in personality characteristics were more satisfied with their relationship (Gonzaga et al., 2010). Similarly, a study of participants at chemistry.com found that "explorers" (curious, creative, adventurous, sexual) were attracted to other "explorers" (H. Fisher, 2009). Being neurotic is also associated with negative relationship outcomes (Burr et al., 2011).

Economic Values, Money Management, and Debt Individuals vary in the degree to which they have money, spend money, and save money. Some are deeply in debt whereas others have no debt. Some carry significant educational debt. The average debt for those with a bachelor's degree is $23,000. Almost half (48%) of those who borrowed money for their undergraduate education report that they are burdened by this debt (Pew Research Center, 2011). Nine percent of 2,922 undergraduates noted that they owed more than a thousand dollars on a credit card (Knox & Hall, 2010). Money becomes an issue in mate selection in that different economic values predict conflict. One undergraduate noted, "There is no way I would get involved with/marry this person as they don't worry about the amount of debt they have."

religious homogamy
tendency for people of similar religious or spiritual philosophies to seek out each other.

Psychological Factors Operative in Partnering

Psychologists have focused on complementary needs, exchanges, parental characteristics, and personality types with regard to mate selection.

5-7a Complementary-Needs Theory

"In spite of the Women's Movement and a lot of assertive friends, I am a shy and dependent person," remarked a transfer student. "My need for dependency is met by Warren, who is the dominant, protective type." The tendency for a submissive person to become involved with a dominant person (one who likes to control the behavior of others) is an example of attraction based on complementary needs.

Complementary-needs theory states that we tend to select mates whose needs are opposite and complementary to our own. Partners can also be drawn to each other on the basis of nurturance versus receptivity. These complementary needs suggest that one person likes to give and take care of another, whereas the other likes to be the benefactor of such care. Other examples of complementary needs are responsibility versus irresponsibility, peacemaker versus troublemaker, and disorder versus order. Helen Fisher (2009), referred to earlier regarding her study of 28,000 participants looking for a partner on chemistry.com, also found evidence of complementary needs. In terms of personality, "directors" (analytical, decisive, focused) sought "negotiators" (introspective, verbal, intuitive) and vice versa.

The idea that mate selection is based on complementary needs was suggested by Winch (1955), who noted that needs can be complementary if they are different (e.g., dominant and submissive) or if the partners have the same need at different levels of intensity. As an example of the latter, two

complementary-needs theory
tendency to select mates whose needs are opposite and complementary to one's own needs.

individuals may have a complementary relationship if they both want to pursue graduate studies but want to earn different degrees. The partners will complement each other if, for instance, one is comfortable with aspiring to a master's degree and approves of the other's commitment to earning a doctorate.

Winch's theory of complementary needs, commonly referred to as "opposites attract," is based on the observation of 25 undergraduate married couples at Northwestern University. Other researchers who have not been able to replicate Winch's study have criticized the findings (Saint, 1994). Jay Leno and his wife, Mavis, have a complementary needs relationship:

> We were, and are, opposites in most every way. Which I love. . . . I don't consider myself to have much of a spiritual side, but Mavis has almost a sixth sense about people and situations. She has deep focus, and I fly off in 20 directions at once. She reads 15 books a week, mostly classic literature. I collect classic car and motorcycle books. She loves European travel; I don't want to go anywhere people won't understand my jokes. (Leno, 1996, pp. 214–215)

Three questions can be raised about the theory of complementary needs:

1. **Couldn't personality needs be met just as easily outside the couple's relationship as through mate selection?** For example,

Couples complement each other in terms of the characteristics they bring to the relationship.

couldn't a person who has the need to be dominant find such fulfillment in a job that involved an authoritative role, such as being a supervisor?

2. **What is a complementary need as opposed to a similar value?** For example, is the desire to achieve at different levels a complementary need or a shared value?

3. **Don't people change as they age?** Could dependent people grow and develop self-confidence so that they might no longer need to be involved with a dominant person? Indeed, such a person might no longer enjoy interacting with a dominant person.

5-7b Exchange Theory

Exchange theory emphasizes that mate selection is based on assessing who offers the greatest rewards at the lowest cost. The following five concepts help to explain the exchange process in mate selection:

1. **Rewards.** Rewards are the behaviors (your partner looking at you with the eyes of love), words (saying "I love you"), resources (being beautiful or handsome; having a car, an apartment, and money), and services (cooking for you, typing for you, taking you to class) your partner provides that you value and that influence you to continue the relationship. Similarly, you provide behaviors, words, resources, and services for your partner that your partner values. Relationships in which the exchange is equal are happiest. Increasingly, men are interested in women who offer "financial independence." "I'm tired of paying for everything," said one of our students. (He obviously has not met those females on campus who work and are economically independent.)

2. **Costs.** Costs are the unpleasant aspects of a relationship. A woman identified the costs associated with being involved with her partner: "He abuses drugs, doesn't have a job, and lives nine hours away." The costs her partner associated with being involved with this woman included "she nags me," "she doesn't like sex," and "she wants her mother to live with us if we marry." Ingoldsby and colleagues (2003) assessed the degree to which various characteristics were associated with reducing one's attractiveness on the marriage market. The most undesirable traits were not being heterosexual, having alcohol or drug problems, having a sexually transmitted disease, and being lazy.

3. **Profit.** Profit occurs when the rewards exceed the costs. Unless the couple described in the preceding paragraph derive a profit from staying together, they are likely to end their relationship and seek someone else with whom there is a higher profit margin.

4. **Loss.** Loss occurs when the costs exceed the rewards.

5. **Alternative.** Is another person currently available who offers a higher profit margin?

Most people have definite ideas about what they are looking for in a mate. The currency used in the marriage market consists of the socially valued characteristics of those involved, such as age, physical characteristics, and economic status. We typically get as much in return for our attributes as we have to offer or trade. Indeed, individuals have a perception of what they are worth on the marriage market and adjust their expectations in terms of the partner they think they can attract (Bredow et al., 2011). An unattractive, drug-abusing high school dropout with no job has little to offer an attractive, drug-free, high-GPA college student. Similarly, the high school student would feel uncomfortable around the college student.

Once you identify a person who offers you a good exchange for what you have to offer, other bargains are made about the conditions of continuing your relationship. The person who has the least interest in continuing the

exchange theory theory that emphasizes that relations are formed and maintained between individuals offering the greatest rewards and least costs to each other.

relationship controls the relationship (Waller & Hill, 1951). This **principle of least interest** is illustrated by the woman who said, "He wants to date me more than I want to date him, so we end up going where I want to go and doing what I want to do." In this case, the woman trades her company for the man's acquiescence to her recreational choices. In effect, the person with the least interest controls the relationship.

5-7c Parental Characteristics

Whereas the complementary-needs and exchange theories of mate selection are relatively recent, Freud suggested that the choice of a love object in adulthood represents a shift in libidinal energy from the first love objects—the parents. **Role (or modeling) theory of mate selection** emphasizes that a son or daughter models the parent of the same sex by selecting a partner similar to the one the parent selected.

This concept means that a man looks for a wife who has characteristics similar to those of his mother and that a woman looks for a husband who is very similar to her father.

5-7d Personality Types

Desired Characteristics of a Potential Mate In a study of 700 undergraduates, both men and women reported that the personality characteristics of being warm, kind, and open and of having a sense of humor were very important to them in selecting a romantic or sexual partner. Indeed, these intrinsic personality characteristics were rated as more important than physical attractiveness or wealth (extrinsic characteristics; Sprecher & Regan, 2002). Similarly, adolescents wanted intrinsic qualities such as intelligence and humor in a romantic partner but looked for physical appearance and high sex drive in a casual partner (no gender-related differences in responses were found; Regan & Joshi, 2003). Sometimes what individuals say they want and what they actually select may be different. Bogg and Ray (2006) noted that, although women in one study said that they consistently preferred to date and marry men who were egalitarian, it was not unusual for them to select ultramasculine men who were sometimes dominant, mysterious, and rebellious.

Assad and colleagues (2007) studied the personality quality of optimism and found that optimistic individuals tended to be involved in satisfying and happy romantic relationships and that a substantial portion of this association was mediated by willingness to be a cooperative problem solver. Hence, selecting a person who is optimistic seems to predict well for the future of the relationship.

In another study, women were significantly more likely than men to identify "having a good job" and "being well-educated" as important attributes in a future mate, whereas significantly more men than women wanted their spouse to be physically attractive (Medora et al., 2002). Toro-Morn and Sprecher (2003) noted that both American and Chinese undergraduate women identified characteristics associated with status (earning potential, wealth) more than physical attractiveness.

Women are also attracted to men who have good manners. In a study of 398 undergraduates, women were significantly more likely than men to report that they wanted to "date or be involved with [only] someone who had good manners," that "manners are very important," and that "the more well mannered the person, the more I like the person" (Zusman et al., 2003). Twenge (2006) noted that good manners are being displayed less often.

Undesired Personality Characteristics of a Potential Mate Several personality factors are predictive of unhappy relationships or divorce. Potential partners who are observed to consistently display these characteristics might be avoided.

1. **Controlling.** The behavior that 60% of a national sample of adult, single women reported as the most serious fault of a man was his being "too controlling" (Edwards, 2000).

2. **Narcissistic.** Individuals who display narcissism view relationships in terms of what they get out

principle of least interest principle stating that the person who has the least interest in a relationship controls the relationship.

role (modeling) theory of mate selection emphasizes that children select partners similar to the one their same-sex parent selected.

of them. When satisfactions wane and alternatives are present, narcissists are the first to go. Because all relationships have difficult times, a narcissist is a high risk for a durable marriage partner.

3. **Poor impulse control.** Lack of impulse control is problematic in relationships because such individuals are less likely to consider the consequences of their actions. For example, to some people, having an affair might seem harmless, but it often has devastating consequences for the partners and their relationship (ask Tiger Woods). For people with poor impulse control, if they feel like having an affair, they often will—and worry about the consequences later.

4. **Hypersensitive.** Hypersensitivity to perceived criticism involves getting hurt easily. Any negative statement or criticism is received with a greater impact than the partner intended. The disadvantage of such hypersensitivity is that a partner may learn not to give feedback for fear of hurting the hypersensitive partner. Such lack of feedback to the hypersensitive partner blocks information about what the person does that upsets the other and what could be done to make things better. The result is a relationship in which the partners can't talk about what is wrong, so the potential for change is limited.

5. **Inflated ego.** An exaggerated sense of oneself is another way of saying a person has a big ego and always wants things to be his or her way. A person with an inflated sense of self may be less likely to consider the other person's opinion in negotiating a conflict and prefer to dictate an outcome. Such disrespect for the partner can be damaging to the relationship.

6. **Perfectionistic.** Individuals who are perfectionists may require perfection of themselves and others. This attitude is associated with relationship problems (Haring et al., 2003).

7. **Insecure.** Feelings of insecurity also compromise marital happiness. Researchers studied the personality trait of attachment and its effect on marriage in 157 couples at two time intervals and found that "insecure participants reported more difficulties in their relationships. . . . [I]n contrast, secure participants reported greater feelings of intimacy in the relationship at both assessments" (Crowell et al., 2002).

© holbox/Shutterstock

8. **Controlled.** Individuals who are controlled by their parents, grandparents, former partner, child, or anyone else compromise the marriage relationship because their allegiance is external to the couple's relationship. Unless the person is able to break free of such control, the ability to make independent decisions will be thwarted, which will both frustrate the spouse and challenge the marriage. An example is a wife whose father dictated that she spend all holidays with him.

9. **Substance abuser.** There is a negative relationship between substance abuse (alcohol and marijuana) and the likelihood of getting married, particularly for females (Blair, 2010). A woman's being labeled as a "drunk" or "stoner" may alienate her from her peers and reduce her chance of being selected as a romantic partner. Wiersma et al. (2011) also noted negative relationship outcomes for the female cohabitant who drinks significantly more than her partner.

In addition to undesirable personality characteristics, Table 5.1 lists some particularly troublesome personality types and how they may negatively impact you if you live with or marry such a person.

Table 5.1
Personality Types Problematic in a Potential Partner

Type	Characteristics	Impact on Partner
Paranoid	Suspicious, distrustful, thin-skinned, defensive	Partners may be accused of everything.
Schizoid	Cold, aloof, solitary, reclusive	Partners may feel that they can never "connect" and that the person is not capable of returning love.
Borderline	Moody, unstable, volatile, unreliable, suicidal, impulsive	Partners will never know what their Jekyll-and-Hyde partner will be like, which could be dangerous.
Antisocial	Deceptive, untrustworthy, conscienceless, remorseless	Such a partner could cheat on, lie to, or steal from a partner and not feel guilty.
Narcissistic	Egotistical, demanding, greedy, selfish	Such a person views partners only in terms of their value. Don't expect such a person to see anything from a partner's point of view; expect such a person to bail in tough times.
Dependent	Helpless, weak, clingy, insecure	Such a person will demand a partner's full time and attention, and other interests will incite jealousy.
Obsessive-compulsive	Rigid, inflexible	Such a person has rigid ideas about how a partner should think and behave and may try to impose them on the partner.

© Cengage Learning 2011

5-8 Sociobiological Factors Operative in Partnering

In contrast to cultural, sociological, and psychological aspects of mate selection, which reflect a social learning assumption, the sociobiological perspective suggests that biological or genetic factors may be operative in mate selection.

5-8a Definition of Sociobiology

Sociobiology suggests a biological basis for all social behavior—including mate selection. Based on Charles Darwin's theory of natural selection, which states that the strongest of the species survive, sociobiology holds that men and women select each other as mates on the basis of their innate concern for producing offspring who are most capable of surviving.

According to sociobiologists, men look for a young, healthy, attractive, sexually conservative woman who will produce healthy children and invest in taking care of the children.

sociobiology
suggests a biological basis for all social behavior.

Women, in contrast, look for an ambitious man with good economic capacity who will invest his resources in her children. Earlier in this chapter, we provided data supporting the idea that men seek attractive women and women seek ambitious, financially successful men.

5-8b Criticisms of the Sociobiological Perspective

The sociobiological explanation for mate selection is controversial. Critics argue that women may show concern for the earning capacity of men because women have been systematically denied access to similar economic resources and selecting a mate with these resources is one of their remaining options. In addition, it is argued that both women and men, when selecting a mate, think about their partners more as companions than as future parents of their offspring.

5-9 Engagement

Engagement moves the relationship of a couple from a private, love-focused experience to a public, parent-involved experience. Family and friends are invited to enjoy the happiness and commitment of the individuals to a future marriage. Unlike casual dating, **engagement** is a time in which the partners are emotionally committed, are sexually monogamous, and are focused on wedding preparations. The engagement period is the last opportunity before marriage to systematically examine the relationship. Johnson and Anderson (2011) studied 610 newly married couples and found that those spouses who were more confident of their decision to marry ended up spending more time together and invested more in their relationship with greater marital satisfaction.

5-9a Asking Specific Questions

Because partners might hesitate to ask for or reveal information that they feel will be met with disapproval during casual dating, the engagement is a time to be specific about the other partner's thoughts, feelings, values, goals, and expectations. The Involved Couple's Inventory (see Chapter 5 Self-Assessment Cards at the end of the text) is designed to help individuals in committed relationships learn more about each other by asking specific questions. Other couples use technology to find out information about their potential partners.

5-9b Visiting Your Partner's Parents

Engaged couples should seize the opportunity to discover the family environment in which each partner was reared and consider the implications for their subsequent marriage. When visiting one partner's parents, the other should observe their standard of living and the way they interact and relate with one another (e.g., level of affection, verbal and nonverbal behavior, marital roles). How does their standard of living compare with that of the visiting partner's family? How does the emotional closeness (or distance) of the partner's family compare with that of the visiting partner's family? Such comparisons are significant because both partners will reflect their respective family or origins. "This is the way we did it in my family" is a phrase that is often repeated. One way to discern how a person will be treated by his or her partner in the future is to observe the way the partner's parent of the same sex treats and interacts with his or her spouse. To know what a partner may be like in the future, one should look at the partner's parent of the same sex. There is a tendency for a man

Couples often benefit from examining their relationship and attending counseling BEFORE they get married.

to become like his father and a woman to become like her mother. A partner's parent of the same sex and the parents' marital relationship are the models of a spouse and a marriage relationship that the person is likely to duplicate.

5-9c Participating in Premarital Education Programs

Various **premarital education programs** (also known as premarital prevention programs, premarital counseling, premarital therapy, and marriage preparation), both academic and religious, are formal, systematized experiences designed to provide information to individuals and to couples about how to have a good relationship. About 30% of couples getting married become involved in some type of premarital education. The greatest predictor of whether a couple will become involved in a premarital education program is the desire and commitment of the female to do so (Duncan et al. 2007). Jackson (2011) developed

engagement
period of time during which committed, monogamous partners focus on wedding preparations and systematically examine their relationship.

premarital education programs
formal, systematized experiences designed to provide information to individuals and couples about how to have a good relationship.

> ## "Marry'd in haste,
> ## we may repent
> ## at leisure."
>
> —WILLIAM CONGREVE, PLAYWRIGHT

and implemented a unique evidenced informed treatment protocol for premarital counseling including six private sessions and two postmarital booster sessions.

Premarital education programs are valuable in that they not only help partners assess the degree to which they are compatible with each other but also provide a context to discuss specific relationship issues. When 325 individuals who attended 12 two-hour sessions provided by Family Wellness instructors were given a pretest and posttest, Devall and colleagues (2009) found significant increases in relationship satisfaction and the ability to resolve conflict. Parents were also less likely to endorse corporal punishment. In effect a 24-hour investment by these respondents paid relationship dividends to the partners in addition to the benefits afforded the children.

PREPARE, a program designed for engaged couples, was assessed by Futris and colleagues (2011) who found that those who participated reported gains in relationship knowledge, feeling confident about their relationship, engaging in positive conflict management behavior, and feeling more satisfied in their relationship.

5-10 Marrying for the Wrong Reasons

Some reasons for getting married are more questionable than others. These reasons include the following:

5-10a Rebound

A rebound marriage results when people marry someone immediately after another person has ended a relationship with them. It is a frantic attempt on their part to reestablish their desirability in their own eyes and in the eyes of the partner who dropped them. One man said, "After she told me she wouldn't marry me, I became desperate. I called up an old girlfriend to see if I could get the relationship going again. We were married within a month. I know it was foolish, but I was very hurt and couldn't stop myself." To marry on the rebound is questionable because the marriage is made in reference to the previous partner and not to the partner being married. In reality, people who engage in rebound marriage are using the person they intend to marry to establish themselves as the "winner" in the previous relationship.

To avoid the negative consequences of marrying on the rebound, wait until the negative memories of a past relationship have been replaced by positive aspects of a current relationship. In other words, marry when the satisfactions of being with the current partner outweigh any feelings of revenge. This adjustment normally takes between 12 and 18 months. In addition, be careful about getting involved in a relationship with someone who is recently divorced or has just emerged from a painful ending of a previous relationship.

5-10b Escape

A person might marry to escape an unhappy home situation in which the parents are oppressive, overbearing, conflictual, alcoholic, and/or abusive. One woman said, "I couldn't wait to get away from home. Ever since my parents divorced, my mother has been drinking and watching me like a hawk. 'Be home early, don't drink, and watch out for those horrible men,' she would always say. I admit it. I married the first guy that would have me. Marriage was my ticket out of there."

Marriage for escape is a bad idea. It is far better to continue the relationship with the partner until mutual love and respect, rather than the desire to escape an unhappy situation, become the dominant forces propelling you toward marriage. In this way, you can evaluate the marital relationship in terms of its own potential and not solely as an alternative to an unhappy situation.

5-10c Unplanned Pregnancy

Getting married just because a partner becomes pregnant is usually a bad idea. Indeed, the decision of whether to marry should be kept separate from decisions about a pregnancy. Adoption, abortion, single parenthood, and unmarried parenthood (the couple can remain together as an unmarried couple and have the baby) are all alternatives to simply deciding to marry if a partner becomes pregnant. Avoiding feelings of being trapped or later feeling that the marriage might not have happened without the pregnancy are two reasons for not rushing into marriage because of pregnancy. Couples who marry when the woman becomes pregnant have an increased chance of divorce.

© Aksenova Natalya/Shutterstock

5-10d Psychological Blackmail

Some individuals get married because their partner insists "I can't live without you" or "I will commit suicide if you leave me." Because the person fears that the partner may commit suicide, the partner agrees to the wedding. The problem with such a marriage is that one partner has learned to manipulate the relationship to get control. Use of such power often creates resentment in the other partner, who feels trapped in the marriage. Escaping from the marriage becomes even more difficult. One way of coping with a psychological blackmail situation is to encourage the person to go with you to a therapist to "discuss the relationship." Once inside the therapy room, you can tell the counselor that you feel pressured to get married because of the suicide threat. Counselors are trained to respond to such a situation.

5-10e Insurance Benefits

In a poll conducted by the Kaiser Family Foundation, a health policy research group, 7% of adults said someone in their household had married in the past year to gain access to insurance. "For today's couples, 'in sickness and in health' may seem less a lover's troth than an actuarial contract. They marry for better or worse, for richer or poorer, for co-pays and deductibles" (Sack, 2008). Although selecting a partner who has resources (which may include health insurance) is not unusual,

selecting a partner for health benefits is yet another matter. Both parties might be cautious if the alliance is more about "benefits" than the relationship.

5-10f Pity

Some partners marry because they feel guilty about terminating a relationship with someone whom they pity. The fiancé of one woman got drunk one Halloween evening and began to light fireworks on the roof of his fraternity house. As he was running away from a Roman candle he had just ignited, he tripped and fell off the roof. He landed on his head and was in a coma for three weeks. A year after the accident, his speech and muscle coordination were still adversely affected. The woman said she did not love him anymore but felt guilty about terminating the relationship now that he had become physically afflicted.

She was ambivalent. She felt marrying her fiancé was her duty, but her feelings were no longer love feelings. Pity may also have a social basis. For example, a partner may fail to achieve a lifetime career goal (e.g., the partner may flunk out of medical school). Regardless of the reason, if one partner loses a limb, becomes brain damaged, or fails in the pursuit of a major goal, keeping the issue of pity separate from the advisability of the marriage is important. The decision to marry should be based on factors other than pity for the partner.

5-10g Filling a Void

A former student in our classes noted that her father died of cancer. She acknowledged that his death created a vacuum, which she felt driven to fill immediately by getting married so that she would have a man in her life. Because she was focused on filling the void, she had paid little attention to the personality characteristics of or her relationship with the man who had asked to marry her.

She reported that she discovered on her wedding night that her new husband had several other girlfriends whom he had no intention of giving up. The marriage was annulled.

In deciding whether to continue or terminate a relationship, listen to what your senses tell you ("Does

> A short courtship does not allow partners to learn about each other's background, values, and goals and does not permit time to observe and scrutinize each other's behavior in a variety of settings.

it feel right?"), listen to your heart ("Do you love this person or do you question whether you love this person?"), and evaluate your similarities ("Are we similar in terms of core values, goals, view of life?"). Also, be realistic. It would be unusual if none of these factors applied to you. Indeed, most people have some negative and some positive indicators before they marry.

5-11 Consider Calling Off the Wedding If . . .

"No matter how far you have gone on the wrong road, turn back" is a Turkish proverb. If your relationship is characterized by the following factors, consider prolonging your engagement and delaying your marriage at least until the most distressing issues have been resolved. Alternatively, break the engagement (which happens in 30% of formal engagements). Indeed, rather than defend a course of action that does not feel right, stop and reverse directions. Breaking an engagement has fewer negative consequences and is less stigmatized than ending a marriage.

5-11a Age 18 or Younger

The strongest predictor of getting divorced is getting married during the teen years.

Individuals who marry at age 18 or younger have three times the risk of divorce than those who delay marriage into their late 20s or early 30s. Teenagers may be more at risk for marrying to escape an unhappy home and may be more likely to engage in impulsive decision making and behavior. Early marriage is also associated with an end to one's education, social isolation from peer networks, early pregnancy or parenting, and locking oneself into a low income. Increasingly, individuals are delaying when they marry. As noted earlier, the median age at first marriage in the United States is almost 29 for men and 27 for women.

Research by Meehan and Negy (2003) on being married while in college revealed higher marital distress among spouses who were also students. In addition, when married college students were compared with single college students, the married students reported more difficulty adjusting to the demands of higher education. The researchers concluded that "these findings suggest that individuals opting to attend college while being married are at risk for compromising their marital happiness and may be jeopardizing their education" (p. 688).

5-11b Known Partner Less Than Two Years

Of 2,922 undergraduates, 27% agreed with the statement "If I were really in love, I would marry someone I had known for only a short time" (Knox & Hall, 2010). Impulsive marriages in which the partners have

© Skip Odonnell/iStockphoto.com

> "Neurotics worry about things that didn't happen in the past instead of worrying like normal people about things that won't happen in the future."
>
> —UNKNOWN

known each other only briefly are associated with a higher-than-average divorce rate. Indeed, partners who date each other for at least two years (25 months to be exact) before getting married report the highest level of marital satisfaction and are less likely to divorce (Huston et al., 2001). However, Dr. Carl Ridley (2009) observed, "I am not sure it is about time but more about attending to self and other needs, desires, preferences and then assessing if one's potential partner can meet these needs or be willing to negotiate so that mutual needs are met. For some, this takes lots of time and for others, not very long."

A short courtship does not allow partners to learn about each other's background, values, and goals and does not permit time to observe and scrutinize each other's behavior in a variety of settings (e.g., with one's close friends, parents, or siblings). Indeed, some individuals may be more prone to falling in love at first sight and to wanting to hurry the partner into a committed love relationship *before* the partner can find out about who they really are. "Let the buyer beware!" If your partner is pressuring you to marry too quickly, tell your partner you need to slow the relationship down and don't want to get married now.

To increase the knowledge you and your partner have about each other, find out each other's answers to the Involved Couple's Inventory, take a five-day "primitive" camping trip, take a 15-mile hike together, wallpaper a small room together, or spend several days together when one partner is sick. If the couple plans to have children, they may want to take care of a 6-month-old together for a weekend. Time should also be spent with each other's parents and friends.

5-11c Abusive Relationship

As we will discuss in Chapter 13, Abuse in Relationships, partners who emotionally and/or physically abuse their partners while dating and living together continue these behaviors in marriage. Abusive lovers become abusive spouses, with predictable negative outcomes. Though extricating oneself from an abusive relationship is difficult before the wedding, it becomes more difficult after marriage and even more difficult once the couple has children.

One characteristic of an abusive partner is his or her attempt to systematically detach the intended spouse from all other relationships ("I don't want you spending time with your family and friends—you should be here with me"). This behavior is a serious red flag for impending relationship doom that should not be overlooked, and one should seek the exit ramp as soon as possible.

5-11d Numerous Significant Differences

Relentless conflict often arises from numerous significant differences. Though all spouses are different from each other in some ways, those who have numerous differences in key areas such as race, religion, social class, education, values, and goals are less likely to report being happy and to have durable relationships. The less couples have in common, the more their marital distress (Amato et al., 2007). As we noted earlier, the more relationship partners have in common, the more likely the relationship will last.

5-11e On-and-Off Relationship

A roller-coaster premarital relationship is predictive of a marital relationship that will follow the same pattern. Partners who break up and get back together several times have developed a pattern in which the dissatisfactions in the relationship become so frustrating that separation becomes the antidote for relief. In courtship, separations are of less social significance than marital separations. "Breaking up" in courtship is called "divorce" in marriage. Couples who routinely break up and get back together should examine the issues that continue to recur in their relationship and attempt to resolve them.

5-11f Parental Disapproval

Parents usually have an opinion of their son's or daughter's mate choice. Mothers disapprove of the mate choice of their daughters if they predict that he will be a lousy father, and fathers disapprove if they predict the suitor will be a poor provider (Dubbs & Buunk, 2010). When student and parent ratings of mate selection traits were compared, parents ranked religion higher than did their offspring, whereas offspring ranked physical attractiveness higher than did parents (Perilloux et al., 2011). Parents were also more focused on earning capacity and education (e.g., college graduate) in their daughter's mate selection than in their son's.

A parent recalled, "I knew when I met the guy it wouldn't work out. I told my daughter and pleaded that she not marry him. She did, and they divorced." Although parental and in-law dissatisfaction is rare (Amato et al., 2007, found that only 13% of parents

© iStockphoto.com

disapprove of the partner their son or daughter plans to marry), such parental predictions (whether positive or negative) often come true. If the predictions are negative, they sometimes contribute to stress and conflict once the couple marries. While individuals can always override parental concerns, they should give serious consideration to those concerns.

5-11g Low Sexual Satisfaction

Sexual satisfaction is linked to relationship satisfaction, love, and commitment. Sprecher (2002) followed 101 dating couples across time and found that low sexual satisfaction (for both women and men) was related to reporting low relationship quality, less love, lower commitment, and breaking up. Hence, couples who are dissatisfied with their sexual relationship might explore ways of improving it (alone or through counseling) or consider the impact of such dissatisfaction on the future of their relationship.

Basically it is time to end a relationship when the gain or advantages of staying together no longer outweigh the pain and disadvantages of staying—the pain of leaving is less than the pain of staying. Of course, all relationships go through periods when the disadvantages outweigh the benefits, so one should not bail out without careful consideration.

STUDY TOOLS ➲ **5**

Ready to study? In this book, you can:

- ➲ Rip out the Chapter Review card in the back of the book to study for exams

- ➲ Take the Self Assessment for this chapter (card in the back of the book) and see where you stand on the vital issues raised in the chapter

Or you can go online to CourseMate at www.cengagebrain.com for these resources:

- ➲ Complete Practice Quizzes to prepare for tests

- ➲ Review Key Terms Flash Cards (online or print)

- ➲ Read about Marriage and Family in the news

- ➲ Play "Beat the Clock" to master concepts

- ➲ Check out Personal Applications

WHY CHOOSE?

Every 4LTR Press solution comes complete with a visually engaging textbook in addition to an interactive eBook. Go to CourseMate for **M&F** to begin using the eBook. Access at **www.cengagebrain.com**

Complete the Speak Up survey in CourseMate at **www.cengagebrain.com**

 Follow us at **www.facebook.com/4ltrpress**

Sexuality in Relationships

"Among **men,** sex sometimes **results in intimacy;** among **women** intimacy sometimes **results in sex.**"

—BARBARA CARTLAND, ENGLISH ROMANCE NOVELIST

SECTIONS

6-1 Sexual Values

6-2 Alternative Sexual Values

6.3 Sexual Double Standard

6-4 Sources of Sexual Values

6-5 Gender Differences in Sexuality

6-6 Pheromones and Sexual Behavior

6-7 Sexuality in Relationships

6-8 Safe Sex: Avoiding Sexually Transmitted Infections

6-9 Sexual Fulfillment: Some Prerequisites

"Like a match, sex ignites and brings life, light, and warmth; it also burns" (Merrill & Knox, 2010). Once sex is introduced into a relationship, the relationship is changed forever, and often in a positive way. Indeed, sexuality is a major relationship aspect and the subject of this chapter. We begin by discussing the definition of sexual values and the various sexual values individuals bring to the relationship table.

6-1 Sexual Values

The following are some examples of choices (based on sexual values) with which individuals in a new relationship are confronted:

- How much sex and how soon in a relationship is appropriate?

- Do you require a condom for vaginal or anal intercourse?

- Do you require a condom and/or dental dam (vaginal barrier) for oral sex?

- Before having sex with someone, do you require that he or she has been recently tested for sexually transmitted infections (STI) and the human immunodeficiency virus (HIV)?

- Do you tell your partner the *actual* number of previous sexual partners?
- Do you tell your partner your sexual fantasies?
- Do you reveal to your partner previous or current same-sex behavior or interests?

Sexual values are moral guidelines for sexual behavior in relationships. Attitudes and values sometimes predict sexual behavior. One's sexual values may be identical to one's sexual behavior. For example, a person who values remaining a virgin or abstaining until marriage will not have sexual intercourse until marriage. But one's sexual behavior does not always correspond with one's sexual values. One explanation for the discrepancy between values and behavior is that a person may engage in a sexual behavior, then decide the behavior was wrong and adopt a sexual value against it.

sexual values moral guidelines for sexual behavior in relationships.

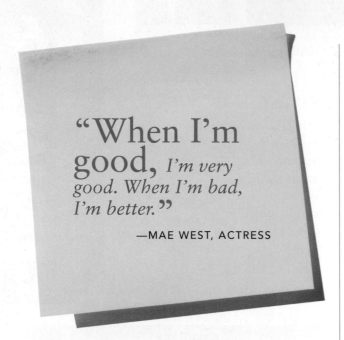

"When I'm good, *I'm very good. When I'm bad, I'm better.*"

—MAE WEST, ACTRESS

6-2 Alternative Sexual Values

At least three sexual values guide choices in sexual behavior: absolutism, relativism, and hedonism. People often have different sexual values over time. For example, as a young adolescent, an individual may have the sexual value of absolutism, but this value typically transitions into the sexual value of relativism. Along the way, the individual may become hedonistic for a period and then move back to relativism.

6-2a Absolutism

Absolutism refers to a belief system based on unconditional allegiance to the authority of religion, tradition, or law. A religious absolutist makes sexual choices on the basis of moral considerations. To make the correct moral choice is to comply with one's religious doctrine. People who are guided by absolutism in their sexual choices have a clear notion of what is right and wrong.

The official creeds of fundamentalist Christian and Islamic religions encourage absolutist sexual values. Intercourse is solely for procreation, and any sexual acts that do not lead to procreation (masturbation, oral sex, homosexuality) are immoral and regarded as sins against God, Allah, self, and

absolutism belief system based on unconditional allegiance to the authority of religion, law, or tradition.

community. Waiting until marriage to have intercourse is also an absolutist sexual value. Of 1,004 undergraduates, 6% of males and 12% of females held this value (Toews & Yazedjian, 2011). True Love Waits is an international campaign designed to challenge teenagers and college students to remain sexually abstinent until marriage. Under this program, created and sponsored by the Baptist Sunday School Board, young people are asked to agree to the absolutist position and sign a commitment to the following: "Believing that true love waits, I make a commitment to God, myself, my family, my friends, my future mate, and my future children to a lifetime of purity including sexual abstinence from this day until the day I enter a biblical marriage relationship" (True Love Waits, 2012).

How effective are the True Love Waits and "virginity pledge" programs in delaying sexual behavior until marriage? Data from the National Longitudinal Study of Adolescent Health revealed that although youth who took the pledge were more likely than other youth to experience a later "sexual debut," have fewer partners, and marry earlier, most eventually engaged in premarital sex, were less likely to use a condom when they first had intercourse, and were more likely to substitute oral and/or anal sex in the place of vaginal sex. There was no significant difference in the occurrence of STIs between "pledgers" and "nonpledgers" (Brucker & Bearman, 2005). The emphasis on virginity may have encouraged the pledgers to engage in noncoital (nonintercourse) sexual activities (e.g., oral sex), which still exposed them to STIs, and to be less likely to seek testing and treatment for STIs. Hollander (2006) also collected national data on adolescents and found that half (more often males and Blacks) who had taken the virginity pledge reported no such commitment a year

© SuperStock/Jupiterimages

later. A study of 1,215 undergraduates confirmed that male pledgers (in contrast to female pledgers) were less likely to remain virgins (Landor & Simons, 2010).

A similar cultural ritual is the Father–Daughter Purity Ball, which encourages teenage daughters to remain a virgin until marriage. These events occur in 48 states, have an evangelical base, and involve fathers with daughters as young as 4 years old who take mutual pledges to be pure. Fathers promise to be faithful to their wives and shun pornography; daughters promise to be pure, which implies the value of absolutism.

Virginity remains a value for some individuals. Fourteen percent of females and 20% of males of 2,922 undergraduates identified absolutism as their sexual value (Knox & Hall, 2010). Boislard and Poulin (2011) were able to predict early virginity loss. In a sample of 402 youths, those who engaged in more antisocial behaviors, had a lot of other-sex friends, and were from a non-intact family were predicted to experience their first intercourse at age 13 or less when compared to those who reported fewer antisocial behaviors, fewer other-sex friends, and being from an intact family.

While virginity is technically defined as never having had a penis enter a vagina, some individuals define themselves as virgins even though they have engaged in oral sex. Of 2,922 university students, 60% agreed with the statement "If you have oral sex, you are still a virgin." Hence, according to these undergraduates, having oral sex with someone is not really having sex (Knox & Hall, 2010). Persons most likely to agree that "oral sex is not sex" are freshmen or sophomores, as well as those self-identifying as religious (Dotson-Blake et al. 2012). Individuals may engage in oral sex rather than sexual intercourse to avoid getting pregnant, to avoid getting an STI, to keep their partner interested, to avoid a bad reputation, and to avoid feeling guilty over having sexual intercourse (Vazonyi & Jenkins, 2010).

Some people regret having had intercourse and want to be a virgin again. Carpenter (2010) discussed the concept of **secondary virginity**—a sexually initiated person's deliberate decision to refrain from intimate encounters for a set period of time and to refer to that decision as a kind of virginity (rather than "mere" abstinence). The decision to return to virginity may be a result of a previous physically painful, emotionally distressing, or romantically

More than 4,000 Father–Daughter Purity Balls occur annually and involve fathers and daughters as young as 4 years old taking mutual pledges to be pure.

disappointing sexual encounter. Of 61 young adults Carpenter interviewed, more than half (women more than men) believed that a person could, under some circumstances, be a virgin more than once. Fifteen people contended that they could resume their virginity in an emotional, psychological, or spiritual sense. Terence Duluca, a 27-year-old, heterosexual, White, Roman Catholic, explained:

> There is a different feeling when you love somebody and when you just care about somebody. So I would have to say if you feel that way, then I guess you could be a virgin again. Christians get born all the time again, so. . . . When there's true love involved, yes, I believe that.

A subcategory of absolutism is **asceticism**. Ascetics believe that giving in to carnal lust is wrong and attempt to rise above the pursuit of sensual pleasure into a life of self-discipline and self-denial. Accordingly, a spiritual life is viewed as the highest good, and self-denial helps one to achieve it. Catholic priests, monks, nuns, and some other celibate people have adopted the sexual value of asceticism.

secondary virginity the conscious decision of a sexually active person to refrain from intimate encounters for a specified period of time.

asceticism the belief that giving in to carnal lusts is wrong and that one must rise above the pursuit of sensual pleasure to a life of self-discipline and self-denial.

6-2b Relativism

In contrast to absolutism, 51% of males and 64% of females of 2,922 undergraduates identified relativism as their sexual value (Knox & Hall, 2010). **Relativism** is a value system emphasizing that sexual decisions should be made in reference to the emotional, security, and commitment aspects of the relationship. Whereas absolutists might feel that having intercourse is wrong for unmarried people, relativists might feel that the moral correctness of sex outside marriage depends on the nature of the relationship. For example, a relativist might say that marital sex between two spouses who are emotionally and physically abusive to each other is not to be preferred over intercourse between two unmarried individuals who love each other, are kind to each other, and are committed to the well-being of each other.

© iStockphoto.com/Maridav

A disadvantage of relativism as a sexual value is the difficulty of making sexual decisions on a relativistic case-by-case basis. Once a person decides that mutual love is the context justifying intercourse, how often and how soon is it appropriate for the person to fall in love? Can love develop after two drinks and two hours of conversation? The freedom that relativism brings to sexual decision making requires responsibility, maturity, and judgment. Though one may feel "in love," "secure," and "committed" at the time first intercourse occurs, only 17% of all first intercourse experiences that women report are with the person they eventually marry (Raley, 2000). A study of 475 young Canadian adults revealed that only 6% of females and 62% of males experienced orgasm during their first sexual experience. Participants whose experiences included

> **relativism** belief system in which sexual decisions are made in reference to the emotional, security, and commitment aspects of the relationship.

The section below contains a possible script that a female who values relativism, and who wants to pace her partner, might use to ensure that their relationship develops before sexual involvement.

"I NEED FOR US TO DELAY HAVING SEX IN OUR RELATIONSHIP"

I need to talk about sex in our relationship. I'm not comfortable talking about this and don't know exactly what I want to say, so I have written it down to try and help me get the words right. I like you, enjoy the time we spend together, and want us to continue seeing each other. The sex is something I need to feel good about to make it good for you, for me, and for us.

I need for us to slow down sexually. I need to feel an emotional connection that goes both ways—we both have very strong emotional feelings for each other. I also need to feel secure that we are going somewhere—that we have a future and that we are committed to each other. We aren't there yet so I need to wait till we get there to be sexual with you.

How long will this take?—I don't know—the general answer is "longer than now." I know this may not be what you have in mind and that you are probably ready now for us to increase the sex. I'm glad that you want it cause I want it too, but I need to feel right about it.

So, for now, let me drive the bus, sexually...when I feel the emotional connection and that we are going somewhere, I'll be the best sex partner you ever had. But let me move us forward.

If this is too slow for you or not what you have in mind, maybe I'm not the girl for you. It is certainly OK for you to tell me you need more and move on. Otherwise, we can still continue to see each other and see where the relationship goes.

So...tell me how you feel and what you want...and don't feel like you need to tell me you love me and want us to get married...ha! I'm not asking that...I'm just asking for time...

© fias100/Shutterstock

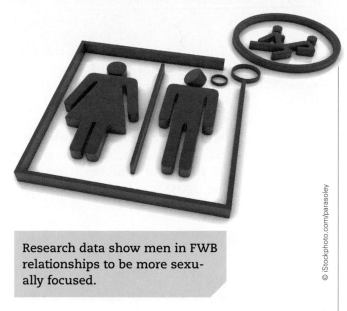

Research data show men in FWB relationships to be more sexually focused.

alcohol or drug use reported fewer positive experiences, and higher sexual regret (Reissing et al., 2012).

Students who become involved in a "friends with benefits" relationship reflect a specific expression of relativism.

6-2c Friends With Benefits

Friends with benefits (FWB) is a relationship of nonromantic friends who also have a sexual relationship. Of 2,922 undergraduates, 39% reported that they have been in an FWB relationship (Knox & Hall, 2010). Males, casual daters, hedonists, nonromantics, the jealous, Blacks, juniors and seniors, those who have had sex without love, and those who regard "financial security" as their top value are more likely to report previous involvement in an FWB relationship (Puentes et al., 2008). In effect, the friends with benefits relationship is primarily sexual and engaged in by nonromantic hedonists who have a pragmatic view of relationships.

To identify how women and men differ in their experience of the FWB relationship, McGinty and colleagues (2007) made three observations of their sample of 170 undergraduates—women in FWB relationships were more emotionally involved than men (63% vs. 36%), while men were more sexually focused than women (44% vs. 14%). Men were also more

> ## "Women want the 'friends' while men want the 'benefits.'"
> —RESULT OF RESEARCH STUDY ON FWB

polyamorous—wanting more than one FWB relationship at the time (35% of men vs. 5% of women).

Concurrent sexual partnerships are those in which the partners have sex with several individuals concurrently. In a study of 783 adults ages 18 to 59, 10% reported that both they and their partners had had other sexual partners during their relationship (Paik, 2010). Men were more likely than women to have been nonmonogamous (17% versus 5%). A serious relationship was the context in which going outside the relationship for sex was least likely. However, if the partners had a casual or friends-with-benefits relationship, the chance of nonmonogamy rose 30% and 44%, respectively.

6.2d Hedonism

Hedonism is the belief that the ultimate value and motivation for human actions lie in the pursuit of pleasure and the avoidance of pain. Twenty-seven percent of males and 16% of females of 2,922 undergraduates identified hedonism as their primary sexual value (Knox & Hall, 2010). Bersamin and colleagues (2012) analyzed data on single, heterosexual college students ($N = 3,907$) ages 18 to 25 from 30 institutions across the United States. The study focused on identity, culture, psychological well-being, and risk behaviors of undergraduates and found that casual sex was negatively associated with psychological well-being and positively associated with psychological distress. There were no gender differences.

friends with benefits (FWB) a relationship between nonromantic friends who also have a sexual relationship.

concurrent sexual partnerships relationships in which the partners have sex with several individuals concurrently.

hedonism belief that the ultimate value and motivation for human actions lie in the pursuit of pleasure and the avoidance of pain.

sexual double standard the view that encourages and accepts sexual expression of men more than women.

6-3 Sexual Double Standard

The **sexual double standard**—the view that encourages and accepts sexual expression of men more than women—is reflected in Table 6.1 (men were about twice as likely to be hedonistic

Table 6.1

Sexual Value by Sex of Undergraduate Respondent

Respondents	Absolutism	Relativism	Hedonism
Male students	14%	51%	27%
Female students	20%	64%	16%

Source: D. Knox & S. Hall, 2010, Relationship and sexual behaviors of a sample of 2,922 university students. Unpublished data collected for this text. Department of Sociology, East Carolina University, and Department of Family and Consumer Sciences, Ball State University.

as women). In another study of 1,004 undergraduates, 59% of the males and 24% of the females reported that they would "engage in casual sex" (Toews & Yazedjian, 2011).

The sexual double standard is also evident in the low disapproval of men having higher numbers of sexual partners but high disapproval of women for having the same number of sexual partners as men. In another study, England and Thomas (2006) noted that the double standard was operative in hooking up. Women who hooked up too often with too many men or had sex too quickly were vulnerable to getting bad reputations. Men who did the same thing got a bad reputation among women, but with fewer stigmas. Men gained status among other men for their sexual exploits. Backstrom and colleagues (2012) noted that a college woman in a hook-up context had to be aggressive about getting cunnilingus, whereas fellatio might be more normative for the male.

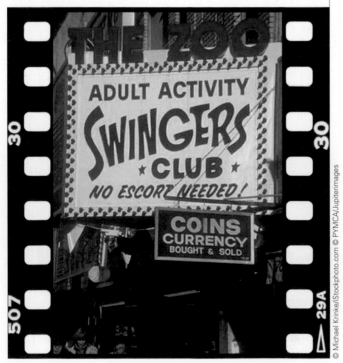

There is also evidence for the double standard in a review of 25 prime-time television programs (Kim et al., 2007). Finally, a study of 689 heterosexual couples revealed that women were disadvantaged in negotiating sexual issues with their partners, which may include the timing of sex and the use of condoms (Greene & Faulkner, 2005).

6-4 Sources of Sexual Values

The sources of one's sexual values are numerous and include one's school, religion, and family, as well as technology, television, social movements, and the Internet. Some schools (e.g., in Texas) promote absolutist sexual values through abstinence education. In the past, schools would lose funding if they discussed contraception, abortion, and similar issues.

Religion is also an important influence. In a sample of 1,289 undergraduates at a large southeastern university, 23% identified themselves as very religious or spiritual, and 50% as somewhat religious or spiritual. In regard to sexual behavior, the researchers found that religiously active young adults were more likely to indicate that sexting and sexual intercourse were inappropriate activities to engage in before being in a committed dating relationship (Miller et al., 2011).

Siblings are also influential in the development of sexual values. Females who have older brothers hold more conservative sexual values. "Those with older brothers in the home may be socialized more strongly to adhere to these traditional standards in line with power dynamics believed to shape and reinforce more submissive gender roles for girls and women" (Kornreich et al., 2003, p. 197).

Reproductive technologies such as birth control pills, the morning-after pill, and condoms influence sexual values by affecting the consequences of behavior. Being able to reduce the risk of pregnancy and HIV infection with the pill and condoms allows one to consider a different value system than if these methods of protection did not exist.

The media is also a source of sexual values. A television advertisement shows an affectionate couple with minimal clothes on in a context where sex could occur. "Be ready for the moment" is the phrase of the announcer, and Levitra, the new quick-start Viagra, is

© Michael Krinke/iStockphoto.com © PYMCA/Jupiterimages

the product for sale. The advertiser used sex to get the attention of the viewer and punch in the product to elicit buying behavior. Ballard and colleagues (2011) studied the sexual socialization of young adults via interviews. Some of the respondents noted the mixed messages in their socialization—for example, the discrepancy between media messages and church messages regarding sexuality: *"They [church members] say abstain, abstain, abstain, but it's so hard when sex is being thrown at you constantly from, you know, watching TV."*

In regard to magazines as media, exposure outcome is more positive. Two researchers assessed sexual health behaviors and magazine reading among 579 undergraduate students. They found that more frequent reading of mainstream magazines was associated with greater sexual health knowledge, safe-sex self-efficacy, and consistency of using contraception (although results varied across sex and magazine genre; Walsh & Ward, 2010).

Social movements such as the women's movement affect sexual values by empowering women with an egalitarian view of sexuality. This view translates into encouraging women to be more assertive about their own sexual needs and giving them the option to experience sex in a variety of contexts (e.g., without love or commitment) without self-deprecation. The net effect is a potential increase in the frequency of recreational, hedonistic sex.

Another influence on sexual values is the Internet, where the sexual content is extensive. The Internet features erotic photos, videos, "live" sex acts, and stripping by webcam sex artists. Individuals can exchange nude photos, have explicit sex dialogue, arrange to have "phone sex" or meet in person, or find a prostitute. Indeed, the adult section of Craig's List was shut down because it featured blatant prostitution.

Khajuraho has the largest group of medieval Hindu and Jain temples in India, famous for their erotic sculpture.

© Frédéric Soltan/Sygma/Corbis

"Why is Viagra like Disneyworld? You have to wait an hour for a three-minute ride."

—Anonymous

© Christian Kargl/ Getty Images

6-5 Gender Differences in Sexuality

Gender differences exist in what men and women believe about sex as well as what they do. In this section, we examine these differences.

6-5a Gender Differences in Sexual Beliefs

Analysis of data from 326 undergraduates on gender differences in beliefs about sex revealed that men were more likely to think that oral sex is not sex, that cybersex is not cheating, that men can't tell if a woman is faking orgasm, and that sex frequency drops in marriage. Meanwhile, women tended to believe that oral sex is sex, that cybersex is cheating, that faking orgasm does occur, and that sex frequency stays high in marriage (Knox et al., 2008). Little wonder there is misunderstanding, frustration, and disappointment between men and women as they include sexuality in their relationship.

6-5b Gender Differences in Sexual Behavior

In national data based on interviews with 3,432 adults, women reported having fewer sexual partners than men (2% versus 5% reported having had five or more sexual partners in the previous year) and reported having orgasm during intercourse less often (29% versus 75%; Michael et al., 1994, pp. 102, 128, 156). In another study of the responses of 5,385 males and 1,038 females who completed a questionnaire online, males were significantly more likely than females to report frequenting strip clubs, paying for sex, having anonymous sex with strangers, and having casual sexual relations (Mathy, 2007).

Pornography use is also higher among males. Motivations include emotional avoidance, excitement seeking, sexual pleasure, and sexual curiosity. The Pornography Consumption Inventory has been used to identify these uses (Reid et al., 2011). Men and women also differ in their motivations for sexual intercourse, with men viewing sex more casually. Earlier, we noted that undergraduate men (in comparison with undergraduate women) were almost twice as likely to report being hedonistic in their sexual values.

Sociobiologists explain males' more casual attitude toward sex, engaging in sex with multiple partners, and being hedonistic as biologically based (that is, due to higher testosterone levels). Social learning theorists, on the other hand, emphasize that the media and peers socialize men to think about and to seek sexual experiences. Men are also accorded social approval and called "studs" for their sexual exploits. Women, on the other hand, are more often punished and labeled "sluts" if they have many sexual partners. Because **social scripts** guide sexual behavior, what individuals think, do, and experience is a reflection of what they have learned (Simon & Gagnon, 1998). These scripts operate at the cultural (e.g., societal norms for sexual conduct—women should not be promiscuous), interpersonal (e.g., sexual desires translated into strategies—man should be aggressive in sexual encounters), and intrapsychic (e.g., sexual dialogues with self that elicit and sustain arousal—"this will be erotic") levels (DeLamater & Hasday, 2007).

There are also differences in the perceptions of foreplay and intercourse. Miller and Byers (2004) compared the reported duration of actual foreplay (men=13 minutes; women=11 minutes) and desired foreplay (men=18 minutes; women=19 minutes), and the reported duration of actual intercourse (men=8 minutes; women=7 minutes) and desired intercourse (men=18 minutes; women=14 minutes) of

social script the identification of the roles in a social situation, the nature of the relationship between the roles, and the expected behaviors of those roles.

Table 6.2
FOREPLAY V. INTERCOURSE

MEN		WOMEN	
FOREPLAY		**FOREPLAY**	
Desired duration	18 min.	Desired duration	19 min.
Actual	13 min.	Actual	11 min.
INTERCOURSE		**INTERCOURSE**	
Desired duration	18 min.	Desired duration	14 min.
Actual	8 min.	Actual	7 min.

Source: Miller, S. A., and E. S. Byers. 2004. Actual and desired duration of foreplay and intercourse: Discordant and misperceptions within heterosexual couples. *The Journal of Sex Research*, 41, 301–309.

Why don't women blink during foreplay?

There isn't time.

> ## "Masturbation—
> ## it's **sex** with someone
> ## I love."
>
> —WOODY ALLEN, DIRECTOR

heterosexual men and women in long-term relationships with each other. These findings suggest that both men and women underestimate their partner's desires for the duration of both foreplay and intercourse; also, they had similar preferences for duration of foreplay, but men wanted longer intercourse than women (see Figure 6.2).

Sexual satisfaction reported by men and women seems to be equal, at least among the French. In a representative sample of 1,002 French respondents (483 men and 519 women) age 35 years, 83% reported relative or full satisfaction with their sex life (Colson et al., 2006).

Mood states typically affect both men and women equally (Lykins et al., 2006). In a study of 663 female college students and 399 male college students, the researchers found that individuals who were depressed were less likely to be interested in engaging in sexual behavior. However, this outcome was not always the case, as about 10% of women and a higher percentage of men reported that they were interested in engaging in sexual behavior in spite of a negative mood state.

Finally, gender differences may also be influenced by ethnic background. Eisenman and Dantzker (2006) surveyed primarily Hispanics and found that women were less permissive and had more negative attitudes than men in regard to oral sex, premarital intercourse, and masturbation. However, in a national sample of all ethnicities, both women and men held positive attitudes about female vibrator use (Herbenick et al., 2011).

6-6 Pheromones and Sexual Behavior

The word *pheromone* comes from the Greek words *pherein*, meaning "to carry," and *hormon*, meaning "to excite." **Pheromones** are "chemical messengers that are emitted into the environment from the body, where they can then activate specific physiological or behavioral responses in other individuals of the same species" (Grammer et al., 2005, p. 136). Pheromones are produced primarily by the apocrine glands located in the armpits and pubic region. The

Sexual Attitudes and Behaviors in Black and White

A survey of 1,915 undergraduate women and 1,111 undergraduate men at four universities revealed that race was the most influential factor differentiating the sexual attitudes and behavior of the sample (Davidson et al., 2008). When Blacks were compared with Whites, the former had more permissive attitudes and were more likely to approve of sexual intercourse with casual, occasional, and regular dating partners. Black people also experienced sexual intercourse earlier and with more lifetime partners. Other researchers have confirmed that sexual behavior of Black people is inconsistent with their close affiliation with religion. One explanation is that the churches attended by Black people are often reluctant to address sexual issues (Uecker, 2008).

© auremar/Shutterstock.com

functions of pheromones include opposite-sex attractants, same-sex repellents, and mother–infant bonding.

Pheromones typically operate without the person's awareness; researchers disagree about whether pheromones do in fact influence human sociosexual behaviors. Although Levin (2004) reviewed the literature on chemical messengers in attraction, the strongest evidence for the effect of hormones

pheromones body scents which activate physiological or behavioral responses in other individuals of the same species.

on sexual behavior is that 38 male volunteers who applied aftershave lotion containing a male hormone reported significant increases in sexual intercourse and sleeping next to a partner when compared with men who had a placebo in their aftershave lotion (Cutler et al., 1998).

6-7 Sexuality in Relationships

Sexuality occurs in a social context that influences its frequency and perceived quality.

6-7a Sexual Relationships Among Never-Married Individuals

Never-married individuals and those not living together report more sexual partners than those who are married or living together. In a nationwide study, 9% of never-married individuals and those not living together reported having had five or more sexual partners in the previous 12 months; 1% of married people and 5% of cohabitants reported the same (Michael et al., 1994).

Unmarried individuals, when compared with married individuals and cohabitants, also reported the lowest level of sexual satisfaction. One third of a sample of people who were not married and not living with anyone reported that they were emotionally satisfied with their sexual relationships. In contrast, 85% of the married and pair-bonded individuals reported emotional satisfaction in their sexual relationships. Hence, although never-married individuals have more sexual partners, they are less emotionally satisfied (Michael et al., 1994).

6-7b Sexual Relationships Among Married Individuals

Marital sex is distinctive for its social legitimacy, declining frequency, and satisfaction (both physical and emotional).

1. **Social legitimacy.** In our society, marital intercourse is the most legitimate form of sexual behavior. Homosexual, premarital, and extramarital intercourse do not have as high a level of social approval as does marital sex. It is not only okay to have intercourse when married, it is expected. People assume that married couples make love and that something is wrong if they do not.

2. **Declining frequency.** Sexual intercourse between spouses occurs about six times a month, which declines in frequency as spouses age. Pregnancy also decreases the frequency of sexual intercourse (Lee et al., 2010). In addition to biological changes due to aging and pregnancy, satiation also contributes to the declining frequency of intercourse between spouses and partners in long-term relationships. Psychologists use the term **satiation** to mean that repeated exposure to a stimulus results in the loss of its ability to reinforce. For example, the first time you listen to a new CD, you derive considerable enjoyment and satisfaction from it. You may play it over and over during the first few days. After a week or so, listening to the same music is no longer new and does not give you the same level of enjoyment that it first did. So it is with intercourse between spouses or long-term partners. The thousandth time that a person has intercourse with the same partner is not as new and exciting as the first few times.

3. **Satisfaction (emotional and physical).** Despite declining frequency and less satisfaction over time, marital sex remains a richly satisfying experience. Contrary to the popular belief that unattached singles have the best sex, married and pair-bonded adults enjoy the most satisfying sexual relationships. In the national sample referred to earlier, 88% of married people said they received great physical pleasure from their sexual lives, and almost 85% said they received great emotional satisfaction (Michael et al. 1994). Individuals least likely to report being physically and emotionally pleased in their sexual relationships are those who are not married, not living with anyone, or not in a stable relationship with one person.

6-7c Sexual Relationships Among Divorced Individuals

Of the almost 2 million people getting divorced, most will have intercourse within one year of being separated from their spouses. The meanings of intercourse

satiation the state in which a stimulus loses its value with repeated exposure.

for separated or divorced individuals vary. For many, intercourse is a way to reestablish—indeed, repair—their crippled self-esteem. Questions such as "What did I do wrong?" "Am I a failure?" and "Is there anybody out there who will love me again?" loom in the minds of divorced people. One way to feel loved, at least temporarily, is through sex. Being held by another and being told that it feels good provides people some evidence that they are desirable. Because divorced people may be particularly vulnerable, they may reach for sexual encounters as if for a lifeboat. "I felt that, as long as someone was having sex with me, I wasn't dead and I did matter," said one recently divorced person.

Because divorced individuals are usually in their mid-30s or older, they may not be as sensitized to the danger of contracting HIV as are people in their 20s. Divorced individuals should always use a condom to lessen the risk of an STI, including HIV infection, and AIDS.

6-8 Safe Sex: Avoiding Sexually Transmitted Infections

The Student Sexual Risks Scale (SSRS) at the back of the book allows you to assess the degree to which you are at risk for contracting an STI, including HIV infection.

One of the negative consequences of unprotected sexual behavior is the risk of contracting an STI. Thirty percent of 1,497 women said that they would have unprotected sex (Foster et al., 2012), and 6% of 1,004 undergraduates reported having an STI (Toews & Yazedjian, 2011). **STI** refers to the general category of sexually transmitted infections such as *Chlamydia*, genital herpes, gonorrhea, and syphilis. The most lethal of all STIs is that caused by the human immunodeficiency virus (**HIV**), which attacks the immune system and can lead to acquired immunodeficiency syndrome (**AIDS**).

6-8a Transmission of HIV and High-Risk Behaviors

HIV can be transmitted in several ways.

1. **Sexual contact.** HIV is found in several body fluids of infected individuals, including blood, semen, and vaginal secretions. During sexual contact with an infected individual, the virus enters a person's bloodstream through the rectum, vagina, penis (an uncircumcised penis is at greater risk because of the greater retention of the partner's fluids), and possibly the mouth during oral sex. Saliva, sweat, and tears are not body fluids through which HIV is transmitted.

2. **Intravenous drug use.** Drug users who are infected with HIV can transmit the virus to other drug users with whom they share needles, syringes, and other drug-related implements.

3. **Blood transfusions.** HIV can be acquired by receiving HIV-infected blood or blood products. Currently, all blood donors are screened, and blood is not accepted from high-risk individuals. Blood that is accepted from donors is tested for the presence of HIV. However, prior to 1985, donor blood was not tested for HIV. Individuals who received blood or blood products prior to 1985 may have been infected with HIV.

4. **Mother–child transmission.** A pregnant woman infected with HIV has a 40% chance of transmitting the virus through the placenta to her unborn

STI sexually transmitted infection.

HIV human immunodeficiency virus, which attacks the immune system and can lead to AIDS.

AIDS acquired immunodeficiency syndrome; the last stage of HIV infection, in which the immune system of a person's body is so weakened that it becomes vulnerable to disease and infection.

child. These babies will initially test positive for HIV as a consequence of having the antibodies from their mother's bloodstream. However, azidothymidine (AZT, alternatively called zidovudine or ZVD) taken by the mother 12 weeks before birth seems to reduce the chance of transmission of HIV to her baby by two thirds. HIV may also be transmitted, although rarely, from mother to infant through breast-feeding.

5. **Organ or tissue transplants and donor semen.** Receiving transplant organs and tissues, as well as receiving semen for artificial insemination, could involve risk of contracting HIV if the donors have not been tested for HIV. Such testing is essential, and recipients should insist on knowing the HIV status of the organ, tissue, or semen donor.

6-8b STI Transmission— The Illusion of Safety in a "Monogamous" Relationship

Most individuals in a serious "monogamous" relationship assume that their partner is faithful and that they are at zero risk for contracting an STI. In a study of 1,341 undergraduates at a large southeastern university, almost 30% (27%) of the males and 20% of the females reported having oral, vaginal, or anal sex outside of a relationship that their partner considered monogamous. People most likely to cheat were men over the age of 20, those who were binge drinkers, members of a fraternity, male NCAA athletes, or nonreligious people. These data suggest the need for educational efforts to encourage undergraduates in committed relationships to reconsider their STI risk and to protect themselves via condom use (Vail-Smith et al., 2010).

6-8c Prevention of HIV and STI Transmission

The safest relationship context for avoiding an STI (including HIV) is marriage, with cohabitation a close second (Hattori & Dodoo, 2007). Of course, the best way to avoid getting an STI is to avoid sexual contact or to have contact only with partners who are not infected. This means restricting your sexual contacts to those who limit their relationships to one person. The person most likely to get an STI has sexual relations with a number of partners or with a partner who has a variety of partners.

Condoms should be used for vaginal, anal, and oral sex and should never be reused. However, in a sample of 2,922 undergraduates, only 25% reported that they always used a condom before having intercourse (Knox & Hall, 2010). Feeling that the partner is disease-free, having had too much alcohol, and believing that "getting an STI won't happen to me" are reasons individuals do not use a condom. Some partners are also forced to have sex or do not feel free to negotiate the use of a condom in their sexual relationship (Heintz & Melendez, 2006). It is important for you to take responsibility for your own safety in regard to your sexual behavior.

Using the condom properly is also important. Putting on a latex or polyurethane condom before the penis touches the partner's body makes passing STIs from one person to another difficult (natural membrane condoms do not block the transmission of STIs). Care should also be taken to withdraw the penis while it is erect to prevent fluid from leaking from the base of the condom into the partner's genital area. If a woman is receiving oral sex, she should wear a dental dam, which will prevent direct contact between the genital area and her partner's mouth.

Sexuality in an age of HIV and STIs demands talking about safer sex issues with a new potential sexual partner. Bringing up the issue of condom use should be perceived as caring for oneself, the partner, and the relationship rather than as a sign of distrust. Some individuals routinely have a condom available, and it is a "given" in any sexual encounter. Figure 6.1 illustrates that one is more likely to contract an STI through high alcohol use, low condom use, and having sex with multiple partners.

Figure 6.1

Risk of Contracting an STI, as Related to Alcohol, Condom Use, and Number of Partners

Amount of alcohol consumed

Use of condom during intercourse or oral sex

Number of sexual partners

Sexual Fulfillment: Some Prerequisites

There are several prerequisites for having a good sexual relationship.

6-9a Self-Knowledge, Self-Esteem, and Health

Sexual fulfillment involves knowledge about yourself and your body. Such information not only makes it easier for you to experience pleasure but also allows you to give accurate information to a partner about pleasing you. It is not possible to teach a partner what you don't know about yourself.

Sexual fulfillment also implies having a positive self-concept. To the degree that you have positive feelings about yourself and your body, you will regard yourself as a person someone else would enjoy touching, being close to, and making love with. If you do not like yourself or your body, you might wonder why anyone else would.

Effective sexual functioning also requires good physical and mental health. This means regular exercise, good nutrition, regular medical checkups, and lack of fatigue. Performance in all areas of life does not have to diminish with age—particularly if people take care of themselves physically.

Good health also implies being aware that some drugs may interfere with sexual performance. Alcohol is the drug most frequently used by American adults. Although a moderate amount of alcohol can help a person become aroused through a lowering of inhibitions, too much alcohol can slow the physiological processes and deaden the senses. Shakespeare may have said it best: "It [alcohol] provokes the desire, but it takes away the performance" (*Macbeth*, 2.3). The result of

an excessive intake of alcohol for women is a reduced chance of orgasm; for men, overindulgence results in a reduced chance of attaining or maintaining an erection.

The reactions to marijuana are less predictable than the reactions to alcohol. Though some individuals report a short-term enhancement effect, others say that marijuana just makes them sleepy. In men, chronic use may decrease sex drive because marijuana may lower testosterone levels.

6-9b A Good Relationship, Positive Motives

A guideline among therapists who work with couples who have sexual problems is to treat the relationship before focusing on the sexual issue. The sexual relationship is part of the larger relationship between the partners, and what happens outside the bedroom in day-to-day interaction has a tremendous influence on what happens inside the bedroom. The statement, "I can't fight with you all day and want to have sex with you at night" illustrates the social context of the sexual experience.

Sexual interaction communicates how the partners are feeling and acts as a barometer for the relationship. Each partner brings to a sexual encounter, sometimes unconsciously, a motive (pleasure, reconciliation, procreation, duty), a psychological state (love, hostility, boredom, excitement), and a physical state (tense, exhausted, relaxed, turned on). The combination of these factors will change from one encounter to another. Tonight one partner may feel aroused and loving and seek pleasure, but the other partner may feel exhausted and have sex only out of a sense of duty. Tomorrow night, both partners may feel relaxed and have sex as a means of expressing their love for each other.

One's motives for a sexual encounter are related to the outcome. When individuals have intercourse out of the desire to enhance personal and relationship pleasure, the personal and interpersonal effect on well-being

Good health = Good sex

© Radius Images/Jupiterimages / © iStockphoto.com/YinYang

is very positive. However, when sexual motives are to avoid conflict, the personal and interpersonal effects do not result in similar positive outcomes (Impett et al., 2005). In a study of 1,002 French adults, sexuality was more synonymous with pleasure (44%) and love (42%) than with procreation, children, or motherhood (8%; Colson et al., 2006).

6-9c An Equal Relationship

In a survey of 27,500 individuals in 29 countries, reported sexual satisfaction was higher where men and women were considered equal. Austria topped the list, with 71% reporting sexual satisfaction; only 26% of those surveyed in Japan reported sexual satisfaction. The United States was among those countries in which a high percentage of the respondents reported sexual satisfaction (Laumann et al., 2006).

6-9d Open Sexual Communication, Sexual Self-Disclosure, and Feedback

Sexually fulfilled partners are comfortable expressing what they enjoy and do not enjoy. Unless both partners communicate their needs, preferences, and expectations to each other, neither is ever sure what the other wants. In essence, the Golden Rule ("Do unto others as you would have them do unto you") is *not* helpful, because what you like may not be the same as what your partner wants.

Sexually fulfilled partners take the guesswork out of their relationship by communicating preferences and giving feedback. Comfort in communicating about sex openly with one's partner varies by culture. Ali (2011) compared Australian and Malaysian couples and found that the former were sexually self-disclosing with their partners, while the latter were

not. She noted that the norms of Malaysian socialization do not allow parents to talk to their children about sex, since doing so is thought to destroy childhood innocence. Even adult married couples do not discuss sex as it is not regarded as proper.

Open sexual communication includes using what some therapists call the touch-and-ask rule. Each touch and caress may include the question, "How does that feel?" Sometimes the feeling is painful. Of 505 women who experienced anal intercourse, 9% reported severe pain during every penetration (Stulhofer & Ajdukovic, 2011).

Guiding and moving the partner's hand or body are also ways of giving feedback. What women and men want each other to know about sexuality is presented in Table 6.3.

6-9e Having Realistic Expectations

To achieve sexual fulfillment, expectations must be realistic. A couple's sexual needs, preferences, and expectations may not coincide. It is unrealistic to assume that your partner will want to have sex with the same frequency and in the same way that you do on all occasions. It may also be unrealistic to expect the level of sexual interest and frequency of sexual interaction in long-term relationships to remain consistently high.

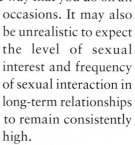

Sexual fulfillment means not asking things of the sexual relationship that it cannot deliver. Failure to develop realistic expectations will result in

Table 6.3
What Women and Men Want Each Other to Know About Sexuality

What Women Want Men to Know About Sex

Women tend to like a loving, gentle, patient, tender, and understanding partner. Rough sexual play can hurt and be a turnoff.

It does not impress women to hear about other women in the man's past.

If men knew what it is like to be pregnant, they would not be so apathetic about birth control.

Most women want more caressing, gentleness, kissing, and talking before and after intercourse.

Some women are sexually attracted to other women, not to men.

Sometimes the woman wants sex even if the man does not. Sometimes she wants to be aggressive without being made to feel that she shouldn't be.

Intercourse can be enjoyable without orgasm.

Many women do not have an orgasm from penetration only; they need direct stimulation of their clitoris by their partner's tongue or finger.

Men should be interested in fulfilling their partner's sexual needs.

Most women prefer to have sex in a monogamous love relationship.

When a woman says no, she means it.

Women do not want men to expect sex every time they are alone with their partner.

Many women enjoy sex in the morning, not just at night.

Sex is *not* everything.

Women need to be lubricated before penetration.

Men should know more about menstruation.

Many women are no more inhibited about sex than men are.

Women do not like men to roll over, go to sleep, or leave right after orgasm.

Intercourse is more of a love relationship than a sex act for some women.

The woman should not always be expected to supply a method of contraception. It is also the man's responsibility.

Men should always have a condom with them and initiate putting it on.

What Men Want Women to Know About Sex

Men do not always want to be the dominant partner; women should be aggressive.

Men want women to enjoy sex totally and not be inhibited.

Men enjoy tender and passionate kissing.

Men really enjoy fellatio and want women to initiate it.

Women need to know a man's erogenous zones.

Men enjoy giving oral sex; it is not bad or unpleasant.

Many men enjoy a lot of romantic foreplay and slow, aggressive sex.

Men cannot keep up intercourse forever. Most men tire more easily than women.

Looks are not everything.

Women should know how to enjoy sex in different ways and different positions.

Women should not expect a man to get a second erection right away.

Many men enjoy sex in the morning.

Pulling the hair on a man's body can hurt.

Many men enjoy sex in a caring, loving, exclusive relationship.

It is frustrating to stop sex play once it has started.

Women should know that not all men are out to have intercourse with them. Some men like to talk and become friends.

© Cengage Learning 2011

frustration and resentment. One's age, health (both mental and physical), sexual dysfunctions of self and partner, and previous sexual experiences will have an effect on one's sexuality and one's sexual relationship and sexual fulfillment (McCabe & Goldhammer, 2012).

6-9f Avoiding Spectatoring

One of the obstacles to sexual functioning is **spectatoring**, which involves mentally observing your sexual performance and that of your partner. When the researchers in one extensive study observed how individuals actually behave during sexual intercourse, they reported a tendency for sexually dysfunctional partners to act as spectators by mentally observing their own and their partners' sexual performance. For example, the man would focus on whether he was having an erection, how complete it was, and whether it would last. He might also watch to see whether his partner was having an orgasm (Masters & Johnson, 1970).

Spectatoring, as Masters and Johnson conceived it, interferes with each partner's sexual enjoyment because it creates anxiety about performance, and anxiety blocks performance. A man who worries about getting an erection reduces his chance of doing so. A woman who is anxious about achieving an orgasm probably will not. The desirable alternative to spectatoring is to relax, focus on and enjoy your own pleasure, and permit yourself to be sexually responsive.

Spectatoring is not limited to sexually dysfunctional couples and is not necessarily associated with psychopathology. It is a reaction to the concern that the performance of one's sexual partner is consistent with expectations. We all probably have engaged in spectatoring to some degree. When such spectatoring is continuous, performance is impaired.

6-9g Debunking Sexual Myths

Sexual fulfillment also means not being victim to sexual myths. Some of the more common myths are that sex equals intercourse and orgasm, that women

Table 6.4
Common Sexual Myths

Masturbation is sick.
Women who love sex are sluts.
Sex education makes children promiscuous.
Sexual behavior usually ends after age 60.
People who enjoy pornography end up committing sexual crimes.
Most "normal" women have orgasms from penile thrusting alone.
Extramarital sex always destroys a marriage.
Extramarital sex will strengthen a marriage.
Simultaneous orgasm with one's partner is the ultimate sexual experience.
My partner should enjoy the same things that I do sexually.
A man cannot have an orgasm unless he has an erection.
Most people know a lot of accurate information about sex.
Using a condom ensures that you won't get HIV.
Most women prefer a partner with a large penis.
Few women masturbate.
Women secretly want to be raped.
An erection is necessary for good sex.
An orgasm is necessary for good sex.

© Cengage Learning 2011

who love sex don't have values, and that the double standard is dead. Another common sexual myth is that the elderly have no interest in sex. In a study of the sexual interests and needs of 563 70-year-olds, 95% reported the continuation of such interests and needs as they aged. Almost 70% (69%) of the married men and 57% of the married women reported continued sexual behavior (Beckman et al., 2006). Believing in various sexual myths makes one vulnerable to sexual dysfunctions (Nobre & Pinto-Gouveia, 2006). Table 6.4 presents some other sexual myths.

spectatoring
mentally observing one's own and one's partner's sexual performance.

STUDY TOOLS **6**

Ready to study? In this book, you can:

- Rip out the Chapter Review card in the back of the book to study for exams

- Take the Self Assessment for this chapter (card in the back of the book) and see where you stand on the vital issues raised in the chapter

Or you can go online to CourseMate at www.cengagebrain.com for these resources:

- Complete Practice Quizzes to prepare for tests

- Review Key Terms Flash Cards (online or print)

- Read about Marriage and Family in the news

- Play "Beat the Clock" to master concepts

- Check out Personal Applications

Marriage Relationships

"Marriage—a word which should be pronounced **'mirage.'"**

—HERBERT SPENCER, PHILOSOPHER

SECTIONS

7-1 Individual Motivations for Marriage

7-2 Societal Functions of Marriage

7-3 Marriage as a Commitment

7-4 Marriage as a Rite of Passage

7-5 Changes After Marriage

7-6 Diversity in Marriage

7-7 Marital Success

The marriage of Barack and Michelle Obama is in the public eye. Campaign manager David Plouffe observed their relationship "up close" and noted differences in the ways they prepared for a major speech.

> Michelle wanted a draft of her speech more than a month out so she could massage it further, get comfortable with it, and practice the delivery. Barack was always crafting his at the eleventh hour. In this regard, Michelle was a concert pianist—disciplined, regimented, methodical—and Barack was a jazz musician, riffing, improvisational, and playing by ear. Both Obamas, it turned out, were clutch performers when the curtain rose. (Plouffe, 2009, pp. 301–302)

In effect, Michelle is very organized, deliberate, and decisive. In contrast, Barack is more of a loosey-goosey type of guy. But their marriage seems to work for them.

The title of this chapter, Marriage Relationships with plural "relationships," confirms that marriages are different. *Diversity* is the term that best describes relationships, marriages, and families today. No longer is there a one-size-fits-all cultural norm of what a relationship, marriage, or family should be. Rather, individuals, couples, and families select their own path. In this chapter, we review the diversity of relationships. We begin by looking at some of the different reasons people marry.

© Shutterstock.com

7-1 Individual Motivations for Marriage

We have defined marriage as a legal contract between two heterosexual adults that regulates their economic and sexual interaction. However, individuals in the United States tend to think of marriage in personal more than legal terms. The following are some of the reasons people give for getting married.

7-1a Love

Many couples view marriage as the ultimate expression of their love for each other—the desire to spend their lives together in a secure, legal, committed relationship. In U.S. society, love is expected to precede marriage—thus, only couples in love consider marriage. Those not in love would be ashamed to admit it.

> "I have **great hopes** that we shall
> **love each other all our lives** as much
> as if we had never married at all."
>
> —LORD BYRON, POET

7-1b Personal Fulfillment

We marry because we feel a sense of personal fulfillment in doing so. We were born into a family (family of origin) and want to create a family of our own (family of procreation). We remain optimistic that our marriage will be a good one. Even if our parents divorced or we have friends who have done so, we feel that our relationship will be different.

7-1c Companionship

Oprah Winfrey once said that lots of people want to ride in her limo, but what she wants is someone who will take the bus when the limo breaks down. One of the motivations for marriage is to enter a relationship with a genuine companion, a person who will take the bus with you when the limo breaks down.

Although marriage does not ensure it, companionship is the greatest expected benefit of marriage in the United States. Companionship has become "the legitimate goal of marriage" (Coontz, 2000, p. 11). Eating meals together is one of the most frequent normative behaviors of spouses. Indeed, although spouses may eat lunch apart, dinner together becomes an expected behavior. **Commensality** is eating with others, and one of the issues spouses negotiate is "who eats with us" (Sobal et al., 2002).

7-1d Parenthood

Most people want to have children. In response to the statement, "Someday, I want to have children," 84% of 2,922 undergraduates (90% of females; 77% of males) answered yes (Knox & Hall, 2010). The amount of time parents spend in rearing their children has increased. Contrary to conventional wisdom, both mothers and fathers report spending greater amounts of time in child-care activities in the late 1990s than in the "family-oriented" 1960s (Sayer et al., 2004).

Although some people are willing to have children outside marriage (in a cohabiting relationship or in no relationship at all), most Americans prefer to have children in a marital context. Previously, a strong norm existed in our society (particularly for White people) that individuals should be married before they have children. This norm has relaxed, with more individuals willing to have children without being married. Indeed, having children has moved from an assumption to a choice (Scott, 2009).

7-1e Economic Security

Married people report higher household incomes than do unmarried people. Indeed, national data from the Health and Retirement Survey revealed that individuals who were not continuously married had significantly lower wealth than those who remained married throughout the life course. Remarriage offsets the negative economic effect of marital dissolution (Wilmoth & Koso, 2002).

Although individuals may be drawn to marriage for the preceding reasons on a conscious level, unconscious motivations may also be operative. Individuals reared in a happy family of origin may seek to duplicate this perceived state of warmth, affection, and sharing. Alternatively, individuals reared in unhappy, abusive, drug-dependent families may inadvertently seek to re-create a similar family because that is what they are familiar with. In addition, individuals are motivated

commensality
eating with others; most spouses eat together and negotiate who joins them.

© Foodpix/Jupiterimages / © Maria Toutoudaki/iStockphoto.com

to marry to alleviate the fear of being alone, to better themselves economically, to avoid birth out of wedlock, and to prove that someone wants them.

Just as most individuals want to marry (regardless of the motivation), most parents want their children to marry. Parents feel that marriage is the context most conducive for their offspring being happy and taken care of. Parents also look forward to grandchildren and a more adult relationship with their own children. Indeed, when their own children have their first child, the "ah-ha" experiences of their offspring begin, and the child and parents have a new role in common—parents.

7-2 Societal Functions of Marriage

As noted in Chapter 1, important societal functions of marriage are to bind a male and female together who will reproduce, provide physical care for their dependent young, and socialize them to be productive members of society who will replace those who die (Murdock, 1949). Marriage helps protect children by giving the state legal leverage to force parents to be responsible to their offspring whether or not they stay married. If couples did not have children, the state would have no interest in regulating marriage.

Additional functions of marriage include regulating sexual behavior (spouses have less exposure to STIs than do singles) and stabilizing adult personalities by providing a companion and "in-house" counselor. In the past, marriage and family have served protective, educational, recreational, economic, and religious functions. However, as these functions have gradually been taken over by police or legal systems, schools, the entertainment industry, workplace, and church or synagogue, only the companionship-intimacy function has remained virtually unchanged.

The emotional support each spouse derives from the other in the marital relationship remains one of the strongest and most basic functions of marriage (Coontz, 2000). In today's social world, which consists mainly of impersonal, secondary relationships, living in a context of mutual emotional support may be particularly important. Indeed, the companionship and intimacy needs of contemporary U.S. marriage have become so strong that many couples consider divorce when they no longer feel "in love" with their partner. Of 2,922 undergraduates, 40% reported that they would divorce their spouse if they no longer loved the spouse (Knox & Hall, 2010).

The very nature of the marriage relationship has also changed from being very traditional or male-dominated to being very modern or egalitarian. A summary of these differences is presented in Table 7.1. Keep in mind that these are stereotypical marriages and that only a small percentage of today's modern marriages have all the traditional or egalitarian characteristics that are listed.

7-3 Marriage as a Commitment

Marriage represents a multilevel commitment—person-to-person, family-to-family, and couple-to-state.

Table 7.1

Traditional Versus Egalitarian Marriages

Traditional Marriage	Egalitarian
There is limited expectation of husband to meet emotional needs of wife and children.	Husband is expected to meet emotional needs of wife and to be involved with children.
Wife is not expected to earn income.	Wife is expected to earn income.
Emphasis is on ritual and roles.	Emphasis is on companionship.
Couples do not live together before marriage.	Couples may live together before marriage.
Wife takes husband's last name.	Wife may keep her maiden name.
Husband is dominant; wife is submissive.	Neither spouse is dominant.
Roles for husband and wife are rigid.	Roles for spouses are flexible.
Husband initiates sex; wife complies.	Either spouse initiates sex.
Wife takes care of children.	Parents share child rearing.
Education is important for husband, not for wife.	Education is important for both spouses.
Husband's career decides family residence.	Career of either spouse determines family residence.

© Cengage Learning 2011

7-3a Person-to-Person Commitment

Commitment is the intent to maintain a relationship. Persons express commitment by telling one another ("I love you and want to spend my life with you"), telling friends ("We have been going together for years and have discussed a June wedding"), doing things for each other ("I'll get the oil changed in your car"), and providing economic resources ("I'll pay off your student loans"). Mutual commitment is associated with higher relationship quality (Weigel, 2010). One partner also tends to mirror the commitment of the other. In a study of 112 couples in marriage counseling, a commitment change in one partner was associated with commitment change in the other (Bartle-Haring, 2010).

7-3b Family-to-Family Commitment

Whereas love is private, marriage is public. Marriage is the second of three times that one's name can be expected to appear in the local newspaper. When individuals marry, the parents and extended kin also become enmeshed. In many societies (e.g., Kenya), the families arrange for the marriage of their offspring, and the groom is expected to pay for his new bride.

How much is a bride worth? In some parts of rural Kenya, premarital negotiations include the determination of **bride wealth**—this is the amount of money a prospective groom will pay to the parents of his bride-to-be. Such a payment is not seen as "buying the woman" but, rather, as compensating the parents for the loss of labor from their daughter. Forms of payment include livestock ("I am worth many cows," said one Kenyan woman), food, and/or money. The man who raises the bride wealth also demonstrates not only that he is ready to care for a wife and children but also that he has the resources to do so (Wilson et al., 2003).

Marriage also involves commitments by each of the marriage partners to the family members of the spouse. For example, married couples are often expected to divide their holiday visits between both sets of parents.

7-3c Couple-to-State Commitment

In addition to making person-to-person and family-to-family commitments, spouses become legally committed to each other according to the laws of the state in which they reside. This means they cannot arbitrarily decide to terminate their own marital agreement.

Just as the state says who can marry (not close relatives, the insane, or the mentally deficient) and when (usually at age 18 or older), legal procedures must be instituted if the spouses want to divorce. The state's interest is that a couple stays married, has children, and takes care of them. Should they divorce, the state will dictate how the parenting is to continue, both physically and economically. Social policies designed to strengthen marriage through divorce law reform reflect the value the state places on stable, committed relationships.

7-4 Marriage as a Rite of Passage

A **rite of passage** is an event that marks the transition from one status to another. Starting school, getting a driver's license, and graduating from high school or college are events that mark major transitions in status (to student, to driver, and to graduate). The wedding itself is another rite of passage that marks the transition from fiancé to spouse. Preceding the wedding is the traditional bachelor party for the soon-to-be groom. Somewhat new on the cultural

commitment an intent to maintain a relationship.

bride wealth the amount of money or goods (e.g., cows) given by the groom or his family to the wife's family for giving her up; also known as *bride price* or *bride payment*.

rite of passage event that marks the transition from one status to another.

landscape is the bachelorette party (sometimes more wild than the bachelor party), which conveys the message of equality (marriage is the end of freedom for both the man and the woman).

7-4a Weddings

The wedding is a rite of passage that is both religious and civil. To the Catholic Church, marriage is a sacrament that implies that the union is both sacred and indissoluble. According to Jewish and most Protestant faiths, marriage is a special bond between the husband and wife sanctified by God, but divorce and remarriage are permitted. Wedding ceremonies still reflect traditional cultural definitions of women as property. For example, the father of the bride usually walks the bride down the aisle and "hands her over to the new husband." In some cultures, the bride is not even present at the time of the actual marriage. For example, in the upper-middle-class Muslim Egyptian wedding, the actual marriage contract signing occurs when the bride is in another room with her mother and sisters. The father of the bride and the new husband sign the actual marriage contract (identifying who is marrying whom, the families they come from, and the names of the two witnesses). The father will then place his hand on the hand of the groom, and the maa'zun, the official presiding, will declare that the marriage has occurred.

That marriage is a public experience is emphasized by weddings in which the couple invites their family and friends to participate. The wedding is a time for the respective families to learn how to cooperate with each other for the benefit of the couple. Conflicts over the number of bridesmaids and ushers, the number of guests to invite, and the place of the wedding are not uncommon.

To obtain a marriage license, some states require the partners to have blood tests to certify that neither has an STI. The document is then taken to the county courthouse, where the couple applies for a marriage license. Two thirds of states require a waiting period between the issuance of the license and the wedding. A member of the clergy marries

80% of couples; the other 20% (primarily in remarriages) go to a justice of the peace, judge, or magistrate.

Brides often wear traditional **artifacts** (concrete symbols that reflect a phenomenon): something old, new, borrowed, and blue. The **"old" wedding artifact** is something that represents the durability of the impending marriage (e.g., an heirloom gold locket). The **"new" wedding artifact**, perhaps in the form of new, unlaundered undergarments, emphasizes the new life to begin. The **"borrowed" wedding artifact** is something that has already been worn by a currently happy bride (e.g., a wedding veil). The **"blue" wedding artifact** represents fidelity (e.g., blue ribbons). When the bride throws her floral bouquet, it signifies the end of girlhood; the rice thrown by the guests at the newly married couple signifies fertility.

Couples now commonly have weddings that are neither religious nor traditional. In the exchange of vows, neither partner may promise to obey the other, and the couple's relationship may be spelled out by the partners rather than by tradition. Vows often include the couple's feelings about equality, individualism, humanism, and openness to change.

In 2012, the average cost of a wedding for a couple getting married for the first time was about $30,000 (www.theknot.com). Campbell, Kaufman, and colleagues (2011) surveyed 610 spouses (married about six years) and found that those who had elaborate weddings reported less present-day satisfaction and commitment, suggesting that individuals who idealize their relationship may enact elaborate weddings, and when their high relational expectations go unmet, satisfaction and commitment decline. Ways in which couples lower the cost of their wedding include marrying any day but Saturday, or marrying off-season (not June) or off-locale (in Mexico or on a Caribbean Island where fewer guests will attend). They

artifact concrete symbol that reflects the existence of a phenomenon.

"old" wedding artifact artifact worn by a bride that symbolizes durability of the impending marriage (e.g., old gold locket).

"new" wedding artifact artifact worn by a bride that symbolizes the new life she is about to begin (e.g., a new undergarment).

"borrowed" wedding artifact artifact worn by a bride that may be a garment or accessory owned by a currently happy bride.

"blue" wedding artifact blue artifact worn by a bride that is symbolic of fidelity.

may also broadcast their wedding over the Internet (http://www.webcastmywedding.net/). Streaming capability means that the couple can get married in Hawaii and have their ceremony beamed back to the states where well-wishers can see the wedding without leaving home.

7-4b Honeymoons

The **honeymoon** is the time in which the couple recovers from the wedding and solidifies their new status as spouses. Thirty-nine percent of 610 spouses (married an average of 6.6 years) reported that they did not take a honeymoon (Campbell, Kaufman et al., 2011). The most common reasons were lack of money (59%) and lack of time to take a honeymoon (39%). Only 6% said they and their partners (5%) were not interested in taking a honeymoon. In another study of 95 spouses who completed a survey about their wedding night, the grooms evaluated the experience higher than did the brides (7.2 to 6.7, respectively, on a 10-point scale). "Just being together" was the best part of the wedding night, with "the accommodations/partner's demeanor" as the worst part (Knox, 2010).

7-5 Changes After Marriage

After the wedding and honeymoon, the new spouses begin to experience changes in their legal, personal, and marital relationship. Some changes are unexpected (Hall & Adams, 2011).

7-5a Legal Changes

Unless the partners have signed a prenuptial agreement specifying that their earnings and property will remain separate, their being married means that each spouse becomes part owner of what the other earns in income and accumulates in property. Although the laws on domestic relations differ from state to state, courts typically award to each spouse half of the assets accumulated during the marriage (even though one of the partners may have contributed a smaller proportion).

For example, if a couple buys a house together, even though one spouse invested more money in the initial purchase, the other will likely be awarded half of the value of the house if they divorce. (Having children complicates the distribution of assets because the house is often awarded to the custodial parent.) In the case of death of a spouse, the remaining spouse is legally entitled to inherit between one third and one half of the partner's estate, unless a will specifies otherwise.

7-5b Personal Changes

New spouses experience an array of personal changes in their lives. One initial consequence of getting married may be an enhanced self-concept. Parents and close friends usually arrange their schedules to participate in your wedding and give gifts to express their approval of you, your partner, and your marriage. In addition, the fact that your spouse loves you and is willing to spend a lifetime with you also communicates that you are a desirable person.

Married people also begin adopting new values and behaviors consistent with the married role. Although new spouses often vow that "marriage won't change me," it does. For example, rather than stay out all night at a party, which is not uncommon for single people who may be looking for a partner, spouses (who are already paired off) tend to go home early. Their roles of spouse, employee, and parent result in their adopting more regular, alcohol- and drug-free hours.

7-5c Friendship Changes

Marriage also affects relationships with friends of the same and other sex. Although time with same-sex

honeymoon time for the new spouses to recover from the wedding and to solidify their relationship.

friends will continue (Hall & Adams, 2011), it will decrease because of the new role demands of the spouses. More time will be spent with other married couples who will become powerful influences on the new couple's relationship. Indeed, couples who have the same friends report increased marital satisfaction.

What spouses give up in friendships, they gain in developing an intimate relationship with each other. However, abandoning one's friends after marriage may be problematic because one's spouse cannot be expected to satisfy all of one's social needs. Because many marriages end in divorce, friendships that have been maintained throughout the marriage can become a vital source of support for a person adjusting to a divorce.

7-5d Relational Changes

Totenhagen and colleagues (2011) studied the variability in the relationships of 328 individuals on seven variables (satisfaction, commitment, closeness, maintenance, love, conflict, ambivalence) and found that there was greater variability for newer couples than for longer term couples. For newly married couples the change can be unsettling. For example, a couple happily married for 45 years spoke to our class and began their presentation with, "Marriage is one of life's biggest disappointments." They spoke of the difference between all the hype and the cultural ideal of what marriage is supposed to be . . . and the reality. One effect of getting married is **disenchantment**—the transition from a state of newness and high expectation to a state of mundaneness tempered by reality. It may not happen in the first few weeks or months of marriage, but disenchantment is inevitable.

Whereas courtship is the anticipation of a life together, marriage is the day-to-day reality of that life together and does not always fit the dream. "Moonlight and roses become daylight and dishes" is an old adage reflecting the realities of marriage. Musick and Bumpass (2012) compared spouses, cohabitants, and singles, and noted that the advantages of being married over not being married tended to dissipate over time.

Disenchantment after marriage is also related to the partners' shifting their focus away from each other to work or children; each partner usually gives and gets less attention in marriage than in courtship. Most college students do not anticipate a nosedive toward disenchantment—of 2,922 respondents, 25% of the males and 17% of the females agreed that "most couples become disenchanted with marriage within five years" (Knox & Hall, 2010). A couple will experience other changes when they marry, such as:

> "**Marriage is like a deck of cards.** *In the beginning all you need is two hearts and a diamond By the end you'll wish you had a freaking club and a spade.*"
>
> —*Unknown*

1. **Loss of freedom.** Single people do as they please. They make up their own rules and answer to no one. Marriage changes that as the expectations of the spouse impact the freedom of the individual. In a study of 1,001 married adults, although 41% said that they missed "nothing" about the single life, 26% reported that they most missed not being able to live by their own rules (Cadden & Merrill, 2007).

2. **More responsibility.** Single people are responsible for themselves. In the study of 1,001 married adults referred to previously, 25% reported that having less responsibility was what they missed most about being single.

3. **Less alone time.** Aside from the few spouses who live apart, most live together. They wake up together, eat their evening meals together, and go to bed together. Each may feel too much togetherness. "This altogether, togetherness thing is something I don't like," said one spouse. In the study referred to previously, 24% reported that "having time alone for myself" was what they missed most about being single (Cadden & Merrill, 2007).

4. **Change in how money is spent.** Entertainment expenses in courtship become allocated to living expenses and setting up a household together in marriage. In the same study,

disenchantment the change in a relationship from a state of newness and high expectation to a state of mundaneness and boredom in the face of reality.

17% reported that they missed managing their own money most (Cadden & Merrill, 2007).

5. **Discovering that one's mate is different from one's date.** Courtship is a context of deception. Marriage is one of reality. Spouses sometimes say, "He (she) is not the person I married." Jay Leno once quipped, "It doesn't matter who you marry since you will wake up to find that you have married someone else."

6. **Sexual changes.** The sexual relationship of the couple also undergoes changes with marriage. Hall and Adams (2011) interviewed 21 recently married childfree couples 3 to 12 months after the wedding. Only those who had not lived together before marriage mentioned unexpected adjustments. "It was also surprising to some (even those who had a previous sexual relationship with one another) that the frequency of sexual encounters was less than expected (or less than they had been before). As one wife explained, sometimes the other pressures and commitments of life got in the way: "We had more sex before we were married. I guess I figured we'd have more, but it's not bad. It's not as often as I figured it would be. But that's because we're busy and tired."

Although married couples may have intercourse less frequently than they did before marriage, marital sex is still the most satisfying of all sexual contexts. Of married people in a national sample, 85% reported that they experienced extreme physical pleasure and extreme emotional satisfaction with their spouses. In contrast, 54% of individuals who were not married or not living with anyone said that they experienced extreme physical pleasure with their partners, and 30% said that they were extremely emotionally satisfied (Michael et al., 1994). In a study of 72 couples just after their weddings, the respondents reported that high sexual satisfaction was associated with high marital satisfaction one year later (Fisher & McNulty, 2008).

7. **Power changes.** The distribution of power changes after marriage and across time. The way wives and husbands perceive and interact with each other continues to change throughout the course of the marriage. Two researchers studied 238 spouses who had been married more than 30 years and observed that (across time) men changed from being patriarchal to collaborating with their wives and that women changed from deferring to their husband's authority to challenging that authority (Huyck & Gutmann, 1992). In effect, men tend to lose power, and women gain power. However, such power changes may not always occur. In abusive relationships, abusive partners may increase the display of power because they fear the partner will try to escape from being controlled.

7-5e Parents and In-Law Changes

Marriage affects relationships with parents. Time spent with parents and extended kin radically increases when a couple has children. Indeed, a major difference between couples with and without children is the amount of time they spend with relatives. Parents and kin rally to help with the newborn and are typically there for birthdays and family celebrations.

Only a minority of spouses (3% to 4%) report that they do not get along with their in-laws (Amato et al., 2007). In a study of 23 daughters-in-law married between 5 and 10 years (with no previous marriages and at least one child from the marriage), the

Interracial partners sometimes experience negative reactions to their relationship. Black people partnered with White people have their blackness and racial identity challenged by other Black people. White people partnered with Black people may lose their White status and have their awareness of whiteness heightened more than ever before. At the same time, one partner is not given full status as a member of the other partner's race (Hill & Thomas, 2000). Other researchers have noted that the pairing of a Black male and a White female is regarded as "less appropriate" than that of a White male and a Black female (Gaines & Leaver, 2002). In the former, the Black male "often is perceived as attaining higher social status (i.e., the white woman is viewed as the black man's 'prize,' stolen from the more deserving white man)" (p. 68). In the latter, when a White male pairs with a Black female, "no fundamental change in power within the American social structure is perceived as taking place" (p. 68). Interracial marriages are more likely to dissolve than same-race marriages (Fu, 2006). Disapproval of cross-racial relationships begins early. Adolescents in one study who were dating cross-racially reported disapproval from peers (Kreaer, 2008).

Black–White interracial marriages are likely to increase—slowly. Not only has White prejudice against Blacks in general declined, but also segregation in school, at work, and in housing has decreased, permitting greater contact between the races.

7-6c Interreligious Marriages

In a survey of 2,922 undergraduates, slightly over a third (33.6%) reported that marrying someone of the same religion was important for them (Knox & Hall, 2010). Of all married couples in the United States, 37% have an interreligious marriage (Pew Research Center, 2008).

Disapproval of cross-racial relationships begins early.

Are people in interreligious marriages less satisfied with their marriages than those who marry someone of the same faith? The answer depends on a number of factors. First, people in marriages in which one or both spouses profess "no religion" tend to report lower levels of marital satisfaction than those in which at least one spouse has a religious tie. People with no religion are often more liberal and less bound by traditional societal norms and values; they feel less constrained to stay married for reasons of social propriety.

The impact of a mixed religious marriage may also depend more on the devoutness of the partners than on the fact that the partners are of different religions. If both spouses are devout in their religious beliefs, they may expect some problems in the relationship (although not necessarily). Less problematic is the relationship in which one spouse is devout but the partner is not. If neither spouse in an interfaith marriage is devout, problems regarding religious differences may be minimal or nonexistent. In their marriage vows, one interfaith couple who married (he Christian, she Jewish) said that they viewed their different religions as an opportunity to strengthen their connections to their respective faiths and to each other. "Our marriage ceremony seeks to celebrate both the Jewish and Christian traditions, just as we plan to in our life together."

7-6d Cross-National Marriages

Of 2,922 undergraduates, 77.3% of the men and 67.4% of the women reported that they would be willing to marry someone from another country (Knox & Hall, 2010). The opportunity to meet someone from another country is increasing as more than 700,000 foreign students are studying at American colleges and universities. Because not enough Americans are going into math and engineering, these foreign students are wanted because they may find the cure for cancer or invent a vaccine for HIV (Marklein, 2008).

Because American students take classes with foreign students, there is the opportunity for dating and romance between the two groups, which may lead to marriage. Some people from foreign countries marry an American citizen to gain citizenship in the United States, but immigration laws now require the marriage to last two years before citizenship is granted. If the marriage ends before two years, the foreigner must prove good faith (that the marriage was not just to gain entry into the country) or he or she will be asked to leave the country.

When the international student is male, more likely than not, his cultural mores will prevail and will clash strongly with his American bride's expectations, especially if the couple should return to his country. One female American student described her experience of marriage to a Pakistani, who violated his parents' wishes by not marrying the bride they had chosen for him in childhood. The marriage produced two children before the four of them returned to Pakistan.

The woman felt that her in-laws did not accept her and were hostile toward her. The in-laws also imposed their religious beliefs on her children and took control of their upbringing. When this situation became intolerable, the woman wanted to return to the United States. Because the children were viewed as being "owned" by their father, she was not allowed to take them with her and was banned from even seeing them. Like many international students, the husband was from a wealthy, high-status family, and the woman was powerless to fight the family. The woman has not seen her children in several years.

Cultural differences do not necessarily cause stress in cross-national marriages; the degree of cultural difference is not necessarily related to the degree of stress. Much of the stress is related to society's intolerance of cross-national marriages, as manifested in attitudes of friends and family. Japan and Korea, for example, place an extraordinarily high value on racial purity. At the other extreme is the racial tolerance evident in Hawaii, where a high level of out-group marriage is normative.

7-6d Military Marriages

Although the war in Iraq is over and U.S. troops are being withdrawn from Afghanistan,

military contract marriage marriage in which a military person and a civilian participate to get more money and benefits from the government.

© Stootsy/Shutterstock

approximately 1.4 million U.S. citizens are active-duty military personnel. Another 814,000 are in the military reserve and National Guard (*Statistical Abstract*, 2012, Table 508). About 60% of military personnel are married or have children or both (NCFR Policy Brief, 2004).

There are three main types of military marriages. In one type, the soldier falls in love with a high school sweetheart, marries the person, and subsequently joins the military. A second type of military marriage consists of those who meet and marry after one of the partners has signed up for the military. This is a typical marriage in which the partners fall in love and one or both of them happens to be in the military. The final and least common military marriage is known as a **military contract marriage** in which a military person marries a civilian to get more money and benefits from the government. For example, a soldier might decide to marry a platonic friend and split the money from the additional housing allowance (which is sometimes a relatively small amount of money and varies depending on geographical location and rank). Other times, the military member keeps the extra money, and the civilian will take the benefit of health insurance. Often, in these types of military marriages, the couple does not reside together. There is no emotional connection because the marriage is mercenary. Military contract marriages are not common, but they do exist.

Military marriages are particularly difficult for women. In their study of military wives whose husbands are deployed, Easterling and Knox (2010) noted:

> Military wives experience life quite differently than civilian wives. For example, they face a unique challenge to fill their roles within their respective military families (e.g., be supportive of the husband whose life belongs to the military and who may constantly be called to the base to deal with a crisis, to attend training missions, etc.). Military wives must also assume the roles of the military husband when he is deployed (hence, she must now do it all since he is no longer there). Military wives must also face some unsettling consequences of being a military wife which include significant barriers to employment (employers are reluctant to hire a military wife since she will move when her husband is trans-

ferred) and education (try starting an MA or Ph.D. program only to learn mid-way your husband is being moved to a new location). Nevertheless, wives cope by finding employment which has a beneficial effect even though looking for employment can have a negative impact on well-being. Volunteering can also have a beneficial effect on the wife's well-being.

Some specific challenges of being in a military marriage include the following:

1. **Traditional sex roles.** Although both men and women are members of the military service, the military has considerably more men than women. In the typical military family, the husband is deployed (sent away), and the wife is expected to "understand" his military obligations and to take care of the family in his absence. Her duties include paying the bills, keeping up the family home, and taking care of the children; a military wife must often play the role of both spouses due to the demands of her husband's military career and obligations. The wife often has to sacrifice her career to follow (or stay behind in the case of deployment) and support her husband in his fulfillment of military duties (Easterling, 2005). In the case of wives or mothers who are deployed, the rare husband is able to switch roles and become Mr. Mom. One military career wife said of her husband, whom she left behind when she was deployed, "What a joke. He found out what taking care of kids and running a family was really like and he was awful. He fed the kids SpaghettiOs for the entire time I was deployed."

Occasionally, both partners are military members, and this can blur traditional sex roles because the woman has already deviated from a traditional "woman's job." Military families in which both spouses are military personnel are rare.

2. **Loss of control—deployment.** Military families have little control over their lives as the specter of deployment is ever-present. Where one of the spouses will be next week and for how long are beyond the control of the spouses. In a survey of 259 military wives (whose husbands had been deployed) who reported feelings of loneliness, fear, and sadness, talking with other military wives who "understood" was the primary mechanism for coping with the husband's deployment. Getting a job, participating in military-sponsored events, and living with a family were also helpful. On the positive side, wives of deployed husbands reported feelings of independence and strength. They were the sole family member available to take care of the house and children, and they rose to the challenge (Easterling & Knox, 2010). Adjusting to the return of the deployed spouse has its own challenges. "You have to learn to dance all over," said one wife (Aducci et al., 2011). A team of researchers observed an increased incidence of spousal violence related to PTSD as a result of having been deployed (Teten et al., 2010).

3. **Infidelity.** Although most spouses are faithful to each other, the context of separation from each other for months (sometimes years) at a time increases the vulnerability of both spouses to infidelity. The double standard may also be operative, whereby "men are expected to have other women when they are away" and "women are expected to remain faithful and be understanding."

Separated spouses try to bridge the time they are apart with Skype, e-mails, and phone calls, but sometimes the loneliness becomes more difficult than anticipated. One enlisted husband said that he returned home after a year-and-a-half deployment to be confronted with the fact that his wife had become involved with someone else. "I absolutely couldn't believe it," he noted. "In retrospect, I think the separation was more difficult for her than for me."

> "A **married man** should forget his **mistakes;** no use two people **remembering** the **same thing.**"
>
> —DUANE DEWEL, JOURNALIST AND STATE LEGISLATOR

4. **Frequent moves and separation from extended family or close friends.** Because military couples are often required to move to a new town, parents no longer have doting grandparents available to help them rear their children. Although other military families become a community of support for each other, the consistency of such support may be lacking. "We moved seven states away from my parents to a town in North Dakota," said one wife. "It was very difficult for me to take care of our three young children with my husband deployed."

   Similar to being separated from parents and siblings is the separation from one's lifelong friends. Although new friendships and new supportive relationships develop within the military community to which the family moves, the relationships are sometimes tenuous and temporary as the new families move on. The result is the absence of a stable, predictable social structure of support, which may result in a feeling of alienation and not belonging in either the military or the civilian community. The more frequent the moves, the more difficult the transition and the more likely the alienation of new military spouses.

5. **Lower marital satisfaction and higher divorce rates.** In a study of 8,056 wives whose husbands were deployed, the researchers found that both the psychological well-being of the spouse and marital satisfaction were at risk (Orthner & Roderick, 2009). Another researcher (Lundquist, 2007) found higher divorce rates in military marriages compared to civilian marriages. Relationship education exposure helps (Allen et al., 2011). The presence of combat stress reaction (CSR) made matters worse. In a study that compared 264 veterans with CSR with 209 veterans who did not experience such stress, results showed that traumatized veterans reported lower levels of marital adjustment and more problems in parental functioning (Solomon et al., 2011).

6. **Employment of spouses.** Employment is beneficial to one's well-being. Military spouses are at a disadvantage when it comes to finding and maintaining careers or even finding a job they can enjoy. Employers in military communities are often hesitant to hire military spouses because they know that the demands placed on them in the absence of the deployed military member can be enormous. They are also aware of frequent moves that military families make and may be reluctant to hire employees for what may be a relatively short period. The result is a disadvantaged wife who has no job and must put her career on hold. Military spouses, when they do find employment, are often underemployed, which can lead to low levels of job satisfaction. Military spouses also are paid less money, on average, than their civilian counterparts with similar qualifications (Easterling, 2005).

In spite of these difficulties, there are enormous benefits to being involved in the military, such as having a stable job (one may get demoted but it is much more difficult to get "fired") and having one's medical bills paid for. In addition, most military families are amazingly resilient. Not only do they anticipate and expect mobilization and deployment as part of their military obligation, they respond with pride. Indeed, some reenlist eagerly and volunteer to return to military life even when retired. One military captain stationed at Fort Bragg, in Fayetteville, North Carolina, noted, "It is part of being an American to defend your country. Somebody's got to do it and I've always been willing to

© AP Photo/The Bulletin, Andy Tullis

do my part." He and his wife made a presentation in my class. She said, "I'm proud that he cares for our country and I support his decision to return to Afghanistan to help as needed. And most military wives that I know feel the same way."

Although military families face great challenges and obstacles, many adopt the philosophy that "whatever doesn't kill us makes us stronger." Facing deployments and frequent moves often forces a military couple to learn to rely on themselves as well as each other. They make it through difficult life events, unique to their lifestyle, which can make day-to-day challenges seem trivial. The strength that is developed within a military marriage through all the challenges they face has the potential to build a strong, resilient marriage.

7-7 Marital Success

Marital success is measured in terms of marital stability and marital happiness. *Stability* refers to how long the spouses have been married and their view on the permanence of the relationship, whereas *marital happiness* refers to more subjective aspects of the relationship such as happiness. In describing marital success, researchers have used the terms *marital satisfaction, marital quality, marital adjustment, lack of distress,* and *marital integration.* Marital success is often measured by asking spouses how happy they are, how often they spend their free time together, how often they agree on various issues, how easily they resolve conflict, how sexually satisfied they are, and how often they have considered separation or divorce.

Researchers have found that asking a single question such as "How happy are you in your marriage?" with answer options of "very happy," "pretty happy," and "not very happy" is just as effective in assessing marital happiness as asking a series of questions (Van Laningham & Johnson, 2009). In addition, a study of marital satisfaction in newlywed couples over a four-year period revealed that marital satisfaction declined across the four years (Lavner & Bradbury, 2010). Steeper declines occurred in those couples where one or both partners evidenced relatively high levels of negative personality, chronic stress, and aggression. Similarly, higher levels of relationship satisfaction were associated with the absence of these characteristics and high levels of positive affect.

Corra and colleagues (2009) analyzed data collected over 30 years (1972–2002) from the General Social Surveys to discover the influence of sex (male or female) and race (White or Black) on the level of reported marital happiness. Findings indicated greater levels of marital happiness among males and White people than among females and Black people. The researchers suggested that males make fewer accommodations in marriage and that White people are not burdened with racism and have greater economic resources.

Wallerstein and Blakeslee (1995) studied 50 financially secure couples in stable (from 10 to 40 years), happy marriages with at least one child. These couples defined marital happiness as feeling respected and cherished. They also regarded their marriages as works in progress that needed continued attention to avoid becoming stale. No couple said that they were happy all the time. Rather, a good marriage is a process. Billingsley and colleagues (1995) interviewed 30 happily married couples who had been wed an average of 32 years and had an average of 2.5 children. The researchers found various characteristics associated with couples who stay together and who enjoy each other. These qualities appear to be the same for both husbands and wives (Amato et al., 2007). Based on these and other studies of couples in stable, happy relationships, the following 15 characteristics emerged:

1. **Personal and emotional commitment to stay married.** Divorce is not considered an option. The spouses are committed to each other for personal reasons rather than societal pressure. In addition, the spouses are committed to maintaining the marriage out of emotional rather than economic need (DeOllos, 2005).

2. **Common interests/positive self-concepts.** Spouses who have similar interests, values, goals, and so on as well as positive self-concepts report higher marital success (Arnold et al., 2011).

3. **Communication/humor.** Gottman and Carrere (2000) studied the communication patterns of couples over an 11-year period and emphasized that those spouses who stay together are five times more likely to lace their arguments with positives ("I'm sorry I hurt your feelings") and to consciously choose to say things to each other that nurture the relationship rather than harm it. Successful spouses also have a sense of humor.

marital success
term for spouses in long-term marriages who are happy.

© iStockphoto.com/iofoto

Indeed, a sense of humor is associated with marital satisfaction across cultures—in the United States, China, Russia, and more (Weisfeld et al., 2011).

4. **Religiosity.** A strong religious orientation and practicing one's religion is associated with being committed to one's marriage (Jorgensen et al., 2011). Religion provides spouses with a strong common value. In addition, religion provides social, spiritual, and emotional support from church members and moral guidance in working out problems.

5. **Trust.** Trust in the partner provides a stable floor of security for the respective partners and their relationship. Neither partner fears that the other partner will leave or become involved in another relationship.

6. **Not materialistic.** Being nonmaterialistic is characteristic of happily married couples (Carroll et al., 2011). Although a couple may live in a nice house and have expensive toys (e.g., a boat and camper), they are tied to nothing. "You can have my things, but don't take away my people," is a phrase from one husband reflecting his feelings about his family.

7. **Role models.** Successfully married couples speak of having positive role models in their parents. Good marriages beget good marriages—good marriages run in families. It is said that the best gift you can give your children is a good marriage.

8. **Sexual desire.** Wilson and Cousins (2005) confirmed that partners' similar rankings of sexual desire are important in predicting long-term relationship success. Earlier, we noted the superiority of marital sex over sex in other relationship contexts in terms of both emotion and physical pleasure.

9. **Equitable relationships.** Amato and colleagues (2007) observed that the decline in traditional gender attitudes and the increase in egalitarian decision making are related to increased happiness in today's couples.

10. **Absence of negative attributions.** Spouses who do not attribute negative motives to their partner's behavior report higher levels of marital satisfaction than spouses who ruminate about negative motives. Dowd and colleagues (2005) studied 127 husbands and 132 wives and found that the absence of negative attributions was associated with higher marital quality.

11. **Forgiveness.** At some time in all marriages, each spouse engages in behavior that may hurt the partner. Forgiveness rather than harboring resentment allows spouses to move forward. Spouses who do not "drop the lowest test score" find that they inadvertently create a failing marriage in which they then must live. McNulty (2008) noted the value of forgiveness, particularly when married to a partner who rarely behaves badly.

12. **Health.** Increasingly, research emphasizes that the quality of family relationships affects family members' health and that the health of family members influences the quality of family relationships and family functioning (Proulx & Snyder, 2009). Indeed, such an association begins early as a dysfunctional family environment can activate the physiological responses to stress, change the brain structurally, and leave children more vulnerable to negative health outcomes. High-conflict spouses also create chronic stress, high blood pressure, and depression.

The family is also the primary socialization unit for physical health in reference to eating nutritious food, avoiding smoking, and getting regular exercise. The father of the author of this text had a high-fat diet, was a chronic smoker, and never exercised. He died of a coronary at age 46.

7-7a Marital Happiness Across Time

Marital happiness has variable patterns over time in relationships. Anderson and colleagues (2010) analyzed longitudinal data of 706 spouses over a 20-year period. Over 90% were in their first marriage; most had two children and 14 years of education. Reported marital happiness, marriage problems, time spent together, and economic hardship were assessed. Five patterns emerged (see Figure 7.1):

1. **High stable 2** (started out happy and remained so across time) = 21.5%

2. **High stable 1** (started out slightly less happy and remained so across time) = 46.1%

3. **Curvilinear** (started out happy, slowly declined, followed by recovery) = 10.6%

4. **Low stable** (started out not too happy and remained so across time) = 18.3%

5. **Low declining** (started out not too happy and declined across time) = 3.6%

The researchers found that couples who start out with a high level of happiness are capable of rebounding if there is a decline. But for those who start out at a low level, the capacity to improve is more limited. These data are in contrast to previous research by Vaillant and Vaillant (1993), who suggested that all couples show a gradual decline and a bounce when the children leave home. This new research shows the complexity of marital happiness patterns over time.

The Healthy Marriage Initiative (http://www.acf.hhs.gov/healthymarriage/) was set up by the U.S. Department of Health and Human Services to provide education and support for relationships, marriages, and families. Research has confirmed the positive effect of such programs not only on the adult couple relationship but also on the parent–child relationship with positive implications for child well-being (Calligas et al., 2010). Family Success in Adams County, a marriage enhancement program, showed positive changes in psychological well-being for participants (Tomkins et al., 2010).

Figure 7.1
Trajectories of Marital Happiness

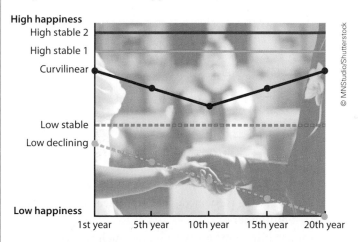

Source: Anderson et al. (2010).

STUDY TOOLS 7

Ready to study? In this book, you can:

- ⊃ Rip out the Chapter Review card in the back of the book to study for exams

- ⊃ Take the Self Assessment for this chapter (card in the back of the book) and see where you stand on the vital issues raised in the chapter

Or you can go online to CourseMate at www.cengagebrain.com for these resources:

- ⊃ Complete Practice Quizzes to prepare for tests

- ⊃ Review Key Terms Flash Cards (online or print)

- ⊃ Read about Marriage and Family in the news

- ⊃ Play "Beat the Clock" to master concepts

- ⊃ Check out Personal Applications

Same-Sex Couples and Families

"You could move."

—ABIGAIL VAN BUREN, "DEAR ABBY," IN RESPONSE TO A READER WHO COMPLAINED THAT A GAY COUPLE WAS MOVING IN ACROSS THE STREET AND WANTED TO KNOW WHAT COULD BE DONE TO IMPROVE THE QUALITY OF THE NEIGHBORHOOD

SECTIONS

8-1 Prevalence of Homosexuality, Bisexuality, and Same-Sex Couples

8-2 Origins of Sexual-Orientation Diversity

8-3 Heterosexism, Homonegativity, Homophobia, and Biphobia

8-4 Gay, Lesbian, Bisexual, and Mixed-Orientation Relationships

8-5 Legal Recognition and Support of Same-Sex Couples and Families

8-6 LGBT Parenting Issues

8-7 Effects of Antigay Bias and Discrimination on Heterosexuals

Aware of prejudice and discrimination against LGBT (lesbian, gay, bisexual, transgender) individuals on campus, a number of colleges and universities have implemented educational interventions with names such as Safe Zone, Safe Space, Safe Harbor, and Safe on Campus. After the training sessions, participants (usually faculty and staff) receive a Safe Zone sticker to put on their door, which indicates they have been through the training and can be counted on to be understanding, supportive, and available for help and advice for those with concerns about sexual orientation and gender identity. The symbol also means that homophobic and heterosexist comments and actions will not be tolerated, but will be addressed in an educational and informative manner. The mere existence of such "safe zone" programs on campus reveals that prejudice and discrimination against LGBT individuals is alive and well. In this chapter we focus on same-sex relationships.

In the United States, homosexuality remains a controversial subject. Although six states in 2012 granted marriage licenses to same-sex partners, other states defined marriage as the exclusive union between a woman and a man. Hate crimes against homosexuals are not uncommon. In 2010, there were 27 murders of LGBTQH (lesbian, gay, bisexual, transgender, queer, HIV-infected individuals; National Coalition of Anti-Violence Programs, 2011). Worldwide, approval varies considerably. The Netherlands, Spain, Belgium, Norway, and South Africa

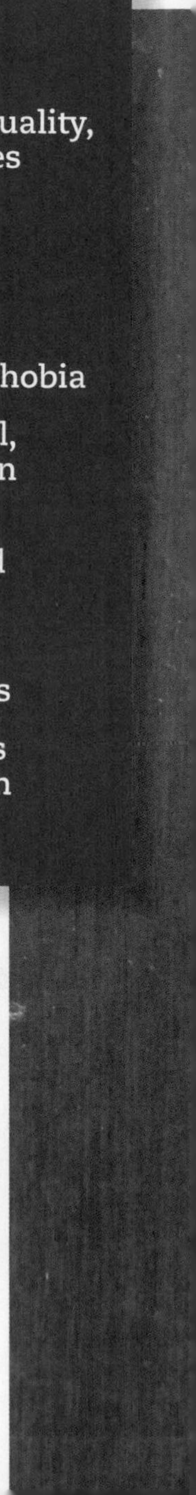

© Chelsea Curry

have granted equal marriage rights to same-sex couples, whereas intense discrimination is the norm in other countries (e.g., Pakistan and Kenya).

In this chapter, we discuss same-sex couples and families—relationships that are, in many ways, similar to heterosexual ones. A major difference, however, is that gay and lesbian couples and families are subjected to **prejudice** and **discrimination**. Although other minority groups also experience prejudice and discrimination, only sexual orientation minorities are denied federal legal marital status and the benefits and responsibilities that go along with marriage (which we discuss later in this chapter). Also, gay couples are sometimes rejected by their own parents, siblings, and other family members. One father told his son, "I'd rather have a dead son than a gay son" (author's files).

Homosexual behavior has existed throughout human history and in most (perhaps all) human societies (Kirkpatrick, 2000). In this chapter, we focus on Western views of sexual diversity that define

prejudice negative attitudes toward others based on differences.

discrimination behavior that denies individuals or groups equality of treatment.

sexual orientation
classification of individuals as heterosexual, bisexual, or homosexual, based on their emotional, cognitive, and sexual attractions and self-identity.

heterosexuality
the predominance of emotional and sexual attraction to individuals of the other sex.

homosexuality
predominance of emotional and sexual attraction to individuals of the same sex.

bisexuality
emotional and sexual attraction to members of both sexes.

lesbian homosexual woman.

gay homosexual woman or man.

lesbigay population
collective term referring to lesbians, gays, and bisexuals.

transgendered
individuals who express their masculinity and femininity in nontraditional ways consistent with their biological sex.

LGBT (GLBT) refers collectively to lesbians, gays, bisexuals, and transgendered individuals.

sexual orientation as a classification of individuals as heterosexual, bisexual, or homosexual, based on their emotional, cognitive, and sexual attractions and self-identity. **Heterosexuality** refers to the predominance of emotional and sexual attraction to individuals of the other sex. **Homosexuality** refers to the predominance of emotional and sexual attraction to individuals of the same sex, and **bisexuality** is emotional and sexual attraction to members of both sexes. The term **lesbian** refers to homosexual women; **gay** can refer to either homosexual women or homosexual men. Lesbians, gays, and bisexuals, sometimes referred to collectively as the **lesbigay population**, are considered part of a larger population referred to as the transgendered community. **Transgendered** individuals are those who express their masculinity and femininity in nontraditional ways consistent with their biological sex. For example, a biological male is not expected to wear a dress. Transgendered

individuals include not only homosexuals and bisexuals but also cross-dressers, transvestites, and transsexuals (see Chapter 2, Gender). Because much of the current literature on the lesbigay population includes other members of the transgendered community, the terms **LGBT** or **GLBT** are often used to refer collectively to lesbians, gays, bisexuals, and transgendered individuals.

The term **queer** is typically used by a male (but it could be used by a female) as a self-identifier to indicate that the person has a sexual orientation other than heterosexual. Traditionally, the term queer was used to denote a gay person, and the connotation was negative. More recently, individuals have begun using the term queer with pride much the same way African Americans called themselves Black during the 1960s civil rights era as part of building ethnic pride and identity. Hence, LGBT people took the term (*queer*) that was used to demean them and started to use it with pride. The term also has shock value, which some people who identify strongly with being queer seem to savor when they introduce themselves as being queer.

Alissa R. King (personal communication, June 1, 2010) of Iowa Central Community College uses queer "as an 'umbrella' term that does not necessarily designate the user's sexual identity."

It's a more inclusive term; thus someone who labels him/herself as "queer" could be gay, lesbian, bisexual, pansexual, trans, intersexed, non-conforming heterosexual, etc. "Queer" is a tricky term to explain because of its ugly history and it is helpful to be aware of all the "possibilities" under that umbrella—i.e., pansexual, intersexed, trans, genderqueer, etc. The clear distinguisher between gay and queer is that if someone identifies as bisexual or pansexual, he or she may not feel as at home using the label "gay" because that implies sexual attraction/behavior with one other person of a specific gender identity, whether it's used to describe same-sex identified men or women. However, "queer" provides a bit more maneuvering room (fluidity) to encompass deviations from same- or other-sex identifications (gay/lesbian or heterosexual)."

Queer theory refers to a movement or theory dating from the early 1990s. Queer theorists want less labeling of sexual orientation and a stronger "anyone can be anything he or she wants" attitude. The theory is that society should support a more fluid range of sexual orientations and that individuals can move through the range as they become more self-aware. For some, queer theory is a very specific subset of gender and human sexuality studies.

BananaStock/Jupiterimages

8-1 Prevalence of Homosexuality, Bisexuality, and Same-Sex Couples

Before looking at prevalence data concerning homosexuality and bisexuality in the United States, it is important to understand the ways in which identifying or classifying individuals as heterosexual, homosexual, gay, lesbian, or bisexual can be problematic.

8-1a Problems Associated With Identifying and Classifying Sexual Orientation

The classification of individuals into sexual orientation categories (e.g., heterosexual, homosexual, bisexual) is problematic for a number of reasons (Savin-Williams, 2006). First, because of the social stigma associated with nonheterosexual identities, many individuals conceal or falsely portray their sexual-orientation identities to avoid prejudice and discrimination.

Second, not all people who are sexually attracted to or have had sexual relations with individuals of the same sex view themselves as homosexual or bisexual. A final difficulty in labeling a person's sexual orientation is that an individual's sexual attractions, behavior, and identity may change across time. For example, in a longitudinal study of 156 lesbian, gay, and bisexual youth, 57% consistently identified as gay or lesbian and 15% consistently identified as bisexual over a one-year period, but 18% transitioned from bisexual to lesbian or gay (Rosario et al., 2006).

Early research on sexual behavior by Kinsey and his colleagues (1948, 1953) found that, although 37% of men

> **"If homosexuality is a disease, let's all call in queer to work: 'Hello. Can't work today, still queer.'"**
>
> —ROBIN TYLER, AUTHOR

and 13% of women had had at least one same-sex sexual experience since adolescence, few of the individuals reported exclusive homosexual behavior. Kinsey suggested that heterosexuality and homosexuality represent two ends of a sexual-orientation continuum and that most individuals are neither entirely homosexual nor entirely heterosexual, but fall somewhere along this continuum. The Heterosexual-Homosexual Rating Scale that Kinsey and his colleagues (1953) developed allows individuals to identify their sexual orientation on a continuum (see Figure 8.1). Very few individuals are exclusively a 0 or a 6, prompting Kinsey to believe that most individuals are bisexual.

Sexual-orientation classification is also complicated by the fact that sexual behavior, attraction, love, desire, and sexual-orientation identity do not always match. For example, "research conducted across different cultures and historical periods (including present-day Western culture) has found that many individuals develop passionate infatuations

queer broad self-identifier term to indicate that the person has a sexual orientation other than heterosexual.

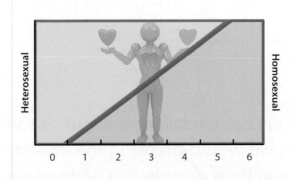

Figure 8.1
Heterosexual-Homosexual Rating Scale

Based on both psychologic reactions and overt experience, individuals rate as follows:
0. Exclusively heterosexual with no homosexual
1. Predominantly heterosexual, only incidentallly homosexual
2. Predominantly heterosexual, but more than incidentally homosexual
3. Equally heterosexual and homosexual
4. Predominantly homosexual, but more than incidentally heterosexual
5. Predominantly homosexual, but incidentally heterosexual
6. Exclusively homosexual

Source: Kinsey et al. (1948), *Sexual Behavior in the Human Male.* Reprinted by permission of the Kinsey Institute for Research in Sex, Gender, and Reproduction, Inc.

with same-gender partners in the absence of same-gender sexual desires ... whereas others experience same-gender sexual desires that never manifest themselves in romantic passion or attachment" (Diamond, 2003, p. 173).

8-1b Prevalence of Homosexuality, Heterosexuality, and Bisexuality

Despite the difficulties inherent in categorizing an individual's sexual orientation, recent data reveal the prevalence of individuals in the United States who identify as lesbian, gay, or bisexual. In a national survey by Mock and Eibach (2011), fewer than 1% (0.80) of 1,370 women and 2% (1.76) of 1,190 men identified themselves as homosexual. Regarding bisexuality, 1.24% of the women and 1.43% of the men in the survey identified as bisexual.

Being bisexual is related to prevalence of contracting sexually transmitted infections. Bisexual women are more likely to report having an STI than are lesbians. This discrepancy occurs because women have a wider area (vulva) to collect sexually transmitted infections and because women collect the male semen, which may also be infected. In regard to whether one's sexual orientation is stable or fluid, Mock and Eibach (2011) collected data 10 years later on the same respondents and found that sexual orientation was stable for the majority. Two percent of heterosexual women and 1% of heterosexual men reported a change in sexual orientation.

8-1c Prevalence of Same-Sex Couple Households

Although U.S. census surveys do not ask about sexual orientation or gender identity, same-sex cohabiting couples may identify themselves as "unmarried partners." Those couples in which both partners are men or both are women are considered to be same-sex couples or households for purposes of research. There are currently over half a million of them (581,300 unmarried same-sex couple households) in the U.S. (*Statistical Abstract of the United States*, 2012, Table 63).

Carpenter and Gates (2008) analyzed data in California and noted that, although 62% of heterosexual couples cohabit, about 40% of gay males and about 60% of lesbians cohabit. Same-sex couples are more likely to live in metropolitan areas than in rural areas. However, the largest proportional increases in the number of same-sex couples self-reporting in 2000 versus 1990 came in rural, sparsely populated states.

Why are data on the numbers of LGBT individuals and couples in the United States relevant? The primary reason is that census numbers on the prevalence of LGBT individuals and couples can influence laws and policies that affect gay individuals and their families. "The more we are counted, the more we count" is the slogan that points out the value  of gay individuals being visible (emphasized in the movie Milk starring Sean Penn as Harvey Milk, a politically active gay man in San Francisco). Visibility and acceptance are increasing. In 2010, 107 openly gay candidates were elected to office ... up a third from 2008. According to a Gallup Poll, almost 70% (67%) say they would vote for a president who is gay (Page, 2011).

© nito/Shutterstock

"The **question** is not what **family form or marriage arrangement** we would prefer in the **abstract** but how we can **help people** in a wide variety of **committed relationships.**"

—STEPHANIE COONTZ, FAMILY HISTORIAN

8-2 Origins of Sexual-Orientation Diversity

Much of the biomedical and psychological research on sexual orientation attempts to identify one or more "causes" of sexual-orientation diversity. The driving question behind this research is this: "Is sexual orientation inborn or is it learned or acquired from environmental influences?" Although a number of factors have been correlated with sexual orientation, including genetics, gender role behavior in childhood, and fraternal birth order, no single theory can explain diversity in sexual orientation.

8-2a Beliefs About What "Causes" Homosexuality

Aside from what "causes" homosexuality, social scientists are interested in what people believe about the "causes" of homosexuality. Most gay people believe that homosexuality is an inherited, inborn trait. In a national study of homosexual men, 90% reported that they believed that they were born with their homosexual orientation; only 4% believed that environmental factors were the sole cause (Lever, 1994).

> No single theory can explain diversity in sexual orientation

Individuals who believe that homosexuality is genetically determined tend to be more accepting of homosexuality and are more likely to be in favor of equal rights for lesbians and gays (Tyagart, 2002). In contrast, "those who believe homosexuals choose their sexual orientation are far less tolerant of gays and lesbians and more likely to feel that homosexuality should be illegal than those who think sexual orientation is not a matter of personal choice" (Rosin & Morin, 1999, p. 8).

Although the terms *sexual preference* and *sexual orientation* are often used interchangeably, the term *sexual orientation* avoids the implication that homosexuality, heterosexuality, and bisexuality are determined voluntarily. Hence, those who believe that sexual orientation is inborn more often use the term *sexual orientation*, and those who think that individuals choose their sexual orientation use *sexual preference* more often.

8-2b Can Homosexuals Change Their Sexual Orientation?

Individuals who believe that homosexual people choose their sexual orientation tend to think that homosexuals can and should change their sexual orientation. Various forms of **reparative therapy** or **conversion therapy** are dedicated to changing homosexuals' sexual orientation. Some religious organizations sponsor "ex-gay ministries," which claim to "cure" homosexuals and transform them into heterosexuals by encouraging them to ask for "forgiveness for their sinful lifestyle," through prayer and other forms of "therapy." In a review of 28 empirically based, peer-reviewed articles, Serovich and colleagues (2008) found them methodologically problematic, which threatens the validity of interpreting available data on this topic.

Parelli (2007) attended private as well as group therapy sessions to change his sexual orientation from homosexuality to heterosexuality. It did not work. In retrospect, he identified seven reasons for the failure of reparative therapy. In effect, he noted that these therapies focus on outward behavioral change and do not acknowledge the inner yearnings:

> By my mid-forties, I was experiencing a chronic need for appropriately affectionate male touch. It was so acute I could think of nothing else. Every cell of my body seemed relationally isolated and emotionally starved. Life was so completely and fatally ebbing out of my being that my internal life-saving system kicked in and put out a high-alert call for help. I desperately needed to be held by loving, human, male arms. (p. 32)

Critics of reparative therapy and ex-gay ministries take a different approach: "It is not gay men and lesbians who need to change . . . but negative attitudes and discrimination against gay people that need to be abolished" (Besen, 2000, p. 7). The National Association for Research and

reparative therapy (conversion therapy) therapy designed to change a person's homosexual orientation to a heterosexual orientation.

Therapy of Homosexuality (NARTH) has been influential in moving public opinion from "gays are sick" to "society is judgmental." The American Psychiatric Association, the American Psychological Association, the American Academy of Pediatrics, the American Counseling Association, the National Association of School Psychologists, the National Association of Social Workers, and the American Medical Association agree that homosexuality is not a mental disorder and needs no cure—that efforts to change sexual orientation do not work and may, in fact, be harmful (Potok, 2005). An extensive review of the ex-gay movement concludes, "There is a growing body of evidence that conversion therapy not only does not work, but also can be extremely harmful, resulting in depression, social isolation from family and friends, low self-esteem, internalized homophobia, and even attempted suicide" (Cianciotto & Cahill, 2006, p. 77). According to the American Psychiatric Association, "clinical experience suggests that any person who seeks conversion therapy may be doing so because of social bias that has resulted in internalized homophobia, and that gay men and lesbians who have accepted their sexual orientation are better adjusted than those who have not done so" (quoted by Holthouse, 2005, p. 14).

heterosexism the denigration and stigmatization of any behavior, person, or relationship that is not heterosexual.

New research brings into question the claim that reparative therapies "never work." Two researchers reported that the majority of 117 men who were dissatisfied with their sexual orientation and who sought sexual orientation change efforts (SOCE) were able to reduce their homosexual feelings and behaviors and increase their heterosexual feelings and behaviors (Karten & Wade, 2010). The primary motivations for their seeking change were religion ("homosexuality is wrong") and emotional dissatisfaction with the homosexual lifestyle. Being married, feeling disconnected from other men prior to seeking help, and feeling able to express nonsexual affection toward other men were the factors predictive of greatest change. Developing nonsexual relationships with same-sex peers was also identified as helpful in change. Jones and Yarhouse (2011) also provided data which suggest that religiously mediated sexual orientation change is possible.

© BananaStock/Jupiter Images

8-3 Heterosexism, Homonegativity, Homophobia, and Biphobia

The United States, along with many other countries throughout the world, is predominantly heterosexist. **Heterosexism** refers to "the institutional and societal reinforcement of heterosexuality as the privileged and powerful norm." Heterosexism is based on the belief that heterosexuality is superior to homosexuality. Of 2,922 undergraduates, 28% agreed with the statement "It is better to be heterosexual than homosexual" (Knox & Hall, 2010). Heterosexism results in prejudice and discrimination against homosexual and bisexual people. The word *prejudice* refers to negative attitudes, whereas *discrimination* refers to behavior that denies equality of treatment for individuals or groups. Before reading further, you may wish to complete the Self-Assessment card at the end of the book, which assesses behaviors toward individuals perceived to be homosexual.

8-3a Homonegativity and Homophobia

The term **homophobia** is commonly used to refer to negative attitudes and emotions toward homosexuality and those who engage in it. Homophobia is not necessarily a clinical phobia (i.e., one involving a compelling desire to avoid the feared object despite recognizing that the fear is unreasonable). Other terms that refer to negative attitudes and emotions toward homosexuality include **homonegativity** and **antigay bias**.

The Sexuality Information and Education Council of the United States (SIECUS) states that "individuals have the right to accept, acknowledge, and live in accordance with their sexual orientation, be they bisexual, heterosexual, gay or lesbian. The legal system should guarantee the civil rights and protection of all people, regardless of sexual orientation. Prejudice/discrimination based on sexual orientation is unethical and immoral" (SIECUS, 2012). Nevertheless, negative attitudes (even by professionals) toward homosexuality continue (Rutledge et al., 2012).

Characteristics associated with positive attitudes toward homosexuals and gay rights include younger age, advanced education, no religious affiliation, liberal political party affiliation, and personal contact with homosexual individuals (Lee & Hicks, 2011). In one study, heterosexual women who kissed other women had more positive attitudes toward homosexuality (Knox et al., 2011). In a study on racial differences in attitudes toward lesbians and gay men (ATLGM), Black students held generally neutral ATLGM whereas White students' attitudes were slightly positive (Whitley et al., 2011).

Negative social meanings associated with homosexuality can affect the self-concepts of LGBT individuals. **Internalized homophobia** is a sense of personal failure and self-hatred among lesbians and gay men resulting from social rejection and the stigmatization of being gay and has been linked to increased risk for depression, substance abuse, anxiety, panic attacks, and suicidal thoughts (Oswalt & Wyatt, 2011; Ricks, 2012; Rubinstein, 2010). Being gay and growing up in a religious context that is antigay may be particularly difficult (Lalicha & McLarena, 2010).

8-3b Biphobia

Just as the term *homophobia* is used to refer to negative attitudes toward homosexuality, gay men, and lesbians, **biphobia** (also referred to as **binegativity**) refers to a parallel set of negative attitudes toward bisexuality and those identified as bisexual. Although heterosexuals (men are less accepting than women; Yost & Thomas, 2011) often reject bisexual-identified individuals, bisexual-identified women and men also face rejection from many homosexual individuals. Thus, bisexuals experience "double discrimination."

Lesbians are a major source of negative views toward bisexuals. In a study of 346 self-identified lesbians, Rust (1993) found that lesbians view bisexuals as less committed to other women than are lesbians, perceive bisexuals as disloyal to lesbians, and resent bisexual women who have close relationships with the "enemy" of lesbians—men. Some negative attitudes toward

homophobia negative (almost phobic) attitudes toward homosexuality.

homonegativity a construct that refers to antigay responses such as negative feelings (fear, disgust, anger), thoughts ("homosexuals are HIV carriers"), and behavior ("homosexuals deserve a beating").

antigay bias any behavior or statement which reflects a negative attitude toward homosexuals.

internalized homophobia a sense of personal failure and self-hatred among lesbians and gay men resulting from the acceptance of negative social attitudes and feelings toward homosexuals.

biphobia (binegativity) refers to a parallel set of negative attitudes toward bisexuality and those identified as bisexual.

bisexual individuals "are based on the belief that bisexual individuals are really lesbian or gay individuals who are in transition or in denial about their true sexual orientation" (Israel & Mohr, 2004, p. 121). Rust (1993) also found that lesbians were likely to be critical of bisexuals if the transition period from identifying as bisexual to identifying as a lesbian took too long. According to this view, bisexuals either fear coming out as lesbian or gay or are trying to maintain heterosexual privilege.

8-4 Gay, Lesbian, Bisexual, and Mixed-Orientation Relationships

Research suggests that gay and lesbian couples tend to be more similar to than different from heterosexual couples (Kurdek, 2005, 2008). In this section, we note the similarities as well as the differences between heterosexual, gay male, and lesbian relationships in regard to relationship satisfaction, conflict and conflict resolution, and monogamy and sexuality. We also look at relationship issues involving bisexual individuals and mixed-orientation couples.

8-4a Relationship Satisfaction

In a review of literature on relationship satisfaction and sexual orientation, Kurdek (1994) concluded, "The most striking finding regarding the factors linked to relationship satisfaction is that they seem to be the same for lesbian couples, gay couples, and heterosexual couples" (p. 251). These factors include having equal power and control, being emotionally expressive, perceiving many attractions and few alternatives to the relationship, placing a high value on attachment, and sharing decision making. Kamen and

I know my roommate is gay. I just wish he would tell me. It would make it easier for me to come out to him.

colleagues (2011) also noted that commitment, trust, and support from one's partner were related to relationship satisfaction in same-sex relationships. In a comparison of relationship quality of cohabitants over a 10-year period involving both partners from 95 lesbian, 92 gay male, and 226 heterosexual couples living without children, and both partners from 312 heterosexual couples living with children, the researcher found that lesbian couples showed the highest levels of relationship quality averaged over all assessments (Kurdek, 2008).

Researchers who studied relationship quality among same-sex couples noted that "in trying to create satisfying and long-lasting intimate relationships, LGBT individuals face all of the same challenges faced by heterosexual couples, as well as a number of distinctive concerns" (Otis et al., 2006, p. 86). These concerns include if, when, and how to disclose their relationships to others and how to develop healthy intimate relationships in the absence of same-sex relationship models.

In one review of research on gay and lesbian relationships, the investigators concluded that the main difference between heterosexual and nonheterosexual relationships is that "[w]hereas heterosexuals enjoy many social and institutional supports for their relationships, gay and lesbian couples are the object of prejudice and discrimination" (Peplau et al., 1996, p. 268). Both gay male and lesbian couples must cope with the stress created by antigay prejudice and discrimination and by "internalized homophobia" or negative self-image and low self-esteem due to being a member of a stigmatized group. Not surprisingly, higher levels of such stress are associated with lower reported levels of relationship quality among LGBT couples (Otis et al., 2006).

Despite the stresses and lack of social and institutional support LGBT individuals experience, gay men and lesbians experience relationship satisfaction at a level that is at least equal to that reported by married heterosexual spouses (Kurdek, 2005). Partners of the same sex enjoy the comfort of having a shared gender perspective, which is often accompanied by a sense of equality in the relationship. For example, contrary to stereotypical beliefs, same-sex couples (male or female) typically do not assign "husband" and "wife" roles in the division of household labor; as well, they are more likely than heterosexual couples to achieve a fair distribution of household labor and at the same time accommodate the different interests, abilities, and work schedules of each partner (Kurdek, 2005). In contrast, division of household labor among heterosexual couples tends to be unequal, with wives doing the majority of such tasks.

Same-sex relationships are not without abuse (Porter & Williams, 2011). In a study of violence in a sample of 284 gay and bisexual men, the researchers found that almost all reported psychological abuse, more than a third reported physical abuse, and 10% reported being forced to have sex. Abuse in gay relationships was less likely to be reported to the police because some gays did not want to be "outed" (Bartholomew et al., 2008).

8-4b Conflict Resolution and Breakup Aftermath

All couples experience conflict in their relationships, and gay and lesbian couples tend to disagree about the same issues that heterosexual couples argue about. In one study, partners from same-sex and heterosexual couples identified the same sources of most conflict in their relationships: finances, affection, sex, being overly critical, driving style, and household tasks (Kurdek, 2004). However, same-sex couples and heterosexual couples tend to differ in how they resolve conflict. Gay and lesbian partners begin their discussions more positively and maintain a more positive tone throughout the discussion than do partners in heterosexual marriages (Gottman et al., 2003). One explanation is that same-sex couples value equality more and are more likely to have equal power and status in the relationship than are heterosexual couples (Gottman et al., 2003). In addition, same-sex couples may have a higher need to reduce conflict because dissolution may involve losing a tight friendship network (Van Eeden-Moorefield et al., 2011). When a breakup occurs, lesbian mothers are likely to maintain cordial communication with the coparent with whom they typically share joint custody (Gartrell et al., 2011).

8-4c Monogamy and Sexuality

Like many heterosexual women, most gay women value stable, monogamous relationships that are emotionally as well as sexually satisfying. Gay and heterosexual women in U.S. society are taught that sexual

© George Doyle/Stockbyte/Getty Images

expression should occur in the context of emotional or romantic involvement.

A common stereotype of gay men is that they prefer casual sexual relationships with multiple partners versus monogamous, long-term relationships. However, although gay men report greater interest in casual sex than do heterosexual men (Gotta et al., 2011), most gay men prefer long-term relationships, and sex outside the primary relationship is usually infrequent and not emotionally involving (Green et al., 1996).

The degree to which gay males engage in casual sexual relationships is better explained by the fact that they are male than by the fact that they are gay. In this regard, gay and straight men have a lot in common: They both tend to have fewer barriers to engaging in casual sex than do women (heterosexual or lesbian). One way that gay men meet partners is through the Internet. Blackwell and Dziegielewski (2012) studied men who seek men for sex on the Internet and noted that these sites promote higher risk sexual activities. "Party and play" (PNP) is one such activity and involves using crystal methamphetamine and having unprotected anal sex. More scholarly inquiry is needed on the extent of this phenomenon.

Such nonuse of condoms results in the high rate of human immunodeficiency virus (HIV) infection and acquired immunodeficiency syndrome (AIDS). Approximately 50,000 new cases of HIV are reported annually, and male-to-male sexual contact is the most common mode of transmission in the United States (29,700 infections annually) (Centers for Disease Control and Prevention, 2012). Women who have sex exclusively with other women have a much lower rate of HIV infection than do men (both gay and straight) and women who have sex with men. Many gay men have lost a love partner to HIV infection or AIDS; some have experienced multiple losses. Those still in relationships with partners who are HIV-positive experience profound changes, such as developing a sense

> "Gay and lesbian people fall in love. We settle down. We commit our lives to one another. We raise our children. We protect them. We try to be good citizens."
>
> —SHEILA KUEHL, CALIFORNIA SENATOR

of urgency to "speed up" their relationship because they may not have much time left together (Palmer & Bor, 2001).

8-4d Relationships of Bisexuals

Individuals who identify as bisexual have the ability to form intimate relationships with both sexes. However, research has found that the majority of bisexual women and men tend toward primary relationships with the other sex (McLean, 2004). Contrary to the common myth that bisexuals are, by definition, nonmonogamous, some bisexuals prefer monogamous relationships (especially in light of the widespread concern about HIV). In another study of 60 bisexual women and men, 25% of the men and 35% of the women were in exclusive relationships; 60% of the men and 53% of the women were in "open" relationships in which both partners agreed to allow each other to have sexual or emotional relationships, or both, with others, often under specific conditions or rules about how such relationships would occur (McLean, 2004). In these "open" relationships, nonmonogamy was not the same as infidelity, and the former did not imply dishonesty. The researcher concluded:

> Despite the stereotypes that claim that bisexuals are deceitful, unfaithful, and untrustworthy in relationships, most of the bisexual men and women I interviewed demonstrated a significant commitment to the principles of trust, honesty, and communication in their intimate relationships and made considerable effort to ensure both theirs [sic] and their partner's needs and desires were catered for within the relationship. (McLean, 2004, p. 96)

Monogamous bisexual women and men find that their erotic attractions can be satisfied through fantasy and their affectional needs through nonsexual friendships (Paul, 1996). Even in a monogamous relationship, "the partner of a bisexual person may feel that

a bisexual person's decision to continue to identify as bisexual . . . is somehow a withholding of full commitment to the relationship. The bisexual person may be perceived as holding onto the possibility of other relationships by maintaining a bisexual identity and, therefore, not fully committed to the relationship" (Ochs, 1996, p. 234). However, this perception overlooks the fact that one's identity is separate from one's choices about relationship involvement or monogamy. Ochs notes that "a heterosexual's ability to establish and maintain a committed relationship with one person is not assumed to falter, even though the person retains a sexual identity as 'heterosexual' and may even admit to feeling attractions to other people despite her or his committed status" (p. 234).

8-4e Mixed-Orientation Relationships

Mixed-orientation couples are those in which one partner is heterosexual and the other partner is gay, lesbian, or bisexual. Up to 2 million gay, lesbian, or bisexual people in the United States have been in heterosexual marriages at some point (Buxton, 2004). Some lesbigay individuals do not develop same-sex attractions and feelings until after they have been married. Others deny, hide, or repress their same-sex desires.

In a study of 20 gay or bisexual men who had disclosed their sexual orientation to their wives, most of the men did not intentionally mislead or deceive their future wives with regard to their sexuality. Rather, they

> Individuals who identify has bisexual have the ability to form intimate relationships with both sexes.

© Chelsea Curry

did not fully grasp their feelings toward men, although they had a vague sense of their same-sex attraction (Pearcey, 2004). The majority of the men in this study (14 of 20) attempted to stay married after disclosure of their sexual orientation to their wives, and nearly half (9 of 20) stayed married for at least three years.

Although gay and lesbian spouses in heterosexual marriages are not sexually attracted to their spouses, they may nevertheless love them. However, that is little consolation to spouses, who, upon learning that their husband or wife is gay, lesbian, or bisexual, often react with shock, disbelief, and anger. The Straight Spouse Network (http://www.straightspouse.org/home.php) provides support to heterosexual spouses or partners, current or former, of LGBT mates.

8-5 Legal Recognition and Support of Same-Sex Couples and Families

As current divorce rates of heterosexuals suggest (4 in 10 marriages end in divorce), maintaining long-term relationships is challenging. However, the challenge is even greater for same-sex couples who lack the many social supports and legal benefits of marriage. A leading researcher and scholar on LGBT issues noted, "[P]erhaps what is most impressive about gay and lesbian couples is . . . that they manage to endure without the benefits of institutionalized supports" (Kurdek, 2005, p. 253). In this section, we discuss laws and policies designed to provide institutionalized support for same-sex couples and families.

8-5a Decriminalization of Sodomy

In the United States, a 2003 Supreme Court decision in Lawrence v. Texas invalidated state laws that criminalized **sodomy**—oral and anal sexual acts. The ruling, which found that sodomy laws were discriminatory and unconstitutional, removed the stigma and criminal branding that sodomy laws had long placed on LGBT individuals. Prior to this historic ruling, sodomy was illegal in 13 states. Sodomy laws, which carried penalties ranging from a $200 fine to 20 years

of imprisonment, were usually not used against heterosexuals but were used primarily against gay men and lesbians. Same-sex sexual behavior is still considered a criminal act in many countries throughout the world.

8-5b Registered Partnerships, Civil Unions, and Domestic Partnerships

Aside from same-sex marriage (which we discuss later in this chapter), other forms of legal recognition of same-sex couples exist in a number of countries throughout the world at the national, state, and/or local levels. In addition, some workplaces recognize same-sex couples for the purposes of employee benefits. Legal recognition of same-sex couples, also referred to as registered partnerships, **civil unions**, or **domestic partnerships**, conveys most but not all the rights and responsibilities of marriage. Half of the partnered lesbians in California are officially registered (Carpenter & Gates, 2008).

State and Local Legal Recognition of Same-Sex Couples There is no federal recognition of same-sex couples in the United States. However, a number of U.S. states allow same-sex couples legal status that entitles them to many of the same rights and

sodomy oral and anal sexual acts.

civil union a pair-bond given legal significance in terms of rights and privileges (more than a domestic partnership and less than a marriage).

domestic partnership a relationship in which individuals who live together and are emotionally and financially interdependent are given some kind of official recognition by a city or corporation so as to receive partner benefits (e.g., health insurance).

responsibilities as married, opposite-sex couples (see Table 8.1). For example, in New Jersey, same-sex couples can apply for a civil union license, which entitles them to all the rights and responsibilities available under state law to married couples. Unlike marriage for heterosexual couples, the rights of partners in same-sex civil unions are not recognized by U.S. federal law, so they do not have the federal protections that go along with civil marriage, and their legal status is not recognized in other states.

The rights and responsibilities granted to domestic partners vary from place to place but may include coverage under a partner's health and pension plan, rights of inheritance and community property, tax benefits, access to housing for married students, child custody and child and spousal support obligations, and mutual responsibility for debts. California provides the broadest array of protections, including eligibility for family leave, other employment and health benefits, the right to sue for wrongful death of a partner or inherit from a partner as next of kin, and access to the stepparent adoption process. Ten states as well as several dozen U.S. municipalities offer domestic partner benefits to the same-sex partners of public employees.

garriage term for relationship of gay individuals who are married or committed.

Recognition of Same-Sex Couples in the Workplace The majority of the largest employers (those employing 5,000 or more) pro-vide benefits for same-sex partners. Of the Fortune 500 companies, most provide health benefits for same-sex partners (Human Rights Campaign, 2012). However, even when companies offer domestic partner benefits to same-sex partners of employees, these benefits are usually taxed as income by the federal government, whereas spousal benefits are not.

> "Anyone who wishes to examine 20 years of peer-reviewed studies on the emotional, cognitive and behavioral outcomes of children of gay and lesbian parents will find not one shred of evidence that children are harmed by their parents' sexual orientation."
>
> —*Carol Trust, executive director, National Association of Social Workers*

8-5c Same-Sex Marriage

As of 2012, seven states and the District of Columbia offered marriage licenses to same-sex couples—Connecticut, Iowa, Massachusetts, New Hampshire, Vermont, Maine, and New York. However, unlike marriages between a man and a woman, the same-sex marriages in these states are not always recognized in other states, nor does the federal government recognize them. Same-sex marriage is still a hotly contested issue in many states. Maine residents, for example, voted to ban same-sex marriage after the courts made same-sex marriage legal.

Attorney Robert Zaleski (2007) coined the term **garriage**. "The 'g' is borrowed from the word gay and connotes the same-sex status of the committed couple. And let the new verb be 'garry,' which would be conjugated in identical fashion with the verb 'marry,' thereby enabling these words to be used interchangeably in conversation." Zaleski also emphasized that the word *garriage* allows heterosexual couples to maintain their uniqueness just as it does for gay couples.

Anti-Gay Marriage Legislation In a national survey, the majority of Americans oppose gay marriage. In a national sample, over half (53%) registered opposition while 39% of U.S. adults were in favor of gay marriage (Pew Research Center, 2011). Among college students, in a sample of

Table 8.1

States that Recognize Same-Sex Relationships

State	Same-Sex Relationship Recognition
California	Recognizes same-sex marriages performed in California before Proposition 8 (defining marriage as man–woman union) was passed November 4, 2008. Ninth Circuit Court overturned.
Connecticut	Grants same-sex marriage licenses to residents.
Hawaii	Offers "reciprocal beneficiary" status to same-sex registered couples.
Iowa	Grants same-sex marriage licenses to residents.
Maine	Grants same-sex marriage licenses to residents.
Massachusetts	Grants same-sex marriage licenses to residents.
New Hampshire	Grants same-sex marriage licenses to residents.
New York and District of Columbia	Grant same-sex marriage licenses to residents.
Vermont	Grants same-sex marriage licenses to residents.

Source: National Conference of State Legislatures (2011).

> ### "I've just concluded that for me personally it is important for me to go ahead and **affirm** that I think **same-sex couples** should be able to **get married.**"
>
> —BARACK OBAMA

2,922 at two universities, 39% were in favor of making same-sex marriage legal (Knox & Hall, 2010). Males are more supportive of lesbian marriage than of gay male marriage (Moskowitz et al., 2010). Where disapproval exists, the primary reason is morality. Gay marriage is viewed as "immoral, a sin, against the Bible." However, support for gay marriage varies by age, with younger adults (ages 18 to 29) more approving.

In 1996, Congress passed and former President Clinton signed the **Defense of Marriage Act (DOMA)**, which states that marriage is a "legal union between one man and one woman" and denies federal recognition of same-sex marriage. In effect, this law allows states to either recognize or not recognize same-sex marriages performed in other states. As of 2012, 37 states had banned gay marriage either through statute or a state constitutional amendment, and 17 states had passed broader antigay family measures that ban other forms of partner recognition in addition to marriage, such as domestic partnerships and civil unions. These broader measures, known as "Super DOMAs," potentially endanger employer-provided domestic partner benefits, joint and second-parent adoptions, health care decision-making proxies, or any policy or document that recognizes the existence of a same-sex partnership (Cahill & Slater, 2004). Some of these "Super DOMAs" ban partner recognition for unmarried heterosexual couples as well.

At the federal level, there have been efforts (under the Bush administration) to amend the U.S. Constitution to define marriage as being between a man and a woman. The Federal Marriage Amendment has been introduced in Congress on several occasions but has failed to reach the required two-thirds majority vote. Should the amendment pass, it would deny marriage and likely civil union and domestic partnership rights to same-sex couples. Passage is not likely.

But should the amendment pass, it would hurt the children in same-sex–couple families. Dr. Kathleen Moltz, an assistant professor at Wayne State University School of Medicine, testified against the passage of an antigay constitutional amendment before the U.S. Senate Judiciary Committee, expressing her fears about how such an amendment would affect her family:

> I don't know what harm . . . a constitutional amendment might cause. I fear that families like mine, with young children, will lose health benefits; will be denied common decencies like hospital visitation when tragedy strikes; will lack the ability to provide support for one another in old age. I fear that my loving, innocent children will face hatred and insults implicitly sanctioned by a law that brands their family as unequal. I know that these sweet children have already been shunned and excluded by people claiming to represent values of decency and compassion. I also know what such an amendment will not do. It will not help couples who are struggling to stay married. It will not assist any impoverished families struggling to make ends meet or to obtain health care for sick children. It will not keep children with their parents when their parents see divorce as their only option. It will not help any single American citizen to live life with more decency, compassion or morality. (Moltz, 2005).

In 2011, the Obama administration announced that the Justice Department will no longer defend federal law that defines marriage as a union between a woman and a man.

Arguments in Favor of Same-Sex Marriage Advocates of same-sex marriage argue that banning or refusing to recognize same-sex marriages granted in other states is a violation of civil rights that denies same-sex couples the many legal and financial benefits that

© glossyplastic/Shutterstock

Defense of Marriage Act (DOMA) legislation passed by Congress denying federal recognition of homosexual marriage and allowing states to ignore same-sex marriages licensed by other states.

are granted to heterosexual married couples. Rights and benefits accorded to married spouses include the following:

- The right to inherit from a spouse who dies without a will

- The benefit of not paying inheritance taxes upon the death of a spouse

- The right to make crucial medical decisions for a spouse and to take care of a seriously ill spouse or a parent of a spouse under current provisions in the federal Family and Medical Leave Act

- The right to collect Social Security survivor benefits

- The right to receive health insurance coverage under a spouse's insurance plan

Other rights bestowed on married (or once-married) partners include assumption of a spouse's pension, bereavement leave, burial determination, domestic violence protection, reduced-rate memberships, divorce protections (such as equitable division of assets and visitation of partner's children), automatic housing lease transfer, and immunity from testifying against a spouse. As noted earlier, same-sex couples are taxed on employer-provided insurance benefits for domestic partners, whereas married spouses receive those benefits tax free. Finally, unlike 17 other countries that recognize same-sex couples for immigration purposes, the United States does not recognize same-sex couples in granting immigration status because such couples are not considered "spouses." Another argument for same-sex marriage is that it would promote relationship stability among gay and lesbian couples: "[T]o the extent that marriage provides status, institutional support, and legitimacy, gay and lesbian couples, if allowed to marry, would likely experience greater relationship stability" (Amato, 2004, p. 963). Indeed, same-sex relationships, like cohabitation relationships, end at a higher rate than marriage relationships (Wagner, 2006).

Recognized marriage, argues Amato, would be beneficial to the children of same-sex parents. Without legal recognition of same-sex families, children living in gay- and lesbian-headed households are denied a range of securities that protect children of heterosexual married couples. These include the right to get health insurance coverage and Social Security survivor benefits from a nonbiological parent. In some cases, children in same-sex households lack the automatic right to continue living with their nonbiological parent should their biological mother or father die (Tobias & Cahill, 2003). It is ironic

that the same pro-marriage groups that stress that children are better off in heterosexual married-couple families disregard the benefits of same-sex marriage to children.

Finally, there are religion-based arguments in support of same-sex marriage. Although many religious leaders teach that homosexuality is sinful and prohibited by God, some religious groups, such as the Quakers and the United Church of Christ (UCC), accept homosexuality, and other groups have made reforms toward increased acceptance of lesbians and gays. In 2005, the UCC became the largest Christian denomination to endorse same-sex marriages. In a sermon titled, "The Christian Case for Gay Marriage," Jack McKinney (2004) interprets Luke 4: "Jesus is saying that one of the most fundamental religious tasks is to stand with those who have been excluded and marginalized. . . . [Jesus] is determined to stand with them, to name them beloved of God, and to dedicate his life to seeing them empowered." McKinney goes on to ask, "Since when has it been immoral for two people to commit themselves to a relationship of mutual love and caring? No, the true immorality around gay marriage rests with the heterosexual majority that denies gays and lesbians more than 1,000 federal rights that come with marriage."

Arguments Against Same-Sex Marriage

Whereas advocates of same-sex marriage argue that they will not be regarded as legitimate families by the larger society so long as same-sex couples cannot be legally married, opponents do not want to legitimize same-sex couples and families. Opponents of same-sex marriage who view homosexuality as unnatural, sick, and/or immoral do not want their children to view homosexuality as socially acceptable.

Opponents of gay marriage also suggest that gay marriage leads to declining marriage rates, increased divorce rates, and increased nonmarital births. However, data in Scandinavia reflect that these trends were occurring 10 years before Scandinavia adopted registered same-sex partnership laws, liberalized alternatives to marriage (such as cohabitation), and expanded exit options (such as no-fault divorce; Pinello, 2008).

Opponents of same-sex marriage commonly argue that such marriages would subvert the stability and integrity of the heterosexual family. However, Sullivan (1997) suggests that homosexuals are already part of heterosexual families:

[Homosexuals] are sons and daughters, brothers and sisters, even mothers and fathers, of heterosexuals. The distinction between "families" and "homosexuals" is, to begin with, empirically false;

and the stability of existing families is closely linked to how homosexuals are treated within them. (p. 147)

Many opponents of same-sex marriage base their opposition on their religious views. In a Pew Research Center national poll, the majority of Catholics and Protestants opposed legalizing same-sex marriage, whereas the majority of secular respondents favored it (Green, 2004). Churches have the right to deny marriage for gay people in their congregations. Legal marriage is a contract between the spouses and the state; it is a civil option that does not require religious sanctioning.

In previous years, opponents of gay marriage have pointed to public opinion polls which showed that the majority of Americans are against same-sex marriage. Although the majority does oppose same-sex marriage, the opposition is decreasing. Recall the 2011 Pew Research Center national poll which found that 53% of U.S. adults oppose legalizing gay marriage, down from 63% in 2004.

8-6 LGBT Parenting Issues

According to the 2010 U.S. census, there are about 600,000 same-sex couples in the United States. A third of lesbians and a fifth of gay men are parents. Over 10 million children are being reared by gay parents. This is a low estimate of children who have gay or lesbian parents, as it does not count children of same-sex couples who did not identify their relationship in the census, those headed by gay or lesbian single parents, or those whose gay parent does not have physical custody but is still actively involved in the child's life.

Many gay and lesbian individuals and couples have children from prior heterosexual relationships or marriages. Some of these individuals married as a "cover" for their homosexuality; others discovered their interest in same-sex relationships after they married. Children with mixed-orientation parents may be raised by a gay or lesbian parent, a gay or lesbian stepparent, a heterosexual parent, and a heterosexual stepparent.

A gay or lesbian individual or couple may have children through the use of assisted reproductive technology, including donor insemination, in vitro fertilization, and surrogate mothers. Others adopt or become foster parents.

Less commonly, some gay fathers are part of an emergent family form known as the hetero-gay family. In a hetero-gay family, a heterosexual mother and a gay father conceive and raise a child together but reside separately.

Antigay views concerning gay parenting include the beliefs that homosexual individuals are unfit to be parents and that children of lesbians and gays will not develop normally and/or will become homosexual. In a sample of 2,922 undergraduates, 21% agreed that "children of gay parents are disadvantaged" (Knox & Hall, 2010). As the following section notes, research findings reveal a positive outcome for the development and well-being of children with gay or lesbian parents.

8-6a Development and Well-Being of Children With Gay or Lesbian Parents

A growing body of research on gay and lesbian parenting supports the conclusion that children of gay and lesbian parents are just as likely to flourish as are children of heterosexual parents. Biblarz and Stacey (2010) compared children from two-parent families with same or different sex co-parents and single-mother with single-father families and found that the strengths typically associated with married mother-father families appeared to the same extent in families with two mothers and potentially in those with two fathers. Hence, children seem to benefit when there are two parents in the household (rather than a single mom or dad), but that the gender of these parents is irrelevant—the structure can be a woman and a man, two women, or two men. Crowl et al. (2008) also reviewed 19 studies on the developmental outcomes and quality of parent-child relationships among children raised by gay and lesbian parents. The researchers confirmed previous studies that children raised by same-sex parents fare equally well when compared with children reared by heterosexual parents. Regardless of family type, adolescents were more likely to show positive adjustment when they perceived more caring from adults and when parents described having close relationships with them. Thus, the qualities of adolescent-parent relationships rather than the sexual orientation of the parents were significantly associated with adolescent adjustment.

Bos and Gartrell (2010) assessed the presence of homophobic stigmatization on the well-being of 39 female and 39 male 17-year-old adolescents whose mothers were lesbian and who were conceived through donor insemination. Forty-one percent reported experiencing stigmatization based on homophobia, and the greater the stigmatization was, the more problem behavior was evident in these adolescents. However, close, positive relationships with their lesbian mothers muted the effect of stigmatization and allowed the adolescents to be resilient. Gay parents, particularly gay male parents, may be particularly sensitive to potential stigmatization and seek gay friendly neighborhoods to rear their children (Goldberg et al. 2012).

In addition, the American Psychological Association (2004) noted that "results of research suggest that lesbian and gay parents are as likely as heterosexual parents to provide supportive and healthy environments for their children" and that "the adjustment, development, and psychological well-being of children [are] unrelated to parental sexual orientation and that the children of lesbian and gay parents are as likely as those of heterosexual parents to flourish." Indeed, Pro-Family Pediatricians cheered when a proposal to ban gay marriage was defeated.

8-6b Discrimination in Child Custody, Visitation, Adoption, and Foster Care

A former student reported that, after she divorced her husband, she became involved in a lesbian relationship. She explained that she would like to be open about her relationship to her family and friends, but she was afraid that if her ex-husband found out that she was in a lesbian relationship, he might take her to court and try to get custody of their children. Although several respected national organizations—including the American Academy of Pediatrics, the Child Welfare League of America, the American Bar Association, the American Medical Association, the American Psychological Association, the American Psychiatric Association, and the National Association of Social Workers—fully support treating gays and lesbians without prejudice in parenting and adoption decisions (Howard, 2006; Landis, 1999), lesbian and gay parents are often discriminated against in child custody, visitation, adoption, and foster care.

Some court judges are biased against lesbian and gay parents in custody and visitation disputes. For example, in 1999, the Mississippi Supreme Court denied custody of a teenage boy by his gay father and instead awarded custody to his heterosexual mother, who remarried into a home "replete with domestic violence and excessive drinking" (Custody and Visitation, 2000, p. 1).

Gay and lesbian individuals and couples who want to adopt children can do so through adoption agencies or through the foster care system in over 20 states and the District of Columbia. For 33 years, Florida forbade adoption by gay and

lesbian people, but the state has reversed its legal position, saying that "the evidence of the suitability of gay parents is extensive." Gay Florida residents may now both adopt children and serve as foster parents. In a national study of first-year freshmen in colleges and universities throughout the United States, three fourths were in favor of gays and lesbians adopting children (Pryor et al., 2011).

Most adoptions by gay people are second-parent adoptions. A **second-parent adoption** (also called coparent adoption) is a legal procedure that allows individuals to adopt their partner's biological or adoptive child without terminating the first parent's legal status as parent. Second-parent adoption gives children in same-sex families the security of having two legal parents. Second-parent adoption potentially benefits a child by:

- Placing legal responsibility on the parent to support the child

- Allowing the child to live with the legal parent in the event that the biological (or original adoptive) parent dies or becomes incapacitated

- Enabling the child to inherit and receive Social Security benefits from the legal parent

- Enabling the child to receive health insurance benefits from the parent's employer

- Giving the legal parent standing to petition for custody or visitation in the event that the parents break up (Clunis & Green, 2003)

Second-parent adoption is not possible when a parent in a same-sex relationship has a child from a previous heterosexual marriage or relationship, unless the former spouse or partner is willing to give up parental rights. Although "third-parent" adoptions have been granted in a small number of jurisdictions, this option is not widely available.

T-shirts for Tolerance

As a junior and senior at Homewood-Flossmoor High School in the suburbs of Chicago, Myka Held played a key role in leading a campaign to promote tolerance of gay and lesbian students. The campaign involved selling T-shirts to students and teachers for them to wear to school on the designated day. The T-shirts say, "gay? fine by me." The goal of wearing the T-shirt is to inform gay people that there are straight people who support them. "I have always supported equal rights for every person and have been disgusted by discrimination and prejudice. As a young Jewish woman, I believe it is my duty to stand up and support minority groups. . . . In my mind, fighting for gay rights is a proxy for fighting for every person's rights" (Held, 2005). Myka Held's T-shirt campaign illustrates that fighting prejudice and discrimination against sexual-orientation minorities is an issue not just for lesbians, gays, and bisexuals but also for all those who value fairness and respect for human beings in all their diversity.

© Jani Bryson/iStockphoto.com

8-7 Effects of Antigay Bias and Discrimination on Heterosexuals

The antigay and heterosexist social climate of our society is often viewed in terms of how it victimizes the gay population. However, heterosexuals are also victimized by heterosexism and antigay prejudice and discrimination. Some of these effects follow:

1. **Heterosexual victims of hate crimes.** Extreme homophobia contributes to instances of violence against homosexuals—acts known as hate crimes. Such crimes include verbal harassment (the most frequent form of hate crime experienced by victims), vandalism, sexual assault and rape, physical assault,

second-parent adoption (also called *coparent adoption*) a legal procedure that allows individuals to adopt their partner's biological or adoptive child without terminating the first parent's legal status as parent.

> "I am a heterosexual. I don't know why
> I'm like this. I was born this way."
>
> —ROSANNE BARR, COMEDIAN

and murder. "Homosexuals are far more likely than any other minority group in the United States to be victimized by violent hate crime" (Potok, 2010, p. 29). Because hate crimes are crimes of perception, victims may not be homosexual; they may just be perceived as being homosexual. The National Coalition of Anti-Violence Programs (2011) reported that, in 2010, heterosexual individuals in the United States were victims of antigay hate crimes, representing 10% of all antigay hate crime victims.

2. **Concern, fear, and grief over well-being of gay or lesbian family members and friends.** Many heterosexual family members and friends of homosexual people experience concern, fear, and grief over the mistreatment of their gay or lesbian friends and/or family members. For example, heterosexual parents who have a gay or lesbian teenager often worry about how the harassment, ridicule, rejection, and violence experienced at school might affect their gay or lesbian child. Will their child drop out of school to escape the harassment, violence, and alienation he or she experiences? Or, will their gay or lesbian child respond to the antigay victimization by turning to drugs or alcohol or by committing suicide? Such fears are not unfounded: Lesbian, gay, and bisexual youth who report high levels of victimization at school also have higher levels of substance use and suicidal thoughts or attempts than heterosexual peers who report high levels of at-school victimization.

Meyer and colleagues (2008) studied the lifetime prevalence of mental disorders and suicide attempts among a diverse group of lesbian, gay, and bisexual individuals and found higher rates of substance abuse among bisexual people than among lesbians and gay men. Also, Latino respondents attempted suicide more often than Caucasian respondents.

Heterosexual individuals also worry about the ways in which their gay, lesbian, and bisexual family members and friends could be discriminated against in the workplace.

To heterosexuals who have lesbian and gay family members and friends, lack of family protections such as health insurance and rights of survivorship for same-sex couples can also be cause for concern. Finally, heterosexuals live with the painful awareness that their gay or lesbian family member or friend is a potential victim of antigay hate crime. Imagine the lifelong grief experienced by heterosexual family members and friends of hate crime murder victims, such as Matthew Shepard, a 21-year-old college student who was brutally beaten to death for no apparent reason other than he was gay. Jamie Nabozny was a gay high school student in Ashland, Wisconsin, who was subjected to relentless harassment and abuse over a four-year period. He attempted suicide and dropped out of school. With his parents' help he sued the school, lost, and appealed in federal court. The school administration was held liable for his mistreatment, and the case was settled for close to a million dollars (see the documentary *Bullied*).

3. **Restriction of intimacy and self-expression.** Because of the antigay social climate, heterosexual individuals, especially males, are hindered in their own self-expression and intimacy in same-sex relationships. Males must be careful about how they hug each other so as not to appear gay. Homophobic scripts also frighten youth who do not conform to gender role expectations, leading some youth to avoid activities—such as arts for boys, athletics for girls—and professions such as elementary education for males. A male student in the author's class revealed that he always wanted to work with young children and had majored in early childhood education. His peers teased him relentlessly about his choice of majors, questioning both his masculinity and his heterosexuality. Eventually, this student changed his major to psychology,

© lenetstar/Shutterstock

which his peers viewed as an acceptable major for a heterosexual male.

4. **Dysfunctional sexual behavior.** Some cases of rape and sexual assault are related to homophobia and compulsory heterosexuality. For example, college men who participate in gang rape, also known as "pulling train," entice each other into the act "by implying that those who do not participate are unmanly or homosexual" (Sanday, 1995, p. 399). Homonegativity also encourages early sexual activity among adolescent men. Adolescent male virgins are often teased by their male peers, who say things like "You mean you don't do it with girls yet? What are you, a fag or something?" Not wanting to be labeled and stigmatized as a "fag," some adolescent boys "prove" their heterosexuality by having sex with girls before they are ready.

5. **School shootings.** Antigay harassment has also been a factor in many of the school shootings in recent years. For example, 15-year-old Charles Andrew Williams fired more than 30 rounds in a San Diego, California, suburban high school, killing two and injuring 13 others. A woman who knew Williams reported that the students had teased him and called him gay.

6. **Loss of rights for individuals in unmarried relationships.** The passage of state constitutional amendments that prohibit same-sex marriage can also result in denial of rights and protections to opposite-sex unmarried couples. For example, Judge Stuart Friedman of Cuyahoga County (Ohio) agreed that a man who was charged with assaulting his girlfriend could not be charged with a domestic violence felony because the Ohio state constitutional amendment granted no such protections to unmarried couples. Some antigay marriage measures also threaten the provision of domestic partnership benefits to unmarried heterosexual couples.

---

2 percent of individuals in the population reporting that they are gay/bisexual >

20 years < maximum jail sentence for sodomy prior to the intervention of the U.S. Supreme Court in 2003

percent of gay teens who attempt suicide > **30%**

1978 < year Harvey Milk was killed

same-sex couples married in Massachusetts the year after it became legal in the state > **>5,000**

number of employers that provided domestic partner health benefits in 2004 > **8,250**

 **8**

Ready to study? In this book, you can:

- ⤷ Rip out the Chapter Review card in the back of the book to study for exams
- ⤷ Take the Self Assessment for this chapter (card in the back of the book) and see where you stand on the vital issues raised in the chapter

Or you can go online to CourseMate at www.cengagebrain.com for these resources:

- ⤷ Complete Practice Quizzes to prepare for tests
- ⤷ Review Key Terms Flash Cards (online or print)
- ⤷ Read about Marriage and Family in the news
- ⤷ Play "Beat the Clock" to master concepts
- ⤷ Check out Personal Applications

Work, Marriage, and Family

"We **worry too much** about something to **live on**—and **too little** about something to **live for.**"

—JIMMY CARTER, FORMER PRESIDENT

SECTIONS

9-1 Effects of Employment on Spouses

9-2 Effects of Employment on Children

9-3 Balancing Work and Family

9-4 Debt

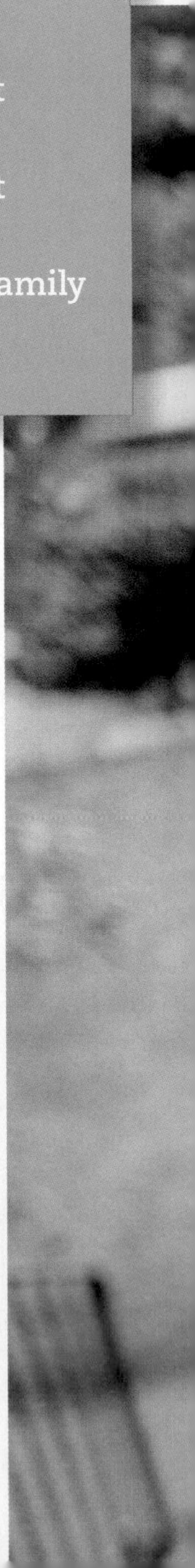

In 2011, Richard James Verone, 59, held up a bank in Gastonia, North Carolina, for one dollar. His goal was to be sent to prison where he would receive free medical care. Verone was unemployed and destitute, and said he didn't know how else to cover the costs of his medical care. His act of desperation illustrates how not being employed affected his well-being. In this chapter we examine how work and money interact with individuals, marriages, and families.

9-1 Effects of Employment on Spouses

We begin with money as power. Gary Rawson (2012), a husband, revealing the truth of money and marriage said: "I make all the big decisions. My wife makes all the small decisions. In 36 years of marriage, I have never made a big decision."

9-1a Money as Power in a Couple's Relationship

Money is associated with power, control, and dominance. Generally, the more money a partner makes, the more power that person has in the relationship (even though spouses typically pool their money; Sonnenberg et al., 2011). Males make considerably more money than females and generally have more power in relationships. The average annual income

© Ryan McVay/Jupiter Images

> "It isn't so much **hard times** are **coming**; the change observed is mostly **soft times going.**"
>
> —GROUCHO MARX, COMEDIAN

of a male with some college education who is working full-time is $52,580, compared with $36,533 for a female with the same education, also working full-time (*Statistical Abstract of the United States, 2012,* Table 702). However, wives typically make the decisions in more areas of the relationship (42% versus 30%; Morin & Cohn, 2008).

In one-fourth of marriages, wives earn higher incomes than their husbands. When a wife earns an income, her power increases in the relationship (Kulik, 2011). In the author's neighborhood, there is a married couple in which the wife recently began to earn an income. Before doing so, her husband's fishing boat was in the protected carport. With her new job and increased power in the relationship, she began to park her car in the carport, and her husband

> ## "If you want to **feel rich,** just **count** the things you **have** that **money can't buy.**"
>
> —CHINESE PROVERB

put his fishing boat underneath the pine trees to the side of the house. Money also provides an employed woman the power to be independent and to leave an unhappy marriage. Indeed, the higher a wife's income, the more likely she is to leave an unhappy relationship (Schoen et al., 2002). Her own income won't induce her to move toward a divorce, but if she is unhappy and has the economic means to do so, she will.

To some individuals, money also means love. While admiring the engagement ring of her friend, a woman said, "What a big diamond! He must really love you." A cultural assumption is that a big diamond equals an expensive diamond and a lot of sacrifice and love. Similar assumptions are often made when gifts are given or received. People tend to spend more money on presents for the people they love, believing that the value of the gift symbolizes the depth of their love. People receiving gifts may make the same assumption. "She must love me more than I thought," mused one man. "I gave her an e-book for Christmas but she gave me an iPad 3. I felt embarrassed."

9-1b Working Wives

Seventy percent of all U.S. wives are in the labor force, most likely when their children are teenagers (wives are ages 35 to 44; *Statistical Abstract of the United States,* 2012, Table 597). The stereotypical family consisting of a husband (who earns the income by working outside the home) and a wife who stays at home (with two or more children) is no longer the norm. Only 13% are "traditional" in the sense of having a breadwinning husband, a stay-at-home wife, and their children (Stone, 2007). Over two thirds of marriages involve two earners. Wives in these marriages who perceive the division of labor to be less equal report lower marital satisfaction (Nazarinia Roy & Britt, 2011).

dual-earner marriage both husband and wife work outside the home to provide economic support for the family.

Because women still take on more child care and household responsibilities than men, women in **dual-earner marriages** (marriages in which both spouses provide significant income to the family unit) are more likely than men to want to be employed part-time rather than full-time. If this option is not possible, many women prefer to work only a portion of the year (the teaching profession, for example, allows employees to work about 10 months and to have 2 months in the summer free). Although many low-wage earners need two incomes to afford basic housing and a minimal standard of living, others have two incomes to afford expensive homes, cars, vacations, and educational opportunities for their children. Whether it makes economic sense is another issue.

Some parents wonder if the money a wife earns by working outside the home is worth the sacrifices to earn it. Not only is the mother away from their children but she must also pay for strangers to care for their children. Sefton (1998) calculated that the value of a stay-at-home mother is $36,000 per year in terms of what a dual-income family spends to pay for all the services that she provides (domestic cleaning, laundry, meal planning and preparation, shopping, providing transportation to activities, taking the children to the doctor, and running errands). Adjusting for changes in the consumer price index, this figure was around $50,000 in 2012. The value of a househusband would be the same. However, because males have higher incomes than females, the loss of income would be greater than for a female.

This $50,000 figure suggests that working outside the home may not be as economically advantageous as one might think—that women may work outside the home for psychological (enhanced self-concept) and social (enlarged social network) benefits. Some may also enjoy a lifestyle that is made possible by earning a significant income from outside employment. One wife noted that the only way she could afford the home she wanted was to help earn the money to pay for it.

© HomeStudio/Shutterstock

The **mommy track** (stopping paid employment to spend time with young children) is another potential cost to those women who want to build a career. Taking time out to rear children in their formative years can derail a career. Noonan and Corcoran (2004) found that female lawyers who took time out for child responsibilities were less likely to make partner and more likely to earn less money if they did make partner. Aware that executive women have found it difficult to reenter the workforce after being on the mommy track, the Harvard Business School created an executive training program for mothers to improve their technical skills and to help them return to the workforce (Rosen, 2006).

9-1c Wives Who "Opt Out"

In *Opting Out?* (Stone, 2007), the author revealed the perspectives of 54 women representing a broad spectrum of professions—doctors, lawyers, scientists, bankers, management consultants, editors, and teachers. **Opting out** involves women leaving their careers and returning home to take care of their children for a variety of reasons. However, two reasons stood out in Stone's survey: (1) husbands who were unavailable or unable to "shoulder significant portions of caregiving and family responsibilities" (p. 68), and (2) employers who had a lot of policies on the books to encourage and support women's parental leave "but not much in the way of making it possible for them to return or stay once they had babies" (p. 119). Part-time employment did not work out for these women because the work was not really "part-time"—the employer kept wanting more hours.

There is disagreement about the wisdom of "opting out." Bennetts (2007) argued that educated women with careers are foolish to leave their lucrative careers and put on their aprons. She emphasized that men cannot be counted on—that they leave the marriage for younger women or die. But Treas et al. (2011) found that homemakers (in 28 countries) are happier than full-time working wives.

© iStockphoto.com/Jason Stitt

9-1d Types of Dual-Career Marriages

A **dual-career marriage** is defined as one in which both spouses pursue careers and maintain a life together that may or may not include dependents. A career is different from a job in that the former usually involves advanced education or training, full-time commitment, night and weekend work "off the clock," and a willingness to relocate. Dual-career couples operate without a person (i.e., a "wife") who stays home to manage the home and care for dependents.

Nevertheless, four types of dual-career marriages are those in which the husband's career takes precedence (HIS/her), the wife's career takes precedence (HER/his), both careers are equal (HIS/HER), or both spouses share a career or work together (THEIR career).

When couples hold traditional gender role attitudes, the husband's career is likely to take precedence (**HIS/her career**). This situation translates into the wife being willing to relocate and to disrupt her career for the advancement of her husband's career. Davis et al. (2012) examined a national sample of married couples during the "great recession" and found that women were more comfortable moving

mommy track stopping paid employment to spend time with young children.

opting out professional women leaving their careers and returning home to care for their children.

dual-career marriage a marriage in which both spouses pursue careers and maintain a life together that may or may not include dependents.

HIS/her career a husband's career is given precedence over a wife's career.

in support of their husbands' careers than men would be for their wives.

Wives who move in reference to their husbands' careers may also have children early, which has an effect on the development of her career. Gordon and Whelan-Berry (2005) interviewed 36 professional women and found that, in 22% of the marriages, the husband's career took precedence. The primary reasons for this arrangement were that the husband earned a higher salary, going where the husband could earn the highest income was easier because the wife could more easily find a job wherever he went (than vice versa), and ego needs (the husband needed to have the dominant career). This arrangement is sometimes at the expense of the wife's career. In a study of educated Papua New Guinean women, Spark (2011) found them reluctant to marry Papua New Guinean men. Indeed, these women avoided marriage because they feared such relationships would destroy their career prospects.

For couples who do not have traditional gender role attitudes, the wife's career may take precedence (**HER/his career**). Of the marriages in the Gordon and Whelan-Berry (2005) study mentioned previously, 19% could be categorized as giving precedence to a wife's career. In such marriages, the husband is willing to relocate and to disrupt his career for his wife's. Such a pattern is also likely to occur when a wife earns considerably more money than her husband. In some cases, the husband who is downsized or who prefers the role of full-time parent becomes "Mr. Mom." A major advantage of this role is the closer relationship/bonding of the father with his children.

When the careers of both the wife and husband are given equal status in the relationship (**HIS/HER career**), they may have a commuter marriage in which they follow their respective careers wherever they lead. Alternatively, Deutsch et al. (2007) surveyed 236 undergraduate senior women to assess their views on husbands, managing children, and work. Most envisioned two egalitarian scenarios in which both spouses would cut back on their careers and/or both would arrange their schedules to devote time to child care. Still other couples may hire domestic child care help so that neither spouse functions in the role of housekeeper. In reality, equal status in HIS/HER careers is not a dominant theme (Stone, 2007).

Finally, some couples have the same career and may work together (**THEIR career**). Some news organizations hire both spouses to travel abroad to cover the same story. These careers are rare. In the following sections, we look at the effects on women, men, their marriage, and their children when a wife is employed outside the home.

9-2 Effects of Employment on Children

Independent of the effect on the wife, husband, and marriage, what is the effect of the wife earning an income outside the home on the children? Individuals disagree on the effects of maternal employment on children.

9-2a Quality Time

Dual-income parents struggle with having "quality time" with their children. The term *quality time* has become synonymous with good parenting. In a study of 220 parents from 110 dual-parent families, the researchers found that "quality time" is defined in different ways (Snyder, 2007). Some parents (structured-planning parents) saw quality time as planning and executing family activities. Mormons set aside "Monday home evenings" as a time to bond, pray, and sing together (thus, quality time). Other parents (child-centered parents) noted that "quality time" occurred when they were having heart-to-heart talks with their children. Still other parents believed that all the time they were with their children was quality time. Glorieux et al. (2011) analyzed time use of couples in a Belgian sample and found that work was the most decisive factor influencing the amount of together time (but work by women was less a threat to couple time).

HER/his career
a wife's career is given precedence over a husband's career.

HIS/HER career
a husband's and wife's careers are given equal precedence.

THEIR career
a career shared by a couple who travel and work together (e.g., journalists).

9-2b Day-Care Considerations

Parents going into or returning to the workforce are intent on finding high-quality day care. Rose and Elicker (2008) surveyed the various characteristics of day care that are important to 355 employed mothers of children under 6 years of age and found that warmth of caregivers, a play-based curriculum, and educational level of caregivers emerge as the first-, second-, and third-most important factors in selecting a day-care center.

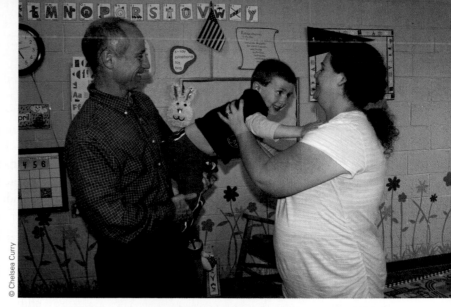

© Chelsea Curry

Quality of Day Care Employed parents are concerned that their children get good-quality care. Their concern is warranted. Cortisol is a steroid hormone that plays critical roles in adaptation to stress. Higher levels of cortisol reflect higher levels of stress experienced by the individual. Gunnar et al. (2010) assessed the cortisol levels of 151 children in full-time, home-based day-care centers and compared these levels to those of children the same age who were at home. Increases were noted in the majority of children (63%) at day care, with 40% classified as a stress response. These increased cortisol levels began in the morning and continued throughout the afternoon.

Vandell et al. (2010) examined the effects of early child care 15 years later and found that higher quality care predicted higher cognitive–academic achievement at age 15, with escalating positive effects at higher levels of quality. High-quality child care also has a positive effect on mothers, who report fewer symptoms of depression (Gordon et al., 2011).

Warash et al. (2005) reviewed the literature on day-care centers and found that the average quality of such centers is mediocre—"unsafe, unsanitary, non-educational, and inadequate in regard to the teacher-child ratio for a classroom." Care for infants was particularly lacking. Of 225 infant or toddler rooms observed, 40% were rated "less than minimal" with regard to hygiene and safety. Because of the low pay and stress of the occupation, the rate of turnover for family child-care providers is very high (estimated at between 33 and 50%; Walker, 2000). However, De Schipper et al. (2008) noted that day-care workers who engage in high-frequency, positive behavior engender secure attachments with the children they work with. Hence, children of such workers don't feel they are on an assembly line but bond with their caretakers.

Parents concerned about the quality of day care their children receive might inquire about the availability of webcams. Some day-care centers offer full-time webcam access so that parents or grandparents can log onto their computers and see the interaction of the day-care worker with their children.

Vincent and Neis (2011) studied the impact of parental work schedules on child academic achievement and reiterated the African proverb that states, "It takes a village to raise a child." Hence, work schedules of parents are sometimes such that an enormous support system needs to be in place. Sun and Sundell (2011) also noted the impact of parental work schedules on children in day care and the effect on increased rates of respiratory tract infections. Ahnert and Lamb (2003) emphasized that attentive, sensitive, loving parents can mitigate any potential negative outcomes in day care. "Home remains the center of children's lives even when children spend considerable amounts of time in child care. . . . Although it might be desirable to limit the amount of time children spend in child care, it is much more important for children to spend as much time as possible with supportive parents" (pp. 1047–1048).

Cost of Day Care Day-care costs are a factor in whether a low-income mother seeks employment, because the cost can absorb her paycheck. Even for dual-earner families, cost is a factor in choosing a day-care center. Day-care costs vary widely—from nothing, where friends trade off taking care of the children, to very expensive institutionalized day care in large cities. A dual-earner metropolitan couple (who use day care for their two children and are planning for them to enter school) identified the costs of institutional day care in the Baltimore area. The father's eye-opening response is on page 175.

9-3 Balancing Work and Family

Work is definitely stressful on individuals, spouses, and relationships. Not only is working longer hours (50 or more hours a week) associated with increased alcohol use for both the male and female worker (Gibb et al., 2012), spouses report feeling greater emotional distance (Lavee & Ben-Ari, 2007). Reducing one's total hours of work may benefit both the spouses and their relationship.

One of the major concerns of employed parents and spouses is how to juggle the demands of work and family. Many family-friendly policies are not family-friendly (Schlehofer, 2012). Women are more likely to experience work as interfering with the family and more likely to resolve the family–work conflict by giving precedence to family (Cinamon, 2006). Kiecolt (2003) examined national data and concluded that employed women with young children are "more likely to find home a haven, rather than finding work a haven" (p. 33). Recall the research by Stone (2007) in regard to women opting out of high-income or high-status work in favor of taking care of their children at home. Nevertheless, the conflict between work and family is substantial, and various strategies are employed to cope with the stress of role overload and role conflict, including (1) the superperson strategy, (2) cognitive restructuring, (3) delegation of responsibility, (4) planning and time management, and (5) role compartmentalization (Stanfield, 1998).

9-3a Superperson Strategy

The **superperson strategy** involves working as hard and as efficiently as possible to meet the demands of work and family. The person who uses the superperson strategy often skips lunch and cuts back on sleep and leisure to have more time available for work. Women are particularly vulnerable because they feel that if they give too much attention to child-care concerns, they will be sidelined into lower paying jobs with no opportunities.

Hochschild (1989) noted that the terms **superwoman** or **supermom** are cultural labels that allow a woman to regard herself as very efficient, bright, and confident. However, Hochschild noted that these labels are a "cultural cover-up" for an overworked and/or frustrated woman. Not only does the woman have a job in the workplace (first shift), she comes home to another set of work demands in the form of house care and child care (second shift). Finally, she has a "third shift" (Hochschild, 1997).

The **third shift** is the expenditure of emotional energy by a spouse or parent in dealing with various issues in family living. Although young children need time and attention, responding to conflicts and problems with teenagers (e.g., dealing with a boyfriend breakup) involves a great deal of emotional energy—the third shift. Opree and Kalmijn (2012) also noted that employed women who take care of aging parents are more likely to report negative changes in their own mental health.

9-3b Cognitive Restructuring

Another strategy used by some women and men experiencing role overload and role conflict is **cognitive restructuring**, which involves viewing a situation in positive terms. Exhausted dual-career earners often justify their time away from their children by focusing on the benefits of their labor—their children live in a nice house in a safe neighborhood and attend the best schools. Whether these outcomes offset the lack of "quality time" may be irrelevant—the beliefs serve simply to justify the two-earner lifestyle.

9-3c Delegation of Responsibility and Limiting Commitments

A third way couples manage the demands of work and family is to delegate responsibility to others for performing certain tasks. Because women tend to bear most of the responsibility for child care (and perform two thirds of the housework; Carriero, 2011), they may ask their partner to contribute more or to take responsibility for these tasks. Although some husbands are involved, cooperative, and contributing, Stone (2007)

---

superperson strategy involves working as hard and as efficiently as possible to meet the demands of work and family

supermom (superwoman) a cultural label that allows a mother who is experiencing role overload to regard herself as particularly efficient, energetic, and confident.

third shift the emotional energy expended by a spouse or parent in dealing with various family issues.

cognitive restructuring viewing a situation in positive terms.

Day Care Costs Are Only the Beginning

There is a sliding scale of costs based on quality of provider, age of child, full-time or part-time. Full-time infant care can be hard to find in this area. Many people put their names on a waiting list as soon as they know they are pregnant. We had our first child on a waiting list for about nine months before we got him into our first choice of providers.

Infant care runs $1,250 per month for high-quality day care. That works out to $15,000 per year—and you thought college was expensive.

The cost goes down when the child turns 2—to around $900–$1,000 per month. In Maryland, this is because the required ratio of teachers to children increases at age 2. This cost break doesn't last long. Preschool programs (more academic in structure than day care) begin at age 3 or 4. When our second child turned 4, he started an academic preschool in September. The school year lasts nine and a half months, and it costs about $14,000. Summer camps or summer day-care costs must be added to this amount to get the true annual cost for the child. My wife and I have budgeted about $30,000 total for our two children to attend private school and summer camps this year.

The news only gets worse when the children get older. A good private school for grades K–5 runs $15,000 per school year. Junior high (grades 6–8) is about $20,000, and private high school here goes for $20,000 to $35,000 per nine-month school year.

Religious private schools run about half the costs above. However, they require that you be a member in good standing in their congregation and, of course, your child undergoes religious indoctrination.

There is always the argument of attending a good public school and thus not having to pay for private school. However, we have found that home prices in the "good school neighborhoods" were out of our price range. Some of the better-performing public schools also now have waiting lists—even if you move into that school's district.

In most urban and suburban areas, cost is secondary to admissions. Getting into any good private school is difficult. In order to get into the good high schools, it's best to be in one of the private elementary or middle schools that serves as a "feeder" school. Of course, to get into the "right" elementary school, you must be in a good feeder kindergarten. And of course to get into the right kindergarten, you have to get into the right feeder preschool. Parents have *a lot* of anxiety about getting into the right preschool, because this can put your child on the path to one of the better private high schools.

Getting into a good private preschool is not just about paying your money and filling out applications. Yes, there are entrance exams. Both of our children underwent the following process to get into preschool: First you must fill out an application. Second, your child's day-care records/transcripts are forwarded for review. If your child gets through this screening, you and your child are called in for a visit. This visit with the child lasts a few hours and your child goes through evaluation for physical, emotional, and academic development. Then you wait several agonizing months to see if your child has been accepted.

Though quality day care is expensive, parents delight in the satisfaction that they are doing what they feel is best for their child.

revealed that 53 respondents in her study opted out of their careers to return home because they lacked help from husbands who either were not at home or did not do much when they were home.

Another form of delegating responsibility involves the decision to reduce one's current responsibilities and not take on additional ones. For example, women and men may give up or limit volunteer responsibilities or commitments. One woman noted that her life was being consumed by the responsibilities of her church; she had to change churches because the demands were relentless. In the realm of paid work, women and men can choose not to become involved in professional activities beyond those that are required.

9-3d Time Management

The use of time management is another strategy for minimizing the conflicting demands of work and family. This method involves prioritizing and making lists of what needs to be done each day. Time management also involves anticipating stressful periods, planning ahead for them, and dividing responsibilities with the spouse. Such division of labor allows each spouse to focus on an activity that needs to be done (grocery shopping, picking up children at day care) and results in a smoothly functioning unit.

Having flexible jobs, supervisor support, and coworker support (Perry-Jenkins et al., 2011) is particularly beneficial for two-earner couples. Being self-employed, telecommuting, or working in academia allows the parents to cooperate on what needs to be done. Alternatively, some dual-earner couples attempt to solve the problem of child care by **shift work**, or having one parent work during the day and the other parent work at night so that one parent can always be with the children. Shift workers often experience sleep deprivation and fatigue, which may make fulfilling domestic roles as a parent or spouse difficult. Similarly, shift work may have a negative effect on a couple's relationship because of their limited time together.

Presser (2000) studied the work schedules of 3,476 married couples and found that recent husbands (married fewer than five years) who had children and who worked at night were six times more likely to divorce than husbands or parents who worked days.

9-3e Role Compartmentalization

Some spouses use **role compartmentalization**—separating the roles of work and home so that they do not think about or dwell on the problems of one when they are at the physical place of the other. Spouses unable to compartmentalize their work and home feel role strain, role conflict, and role overload, with the result that their efficiency drops in both spheres. Since working mothers are more likely to prioritize the role of parent than working men, they are more likely to struggle with this issue.

9-4 Debt

Soaring gas prices, home foreclosures, job loss, inflation, and health care costs result in families being unable to pay their bills and accumulating more debt. Dew (2008) analyzed data on 1,078 couples and found that couples in debt spend less time together and argue more over money, both of which are associated with decreases in marital satisfaction. Separated, divorced, and never married individuals are most vulnerable to increasing debt (Caputo 2012). Parents today are caught in the "parent trap" where they must make choices such as saving for retirement or paying for their offspring's college education.

shift work having one parent work during the day and the other parent work at night so that one parent can always be with the children.

role compartmentalization separating the roles of work and home so that an individual does not dwell on the problems of one role while physically being at the place of the other role.

© Fuse/Jupiter Images

> "Money is the most egalitarian force in society. It confers power on whoever holds it."
>
> —ROGER STARR, COLUMNIST, EDITOR

9-4a Income Distribution and Poverty

A deep recession beginning in mid-2008 and a slow recovery (see Table 9.1) have been reflected in job lay-offs, housing foreclosures, and couples fearful for their economic future. The consequences have been catastrophic and include increased homicides in the family (e.g., murder by a spouse; Diem & Pizarro, 2010). But consequences have also been positive. Etzioni (2011) noted that the Great Recession has caused individuals to become less consumerist oriented and more rich in their social relations and transcendental activities (religious, contemplative). Etzioni asked if this "new normal" would be associated with greater acceptance—acknowledging that getting and keeping a job will be more difficult and that salary raises may not occur for several years in a row.

What is the definition of poverty in terms of actual dollars in the United States? Table 9.2 reflects the Department of Health and Human Services' various poverty level guidelines by size of family and location. A significant proportion of families in the United States continue to be characterized by unemployment and low wages.

Poverty has traditionally been defined as the lack of resources necessary for material well-being—most

Table 9.1
Distribution of Income in U.S. Families in 2009

Income of Family	Percentage of Families
Less than $15,000	8.7
$15,000–$24,999	9.1
$25,000–$34,999	10.0
$35,000–$49,999	13.8
$50,000–$74,999	19.4
$75,000–$99,000	15.5
$100,000 or more	25.6

Source: *Statistical Abstract of the United States* (2012, 131st ed.), Table 696 (Washington, DC: U.S. Census Bureau).

Table 9.2
2012 DHHS Poverty Guidelines

People in Family or Household	48 Contiguous States and DC	Alaska	Hawaii
1	$11,170	$13,970	$12,860
2	15,130	18,920	17,410
3	19,090	23,870	21,960
4	23,050	28,820	26,510
5	27,010	33,770	31,060
6	30,970	38,720	35,610
7	34,930	43,670	40,160
8	38,890	48,620	44,170
For each additional person, add	3,960	4,950	4,550

Source: 2012 Poverty Guidelines, 77 Fed. Reg. 4034–4035 (Jan. 26, 2012), http://aspe.hhs.gov/poverty/12fedreg.shtml.

importantly food and water, but also housing, land, and health care. This lack of resources that leads to hunger and physical deprivation is known as **absolute poverty**. In contrast, **relative poverty** refers to a deficiency in material and economic resources compared with some other group of people. Although many lower income Americans, for example, have resources and a level of material well-being that millions of people living in absolute poverty can only dream of (e.g., those in third world countries), they are relatively poor compared with the American middle and upper classes.

One of the factors driving Americans into poverty is the cost of medical care. Indeed, the primary cause of bankruptcy is the inability to pay hospital bills. Compounding the problem is the inability to afford health-care insurance.

How much it costs to live varies by country. In rural Mexico, the costs of food and housing are substantially below the costs in rural America. For example, lunch in the former may cost less than a dollar whereas lunch in the latter would more likely be three or four times as much. Similarly, due to the low value of the dollar in Europe, a Big Mac in America costs around $3.50 but is $5.13 in Paris (in 2012).

poverty the lack of resources necessary for material well-being.

absolute poverty the lack of resources that leads to hunger and physical deprivation.

relative poverty a deficiency in material and economic resources compared with some other population.

9-4b Effects of Poverty on Marriages and Families

Poverty is devastating to couples and families. Those living in poverty have poorer physical and mental health, report lower personal and marital satisfaction, and die sooner. The anxiety over lack of money may result in relationship conflict. Money is the most common problem that couples report. Money is such a profound issue in marriage because of its symbolic significance (e.g., power and control) and because individuals do something daily in reference to money—spend it, save it, or worry about it (Stanley & Einhorn, 2007). The potential for conflict is endless. Zimmerman (2012) found that couples who participated in a money management course (Dave Ramsey's Financial Peace) reported an increase in relationship quality. The researcher attributed this to increased communication about financial issues.

The stresses associated with low income also contribute to substance abuse, domestic violence, child abuse and neglect, divorce, and questionable parenting practices. For example, economic stress is associated with greater marital discord, and couples with incomes less than $25,000 are 30% more likely to divorce than couples with incomes greater than $50,000 (Whitehead & Popenoe, 2004). Emotional well-being is the frequency and intensity of joy, stress, sadness, anger, and affection that make one's life pleasant or unpleasant. Emotional well-being rises with income and tops out at $75,000 a year. Increases above this figure are not associated with increases in satisfaction (Kahneman & Deaton, 2010). Indeed, materialistic values, even when jointly shared between spouses, are associated with lower marital quality (Carroll et al., 2011).

Another family problem associated with poverty is teenage pregnancy. Although teen pregnancies in general are decreasing (Pazol et al., 2011), poor adolescent girls are more likely to have babies as teenagers or to become young single mothers. Early childbearing is associated with numerous problems, such as increased risk of having premature babies or babies of low birth weight, dropping out of school, and earning less money as a result of lack of academic achievement.

9-4c Global Inequality

Global economic inequality has reached unprecedented levels. The most comprehensive study on the world distribution of household wealth offers the following facts on wealth inequality worldwide:

- The richest 1% of adults in the world own 40% of global household wealth; the richest 2% of adults own more than half of global wealth; and the richest 10% of adults own 85% of total global wealth.

- The poorest half of the world's adult population owns barely 1% of global wealth.

- Households with assets of $2,200 per adult are in the top half of the world wealth distribution; assets of $61,000 per adult place a household in the top 10%, and assets of more than $500,000 per adult place a household in the richest 1% worldwide.

- Although North America has only 6% of the world's adult population, it accounts for one third (34%) of all household wealth worldwide. More than one third (37%) of the richest 1% of individuals in the world reside in the United States.

- The degree of wealth inequality in the world is as if one person in a group of 10 takes 99% of the total pie and the other nine people in the group share the remaining 1%. (Davies et al., 2006)

© iStockphoto.com/Mark Evans

Sheriff's Deputy Rick Ferguson supervises as an eviction team removes household items from a house foreclosed upon in Lafayette, Colorado. The owners had stopped making their mortgage payments, and the bank foreclosed on the house with a court order. Due to an unusually high unemployment rate, many Americans are falling farther behind in their rent and mortgage payments, resulting in more evictions and foreclosures.

9-4d Being Wise About Credit

One way to keep from slipping deeper into debt or poverty is to use credit wisely. The "free" credit cards that college students receive in the mail are Trojan horses and can plunge them into massive debt from which recovering will take years.

You use credit when you take an item home from the store today and pay for it later. The amount you pay later will depend on the arrangement you make with the seller. Suppose you want to buy a flat panel, 42-inch, high-definition television that sells for $2,000. Unless you pay cash, the seller will set up one of three types of credit accounts with you: installment, revolving charge, or open charge.

Under the **installment plan**, you sign a contract to pay for the item in monthly installments. You and the seller negotiate the period of time over which the payments will be spread and the amount you will pay each month. The seller adds a finance charge to the cash price of the television set and remains the legal owner of the set until you have made your last payment. Most department stores, appliance and furniture stores, and automobile dealers offer installment credit. The total cost of buying the $2,000 high-definition TV will actually be $2,175 as shown in Table 9.3.

Instead of buying the flat panel, $2,000 high-definition television set on the installment plan, you might want to buy it on the **revolving charge** plan. Most credit cards, such as Visa and MasterCard, represent revolving charge accounts that permit you to buy on credit up to a stated amount during each month. At the end of the month, you may pay the total amount you owe, any amount over the stated minimum payment due, or the minimum payment. If you choose to pay less than the full amount, the cost of the credit on the unpaid amount is approximately 1.5% per month, or 18% per year. Avoid these finance charges by paying off the credit card monthly.

You can also purchase items on an **open charge** (30-day) account. Under this system, you agree to pay in full within 30 days. Because this type of account has no direct service charge or interest, the television set would cost only the purchase price. For example, Sears and JC Penney offer open charge accounts. If you do not pay the full amount in 30 days, a finance charge is placed on the remaining balance.

Use of all credit is costly in terms of paying interest on the money. For a condo that costs $81,900, if the buyer pays 5.75% interest over a 30-year period, a total of $171.195.07 will be paid to the seller—this includes $82, 295.07 in interest. The point is to avoid paying interest by paying off the principle as soon as possible.

Table 9.3
Calculating the Cost of Installment Credit

Amount to be financed	
Cash price	$2,000
Amount to be paid	
Monthly payments	$75
X number of payments	x 29
Total amount to be repaid	$2,175

© Cengage Learning 2011

9-4e Credit Rating and Identity Theft

When you apply for a loan, the lender will seek a credit report from credit bureaus such as Equifax, TransUnion, and Experian. In effect, you will have a "credit score"—also referred to as a NextGen score or a FICO (Fair Isaac Credit Company) score—calculated as follows:

35 percent based on late payments, bankruptcies, judgments

installment plan repayment plan whereby you sign a contract to pay for an item with regular installments over an agreed-upon period.

revolving charge the repayment plan whereby you may pay the total amount you owe, any amount over the stated minimum payment due, or the minimum payment.

open charge repayment plan whereby you agree to pay the amount owed in full within the agreed amount of time.

30 percent based on current debts

15 percent based on how long accounts have been opened and established

10 percent based on type of credit (credit cards, loan for house or car)

10 percent based on applications for new credit or inquiries

As these percentages indicate, the way to improve your credit is to make payments on time and reduce your current debt. Your score will range from 620 to 850 (a score of 750 to 850 is excellent; a score of 660 to 749 is good; a score of 620 to 659 is fair; a score of 400 to 619 is poor). The higher your score, the lower your rate of interest. For example, on a $150,000 30-year, fixed-rate mortgage, a score above 760 would result in a 5.5% interest rate with a monthly payment of $852. In contrast, a score of 639 and below would result in an interest rate of 7.09%, with a monthly payment of $1,007 (see http://www.myfico.com/ or type in "credit report" on www.google.com). Taking all credit cards to the limit can lead to financial trouble including eventual bankruptcy.

identity theft one person using the Social Security number, address, and bank account numbers of another, posing as that person to make purchases.

When someone poses as you and uses your credit history to buy goods and services, that person has stolen your identity. More than 10 million Americans are victims of **identity theft**, which can destroy your credit, plunge you into debt, and keep you awake at night with lawsuits from creditors. Identity theft happens when someone gets access to personal information such as your Social Security number, credit card number, or bank account number and goes online to pose as you to buy items or services. Identity theft is the number one fraud complaint in the United States.

Safeguards include (1) never giving such information over the phone or online unless you initiate the contact, and (2) shredding bank and credit card statements and preapproved credit card offers. (A shredder can cost as little as $20.) Also, avoid paying your bills by putting an envelope in your mailbox with the flag up—use a locked box or the post office. Finally, check your credit reports, scrutinize your bank statements, and guard your personal identification number (PIN) at automatic teller machines (ATMs). If you use the Internet, protect your safety by installing firewall software.

9-4f Discussing Debt and Money: Do It Now

Sex, religion, and money are sensitive issues in polite society, and many couples may shy away from discussing such details (Shapiro, 2007). Indeed, once partners define themselves as being in a serious relationship, Shapiro (a marriage and family therapist) recommends that they talk about how they feel about running up debt on their credit card or incurring late charges, and how much they are bothered by debt. The issue becomes

Money can be a central issue in a relationship, so couples should talk about how they feel about money and debt.

relevant because the debt of one partner may become the debt of the other if the relationship continues. If John thinks nothing of charging a high-definition TV on his third credit card (because the other two have been maxed out), Mary might legitimately be concerned. The following are other issues that couples might discuss:

- At the time of engagement, what do the partners feel about a ring? Should one be bought? How much should be spent on it? And what is the symbolic meaning of the "size of the ring"—does a smaller, less expensive ring mean less love and less commitment?

- When the couple has their first child, will the wife stop working? What is the implication in terms of her access to money? Does she have to ask for money, or is "his" money deposited in an account so that both have access?

- As children get older, what are the feelings of the spouses in regard to paying for college versus obtaining student loans? What about sending money to parents who may need financial help with health-care bills (including a nursing home for one or both sets of parents)?

- As retirement comes, will the couple save their money or travel around the world?

At each stage of the family life cycle, financial decisions can cause very deep-seated feelings about money to surface. Couples should also be aware of the inequities in income between women and men in regard to levels of education. Although the slogan "the more you learn, the more you earn" is true, gender inequities continue. Table 9.4 reflects the discrepancy of income for males and females at various educational levels.

Table 9.4

Women's and Men's Median Incomes with Similar Education

	Bachelor's Degree	Master's Degree	Doctoral Degree
Men	$54,091	$69,825	$89,845
Women	$35,972	$50,576	$65,587

Source: *Statistical Abstract of the United States* (2012, 131st ed.), Table 70 (Washington, DC: U.S. Census Bureau).

STUDY TOOLS 9

Ready to study? In this book, you can:

- Rip out the Chapter Review card in the back of the book to study for exams

- Take the Self Assessment for this chapter (card in the back of the book) and see where you stand on the vital issues raised in the chapter

Or you can go online to CourseMate at www.cengagebrain.com for these resources:

- Complete Practice Quizzes to prepare for tests

- Review Key Terms Flash Cards (online or print)

- Read about Marriage and Family in the news

- Play "Beat the Clock" to master concepts

- Check out Personal Applications

Planning for Children

"Love is a fourteen letter word—
family planning."

—PLANNED PARENTHOOD

SECTIONS

10-1 Do You Want to Have Children?

10-2 How Many Children Do You Want?

10-3 Teenage Motherhood

10-4 Infertility

10-5 Planning for Adoption

10-6 Foster Parenting

10-7 Abortion

Although over 4 million babies are born in the United States annually, the cultural message on having children is mixed. Although young married individuals are encouraged to "have fun and travel before they begin their family" and to "strap on their seat belts when their children become teenagers," spouses are also socialized to believe that "marriage has no real meaning without children." However, having children continues to be a major goal of young adults. Among youth between the ages of 18 and 29, almost three fourths (74%) in a Pew Research Center report noted that they wanted to have children. And most said that "being a good parent" was more important than "having a successful marriage" (52% and 30%, respectively) (Wang & Taylor, 2011).

Planning children, or failing to do so, is a major societal issue. Planning when to become pregnant has benefits for both the mother and the child. Having several children at short intervals increases the chances of premature birth, infectious disease, and death of the mother or the baby. Would-be parents can minimize such risks by planning fewer children with longer intervals between. Women who plan their pregnancies can also modify their behaviors (e.g., stop smoking cigarettes and drinking alcohol) and seek preconception care from a health-care practitioner to maximize their chances of having healthy pregnancies and babies. Partners

© rSnapshotPhotos/Shutterstock.com

who plan their children also benefit from family planning by pacing the financial demands of their offspring. Having children four years apart helps to avoid having more than one child in college at the same time. Conscientious family planning will also help to reduce the number of unwanted pregnancies. One third (34%) of the births in the United States are unintended (Wildsmith et al., 2010). One researcher asked 192 women how they would feel if they learned they were pregnant. Nine percent reported that they would feel like they were dying (Schwarz et al., 2008).

Your choices in regard to children and contraception have important effects on your happiness, lifestyle, and resources. These choices are influenced by social and cultural factors that may operate without your awareness. We now discuss these influences.

10-1 Do You Want to Have Children?

Beyond a biological drive to reproduce, societies socialize their members to have children. This section examines the social influences that motivate individuals to have children, the lifestyle changes that result from such a choice, and the costs of rearing children.

10-1a Social Influences Motivating Individuals to Have Children

Our society tends to encourage childbearing, an attitude known as **pronatalism**. Our family, friends, religion, and government help to develop positive attitudes toward parenthood. Cultural observances also function to reinforce these attitudes.

Family Our experience of being reared in families encourages us to have families of our own. Our parents are our models. They married; we marry. They had children; we have children. Some parents exert a much more active influence. "I'm 73 and don't have much time. Will I ever see a grandchild?" asked the mother of an only child.

Friends Our friends who have children influence us to do likewise. After sharing an enjoyable weekend with friends who had a little girl, one husband wrote to the host and hostess, "Lucy and I are always affected by Karen—she is such a good child to have around. We haven't made up our minds yet, but our desire to have a child of our own always increases after

pronatalism view that encourages having children.

It's a boy!

we leave your home." This couple became parents 16 months later.

Religion Religion is a powerful influence on the decision to have children. Catholics are taught that having children is the basic purpose of marriage and gives meaning to the union. Mormonism and Judaism also have a strong family orientation.

Race Hispanics have the highest fertility rate of any racial or ethnic category.

Government The tax structures that our federal and state governments impose support parenthood. Married couples without children pay higher taxes than couples with children, although the reduction in taxes is not sufficient to offset the cost of rearing a child and is not large enough to be a primary inducement to have children.

Economy Times of affluence are associated with a high birth rate. The postwar expansion of the 1950s resulted in the oft-noted "baby boom" generation. Similarly, couples are less likely to decide to have a child during economically depressed times. Finally, the decision to have two wage earners in a marriage occurred at the same time that couples decided to have fewer children (everything is more expensive).

Cultural Observances Our society reaffirms its approval of parents every year by identifying special days for Mom and Dad. Each year on Mother's Day and Father's Day (and now Grandparents' Day), parenthood is celebrated across the nation with cards, gifts, and embraces. People choosing not to have children have no cultural counterpart (e.g., Child-free Day). In addition to influencing individuals to have children, society and culture also influence feelings about the age parents should be when they have children. Recently, couples have been having children at later ages.

"Now the thing about **having a baby**—
and I can't be the **first person** to have noticed this—
is that thereafter **you have it**."

—JEAN KERR, AUTHOR AND PLAYWRIGHT

10-1b Individual Motivations for Having Children

Individual motivations, as well as social influences, play an important role in the decision to have children. Some of these inducements are conscious, as in the desire to love and to be loved by one's own child, companionship, and the desire to be personally fulfilled as an adult by having a child. Some people also want to recapture their own childhood and youth by having a child. Unconscious motivations for parenthood may also be operative. Examples include wanting a child to avoid career tracking and to gain the acceptance and approval of one's parents and peers. Teenagers sometimes want to have a child to have someone to love them.

10-1c Lifestyle Changes and Economic Costs of Parenthood

Although becoming a parent has numerous potential positive outcomes, parenting also has drawbacks. Every parent knows that parenthood involves difficulties as well as joys. Some of the difficulties associated with parenthood are discussed next.

Lifestyle Changes Becoming a parent often involves changes in lifestyle. Daily living routines become focused around the needs of the children. Living arrangements change to provide space for another person in the household. Some parents change their work schedule to allow them to be home more. Food shopping and menus change to accommodate the appetites of children. A major lifestyle change is the loss of freedom of activity and flexibility in one's personal schedule. Lifestyle changes are particularly dramatic for women. The time and effort required to be pregnant and rear children often compete with the time and energy needed to finish one's education. Building a career is also negatively impacted by the birth of children. Parents learn quickly that both being involved, on-the-spot parents and climbing the career ladder are difficult. The careers of women may suffer most.

Financial Costs Meeting the financial obligations of parenthood is difficult for many parents. The costs begin with prenatal care and continue at childbirth. For an uncomplicated vaginal delivery, with a two-day hospital stay, the cost may total $10,550, whereas a cesarean section birth may cost $14,770. The annual cost of a child less than 2 years old for middle-income parents ($48,319 to $81,340)—which includes housing ($4,230), food ($1,350), transportation ($1,466), clothing ($432), health care ($823), child care ($2,110), and miscellaneous ($1,149)—is $11,560. For a 15- to 17-year-old, the cost is $13,327.45 (*Statistical Abstract of the United States*, 2012). These costs do not include the wages lost when a parent drops out of the workforce to provide child care. But in spite of the costs children incur, most people look forward to having children.

Most parents anticipate their children attending college. The price varies depending on whether a child attends a public or a private college, and websites such as collegeboard.com identify the cost of a specific college. On average, the estimated annual cost in 2013 for a child attending a four-year public college in the state of residence is around $15,971 (including tuition, board, dorm); for a private college the cost is around $42,504 annually (http://www.csgnetwork.com/educostcalc.html).

© Aaron Amat/Shutterstock

How Many Children Do You Want?

"The reproductive imperative of marriage is over" (Haag, 2011). Couples can now choose not to have children or to have them in or out of marriage. Most decide to have children, and most inside marriage.

10-2a Child-free Marriage

Procreative liberty is the freedom to decide whether or not to have children. More women are deciding not to have children or to have fewer children. The White House Study on Women in America confirmed that about 18% of women ages 40 to 44 have never had a child—this percentage has almost doubled since 1976 (Department of Commerce et al., 2011). About 44% of these child-free women have chosen not to have children (Smock & Greenland, 2010).

The intentionally child-free may be viewed with suspicion ("they are selfish"), avoidance ("since they don't have children they won't like us or support our family values"), discomfort ("what would I have in common with these people?"), rejection ("they are wrong not to want children"; "I don't want to spend time with these people"), and pity ("they don't know what they are missing"; Scott, 2009). Sanders (2012) noted that men reared in stepfamilies were the least likely to want children.

Stereotypes about couples who deliberately elect not to have children include that they don't like kids, are immature, and are not fulfilled because they don't have a child to make their lives "complete." The reality is that such individuals may enjoy children, and some deliberately choose careers to work with them (e.g., elementary school teacher). But they don't want the full-time emotional and economic responsibility of having their own children.

Some people simply do not like children. Aspects of our society reflect **antinatalism** (a perspective against children).

procreative liberty the freedom to decide whether or not to have children.

antinatalism opposition to having children.

Indeed, there is a continuous fight for corporations to implement or enforce family policies (from family leaves to flex time to on-site day care). Profit and money—not children—are priorities. In addition, although people are generally tolerant of their own children, they often exhibit antinatalistic behavior in reference to the children of others. Notice the unwillingness of some individuals to sit next to a child on an airplane.

10-2b One Child

Only 3% of adults view one child as the ideal family size (Sandler, 2010). Those who have only one child may do so because they want the experience of parenthood without children markedly interfering with their lifestyle and careers. Still others have an only child because of the difficulty in pregnancy or birthing the child. One mother said, "I threw up every day for nine months including on the delivery table." Another said, "I was torn up giving birth to my child." Still another mother said, "It took two years for my body to recover. Once is enough for me." There are also those who have only one child because they can't get pregnant a second time.

10-2c Two Children

The most preferred family size in the United States (for non-Hispanic White women) is the two-child family (1.9 to be exact!). Reasons for this preference include feeling that a family is "not complete" without two children, having a companion for the first child, having a child of each sex, and repeating the positive experience of parenthood enjoyed with the first child. Some couples may not want to "put all their eggs in one basket." They may fear that, if they have only one child and that child dies or turns out to be disappointing, they will not have another opportunity to enjoy parenting.

10-2d Three Children

Religion is a strong influence in the number of children a couple has. Twenty percent of Mormons and 15% of Muslims have at least three children (Pew Research Center, 2008). In addition to religious influences, couples are more likely to have a third child, and to do so quickly, if they already have two girls rather than two boys. They are least likely to bear a

© Creatas/Jupiter Images

third child if they already have a boy and a girl. Some individuals may want three children because they enjoy children and feel that "three is better than two." In some instances, a couple that has two children may simply want another child because they enjoy parenting and have the resources to do so.

Having a third child creates a "middle child." This child is sometimes neglected because parents of three children may focus more on the "baby" and the firstborn than on the child in between. However, an advantage to being a middle child is the chance to experience both a younger and an older sibling. Each additional child also has a negative effect on the existing children by reducing the amount of parental time available to the other children. The economic resources for each child are also affected by each subsequent child.

Hispanics are more likely to want larger families than are White or African American people. Larger families have complex interactional patterns and different values. The addition of each subsequent child dramatically increases the possible relationships in the family. For example, in a one-child family, four interpersonal relationships are possible: mother–father, mother–child, father–child, and father–mother–child. In a family of four, 11 relationships are possible; in a family of five, 26; and in a family of six, 57.

"*If you want children to keep their feet on the ground, put some responsibility on their shoulders.*"

—*Abigail Van Buren ("Dear Abby")*

10-2e Four Children—New Standard for the Affluent?

Among affluent couples, four children may be the new norm (Smith, 2007). Fueled by competitive career moms who have opted out of the workforce and who find themselves in suburbia surrounded by other moms with resources and time on their hands, having a large family is being reconsidered. A pattern has emerged called **competitive birthing**, where "keeping up with the Joneses" now means having the same number of kids. Subsequent research will need to confirm whether the pattern is widespread.

10-3 Teenage Motherhood

Reasons for teenagers having a child include not being socialized as to the importance of contraception, having limited parental supervision, and perceiving few alternatives to parenthood. Indeed, motherhood may be one of the only remaining meaningful roles available to them. In addition, some teenagers feel lonely and unloved and have a baby to create a sense of being needed and wanted. In contrast, in Sweden, eligibility requirements for welfare payments make it almost necessary to complete an education and get a job before becoming a parent.

competitive birthing pattern in which a woman will want to have the same number of children as her peers.

10-3a Problems Associated With Teenage Motherhood

Teenage parenthood is associated with various negative consequences, including the following:

1. **Stigmatization and marginalization.** Because teen mothers resist the typical life trajectory of their middle-class peers, they are often stigmatized and marginalized (Wilson & Huntington, 2006). In effect, they are a threat to societal goals of economic growth through higher education and increased female workforce participation. In spite of such stigmatization and marginalization, these teen mothers are very resilient: They invest in the "good" mother identity, maintain kin relations, and prioritize the mother–child dyad in their life (McDermott & Graham, 2005). One researcher interviewed 33 young women who were mothers before the age of 21 and discovered three themes common to their experience of teenage motherhood—"hardship and reward," "growing up and responsibility," and "doing things differently" (Rolfe, 2008, p. 299).

2. **Poverty among single teen mothers and their children.** Many teen mothers are unwed. Livermore and Powers (2006) studied a sample of 336 unwed mothers and found them plagued with financial stress; 18.5% had difficulty providing food for themselves and their children, had their electricity cut off for nonpayment (19.7%), and had no medical care for their children (18.2%). Almost half (47%) reported experiencing "one or more financial stressors" (p. 6).

3. **Poor health habits.** Teenage unmarried mothers are less likely to seek prenatal care and more likely than older and married women to smoke, drink alcohol, and take other drugs. These factors have an adverse effect on the health of the baby. Indeed, babies born to unmarried teenage mothers are more likely to have low birth weights (less than 5 pounds, 5 ounces) and to be born prematurely. Children of teenage unmarried mothers are also more likely to be developmentally delayed. These outcomes are largely a result of the association between teenage unmarried childbearing and persistent poverty.

4. **Lower academic achievement.** Poor academic achievement is both a contributing factor to and a potential outcome of teenage parenthood. Some studies note that between 30% and 70% of teen mothers drop out of high school before graduation (the schools may push them out or they may feel the stress of rearing a baby and going to school is overwhelming). Mollborn (2007) confirmed that teen parenthood diminished the chance that the mother would complete high school.

   Zachry (2005) interviewed 19 teen mothers and noted that, although all dropped out of school, each evidenced a new appreciation for education as a way of providing a better future for their child. Wendy, one of the mothers, said, "I want to better my education for my kids, and myself . . . because I'm their role model. And they're only gonna learn from what they see from me" (p. 2566).

   Children of teen mothers may be disadvantaged. Lipman et al. (2011) analyzed data from the Ontario Child Health Study (OCHS) and found that being born to a teen mother was associated with poorer educational achievement, personal income, and life satisfaction.

infertility the inability to achieve a pregnancy after at least one year of regular sexual relations without birth control, or the inability to carry a pregnancy to a live birth.

© iStockphoto.com/Josh Finehults

10-4 Infertility

Infertility is defined as the inability to achieve a pregnancy after at least one year of regular sexual relations without birth control, or the inability to carry a pregnancy to a live birth. Different types of infertility include the following:

"There is also an epidemic of infertility in this country. There are more women who have put off childbearing in favor of their professional lives."

—Iris Chang, historian

1. **Primary infertility.** The woman has never conceived even though she wants to and has had regular sexual relations for the past 12 months.

2. **Secondary infertility.** The woman has previously conceived but is currently unable to do so even though she wants to and has had regular sexual relations for the past 12 months.

3. **Pregnancy wastage.** The woman has been able to conceive but has been unable to produce a live birth.

10-4a Causes of Infertility

Although popular usage does not differentiate between the terms *fertilization* and the *beginning of pregnancy*, **fertilization** or **conception** refers to the fusion of the egg and sperm, whereas **pregnancy** is not considered to begin until five to seven days later, when the fertilized egg is implanted (typically in the uterine wall). Hence, not all fertilizations result in a pregnancy. An estimated 30 to 40% of conceptions are lost prior to or during implantation.

Forty percent of infertility problems are attributed to the woman, 40% to the man, and 20% to both of them. Some of the more common causes of infertility in men include low sperm production, poor semen motility, effects of STIs (such as chlamydia, gonorrhea, and syphilis), and interference with passage of sperm through the genital ducts due to an enlarged prostate. The causes of infertility in women include blocked fallopian tubes, endocrine imbalance that prevents ovulation, dysfunctional ovaries, chemically hostile cervical mucus that may kill sperm, and effects of STIs. Brandes et al. (2011) noted that unexplained infertility is one of the most common diagnoses in fertility care and is associated with a high probability of achieving a pregnancy—most spontaneously. Schmidt et al. (2012) emphasized that delaying pregnancy until age 30 is associated with more difficulty getting pregnant; delaying pregnancy until age 35

and beyond is associated with higher risk of preterm births and stillbirths.

An at-home fertility kit, **Fertell**, allows women to measure their egg quality. The test takes 30 minutes and involves a urine stick. The same kit allows men to measure the concentration of motile sperm. Men provide a sample of sperm (e.g., via masturbation) that swim through a solution similar to cervical mucus. This procedure takes about 80 minutes. Fertell has been approved by the Food and Drug Administration (FDA), no prescription is necessary, and the kit costs around $100.

Being infertile (for the woman) may have a negative lifetime effect. Wirtberg et al. (2007) interviewed 14 Swedish women 20 years after their infertility treatment and found that childlessness had had a major impact on all the women's lives and remained a major life theme. The effects were both personal (sad) and interpersonal (half were separated and all reported negative effects on their sex lives).

10-4b Assisted Reproductive Technology

A number of technological innovations are available to assist women and couples in becoming pregnant. These include hormonal therapy, artificial insemination, ovum transfer, in vitro fertilization, gamete intrafallopian transfer, and zygote intrafallopian transfer.

Hormone Therapy Drug therapies are often used to treat hormonal imbalances, induce ovulation, and correct problems in the luteal phase of the menstrual cycle. Frequently used drugs include Clomid, Pergonal, and human chorionic gonadotropin (HCG), a hormone extracted from human placenta. These drugs stimulate the ovary to ripen and release an egg. Although they are fairly effective in stimulating ovulation, hyperstimulation can occur, which may result in permanent damage to the ovaries.

Hormone therapy also increases the likelihood that

fertilization (conception) the fusion of the egg and sperm.

pregnancy a condition that begins five to seven days after conception, when the fertilized egg is implanted (typically in the uterine wall).

Fertell an at-home fertility kit that allows women to measure the level of their follicle-stimulating hormone on the third day of their menstrual cycle and men to measure the concentration of motile sperm.

multiple eggs will be released, resulting in multiple births. The increase in triplets and higher order multiple births over the past decade in the United States is largely attributed to the increased use of ovulation-inducing drugs for treating infertility. Infants of higher order multiple births are at greater risk of having low birth weight, and their mortality rates are higher. Mortality rates have improved for these babies, but the low birth weight survivors may need extensive neonatal medical and social services.

Artificial Insemination When the sperm of the male partner are low in count or motility, sperm from several ejaculations may be pooled and placed directly into the cervix. This procedure is known as *artificial insemination by husband* (AIH). When sperm from someone other than the woman's partner are used to fertilize a woman, the technique is referred to as *artificial insemination by donor* (AID).

Lesbians who want to become pregnant may use sperm from a friend or from a sperm bank (some sperm banks cater exclusively to lesbians). Regardless of the source of the sperm, it should be screened for genetic abnormalities and STIs, quarantined for 180 days, and retested for human immunodeficiency virus (HIV); also, the donor should be younger than 50 to diminish hazards related to aging. These precautions are not routinely taken—let the buyer beware.

How do children from donor sperm feel about their fathers? A team of researchers (Scheib et al., 2005) studied 29 individuals (41% from lesbian couples, 38% from single women, and 21% from heterosexual couples) and found that most (75%) always knew about their origin and were comfortable with it. All but one reported a neutral to positive relationship with their birth mother. Most (80%) indicated a moderate interest in learning more about the donor. No youths reported wanting money, and only 7% reported wanting a father–child relationship. Berger and Paul (2008) studied the effects of disclosing or not disclosing to the child that he or she is from a donor sperm. The results were inconclusive but favored disclosure.

Artificial Insemination of a Surrogate Mother In some instances, artificial insemination does not help a woman get pregnant. (Her fallopian tubes may be blocked, or her cervical mucus may be hostile to sperm.) The couple that still wants a child and has decided against adoption may consider parenthood through a surrogate mother. There are two types of surrogate mothers. One is the contracted surrogate mother who supplies the egg, is impregnated with the male partner's sperm, carries the child to term, and gives the baby to the man and his partner. A second type is the surrogate mother who carries to term a baby to whom she is not genetically related (a fertilized egg from the "infertile couple" who can't carry a baby to term is implanted in her uterus). As with AID, the motivation of the prospective parents is to have a child that is genetically related to at least one of them. For the surrogate mother, the primary motivation is to help childless couples achieve their aspirations of parenthood and to make money. Although some American women are willing to "rent their wombs," women in India also provide this service. The cost for a woman in India to carry a baby to term is $22,000 to $35,000 (http://www.medicaltourismco.com/assisted-reproduction-fertility/low-cost-surrogacy-india.php).

California is one of 12 states in which entering into an arrangement with a surrogate mother is legal. The fee for the surrogate mother, travel, the hospital, lawyers, and so on is between $40,000 and $65,000. Surrogate mothers typically have their own children, making giving up a child that they carried easier. For information about the legality of surrogacy in your state, see http://www.surrogacy.com/legals/map.html.

In Vitro Fertilization About 2 million couples cannot have a baby because the woman's fallopian tubes are blocked or damaged, preventing the passage of eggs to the uterus. In some cases, blocked tubes can be opened via laser surgery or by inflating a tiny balloon within the clogged passage. When these procedures are not successful (or when the woman decides to avoid invasive tests and exploratory surgery), *in vitro* (meaning "in glass") *fertilization* (IVF), also known as test-tube fertilization, is an alternative.

Using a laparoscope (a narrow, telescope-like instrument inserted through an incision just below the woman's naval to view tubes and ovaries), the physician is able to see a mature egg as it is released from the woman's ovary. The time of release can be predicted accurately within two hours. When the egg emerges, the physician uses an aspirator to remove the egg, placing it in a small tube containing stabilizing fluid. The egg is taken to the laboratory, put in a culture petri dish, kept at a certain temperature and acidity level, and surrounded by sperm from the woman's partner (or donor). After one of these sperm fertilizes the egg, the egg divides and is implanted by the physician in the wall of the woman's uterus. Usually, several eggs are implanted in the hope one will survive. This was the case of Nadya Suleman, who ended up giving birth to eight babies. Eight embryos were transferred into her body in 2008 at Duke University's in vitro fertilization program with the thought that some would not survive . . . all did (Rochman, 2009).

Some couples want to ensure the sex of their baby. In a procedure called "family balancing" because couples that already have several children of one sex often use it, the eggs of a woman are fertilized, and the sex of the embryos at three and eight days old is identified. Only those of the desired sex are then implanted in the woman's uterus.

Alternatively, the Y chromosome of the male sperm can be identified and implanted. The procedure is accurate 75% of the time for producing a boy baby and 90% of the time for a girl baby. The Genetics and IVF Institute in Fairfax, Virginia, specializes in the sperm sorting technique (see http://www.givf.com/).

> The typical success rate (live birth) for infertile couples who seek help in one of the 450 fertility clinics is 28%

Occasionally, some fertilized eggs are frozen and implanted at a later time, if necessary. This procedure is known as **cryopreservation**. Separated or divorced couples may disagree over who owns the frozen embryos, and the legal system is still wrestling with the fate of unused embryos, sperm, or ova after a divorce or death.

Ovum Transfer In conjunction with in vitro fertilization is ovum transfer, also referred to as embryo transfer. In this procedure, an egg is donated, fertilized in vitro with the husband's sperm, and then transferred to his wife. Alternatively, a physician places the sperm of the male partner in a surrogate woman. After about five days, her uterus is flushed out (endometrial lavage), and the contents are analyzed under a microscope to identify the presence of a fertilized ovum.

The fertilized ovum is then inserted into the uterus of the otherwise infertile partner. Although the embryo can also be frozen and implanted at another time, fresh embryos are more likely to result in successful implantation. Infertile couples that opt for ovum transfer do so because the baby will be biologically related to at least one of them (the father) and the partner will have the experience of pregnancy and childbirth. As noted earlier, the surrogate woman participates out of her desire to help an infertile couple and/or to make money.

Other Reproductive Technologies A major problem with in vitro fertilization is that only about 15 to 20% of the fertilized eggs will implant on the uterine wall. To improve this implant percentage (to between 40% and 50%), physicians place the egg and the sperm directly into the fallopian tube, where they meet and fertilize. Then the fertilized egg travels down into the uterus and implants.

Because the term for sperm and egg together is *gamete,* this procedure is called *gamete intrafallopian transfer,* or GIFT. This procedure, as well as in vitro fertilization, is not without psychological costs to the couple.

Gestational surrogacy, another technique, involves fertilization in vitro of a woman's ovum and transfer to a surrogate. Trigametic IVF also involves the use of sperm in which the genetic material of another person has been inserted. This technique allows lesbian couples to have a child genetically related to both women. Infertile couples hoping to get pregnant through one of the more than 450 in vitro fertilization clinics should make informed choices by asking questions such as, "What is the center's pregnancy rate for women with a similar diagnosis?"

What percentage of these women has a live birth? According to the Centers for Disease Control and Prevention, the typical success rate (live birth) for infertile couples who seek help in one of the 450 fertility clinics is 28% (Lee, 2006). Beginning assisted reproductive technology as early after infertility is suspected is important. Wang et al. (2008) analyzed data on 36,412 patients to assess success of actual births for infertile women using assisted reproductive technology and found that, for women age 30 and above, each additional year in age was associated with an 11% reduction in the chance of achieving pregnancy and a 13% reduction

cryopreservation procedure by which fertilized eggs are frozen and implanted at a later time.

in the chance of a live delivery. If women age 35 years or older would have had their first treatment one year earlier, 15% more live deliveries would be expected.

Finally, the cost of treating infertility is enormous. Katz et al. (2011) examined the costs for 398 women in eight infertility practices over an 18-month period. The cost of a successful outcome (delivery or ongoing pregnancy by 18 months of treatment) for IVF was $61,377. And, women could spend this amount or more and still not have a baby or pregnancy.

10-5 Planning for Adoption

Angelina Jolie and Brad Pitt are celebrities who have given national visibility to adopting children. They are not alone in their desire to adopt children. The various routes to adoption are public (children from the child welfare system), private agency (children placed with nonrelatives through agencies), independent adoption (children placed directly by birth parents or through an intermediary such as a physician or attorney), kinship (children placed in a family member's home), and stepparent (children adopted by a spouse). Motives for adopting a child include an inability to have a biological child (infertility), a desire to give an otherwise unwanted child a permanent loving home, or a desire to avoid contributing to overpopulation by having more biological children. Some couples may seek adoption for all of these motives. Adoption is actually quite rare. Just over 1% of 18- to 44-year-old women reported having adopted a child (15% are children from other countries; Smock & Greenland, 2010).

10-5a Demographic Characteristics of People Seeking to Adopt a Child

Whereas those who typically adopt are White, educated, and high income, adoptees are being increasingly placed in nontraditional families including older, gay, and single individuals. Sixteen states have taken steps to ban adoption by gay couples on the grounds that, because "marriage" is "heterosexual marriage," children do not belong in homosexual relationships (Stone, 2006). Leung et al. (2005) compared children adopted or reared by gay or lesbian parents with those adopted or raised by heterosexual parents. The researchers found no negative effects when the adoptive parents were gay or lesbian. Waterman et al. (2011) found that parents reported high adoption satisfaction despite ongoing behavioral problems with a third of the children in a sample of children who had been adopted from foster homes.

10-5b Characteristics of Children Available for Adoption

Adoptees in the highest demand are healthy, White infants. Those who are older, of a racial or ethnic group different from that of the adoptive parents, of a sibling group, or with physical or developmental disabilities have been difficult to place. Flower Kim (2003) noted that, because the waiting period for a healthy, White infant is from 5 to 10 years, couples are increasingly open to cross-racial adoptions. Of the 1.6 million adopted children younger than 18 living in U.S. households, the percentages adopted from other countries are as follows: 24% from Korea; 11% from China; 10% from Russia; and 9% from Mexico. International or cross-racial adoptions may complicate the adoptive child's identity. Children adopted after infancy may also experience developmental delays, attachment disturbances, and posttraumatic stress

© iStockphoto.com/Nathan Gleave/© iStockphoto.com

disorder (Nickman et al., 2005). Baden and Wiley (2007) reviewed the literature on adoptees as adults and found that the mental health of most was on par with those who were not adopted. However, a small subset of the population showed concerns that may warrant therapeutic intervention.

10-5c Costs of Adoption

Adopting from the U.S. foster care system is generally the least expensive type of adoption, usually involving little or no cost, and states often provide subsidies to adoptive parents. However, a couple can become foster care parents to a child and become emotionally bonded with the child, and then the birth parents might request their child back. Stepparent and kinship adoptions are also inexpensive and have less risk of the child being withdrawn. Agency and private adoptions can range from $5,000 to $40,000 or more, depending on travel expenses, birth mother expenses, and requirements in the state. International adoptions can range from $7,000 to $30,000 (see http://costs.adoption.com/).

10-5d Transracial Adoption

Transracial adoption is defined as the practice of adopting children of a race different from that of the parents—for example, a White couple adopting a Korean or African American child. In a study on transracial adoption attitudes of college students, the scores of the 188 respondents reflected overwhelmingly positive attitudes toward transracial adoption. Overall, women, people willing to adopt a child at all, interracially experienced daters, and those open to interracial dating were more willing to adopt transracially than were men, people rejecting adoption as an optional route to parenthood, people with no previous interracial dating experience, and people closed to interracial dating (Ross et al., 2003).

Ethiopia has become a unique country from which to adopt a child. Not only is the adoption time shorter (four months) and less expensive, but the children there are also psychologically very healthy. "You don't hear crying babies [in the orphanages] . . . they are picked up immediately" (Gross & Connors, 2007, p. A16). In addition, "adoption families are encouraged to meet birth families and visit the villages where the children are raised." (p. A16). Ethiopian adoptions have received considerable visibility in the United States due to the involvement of celebrity Angelina Jolie, who adopted there.

Transracial adoptions are controversial. Kennedy (2003) noted, "Whites who seek to adopt black children are widely regarded with suspicion. Are they ideologues, more interested in making a political point than in actually being parents?" (p. 447). Another controversy is whether it is beneficial for children to be adopted by parents of the same racial background. In regard to the adoption of African American children by same-race parents, the National Association of Black Social Workers (NABSW) passed a resolution against transracial adoptions, citing that such adoptions prevented Black children from developing a positive sense of themselves "that would be necessary to cope with racism and prejudice that would eventually occur" (Hollingsworth, 1997, p. 44).

The counterargument is that a healthy self-concept, an appreciation for one's racial heritage, and skills for coping with racism or prejudice can be learned in a variety of contexts. Legal restrictions on transracial adoptions have disappeared, and social approval for transracial adoptions is increasing. However, a substantial number of studies conclude that "same-race placements are preferable and that special measures should be taken to facilitate such placements, even if it means delaying some adoptions" (Kennedy, 2003, p. 469).

One 26-year-old Black female was asked how she felt about being reared by White parents and replied, "Again, they are my family and I love them, but I am black. I have to deal with my reality as a black woman" (Simon & Roorda, 2000, p. 41). A Black man reared in a White home advised White parents considering a transracial adoption to "[m]ake sure they have the influence of blacks in their lives; even if they have to go out and make friends with black families—it's a must" (p. 25). Indeed, Huh and Reid (2000) found that positive adjustment by adoptees was associated with participation in the cultural activities of the race of the parents who adopted them. Thomas and Tessler (2007) found that American parents intent on keeping the Chinese cultural heritage of their adopted child alive take specific steps (e.g., establishing friendships with Chinese adults and families).

10-5e Open Versus Closed Adoptions

Another controversy is whether adopted children should be

transracial adoption the practice of parents adopting children of another race.

Another controversy is whether adopted children should be allowed to obtain information about their biological parents.

heterosexual couples who were involved in an open adoption. Although there were some tensions with the birth parents over time, most of the 45 adoptive couples reported satisfying relationships.

10-6 Foster Parenting

Some individuals seek the role of parent via foster parenting. A **foster parent**, also known as a *family caregiver*, is neither a biological nor an adoptive parent but is a person who takes care of and fosters a child taken into custody. A foster parent has made a contract with the state for the service, has judicial status, and is reimbursed by the state. Foster parents are screened for previous arrest records and child abuse and neglect. Foster parents are licensed by the state, and some states require a "foster parent orientation" program. Rhode Island, for example, provides a 27-hour course. Brown (2008) asked 63 foster parents what they needed that would allow them to have a successful foster parenting experience. Participants reported that they needed the right personality (e.g., patience and nurturance), information about the foster child, a good relationship with the fostering agency, linkages to other foster families, and supportive immediate and extended families. Other research has found the need for formal foster parent organizations.

Children placed in foster care have typically been removed from parents who are abusive, who are substance abusers, and/or who are mentally incompetent. The goal of placing children in foster care is to improve their living conditions and then either return them to their family of origin or find a more permanent adoptive or foster home. Some couples become foster parents in the hope of being able to adopt a child that is placed in their custody. Meyer et al. (2010) studied the conditions under which parental rights to children in foster care are terminated and found that parents who were incarcerated and who had mental health problems were the most vulnerable.

Due to longer delays for foreign adoptions (e.g., it typically takes three years to complete a foreign adoption) and fewer domestic infants, more couples adopt a foster child. Thirty-nine percent are available within one year (Huggins & Gelles, 2011). Tax credits are available for up to $11,650 for adopting a child with special needs (Block, 2008).

allowed to obtain information about their biological parents. In general, there are considerable benefits to having an open adoption, especially the opportunity for the biological parent(s) to stay involved in the child's life. Adoptees learn early that they are adopted and who their biological parents are. Birth parents are more likely to avoid regret and to be able to stay in contact with their child. Adoptive parents have information about the genetic background of their adopted child. Ge et al. (2008) studied birth mothers and adoptive parents and found that increased openness between the two sets of parents was positively associated with greater satisfaction for both birth mothers and adoptive parents. Goldberg et al. (2011) studied lesbian, gay, and

foster parent (family caregiver) a person who either alone or with a spouse takes care of and fosters a child taken into custody.

© Waschnig/Shutterstock.com

10-7 Abortion

Among American women, half will have an unintended pregnancy and 30% will have an abortion. About 60% of women having an abortion are in their 20s and unmarried (Guttmacher Institute, 2012). An abortion may be either an **induced abortion**, which is the deliberate termination of a pregnancy through chemical or surgical means, or a **spontaneous abortion (miscarriage)**, which is the unintended termination of a pregnancy. (Geller et al. (2010) emphasized that miscarriage is often a significant loss, provoking both depression and anxiety, and that health-care professionals are often oblivious to its treatment. Abortion is legal in the United States.

10-7a Incidence of Abortion

About 1.2 million abortions are performed annually in the United States. Although the number of abortions has been increasing among the poor (lower access to health care and health education), there has been a decrease among higher income women (increased acceptability of having a child without a partner, increased use of contraception). Ninety percent of abortions occur within the first three months of pregnancy (Guttmacher Institute, 2012). The **abortion rate** (the number of abortions per 1,000 women ages 15 to 44) increased 1% between 2005 and 2008, from 19.4 to 19.6 abortions per 1,000 women ages15 to 44; the total number of abortion providers was virtually unchanged (Jones & Kooistra, 2011). About 40% of abortions are repeat abortions (Ames & Norman, 2012).

The **abortion ratio** refers to the number of abortions per 1,000 live births. Abortion is affected by the need for parental consent and parental notification. **Parental consent** means that a woman needs permission from a parent to get an abortion if she is under a certain age, usually 18. **Parental notification** means that a woman has to tell a parent she is getting an abortion if she is under a certain age, usually 18, but she doesn't need parental permission. Laws vary by state. See Table 10.1 or call the National Abortion Federation Hotline at 1-800-772-9100 to find out the laws in your state.

10-7b Reasons for an Abortion

A team of researchers (Finer et al., 2005) surveyed 1,209 women who reported having had an abortion. The most frequently cited reasons were that having a child would interfere with a woman's education, work, or ability to care for dependents (74%), that she could not afford a baby now (73%), and that she did not want to be a single mother or was having relationship problems (48%). Nearly 4 in 10 women said they had completed their childbearing, and almost one third of the women were not ready to have a child. Fewer than 1% said their parents' or partner's desire for them to have an abortion was the most important reason.

Abortions performed to protect the life or health of the woman are called **therapeutic abortions**. However, there is disagreement over this definition. "Some physicians argue that an abortion is therapeutic if it prevents or alleviates a serious physical or mental illness, or even if it alleviates temporary emotional upsets. In short, the health of the

induced abortion the deliberate termination of a pregnancy through chemical or surgical means.

spontaneous abortion (miscarriage) an unintended termination of a pregnancy.

abortion rate the number of abortions per 1,000 women ages 15 to 44.

abortion ratio the number of abortions per 1,000 live births.

parental consent woman needs permission from parent to get an abortion if she is under a certain age, usually 18.

parental notification woman is required to tell parents she is getting an abortion if she is under a certain age, usually 18, but she does not need parental permission.

therapeutic abortion an abortion performed to protect the life or health of a woman.

© Larry W. Smith/epa/Corbis

Table 10.1
Parental Consent by State

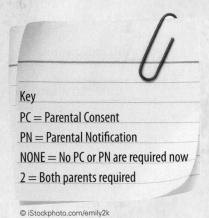

Key

PC = Parental Consent

PN = Parental Notification

NONE = No PC or PN are required now

2 = Both parents required

© iStockphoto.com/emily2k

State	Status	Comments
Alabama	PC	
Alaska	NONE	New law since December 2010
Arizona	PC	Notarized written consent required
Arkansas	PC	Notarized written consent required
California	NONE	PC law stopped by court
Colorado	PN	
Connecticut	NONE	
Delaware	PN	Applies only to girls under 16 years old; notice may also be to grandparent or counselor; doctor can bypass
District of Columbia	NONE	
Florida	PN	
Georgia	PN	
Hawaii	NONE	
Idaho	PC	
Illinois	NONE	PN law stopped by court
Indiana	PC	
Iowa	PN	Also allows consent of grandparent instead
Kansas	PC2	Notarized written consent required
Kentucky	PC	
Louisiana	PC	Notarized written consent required
Maine	NONE	
Maryland	PN	Doctor can bypass
Massachusetts	PC	
Michigan	PC	
Minnesota	PN2	
Mississippi	PC2	
Missouri	PC	
Montana	NONE	PN stopped by court
Nebraska	PN	Notarized written consent required
Nevada	NONE	PN stopped by court
New Hampshire	PN	
New Jersey	NONE	PN law stopped by court
New Mexico	NONE	PC law stopped by court
New York	NONE	
North Carolina	PC	Allows for consent by grandparent instead of parent
North Dakota	PC2	
Ohio	PC	
Oklahoma	PN and PC	
Oregon	NONE	
Pennsylvania	PC	
Rhode Island	PC	
South Carolina	PC	Women under 17 years old; also allows for consent by grandparent
South Dakota	PN	
Tennessee	PC	
Texas	PN and PC	
Utah	PN and PC	
Virginia	PC	Also allows consent of grandparent instead
Vermont	NONE	
Washington	NONE	
West Virginia	PN	Doctor can bypass
Wisconsin	PC	Also allows other family members over 25 to consent; doctor can bypass
Wyoming	PN and PC	

Source: Adapted from "Resources: Parental Consent," Coalition for Positive Sexuality. Retrieved January 13, 2012, from http://www.positive.org/Resources/consent.html.

© Background Land/Shutterstock

pregnant woman is given such a broad definition that a very large number of abortions can be classified as therapeutic" (Garrett et al., 2001, p. 218).

Some women with multifetal pregnancies (a common outcome of the use of fertility drugs) may have a procedure called *transabdominal first-trimester selective termination*. In this procedure, the lives of some fetuses are terminated to increase the chance of survival for the others or to minimize the health risks associated with multifetal pregnancy for the woman. For example, a woman carrying five fetuses may elect to abort three of them to minimize the health risks to the other two.

10-7c Pro-Life and Pro-Choice Abortion Positions

The dichotomy of attitudes toward abortion is reflected in two opposing groups of abortion activists. Individuals and groups who oppose abortion are commonly referred to as "pro-life" or "antiabortion."

Pro-Life Of 2,922 undergraduates at two large universities, 22% reported that abortion was not acceptable under certain conditions (Knox & Hall, 2010). Pro-life groups favor abortion regulation policies or a complete ban on abortion. They essentially believe the following:

- The unborn fetus has a right to live and that right should be protected.

- Abortion is a violent and immoral solution to unintended pregnancy.

- The life of an unborn fetus is sacred and should be protected, even at the cost of individual difficulties for the pregnant woman.

Individuals who are over the age of 44, female, mothers of three or more children, married to white-collar workers, affiliated with a religion, and Catholic are most likely to be pro-life (Begue, 2001). Pro-life individuals emphasize the sanctity of human life and the moral obligation to protect it. The unborn fetus cannot protect itself so is literally dependent on others for life. Naomi Judd noted that if she had had an abortion she would have deprived the world of one of its greatest singers—Wynonna Judd.

Pro-Choice In the sample of 2,922 students referred to earlier, 52% reported that "abortion is acceptable under certain conditions" (Knox & Hall, 2010). Pro-choice advocates support the legal availability of abortion for all women. They essentially believe the following:

- Freedom of choice is a central value—the woman has a right to determine what happens to her own body.

- Those who must personally bear the burden of their moral choices ought to have the right to make these choices.

- Procreation choices must be free of governmental control.

People most likely to be pro-choice are female, are mothers of one or two children, have some college education, are employed, and have an annual income of more than $50,000. Although many self-proclaimed feminists and women's organizations, such as the National Organization for Women (NOW), have been active in promoting abortion rights, not all feminists are pro-choice.

10-7d Physical Effects of Abortion

Part of the debate over abortion is related to the presumed effects of abortion. In regard to the physical effects, legal abortions, performed under safe medical conditions, are safer than continuing the pregnancy. The earlier in the pregnancy the abortion is performed, the safer it is. Vacuum aspiration, a frequently used method in early pregnancy, does not increase the risks to future childbearing. However, late-term abortions do increase the risks of subsequent miscarriages, premature deliveries, and babies of low birth weight.

Postabortion complications include the possibility of incomplete abortion, which occurs when the initial procedure misses the fetus and must be repeated. Other possible complications include uterine infection; excessive bleeding; perforation or laceration of the uterus, bowel, or adjacent organs; and an adverse reaction to a medication or anesthetic. After having an abortion, women are advised to expect bleeding (usually not heavy) for up to two weeks and to return to their health-care provider 30 days after the abortion to check that all is well.

10-7e Psychological Effects of Abortion

Of equal concern are the psychological effects of abortion. The American Psychological Association reviewed all outcome studies on the mental health effects of abortion and issued the following statement:

Based on our comprehensive review and evaluation of the empirical literature published in peer-reviewed journals since 1989, this Task Force on Mental Health and Abortion concludes that the most methodologically sound research indicates that among women who have a single, legal, first-trimester abortion of an unplanned pregnancy for nontherapeutic reasons, the relative risks of mental health problems are no greater than the risks among women who deliver an unplanned pregnancy. (Major et al., 2008, p. 71)

Steinberg and Russo (2008) also looked at national data and did not find a significant relationship between first pregnancy abortion and subsequent rates of generalized anxiety disorder, social anxiety, or posttraumatic stress disorder.

10-7f Postabortion Attitudes of Men

Researchers Kero and Lalos (2004) conducted interviews with men 4 and 12 months after their partners had had an abortion. Overwhelmingly, the men (at both time periods) were happy with the decision of their partners to have an abortion. More than half accompanied their partner to the abortion clinic (which they found less than welcoming); about a third were not using contraception a year later.

© Rehan Qureshi/Shutterstock

STUDY TOOLS **10**

Ready to study? In this book, you can:

- ⮑ Rip out the Chapter Review card in the back of the book to study for exams

- ⮑ Take the Self Assessment for this chapter (card in the back of the book) and see where you stand on the vital issues raised in the chapter

Or you can go online to CourseMate at www.cengagebrain.com for these resources:

- ⮑ Complete Practice Quizzes to prepare for tests

- ⮑ Review Key Terms Flash Cards (online or print)

- ⮑ Read about Marriage and Family in the news

- ⮑ Play "Beat the Clock" to master concepts

- ⮑ Check out Personal Applications

4LTR Press solutions are designed for today's learners through the continuous feedback of students like you. Tell us what you think about **M&F** and help us improve the learning experience for future students.

YOUR FEEDBACK MATTERS.

Complete the Speak Up survey in CourseMate at www.cengagebrain.com

 Follow us at www.facebook.com/4ltrpress

Parenting

"When bringing up **children,** spend on them **half** as much **money** and **twice** as much **time.**"

—LAURENCE PETER, EDUCATOR

SECTIONS

11-1 Roles Involved in Parenting

11-2 Choices Perspective of Parenting

11-3 Transition to Parenthood

11-4 Parenthood: Some Facts

11-5 Principles of Effective Parenting

11-6 Single-Parenting Issues

I didn't have any talent to be a father," said the late Paul Newman of his relationship with his six children (three each by two wives; Levy 2009, p. 256). "The process of really connecting is very long and painful for me. . . . I sometimes have a hard time talking because I have a hard time talking to anybody" (p. 256). Parenting was also difficult for his Academy Award–winning wife, Joanne Woodward. "I don't like children. . . . I like my own children; I occasionally like other people's children. But I don't like babies per se" (p. 257). Although some parents find the meaning of life in having and rearing children and others regard the role as more of a burden, the majority of parents fall somewhere in between. We now examine what is involved in parenting.

11-1 Roles Involved in Parenting

parenting defined in terms of roles including caregiver, emotional resource, teacher, and economic resource.

Although finding one definition of **parenting** is difficult, there is general agreement about the various roles parents play in the lives of their children. New parents assume at least seven roles:

1. **Caregiver.** A major role of parents is the physical care of their children. From the moment of birth, when infants draw their first breath, parents stand ready to provide nourishment (milk), cleanliness (diapers), and temperature control (warm blanket).

The need for such sustained care in terms of a place to live and eat continues into adulthood, as one fourth of 18 to 34 year olds have moved back in with their parents after living on their own (Parker, 2012). These **boomerang generation** children return primarily for economic reasons.

2. **Emotional resource.** Beyond providing physical care, parents are sensitive to the emotional needs of children in terms of their need to belong, to be loved, and to develop positive self-concepts. In hugging, holding, and kissing an infant, parents not only express their love for the infant but also reflect an awareness that such display of emotion is good for the child's sense of self-worth. Children exhibit increased emotional insecurity when their parents are in conflict or are depressed (Kouros et al., 2008). The family context

boomerang generation 18 to 34 year old young adults who have moved back in with their parents after having lived on their own.

is the emotional context for children. Strife or depression in this context does not occur without a negative effect on the children.

3. **Teacher.** All parents think they have a philosophy of life or set of principles their children will benefit from. Parents later discover that their children may not be interested in their religion or philosophy—indeed, they may rebel against it. This possibility does not deter parents from their role as teacher. Children are forever learning from their parents, more often by observing their behavior.

Parents also feel that their role is made more difficult by an increasingly liberal society. A sample of 2,020 Americans noted that one of the biggest problems confronting parents today

What the world needs is not romantic lovers but husbands and wives who willingly give their time and attention to their children.

—Margaret Mead, anthropologist

is the societal influence on their children. Challenges include drugs and alcohol; peer pressure; TV, Internet, and movies; and crime and gangs (Pew Research Center, 2007).

Parents may also teach without awareness. In one study, parents (whether together or divorced) who reported high conflict in their marriage tended to have children who as young adults reported high conflict and low quality in their own romantic relationships (Cui et al., 2008).

4. **Economic resource.** New parents are also acutely aware of the costs for medical care, food, and clothes for infants and seek ways to ensure that such resources are available to their children. Working longer hours, taking second jobs, and cutting back on leisure expenditures are attempts to ensure that money is available to meet the needs of the child. Sometimes the pursuit of money for the family has a negative consequence for children. Two researchers investigated the effects of parents' working schedules on the time they devoted to their children and confirmed that the more parents worked, the less time they spent with their children (Rapoport & Le Bourdais, 2008). In view of extensive work schedules, parents are under pressure to spend "quality time" with their children, and it is implied that putting children in day care robs children of this time. However, Booth et al. (2002) compared children in day care with those in home care in terms of time the mother and child spent together per week. Although the mothers of children in day care spent less time with their children than the mothers who cared for their children at home, the researchers concluded that the "groups did not differ in the quality of mother–infant interaction" and that the difference in the "quality of the mother–infant interaction may be smaller than anticipated" (p. 16).

Parents provide an economic resource for their children by providing free room and board for them. Some young adults continue to live with their parents well into adulthood and may return at other times such as following a divorce, job loss, and a need to return to school.

5. **Protector.** Parents also feel the need to protect their children from harm. This role may begin in pregnancy. Researchers interviewed 1,451 women about their smoking behavior during pregnancy.

MAMA SAYS

Things your mother taught you (Internet humor):

1. My mother taught me to *appreciate a job well done*: "If you're going to kill each other, do it outside. I just finished cleaning."

2. My mother taught me *religion*: "You better pray that will come out of the carpet."

3. My mother taught me about *time travel*: "If you don't straighten up, I'm going to knock you into the middle of next week!"

4. My mother taught me *logic*: "Because I said so, that's why."

5. My mother taught me *more logic*: "If you fall out of that swing and break your neck, you're not going to the store with me."

6. My mother taught me *foresight*: "Make sure you wear clean underwear, in case you're in an accident."

7. My mother taught me *irony*: "Keep crying, and I'll give you something to cry about."

8. My mother taught me about the science of *osmosis*: "Shut your mouth and eat your supper."

9. My mother taught me about *contortionism*: "Will you look at that dirt on the back of your neck!"

10. My mother taught me about *stamina*: "You'll sit there until all that spinach is gone."

11. My mother taught me about *weather*: "This room of yours looks as if a tornado went through it."

12. My mother taught me about *hypocrisy*: "If I told you once, I've told you a million times. Don't exaggerate!"

13. My mother taught me the *circle of life*: "I brought you into this world, and I can take you out."

14. My mother taught me about *behavior modification*: "Stop acting like your father!"

15. My mother taught me about *envy*: "There are millions of less fortunate children in this world who don't have wonderful parents like you do."

16. My mother taught me about *anticipation*: "Just wait until we get home."

17. My mother taught me about *receiving*: "You are going to get it when you get home!"

18. My mother taught me *medical science*: "If you don't stop crossing your eyes, they are going to get stuck that way."

19. My mother taught me about *ESP*: "Put your sweater on; don't you think I know when you are cold?"

20. My mother taught me about my *roots*: "Shut that door behind you. Do you think you were born in a barn?"

✓ educate

✓ control

✓ remove risk

© Robyn Mackenzie/iStockphoto.com

Although 89% reduced their smoking during pregnancy, 25% stopped smoking during pregnancy (Castrucci et al., 2006).

Other expressions of the protective role include insisting that children wear seat belts, protecting them from violence or nudity in the media, and protecting them from strangers—including those they meet on the Internet. In a Dutch study of 1,796 adolescents, 17% had real-life contacts with someone they had met on the Internet—their parents were aware only 30% of the time (Van den Heuvel et al., 2012).

Researchers identified three principal strategies parents use to protect their children—educate, control, and remove risk (Diamond et al., 2006). The strategy used depended on the age and temperament of the child. For example, some parents felt that protecting their children from certain television content was important.

Some parents feel that protecting their children from harm implies appropriate

oppositional defiant disorder disorder in which children fail to comply with requests of authority figures.

discipline for inappropriate behavior. Galambos et al. (2003) noted that "parents' firm behavioral control seemed to halt the upward trajectory in externalizing problems among adolescents with deviant peers." Harris-McKoy and Cui (2011) confirmed that controlling adolescent behavior—monitoring who adolescents associate with, the television programs they watch, and how much unsupervised time they are allowed outside the house—is associated with a decreased chance that the child will be involved in delinquent behavior.

Researchers compared clinically referred boys and girls (ages 6 to 11) diagnosed with **oppositional defiant disorder** (children do not comply with requests of authority figures) to a matched sample of healthy control children and found that the former had greater exposure to delinquent peers (Kolko et al., 2008). Hence, parents who monitor their children's peer relationships and minimize their children's exposure to delinquent models are making a wise time investment.

Increasingly, parents are joining the technological age and learning how to text message. In their role as protector, this ability allows parents to text message their children to tell them to come home, to phone home, or to work out a logistical problem—"meet me at the food court in the mall." Children can also use text messaging to let parents know that they arrived safely at a destination, when they need to be picked up, or when they will be home.

6. **Health promoter.** Parents are major agents in promoting healthy food choices, responsible drinking, not using drugs, safe sexual behavior, safe driving, and ending smoking behavior. Knog et al. (2012) observed that parents who were most successful in getting their adolescents to stop smoking were positive models (they did not smoke themselves) and disapproved of their adolescents' smoking. Baltazar et al. (2011) found that having a close relationship with children in 7th, 8th, and 9th grades was associated with less smoking, drinking, and using inhalants.

7. **Ritual bearer.** To build a sense of family cohesiveness, parents often foster rituals to bind members together in emotion and in memory. Prayer at meals and before bedtime, birthday celebrations, and vacations at the same place (beach, mountains, and so on) provide predictable times of togetherness and sharing.

11-2 Choices Perspective of Parenting

Although both genetic and environmental factors are at work, the choices parents make have a dramatic impact on their children. In this section, we review the nature of parental choices and some of the basic choices parents make.

11-2a Nature of Parenting Choices

Parents might keep the following points in mind when they make choices about how to rear their children.

1. **Not to make a parental decision is to make a decision.** Parents are constantly making choices even when they think they are not doing so. When a child is impolite and the parent does not provide feedback and encourage polite behavior, the parent has chosen to teach the child that being impolite is acceptable. When a child makes a promise ("I'll text you when I get to my friend's house") and does not do as promised, the parent has chosen to allow the child to not take commitments seriously. Hence, parents cannot choose not to make choices in their parenting, because their inactivity is a choice that has as much impact as a deliberate decision to reinforce politeness and responsibility.

2. **All parental choices involve trade-offs.** Parents are also continually making trade-offs in the parenting choices they make. The decision to take on a second job or to work overtime to afford the larger house will come at the price of having less time to spend with one's children and being more exhausted when such time is available. The choice to enroll one's child in the highest-quality day care (which may also be the most expensive) will mean less money for family vacations. The choice to have an additional child will provide siblings for the existing children but will mean less time and fewer resources for those children. Parents should increase their awareness that no choice is without a trade-off and should evaluate the costs and benefits of making such decisions.

3. **Reframe "regretful" parental decisions.** All parents regret a previous parental decision (e.g., they should have held their child back a year in school, or not done so; they should have intervened in a bad peer relationship; they should have handled their child's drug use differently). Whatever the issue, parents chide themselves for their mistakes. Rather than berate themselves as parents, they might emphasize the positive outcome of their choices: not holding the child back made the child the "first" to experience some things among his or her peers; they made the best decision they could at the time; and so on. Children might also be encouraged to view their own decisions positively.

11-2b Five Basic Parenting Choices

The five basic choices parents make include deciding (1) whether to have a child, (2) the number of children, (3) the interval between children, (4) the method of discipline and guidance, and (5) the degree to which they will be invested in the role of parent. Though all of these decisions are important, the relative importance one places on parenting as opposed to one's career will have implications for the parents, their children, and their children's children. Parents continually make choices in reference to their children, including whether their children will sleep with them in the "family bed."

11-3 Transition to Parenthood

The **transition to parenthood** refers to that period from the beginning of pregnancy through the first few months after the birth of a baby. The mother, the father, and the two of them as a couple undergo changes and adaptations during this period.

11-3a Transition to Motherhood

In a study of 1,035 working adults ages 18 and older, 37% of moms and 16% of dads reported spending more than 8 hours a day parenting (Huggins & Ward,

transition to parenthood period from the beginning of pregnancy through the first few months after the birth of a baby, during which the mother and father undergo changes.

2011). These data reflect that women typically devote more time to parenting. Michele Obama emphasizes that motherhood is her priority. She noted in an interview, "Like any mother, I am just hoping that I don't mess them up," which for her "means getting them out of the White House and paying attention to their lives given the fact that they are in the public spotlight." Mrs. Obama notes that their father's job is the last topic on the menu. "It's sitting down at the dinner table and having Barack's day be the last thing anyone really cares about."

Although childbirth is sometimes thought of as a painful ordeal, some women describe the experience as fantastic, joyful, and unsurpassed. A strong emotional bond between the mother and her baby usually develops early, and both the mother and infant resist separation.

Sociobiologists suggest that the attachment between a mother and her offspring has a biological basis (one of survival). The mother alone carries the fetus in her body for nine months, lactates to provide milk, and, during the expulsive stage of labor, produces **oxytocin**, a hormone from the pituitary gland that has been associated with the onset of maternal behavior in lower animals.

Not all mothers feel joyous after childbirth. Naomi Wolf uses the term "the conspiracy of silence" to note motherhood is "a job that sucks 80 percent of the time" (quoted in Haag, 2011, 83). Some mothers don't bond immediately and feel overworked, exhausted, mild depression, irritability, crying, loss of appetite, and difficulty in sleeping. Many new mothers experience **baby blues**—transitory symptoms of depression 24 to 48 hours after the baby is born. A few, about 10%, experience postpartum depression—a more severe reaction than baby blues.

Postpartum depression is believed to be a result of the numerous physiological and psychological changes that occur during pregnancy, labor, and delivery. Although the woman may become depressed during pregnancy or delivery, she more often experiences these feelings within the first month after returning home with her baby (sometimes the woman does not experience postpartum depression until a couple of years later). Most women recover within a short time; some (between 5 and 10%) become suicidal (Pinheiro et al., 2008).

Gelabert et al. (2012) looked at various personality traits that were associated with postpartum depression. They found that women who were perfectionistic (characterized by concern over mistakes, personal standards, parental expectations, parental criticism, doubt about actions and organization) were more likely to report

oxytocin a hormone released from the pituitary gland during the expulsive stage of labor that has been associated with the onset of maternal behavior in lower animals.

baby blues transitory symptoms of depression twenty-four to forty-eight hours after the baby is born.

postpartum depression a severe reaction following the birth of a baby, which occurs in reference to a complicated delivery as well as numerous physiological and psychological changes; usually in the first month after birth but can be experienced after a couple of years have passed.

© Charlene Johnson

postpartum depression. When compared with a control group, the prevalence of high-perfectionism was higher in the postpartum depression group than in the control group (34% versus 11%). In addition, the high concern over mistakes dimension increased over fourfold the odds of major depression in the postpartum period. New babies create a context of uncertainty—just the lack of normative stability that can unnerve the new mother.

To minimize baby blues and postpartum depression, antidepressants such as Zoloft and Prozac are used. Celebrities such as Brooke Shields have appeared on various talk shows to increase awareness about the issue of postpartum depression, its physiological basis, and the value of medication.

As with mothers, fathers may also experience depression following the birth of a baby. Qing et al. (2011) examined the postnatal reactions of 378 pairs of mothers and fathers and found that some parents of both genders reported postnatal depression (15% of mothers and 13% of fathers). The preference for a male baby was associated with the fathers' depression.

Postpartum psychosis, a reaction in which a woman wants to harm her baby, is experienced by only one or two women per 1,000 births (British Columbia Reproductive Mental Health Program, 2005). While having misgivings about a new infant is normal, the parent who wants to harm the infant should make these feelings known to the partner, a close friend, and a professional.

Hammarberg et al. (2008) assessed the different parenting experiences of those who had difficulty getting pregnant or who used assisted reproductive technology (ART) compared to those who did not use ART. The researchers concluded that, although the evidence is inconclusive, those couples who become pregnant via ART may idealize parenthood, and this outlook might then hinder adjustment and the development of a confident parental identity. Is transition to motherhood similar for lesbian and heterosexual mothers? Not according to Cornelius-Cozzi (2002), who interviewed lesbian mothers and found that the egalitarian norm of the lesbian relationship had been altered; for example, the biological mother became the primary caregiver, and the coparent, who often heard the biological mother refer to the child as "her child," suffered a lack of validation.

11-3b Transition to Fatherhood

Mothers are typically disappointed in the amount of time the father helps with the new baby (Biehle and Michelson 2012). Schindler (2010) found that fathers' engagement in parenting and financial contributions to the family predicted improvements in their psychological well-being. Hence, fathers benefit from active involvement with their children. Schoppe-Sullivan et al. (2008) emphasized that mothers are the "gatekeepers" of the father's involvement with his children. A father may be involved or not involved with his children to the degree that a mother encourages or discourages a father's involvement. The **gatekeeper role** is particularly pronounced after a divorce in which the mother receives custody of the children (the role of the father may be severely limited).

The importance of the father in the lives of his children is enormous and goes beyond his economic contribution (McClain 2011; Bronte-Tinkew et al., 2008; Flouri & Buchanan, 2003). While the role of father is not clearly defined and positive models are lacking (Ready et al. 2011), children who have a father who maintains active involvement in their lives tend to:

- Make good grades
- Be less involved in crime
- Have good health/self-concept
- Have a strong work ethic
- Have durable marriages
- Have a strong moral conscience
- Have higher life satisfaction
- Have higher incomes as adults
- Have higher education levels
- Have two biological parents at home
- Have stable jobs
- Have fewer premarital births
- Have lower incidences of child sex abuse
- Exhibit fewer anorectic symptoms

Daughters may have an extra benefit of a close relationship with fathers. Byrd-Craven et al. (2012) noted that such a relationship was associated with the daughters having lower stress levels which assisted them coping with problems, managing interpersonal relationships, etc. Fathers' involvement is predicted by the quality of the parents' romantic relationship (Gavin et al., 2002). Fathers who are emotionally and physically involved with the mother tend to take an active role in the child's life.

postpartum psychosis a reaction (rare) following the birth of a woman's baby where she wants to harm her baby.

gatekeeper role term used to refer to the influence or control of the mother on the father's involvement and relationship with his children.

Researchers have noted the inadequacy of the stereotype of the "uninvolved" African American father created by data showing that they often do not live in the household with the mother and have sought to redefine father presence in the context of children's feelings of closeness to their father as well as the frequency of father visitation (Thomas et al., 2008). Their findings confirmed that a considerable portion of African American nonresident fathers visit their children on a daily or weekly basis. In addition, African American adult children with nonresident fathers often feel significantly closer to their fathers than do their White peers. Finally, African American adult children were more likely than their White peers to believe that their mothers supported their relationship with their father and to have positive perceptions of their parents' relationship.

11-3c Transition from a Couple to a Family

Research consistently reveals that having a child has a negative effect on marital happiness. Researchers interviewed 137 couples before the birth of their first child and then at 3-, 12-, and 24-month periods. The spouses during the respective time periods consistently reported depression and adjustment through 24 months postpartum (Bost et al., 2002). In another study of 148 samples representing 47,692 individuals on the effect children have on marital satisfaction, Twenge et al. (2003) found that (1) parents (both women and men) reported lower marital satisfaction than nonparents; (2) mothers of infants reported the most significant drop in marital satisfaction; (3) the higher the number of children, the lower the marital satisfaction; and (4) the factors in depressed marital satisfaction were conflict and loss of freedom. Goldberg et al. (2010) found that a decrease in relationship quality across the first year of parenthood occurs regardless of whether the child is biological or adopted and regardless of whether the parents are heterosexual or homosexual. Women experience the steeper decline.

For parents who experience a pattern of decreased happiness, it bottoms out during the teen years. Facer and Day (2004) found that adolescent problem behavior, particularly that of a daughter, is associated with increases in marital conflict. Of even greater impact was the parents' perception of the child's emotional state. Parents who viewed their children as "happy" were less maritally affected by their adolescent's negative behavior.

Regardless of the negative effect children may have on marital happiness, spouses report more commitment to their relationship once they have children (Stanley & Markman, 1992). Figure 11.1 illustrates that the more children a couple has, the more likely the couple will stay married. A primary reason for this increased commitment is the desire on the part of both parents to provide a stable family context for their children. In addition, parents of dependent children may keep their marriage together to maintain a higher standard of living for their children. Finally, people (especially mothers) with small children feel more pressure to stay married (if the partner provides sufficient economic resources) regardless of how unhappy they may be. Hence, though children may decrease happiness, they increase stability because pressure exists to stay together.

Figure 11.1
Percentage of Couples Getting Divorced by Number of Children

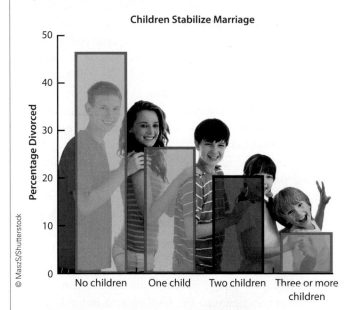

© MaszS/Shutterstock

11-4 Parenthood: Some Facts

Parenting is only one stage in an individual's or couple's life (children typically live with an individual 30% of that person's life and with a couple 40% of their marriage). Parenting involves responding to the varying needs of children as they grow up, and parents require help from family and friends in rearing their children.

Some additional facts of parenthood follow.

11-4a Views of Children Differ Historically

Whereas children of today are thought of as dependent, playful, and adventurous, they were viewed quite differently in the past (Mayall, 2002). Indeed, the concept of childhood, like gender, is a social construct rather than a fixed life stage. From the 13th through the 16th centuries, children were viewed as innocent, sweet, and a source of amusement for adults. From the 16th through the 18th centuries, they were viewed in need of discipline and moral training. In the 19th century, whippings were routine as a means of breaking children's spirits and bringing them to submission. Although remnants of both the innocent and moralistic views of children exist today, the lives of children are greatly improved. Child labor laws protect children from early forced labor, education laws ensure a basic education, and modern medicine has been able to increase the life span of children.

Children of today are the focus of parental attention. In some families, everything they do is "fantastic" and deserves a gold star. The result is a generation of young adults who feel that they are special, who feel that they deserve to be catered to—they are entitled. Nelson (2010) noted that some parents have become "helicopter parents" (also referred to as hovercrafts and PFH—Parents From Hell) in that they are constantly hovering at school and in the workplace to ensure their child's "success." The workplace has become the new field where parents negotiate the benefits and salaries of

their children. However, employers may not appreciate the tampering, and the parents risk hampering their child's development of life skills. Hersh et al. (2011) found that children of helicopter parents are more likely to put off getting married.

11-4b Each Child Is Unique

Children differ in their genetic makeup, physiological wiring, intelligence, tolerance for stress, capacity to learn, comfort in social situations, and interests. Parents soon become aware of the uniqueness of each child—of the child's difference from every other child they know. Parents of two or more children are often amazed at how children who have the same parents can be so different.

Children also differ in their academic capabilities. Increasingly, children are being diagnosed with ADHD, which has implications for their performance in school. Mental health and physical differences also present challenges to parents. Fear that their child will be bullied for being different is an increasing concern.

Although parents often contend, "We treat our children equally," Tucker et al. (2003) found that parents treat children differently, with firstborns usually receiving more privileges than children born later. Suitor and Pillemer (2007) found that elderly mothers reported that they tended to establish a closer relationship with the firstborn child whom they are more likely to call on in later life when there is a crisis.

11-4c Birth Order Effects on Personality

Psychologist Frank Sulloway (1996) examined how a child's birth order influenced the development of various personality characteristics. His thesis is that children with siblings develop different strategies to maximize parental investment in them. For example, children can promote parental favor directly by "helping and obeying parents" (p. 67). Sulloway identified the following personality characteristics that have their basis in a child's position in the family:

"Having a child is like throwing a hand grenade into a marriage."

—Nora Ephron, writer, producer, and director

1. **Conforming or traditional.** Firstborns are the first on the scene with parents and always have the "inside track." They want to stay that way, so they are traditional and conforming to their parent's expectations.

2. **Experimental or adventurous.** Children born later learn quickly that they enter an existing family constellation where everyone is bigger and stronger. They cannot depend on having established territory so must excel in ways different from the firstborn. They are open to experience, adventurousness, and trying new things because their status is not already assured.

3. **Neurotic or emotionally unstable.** Because firstborns are "dethroned" by younger children to whom parents have to divert their attention, they tend to be more jealous, anxious, and fearful. Children born later are never number one to begin with so do not experience this trauma.

As support for his ideas, Sulloway (1996) cited 196 controlled birth order studies and contended that, although there are exceptions, one's position in the family is a factor influencing personality outcomes. Nevertheless, he acknowledged that researchers disagree on the effects of birth order on the personalities of children and that birth order research is incomplete in that it does not consider the position of each child in families that vary in size, gender, and number. Indeed, Sulloway (2007) has continued to conduct research, the findings of which do not always support his prediction. For example, he examined intelligence and birth order among 241,310 Norwegian 18- and 19-year-olds and found no relationship. He recommended that researchers look to how a child was reared rather than birth order for understanding a child's IQ in the family.

11-4d Parents Are Only One Influence in a Child's Development

Although parents often take the credit—and the blame—for the way their children turn out, they are only one among many influences on a child's development. Although parents are the first significant influence, peer influence becomes increasingly important during adolescence. For example, Ali and Dwyer (2010) studied a nationally representative sample of adolescents and found that having peers who drank alcohol was related to the adolescents themselves drinking alcohol.

Siblings also have an important and sometimes lasting effect on each other's development. Siblings are social mirrors and models (depending on the age) for each other. They may also be sources of competition and can be jealous of each other.

Teachers are also significant influences in the development of a child's values. Some parents send their children to religious schools to ensure that they will have teachers with conservative religious values. Selecting this structure for a child's education may continue into college.

Media in the form of television—replete with MTV and "parental discretion advised" movies—are a major source of language, values, and lifestyles for children that may be different from those of the parents. Parents are also concerned about the violence to which television and movies expose their children.

Another influence of concern to parents is the Internet. Though parents may encourage their children to conduct research and write term papers using the Internet, they may fear their children are accessing pornography and related sex sites. Parental supervision of teens on the Internet and teens' right to privacy remain potential sources of conflict.

"If I had my child to raise all over again, I'd build self-esteem first, and the house later. I'd finger-paint more, and point the finger less. I would do less correcting and more connecting."

—DIANE LOOMANS, FROM "IF I HAD MY CHILD TO RAISE OVER AGAIN"

11-4e Parenting Styles Differ

Diana Baumrind (1966) developed a typology of parenting styles that has become classic in the study of parenting. She noted that parenting behavior has two dimensions: responsiveness and demandingness. **Responsiveness** refers to the extent to which parents respond to and meet the needs of their children. In other words, how supportive are the parents? Warmth, reciprocity, person-centered communication, and attachment are all aspects of responsiveness. **Demandingness**, on the other hand, is the manner in which parents place demands on children in regard to expectations and discipline. How much control do they exert over their children? Monitoring and confrontation are also aspects of demandingness. Categorizing parents in terms of their responsiveness and their demandingness creates four categories of parenting styles: permissive (also known as indulgent), authoritarian, authoritative, and uninvolved.

1. Permissive parents are high on responsiveness and low on demandingness. They are very lenient and allow their children to largely regulate their own behavior.

2. Authoritarian parents are high on demandingness and low in responsiveness. They feel that children should obey their parents no matter what, and they provide a great deal of structure in the child's world.

3. Authoritative parents are both demanding and responsive. This style offers a balance of warmth and control and is associated with the most positive outcome for children (fewer behavior problems; Tan et al., 2012). Walcheski and Bredehoft (2010) gave examples of the authoritative parenting style—parents telling the child their expectations of the child's behavior before the child engages in the activity, giving the child reasons why rules should be obeyed, talking with the child when he or she has misbehaved, and explaining consequences. Chen et al. (2012) found that the authoritative parenting style was the type most commonly used by Chinese parents.

4. Uninvolved parents are low in responsiveness and demandingness. These parents are not invested in their children's lives.

McKinney and Renk (2008) identified the differences between maternal and paternal parenting styles, with mothers tending to be authoritative and fathers tending to be authoritarian. Mothers and fathers also use different parenting styles for their sons and daughters, with fathers being more permissive with their sons than with their daughters. Overall, this study emphasizes the *importance* of examining the different parenting styles of parents and adolescent outcome and suggests that having one authoritative parent may be a protective factor for late adolescence. Recall that the authoritative parenting style is the combination of warmth, guidelines, and discipline—"I love you but you need to be in by midnight or lose privileges of having a cell phone and a car."

responsiveness refers to the extent to which parents respond to and meet the needs of their children.

demandingness the manner in which parents place demands on children in regard to expectations and discipline.

11-5 Principles of Effective Parenting

Numerous principles are involved in being an effective parent (Keim & Jacobson, 2011). We begin with the most important of these, which

© Andrea Gingerich/iStockphoto.com / © Greg Nicholas/iStockphoto.com

involves giving time and love to children as well as praising and encouraging them.

11-5a Give Time, Love, Praise, and Encouragement

Children most need to feel that they are worth spending time with and that someone loves them. Because children depend first on their parents for the development of their sense of emotional security, it is critical that parents provide a warm emotional context in which the children can develop. Feeling loved as an infant also affects one's capacity to become involved in adult love relationships.

As children mature, positive reinforcement for prosocial behavior also helps to encourage desirable behavior and a positive self-concept. Instead of focusing only on correcting or reprimanding bad behavior, parents should frequently comment on and reinforce good behavior. Comments such as "I like the way you shared your toys," "You asked so politely," and "You did such a good job cleaning your room" help to reinforce positive social behavior and may enhance a child's self-concept. However, parents need to be careful not to overpraise their children, as too much praise may lead to the child's striving to please others rather than trying to please herself or himself.

Praise focuses on other people's judgments of a child's actions, whereas encouragement focuses more on the child's efforts. For example, telling a child who brings you their painting, "I love your picture; it is the best one that I have ever seen," is not

as effective in building the child's confidence as saying, "You worked really hard on your painting. I notice that you used lots of different colors." Some parents feel that rewarding positive behavior is not a good idea.

11-5b Be Realistic

Parents should rid themselves of the illusion of childhood honesty. Talwar and Lee (2008) confirmed that children lie. In an experiment, 138 children (3 to 8 years) were told not to peek at a toy—82% peeked in the experimenter's absence, and 64% lied about their transgression.

11-5c Avoid Overindulgence

Overindulgence may be defined as more than just giving children too much. It includes overnurturing and providing too little structure. Using this definition, a study of 466 participants revealed that those who are overindulged tend to hold materialist values for success, are not able to delay gratification, and are less grateful for things and to others. Indeed, not being overindulged promotes the ability to delay gratification, be grateful, and experience a view (nonmaterialistic) that is associated with happiness (Slinger & Bredehoft, 2010). The book *How Much Is Enough?* emphasizes attentiveness to avoiding overindulgence (Dawson et al., 2003).

Parents typically overindulge because they feel guilty or because they did not have certain material goods in their own youth. In a study designed to identify who overindulged, a researcher found mothers were four times more likely to overindulge than fathers (Clarke, 2004). The result of overindulgence is that children grow up without consequences, and they avoid real jobs where employers expect them to show up at 8:00 a.m. Corporate employers have discovered that Generation Y youth (79.8 million of them born between 1977 and 1995) are no longer anxious to work or appreciative of getting a job. Rather, they want to know their benefits, and they make it clear they have no intention of working 60 hours a week.

> *"Obstinacy in children is like a kite; it is kept up just as long as we pull against it."*
> —Marlene Cox, coach/mentor

11-5d Monitor Activities and Drug Use

Abundant research suggests that parents who monitor their children and teens and know where their children are, who they are

© Jodi Matthews/iStockphoto.com

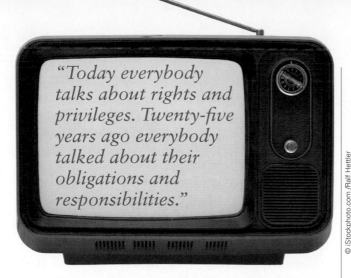

"Today everybody talks about rights and privileges. Twenty-five years ago everybody talked about their obligations and responsibilities."

—Lou Holtz, football television commentator

with, and what they are doing are less likely to report that their adolescents receive low grades or are engaged in early sexual activity, delinquent behavior, and alcohol or drug use (Crosnoe & Cavanagh, 2010). Regarding alcohol use, Crutzen et al. (2012) assessed the effects of parental approval of children's drinking alcohol at home on subsequent drinking behavior of the children. In a study (which took place in the Netherlands) of 1,500 primary school children, those children who thought they were not allowed to drink at home were more likely to use alcohol. Hence, children who were allowed to drink alcohol around their parents were less likely to consume alcohol when they were away from their parents.

Parents who drank alcohol under age and who used marijuana or other drugs wonder how to go about encouraging their own children to be responsible alcohol users and drug free. Drugfree.org has some recommendations for parents, including being honest with their children about previous alcohol and drug use, making clear that they do not want their children to use alcohol or drugs, and explaining that although not all alcohol or drug use leads to negative consequences, staying clear of such possibilities is the best course of action.

11-5e Set Limits and Discipline Children for Inappropriate Behavior

The goal of limits and discipline is self-control. Parents want their children to be able to control their own behavior and to make good decisions without their parents.

Children engage in behavior in reference to consequences. When parents express approval to their children for showing acceptable behavior and punish them for being disobedient (e.g., by imposing a time out or withdrawing privileges), children are likely to behave appropriately. Unless parents tell their children that they appreciate their honesty and nonaggressive behavior and provide negative consequences for lying, stealing, and hitting, their children can grow up to be dishonest, to steal, and to be inappropriately aggressive. Children learn what they are taught.

Time-out (a noncorporal form of punishment that involves removing the child from a context of reinforcement to a place of isolation for one minute for each year of the child's age) has been shown to be an effective consequence for inappropriate behavior. Withdrawal of privileges (watching television, playing with friends), pointing out the logical consequences of the misbehavior ("you were late; we won't go"), and positive language ("I know you meant well but . . .") are also effective methods of guiding children's behavior.

Jordan and Curtner-Smith (2011) studied a sample of 374 undergraduates and found that mothers (the primary disciplinarians) who relied on corporal punishment as a method of discipline incurred negative feelings in the child, which affected the child's attachment to the mother. African American children were at greater risk for corporal punishment and negative attachment.

11-5f Provide Security

Predictable responses from parents, a familiar bedroom or playroom, and an established routine help to encourage a feeling of security in children. Security provides children with the needed self-assurance to venture beyond the family. If the outside world becomes too frightening or difficult, a child can return to the safety of the family for support. Knowing it is always possible to return to an accepting environment enables a child to become more involved with the world beyond the family.

11-5g Encourage Responsibility

Giving children increased responsibility encourages the autonomy and independence they need to be assertive and self-governing. Giving children more responsibility as they grow older can take the form of encouraging them to choose healthy snacks and letting them decide what to wear and when to return from playing with a friend (of course, the parents should praise appropriate choices).

time-out a noncorporal form of punishment that involves removing the child from a context of reinforcement to a place of isolation

Children who are not given any control over, and responsibility for, their own lives remain dependent on others. Successful parents can be defined in terms of their ability to rear children who can function as independent adults. A dependent child is a vulnerable child.

Some American children remain in their parents' house into their 20s and 30s. They do so primarily because it is cheaper. Ward and Spitze (2007) analyzed data from the National Survey of Families and Households in regard to children ages 18 and older who lived with their parents. Findings revealed that although disagreements between parents and children increased, the quality of parent–child or husband–wife relations did not change.

11-5h Teach Emotional Competence

Wilson et al. (2012) emphasized the importance of teaching children **emotional competence**—experience emotion, express emotion, and regulate emotion. Being able to label when one is happy or sad (experience emotion), to express emotion ("I love you"), and to regulate emotion (e.g. anger) assists children in "getting in touch with their feelings" and being empathetic with others. Wilson et al. (2012) reported on "Tuning in the Kids," a training program for parents to learn how to teach their children to be emotionally competent. Follow up data on parents who took the six-session, two-hour-a-week program revealed positive outcomes and changes.

emotional competence capacity to experience emotion, express emotion, and regulate emotion.

menarche first menstruation signaling a woman's fertility.

nature-deficit disorder children who are not encouraged and who have little opportunity to have direct contact with nature; instead they are overscheduled with activities such as soccer, swimming, etc.

11-5i Provide Sex Education and Teach Nonviolence

Wilson et al. (2010) found that although parents value talking with their children about sex, they often do not do so because they feel their children are "too young" or they don't know how to talk with their children about sex. Those who found it easiest to talk about sex with their children reported very positive parent–child relationships and talked about sex when their children were young. Hence, there are definite benefits when parents talk with their children about sex.

In regard to adolescent females experiencing their first period, Lee (2008) reported that, of 155 young women, most said that their mothers were supportive and emotionally engaged with them regarding **menarche**. These emotionally connected mothers were, for the most part, able to mitigate feelings of shame and humiliation associated with the onset of menstruation in contemporary culture.

In addition to teaching children about sex, effective parenting involves teaching children to be nonviolent—to handle anger, stress, and conflict with others in productive ways. Researchers Knox et al. (2011) found that children of parents who take the eight-week ACT Raising Safe Kids Parenting Program evidence fewer behavior problems and more skill in managing conflict. For example, children learn it is okay to feel anger but not to express violence.

11-5j Establish Norm of Forgiveness

Carr and Wang (2012) emphasized the importance of forgiveness in family relationships and the fact that it is a complex, time-involved process rather than a one-time cognitive event (e.g. "I forgive you"). Respondents revealed in interviews with the researchers that forgiveness involved head over heart (finding ways to explain the transgression), time (sometimes months and years), and distance (giving the relationship a rest and coming back with renewed understanding). Former governor of New York Elliott Spitzer was asked if there was anything positive about his fall from grace after he was caught having sex with prostitutes. He replied that the depth of his wife's forgiveness was beyond what one considers possible.

11-5k Keep Children Connected With Nature

Louv (2006) used the term **nature-deficit disorder** to describe the result of encouraging children to

© Paul Matthew Photography/Shutterstock

© David Knox

avoid direct contact with nature—playing in the woods, wading through a stream, catching tadpoles—by means of overscheduling them with planned "activities" such as swimming, tennis, gymnastics, soccer, and piano lessons, which leaves little time for anything else. He recommended that children need such contact with nature just as they need good nutrition and adequate sleep.

11-5| Respond to the Teen Years Creatively

Research has revealed that teenagers have unique brains with lower amounts of dopamine, which may disrupt their reward function and make them less responsive to social stimuli (Forbes & Dahl, 2012). Teenagers are more likely to defy authority, act rebellious, and engage in risky behavior. Other teens are more likely to smoke, drink, drive recklessly, take dares, and engage in sexual behavior (Becker, 2010).

Conflicts between parents and teenagers often revolve around money and independence. The desires for a cell phone, iPod, iPad, and flat-screen TV can outstrip the budget of many parents. Teens also increasingly want more freedom. However, neither of these issues needs to result in conflicts. When they do, the effect on the parent–child relationship may be inconsequential. One parent tells his children, "I'm just being the parent, and you're just being who you are; it is okay for us to disagree—but you can't go."

Sometimes teenagers present challenges with which the parents feel unable to cope. Aside from monitoring their behavior closely, family therapy may be helpful. Two major goals of such therapy are to increase the emotional bond between the parents and the teenagers and to encourage positive consequences for desirable behavior (e.g., concert tickets for good grades) and negative consequences for undesirable behavior (e.g., loss of car privileges for getting a speeding ticket).

11-6 Single-Parenting Issues

Forty percent of births in the United States are to unmarried mothers. By adolescence, 20% of children have no contact with their father (Doherty & Craft, 2011). Some children may start out with a father figure in their lives, but over time, there is less involvement by the father.

Distinguishing between a single-parent "family" and a single-parent "household" is important. A single-parent family is one in which there is only one parent—the other parent is completely out of the child's life through death or complete abandonment or as a result of sperm donation, and no contact is ever made. In contrast, a single-parent household is one in which one parent typically has primary custody of the child or children, but the parent living out of the house is still a part of the child's life. This arrangement is also referred to as a binuclear family. In most divorce cases in which the mother has primary physical custody of the child, the child lives in a single-parent household because the child is still connected to the father, who remains part of the child's life. In cases in which one parent has died, the child or children live with the surviving parent in a single-parent family because there is only one parent.

11-6a Single Mothers by Choice

Single parents enter their role through divorce or separation, widowhood, adoption, or deliberate choice to rear a child or children alone. Jodie Foster, an Academy Award–winning actress, has elected to have children without a husband. She now has two children and smiles when asked, "Who's the father?" The implication is that choosing to have a single-parent family is a viable option. Sandra Bullock also adopted a child as a single mother. An organization for women who want children and who may or may not marry is Single Mothers by Choice.

Bock (2000) noted that single mothers by choice are, for the most part, in the middle to upper class, mature, well-employed, politically aware, and dedicated to motherhood. Interviews with 26 single mothers by choice revealed their struggle to avoid stigmatization and to seek legitimization for their choice. Most felt that their age (older), sense of responsibility, maturity, and

PUT OUT THE FIRE

Keeping conflicts with teens to a low level

The following suggestions can help to keep conflicts with teenagers at a low level:

1. Catch them doing what you like rather than criticizing them for what you don't like. Adolescents are like everyone else—they don't like to be criticized, but they do like to be noticed for what they do that is good.

2. Be direct when necessary. Though parents may want to ignore some behaviors of their children, addressing some issues directly may also be effective. Regarding the avoidance of STIs or HIV infection and pregnancy, Dr. Louise Sammons tells her teenagers, "It is utterly imperative to require that any potential sex partner produce a certificate indicating no STIs or HIV infection and to require that a condom or dental dam be used before intercourse or oral sex" (personal communication, 2008).

3. Provide information rather than answers. When teens are confronted with a problem, try to avoid making a decision for them. Rather, provide information on which they may base a decision. What courses to take in high school and what college to apply for are decisions that might be made primarily by the adolescent. The role of the parent might best be to listen. The website Radical Parenting (http://www.radicalparenting.com/) focuses on what teens are thinking (and what parents want to know).

4. Be tolerant of high activity levels. Some teenagers are constantly listening to loud music, going to each other's homes, and talking on cell phones for long periods. Parents often want to sit in their easy chairs and be quiet. Recognizing that it is not realistic to expect teenagers to be quiet and sedentary may be helpful in tolerating their disruptions.

5. Engage in some activity with your teenagers. Whether renting a DVD, eating a pizza, or taking a camping trip, structuring some activities with your teenagers is important. Such activities permit a context in which to communicate with them.

© Inspirestock/Jupiterimages/© Hemera Technologies/Jupiterimages

fiscal capability justified their choice. Their self-concepts were those of competent, ethical, mainstream mothers.

11-6b Challenges Faced by Single Parents

The single-parent lifestyle involves numerous challenges. See http://singleparent.lifetips.com/ for some interesting tips. Challenges associated with being a single parent include the following.

1. **Responding to the demands of parenting with limited help.** Perhaps the greatest challenge for single parents is to take care of the physical, emotional, and disciplinary needs of their children—alone. Solem et al. (2011) noted that single parents more often are less educated, are less likely to be employed, and have less social support and that these factors weigh against building a protective frame around a child, resulting in more behavioral problems.

2. **Meeting adult emotional needs.** Single parents have emotional needs of their own that children are often incapable of satisfying. The unmet need to share an emotional relationship with an adult can weigh heavily on a single parent. Many single women solve this problem by reaching out to their own parents, extended kin, and a network of friends.

3. **Meeting adult sexual needs.** Some single parents regard their parental role as interfering with their sexual relationships. They may be concerned that their children will find out if they have a sexual encounter at home or be frustrated if they have to go away from home to enjoy a sexual relationship. Some choices with which they are confronted include, "Do I wait until my children are asleep and then ask my lover to leave before morning?" or "Do I openly acknowledge my lover's presence in my life to my children and ask them not to tell anybody?" and "Suppose my kids get attached to my lover, who may not be a permanent part of our lives?"

4. **Coping with the lack of money.** The median income of a single-woman householder is $29,770; it is $71,627 for a married couple (*Statistical Abstract of the United States*, 2012, Table 698).

5. **Ensuring guardianship.** If the other parent is completely out of the child's life, the single parent needs to appoint a guardian to take care of the child in the event of the parent's death or disability.

6. **Obtaining prenatal care.** Children born to single-parent mothers are likely to be born prematurely and to have low birth weight (Mashoa et al., 2010) Reasons include a lack of funds (no partner with economic resources is available) and the lack of social support for the pregnancy or the working conditions of the mothers, all of which result in less prenatal care for their babies.

7. **Coping with the absence of a father.** Another consequence for children of single-parent mothers is that they often do not have the opportunity to develop an emotionally supportive relationship with their father. Barack Obama noted in one of his campaign speeches, "I know what it is like to grow up without a father." The late comedian Rodney Dangerfield said that in his entire life he spent an average of two hours a year with his father. Shook et al. (2010) noted that when single mothers receive financial support from the father who is also active in the role of coparent, child competence increased. When neither of these behaviors occurred, child maladjustment was more likely.

8. **Avoiding negative life outcomes for the child in a single-parent family.** Researcher Sara McLanahan, herself a single mother, set out to prove that children reared by single parents were just as well off as those reared by two parents. McLanahan's data on 35,000 children of single parents led her to a different conclusion—children of only one parent were twice as likely as those reared by two married parents to drop out of high school, get pregnant before marriage, have drinking problems, and experience a host of other difficulties, including getting divorced themselves (McLanahan, 1991; McLanahan & Booth, 1989). Lack of supervision, fewer economic resources, and less extended-family support were among the culprits. Other research suggests that negative outcomes are reduced or eliminated when income levels remain stable (Pong & Dong, 2000).

Though the risk of negative outcomes is higher for children in single-parent homes, most are happy and well-adjusted. Benefits to single parents themselves include a sense of pride and self-esteem that results from being independent.

STUDY TOOLS 11

Ready to study? In this book, you can:

- ⊃ Rip out the Chapter Review card in the back of the book to study for exams

- ⊃ Take the Self Assessment for this chapter (card in the back of the book) and see where you stand on the vital issues raised in the chapter

Or you can go online to CourseMate at www.cengagebrain.com for these resources:

- ⊃ Complete Practice Quizzes to prepare for tests

- ⊃ Review Key Terms Flash Cards (online or print)

- ⊃ Read about Marriage and Family in the news

- ⊃ Play "Beat the Clock" to master concepts

- ⊃ Check out Personal Applications

Stress and Crisis in Relationships

"Life is not about waiting for the **storm to pass . . .** it is about **learning** to **dance** in the **rain."**

—UNKNOWN

SECTIONS

12-1 Personal Stress and Crisis Events

12-2 Positive Stress-Management Strategies

12-3 Harmful Stress-Management Strategies

12-4 Family Crisis Examples

A married couple enjoyed their respective careers and looked forward to retirement and travel. Because they both had lucrative careers, they put off retirement. When the husband reached 55, he suddenly developed MS and was in a wheelchair by age 60. Their traveling days came to an abrupt end; he died at age 63. The wife noted that a personal, marital, or family crisis can come at any time and that one should not put off living because there is no guarantee of tomorrow. In this chapter we review crisis events that alter individuals and relationships. Our focus is not only about experiencing these events but coping with them over time.

12-1 Personal Stress and Crisis Events

In this section, we review the definitions of crisis and stressful events, the characteristics of resilient families, and a framework for viewing a family's reaction to a crisis event.

12-1a Definitions of Stress and Crisis Events

stress a nonspecific response of the body to demands made on it.

Stress is a reaction of the body to substantial or unusual demands (physical, environmental, or interpersonal). Stress is often accompanied by irritability, high blood pressure, and depression (Barton & Kirtley,

2012). Stress also has an effect on a person's relationships and sex life. Bodenmann et al. (2010) found that stress in daily life was associated with a decrease in relationship satisfaction and lower levels of sexual activity.

Stress is a process rather than a state. For example, a person will experience different levels of stress throughout a divorce—acknowledging that one's marriage is over, telling the children, leaving the family residence, getting the final decree, and seeing one's ex may all result in varying levels of stress.

> "Stress is an **ignorant state.** It believes that **everything** is an **emergency.**"
>
> —NATALIE GOLDBERG, *WILD MIND*

A **crisis** is a situation that requires changes in normal patterns of behavior. A family crisis is a situation that upsets the normal functioning of the family and requires a new set of responses to the stressor. Sources of stress and crises can be external, such as the hurricanes that annually hit our coasts or devastating tornadoes in the spring. Other examples of an external crisis are economic recession, downsizing, or military deployment to Afghanistan. Stress and crisis events may also be induced internally (e.g., alcoholism, an extramarital affair, or Alzheimer's disease in a spouse or parents). Peilian et al. (2011) studied stressful life events of 1,749 participants in seven Chinese cities and found that crisis events reduced marital satisfaction. (The rapid assault of crisis events can be a challenge for individuals and couples. Getting news of one's ill health, losing one's job, and having an aging parent who needs full-time care with no siblings to help can produce overload.)

Stressors or crises may also be categorized as expected or unexpected. Examples of expected family stressors include the need to care for aging parents and the death of one's parents. Unexpected stressors include contracting human immunodeficiency virus (HIV), having a miscarriage, or experiencing the suicide of one's teenager.

Both stress and crisis are normal parts of family life and sometimes reflect a developmental sequence. Pregnancy, childbirth, job change or loss, children's leaving home, retirement, and widowhood are all stressful and predictable for most couples and families. Crisis events may have a cumulative effect: the greater the number in rapid succession, the greater the stress.

crisis a sharp change for which typical patterns of coping are not adequate and new patterns must be developed.

resiliency the ability of a family to respond to a crisis in a positive way.

family resilience the successful coping of family members under adversity that enables them to flourish with warmth, support, and cohesion.

12-1b Resilient Families

Just as the types of stress and crisis events vary, individuals and families vary in their abilities to respond successfully to crisis events. **Resiliency** refers to a family's strengths and ability to respond to a crisis in a positive way. Black and Lobo (2008) defined **family resilience** as the successful coping of family members under adversity that enables them to flourish with warmth, support, and cohesion. The key factors that promote family resiliency include positive outlook, spirituality, flexibility, communication, financial management, shared family recreation, routines or rituals, and support networks. A family's ability to bounce back from a crisis (from loss of one's job to the death of a family member) reflects its level of resiliency. Resiliency may also be related to individuals' perceptions of the degree to which they are in control of their destiny.

12-2 Positive Stress-Management Strategies

Researchers Burr and Klein (1994) administered an 80-item questionnaire to 78 adults to assess how families experiencing various stressors, such as bankruptcy, infertility, a disabled child, and a troubled teen, used various coping strategies and how useful they felt these strategies were. In the following sections, we detail some helpful stress-management strategies.

12-2a Changing Basic Values and Perspective

The strategy that the highest percentage of respondents reported as being helpful was changing basic values as a result of the crisis situation. Survivors of

"Our remedies oft in ourselves do lie."
—*Shakespeare,* All's Well That Ends Well

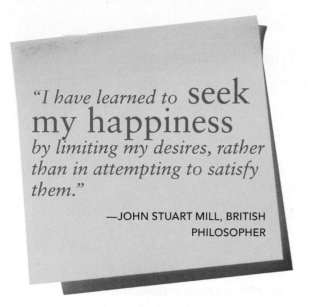

hurricanes and tornadoes who focus on the sparing of their lives rather than the loss of their home or material possessions rebound more quickly. Sharpe and Curran (2006) confirmed that finding positive meaning in a crisis situation is associated with positive adjustment to the situation. Similarly, Waller (2008) studied new parents at two time frames (when the child was 1 and 4), and noted that whether they survived the crisis of having a child was a function of their perception. She found that "parents in stable unions framed tensions as manageable within the context of a relationship they perceived to be moving forward, whereas those in unstable unions viewed tensions as intolerable in relationships they considered volatile." Positive cognitive functioning is associated with keeping the crisis in perspective and moving on (Phillips et al., 2012).

Some crisis events provide an opportunity for positive growth. In responding to the crisis of bankruptcy, people may reevaluate the importance of money and conclude that relationships are more important. In coping with unemployment, people may decide that the amount of time they spend with family members is more valuable than the amount of time they spend making money. Buddhists have the saying, "Pain is inevitable; suffering is not." This is another way of emphasizing that how one views a situation, not the situation itself, determines its impact on you.

12-2b **Exercise**

The Centers for Disease Control and Prevention (CDC) and the American College of Sports Medicine (ACSM) recommend that people ages 6 years and older engage regularly, preferably daily, in light to moderate physical activity for at least 30 minutes at a time. Tetlie et al. (2008) confirmed that a structured exercise program lasting 8 to 12 weeks was associated with individuals reporting improved feelings of well-being. In addition, Taliaferro et al. (2008) noted that vigorous exercise and involvement in sports are associated with lower rates of suicide among adolescents. Exercise is the most important of all behaviors for good health (see http://www.youtube.com /watch?feature=player_embedded&v=aUaInS6HIGo).

12-2c **Friends and Relatives**

A network of relationships is associated with successful coping with various life transitions. News media covering hurricanes, earthquakes, and tsunamis emphasize that, as long as their family is alive and together, individuals view the loss of possessions as irrelevant.

12-2d **Love**

A love relationship also helps individuals cope with stress. Being emotionally involved with another and sharing the experience with that person helps to insulate individuals from being devastated by a crisis event. Love is also viewed as helping resolve relationship problems. Over 85% (85.9%) of undergraduate males and 72.5% of undergraduate females agreed with the statement, "If you love someone enough, you will be able to resolve your problems with that person" (Dotson-Blake et al., 2010).

© Radius Images/Jupiterimages

12-2e Religion and Spirituality

Ellison et al. (2011) examined the role of religion in marital satisfaction and coping with stress. They found that **sanctification** (viewing the marriage as having divine character or significance) is associated with both predicting positive marital quality and providing a buffer for financial stress and general stress on the marriage. Green and Elliott (2010) also noted that those who identify as religious report better health and marital satisfaction. Not only does religion provide a rationale for one's plight ("It is God's will"), but it also offers a mechanism to ask for help. Religion is also a social institution that connects one to others who may offer both empathy and help.

12-2f Humor

The role of humor in coping with stress has been the focus of research (Doosje et al., 2012). Humor is related to lower anxiety and a happier mood and is more effective than aerobic exercise or listening to music for relieving stress. Just sitting quietly seems to have no effect on lowering anxiety (Szabo et al., 2005). Smedema et al. (2010) confirmed that humor was a positive mechanism for coping with stressful events.

12-2g Sleep

Getting an adequate amount of sleep is also associated with lower stress levels. Even midday naps are associated with positive functioning, particularly memory and cognitive function (Pietrzak et al., 2010). Indeed, adequate sleep helps one to cope with crisis events.

12-2h Biofeedback

sanctification
viewing the marriage as having divine character or significance.

biofeedback
a process in which information that is relayed back to the brain enables people to change their biological state.

Biofeedback is a process in which information that is relayed back to the brain enables people to change their biological state. Biofeedback treatment teaches a person to influence biological responses such as heart rate, nervous system arousal, muscle contractions, and even brain-wave functioning. Biofeedback is used at about 1,500 clinics and treatment centers worldwide. A typical session lasts about an

> ### "The only way you can hurt your body is not use it. Inactivity is the killer—it is never too late."
>
> —JACK LALANNE, FITNESS GURU WHO DIED AT AGE 96

hour and costs $60 to $150. The following are several types of biofeedback:

1. **Electromyographic (EMG) biofeedback.** EMG measures electrical activity created by muscle contractions and is often used for relaxation training and for stress and pain management.

2. **Thermal or temperature biofeedback.** Because stress causes blood vessels in the fingers to constrict, reducing blood flow and leading to cooling, thermal biofeedback uses a temperature sensor to detect changes in temperature of the fingertips or toes. This biofeedback trains people to quiet the nervous system arousal mechanisms that produce hand and/or foot cooling and is often used for stress, anxiety, and pain management.

3. **Galvanic skin response (GSR) biofeedback.** GSR utilizes a finger electrode to measure sweat gland activity. This measure is very useful for relaxation and stress management training and is also used in the treatment of attention deficit/hyperactivity disorder.

4. **Neurofeedback.** Also called *neurobiofeedback* or *EEG (electroencephalogram) biofeedback*, neurofeedback may be particularly helpful for individuals coping with a crisis. It trains people to enhance their brain-wave functioning and has been found to be effective in treating a wide range of conditions, including anxiety, stress, depression, tension and migraine headaches, addictions, and high blood pressure.

Because neurofeedback is the fastest-growing field in biofeedback, we take a closer look at this treatment modality. Neurofeedback involves a series of sessions in which a client sits in a comfortable chair facing a specialized game computer. Small sensors are placed on the scalp to detect brain-wave activity and transmit this information to

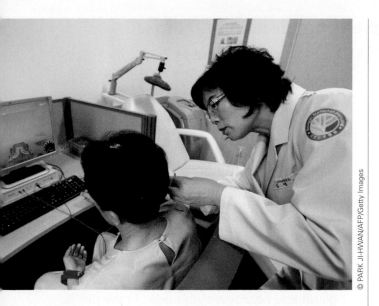
© PARK JI-HWAN/AFP/Getty Images

the computer. The neurofeedback therapist (in the same room) also sits in front of a computer that displays the client's brain-wave patterns in the form of an electroencephalogram (EEG). After a clinical assessment of the client's functioning, the therapist determines what kinds of brain-wave patterns are optimal for the client. During neurofeedback sessions, clients learn to produce desirable brain waves by controlling a computerized game or task, similar to playing a video game, but the client's brain waves—instead of a joystick—control the game.

Neurofeedback is like an exercise of the brain, helping it to become more flexible and effective. Unlike body exercise, which will lose its benefits over time when training is stopped, brain-wave training generally does not diminish. Once the brain is trained to function in its optimal state (which may take an average of 20 to 25 sessions), it generally remains in this more healthy state. Neurofeedback therapists liken the process to that of learning to ride a bicycle: Once you learn to ride a bike, you can do so even if you have not ridden in years. As one neurofeedback therapist explains, "Clients speak often of their disorders—panic attacks, chronic pain, etc.—as if they were stuck in a certain pattern of response. Consistently in clinical practice, EEG biofeedback helps 'unstick' people from these unhealthy response patterns" (J. Carlson-Catalano, personal communication, June 9, 2003).

12-2i Deep Muscle Relaxation

Tensing and relaxing one's muscles have been associated with an improved state of relaxation. Calling the activity abbreviated progressive muscle relaxation (APMR), Termini (2006) found that the cognitive benefits were particularly evident; people who tensed and relaxed various muscle groups noticed a mental relaxation more than a physical relaxation.

12-2j Education

Sometimes becoming informed about a family problem helps individuals to cope with the problem. Friedrich et al. (2008) studied how siblings cope with the fact that a brother or sister is schizophrenic. Education and family support were the primary coping mechanisms. Becoming informed about schizophrenia helped siblings understand that the "parents were not to blame."

12-2k Pets

Hughes (2011) emphasized that animals are associated with reducing blood pressure and stress, preventing heart disease, and fighting depression. Veterinary practices are encouraged to increase the visibility of this connection so that more individuals might benefit.

12-3 Harmful Stress-Management Strategies

Some coping strategies not only are ineffective for resolving family problems but also add to the family's stress by making the problems worse. Respondents in the Burr and Klein (1994) research identified several strategies they regarded as harmful to overall family functioning. These included keeping feelings inside, taking out frustrations on or blaming others, and denying or avoiding the problem.

Burr and Klein's research also suggests that women and men differ in their perceptions of the usefulness of various coping strategies. Women were more likely than men to view as helpful such strategies as sharing concerns with relatives and friends, becoming more involved in religion, and expressing emotions. Men were more likely than women to use potentially harmful strategies such as using alcohol, keeping feelings inside, or keeping others from knowing how bad the situation was.

"But do not distress yourself with dark imaginings.
Many fears are born of fatigue and loneliness."

—MAX EHRMANN, FROM *DESIDERATA*

12-4 Family Crisis Examples

Some of the more common crisis events that spouses and families face include physical illness, mental challenges, an extramarital affair, unemployment, substance abuse, and death.

12-4a Physical Illness and Disability

Absence of physical illness is important for individuals to define themselves as being healthy overall. Awareness that one has a disease that is life threatening increases one's stress level. The following example of coping with prostate cancer makes the point.

Reacting to Prostate Cancer: One Husband's Experience For men, prostate cancer is an example of a medical issue that can rock the foundation of their personal and marital well-being. A student described his experience with prostate cancer as follows:

> Since my father had prostate cancer, I was warned that it is genetic and to be alert as I reached age 50. At the age of 56, I noticed that I was getting up more frequently at night to urinate. My doc said it was probably just one of my usual prostate infections, and he prescribed the usual antibiotic. When the infection did not subside, a urologist did a transrectal ultrasound (TRUS) needle biopsy. The TRUS gives the urologist an image of the prostate while he takes about ten tissue samples from the prostate with a thin, hollow needle.
>
> Two weeks later I learned the bad news (I had prostate cancer) and the good news (it had not spread). Since the prostate is very close to the spinal column, failure to act quickly can allow time for cancer cells to spread from the prostate to the bones. My urologist outlined a number of treatment options, including traditional surgery, laparoscopic surgery,

radiation treatment, implantation of radioactive "seeds," cryotherapy (in which liquid nitrogen is used to freeze and kill prostate cancer cells), and hormone therapy, which blocks production of the male sex hormones that stimulate growth of prostate cancer cells. He recommended traditional surgery ("radical retropubic prostatectomy"), in which an incision is made between the belly button and the pubic bone to remove the prostate gland and nearby lymph nodes in the pelvis. This surgery is generally considered the "gold standard" when the disease is detected early. Within three weeks I had the surgery.

> Every patient awakes from radical prostate surgery with urinary incontinence and impotence—which can continue for a year, two years, or forever. Such patients also awake from surgery hoping that they are cancer-free. This is determined by laboratory analysis of tissue samples taken during surgery. The patient waits for a period of about two weeks, hoping to hear the medical term "negative margins" from his doctor. That finding means that the cancer cells were confined to the prostate and did not spread past the margins of the prostate. A finding of negative margins should be accompanied by a PSA (prostate specific antigen) score of zero, confirming that the body no longer detects the presence of cancer cells. I cannot describe the feeling of relief that accompanies such a report, and I am very fortunate to have heard those words used in my case.
>
> The psychological effects have been devastating—more for me than for my partner. I have only been intimate with one woman—my wife—and having intercourse with her was one of the greatest pleasures in my life. For a year, I was left with no erection and an inability to have an orgasm. Afterwards, I was able to have an erection (via a self-injection of Alprostadil) and an attenuated (weakened) orgasm. Dealing with urinary incontinence (I

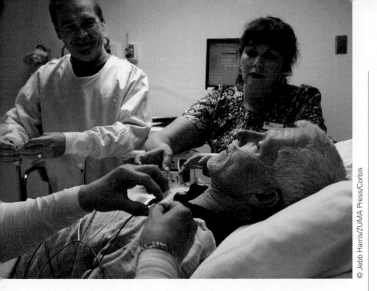
© Jebb Harris/ZUMA Press/Corbis

refer to myself as Mr. Drippy) is a "wish it were otherwise" on my psyche.

When faced with the decision to live or die, the choice for most of us is clear. In my case, I am alive, cancer-free, and enjoying the love of my life (now in our 44th year together). (Reprinted with permission from the author.)

In addition to prostate cancer, chronic degenerative diseases (autoimmune dysfunction, rheumatoid arthritis, lupus, Crohn's disease, and chronic fatigue syndrome) can challenge a mate's and couple's ability to cope. These illnesses are particularly distressing because conventional medicine has little to offer besides pain medication. For example, spouses with chronic fatigue syndrome may experience financial consequences ("I could no longer meet the demands of my job so I quit"), gender role loss ("I couldn't cook for my family" or "I was no longer a provider"), and changed perceptions by their children ("They have seen me sick for so long they no longer ask me to do anything"). The problem is compounded because insurance companies will not pay for nontraditional treatments (e.g., integrative medicine).

In those cases in which the illness is fatal, **palliative care** is helpful. This term describes the health care for the individual who has a life-threatening illness (focusing on relief of pain and suffering) and support for the individual and his or her loved ones. Such care may involve the person's physician or a palliative care specialist who works with the physician, nurse, social worker, and chaplain. Pharmacists or rehabilitation specialists may also be involved. The goals of such care are to approach the end of life with planning (how long should life be sustained on machines?) and forethought to relieve pain and provide closure.

Another physical issue with which some parents cope is that of their children being overweight. Body mass index (BMI) is calculated as weight in kilograms divided by height in meters squared; overweight is indicated by a BMI of 25.0 to 29.9, and obesity by 30.0 or higher. Moens and Braet (2012) emphasized the family as the context in which children learn healthy eating habits. They conducted a study in which 50 families of overweight children (ages 6 to 12) were randomly allocated to a parent-led intervention group (cognitive behavioral training) or to a waiting list control group. Results demonstrated that children in the intervention group showed a significantly greater decrease in BMI over a six-month period. Parents reported significant positive changes in children's eating. Bringing children's eating under control is particularly important in light of the prejudice and discrimination that are displayed toward overweight children. Such children are targets for ridicule and bullying at school with devastating consequences for their self-concept.

12-4b Mental Illness

In a national survey of 9,282 respondents (Mahoney, 2005), the reported psychological problems and their percentages of lifetime incidence included: anxiety (29%), impulse control disorder (25%), mood disorder (21%), and substance abuse (15%). Depression among the elderly is higher than among younger age groups (particularly if older people can no longer take care of themselves and their social support is lacking; Ciro et al., 2012). Depression is not uncommon among college students. Field et al. (2012) found an unusually high percentage of depression (52%) among the 238 respondents who also reported anxiety, intrusive thoughts, and sleep disturbances.

Insel (2008) noted the enormous economic costs of serious mental illness, including a high rate of emergency room care (e.g., suicide attempts), a high prevalence of pulmonary disease (people with serious mental illness smoke 44% of all cigarettes in the United States), and early mortality (a loss of 13 to 32 years).

The toll of mental illness on a relationship can be immense. A major initial attraction of partners to each other includes intellectual and emotional qualities. Butterworth and Rodgers (2008) surveyed 3,230 couples to assess the degree to which mental illness of a spouse or spouses affects divorce and found that couples in which either men or women reported mental health problems had higher rates

palliative care health care focused on the relief of pain and suffering of the individual who has a life-threatening illness and support for them and their loved ones.

of marital disruption than did couples in which neither spouse experienced mental health problems. For couples in which both spouses reported mental health problems, rates of marital disruption reflected the additive combination of each spouse's separate risk. Mental illness is also a disadvantage for those seeking remarriage. Teitler and Reichman (2008) found that unmarried mothers with mental illness were about two thirds as likely as mothers without mental illness to marry.

Children may also have difficulty being attentive and processing information. An example is attention deficit/hyperactivity disorder (8.6% of children, ages 3 to 17; *Statistical Abstract of the United States,* 2012, Table 188). Children diagnosed with ADHD can be stressful to spouses and limit their coping capacity, which may put an enormous strain on their marriage.

The reverse is also true; children must learn how to cope with the mental illness of their parents. Mordoch and Hall (2008) studied 22 children between 6 and 16 years of age, who were living part- or full-time with a parent with depression, schizophrenia, or bipolar illness. They found that the children learned to maintain connections with their parents by creating and keeping a safe distance between themselves and their parents so as not to be engulfed by their parents' mental illnesses.

12-4c Middle-Age Crazy (Midlife Crisis)

MIDLIFE CRISIS AHEAD
© Sam72/Shutterstock

The stereotypical explanation for 45-year-old people who buy convertible sports cars, have affairs, marry 20-year-olds, or adopt a baby is that they are "having a midlife crisis." The label conveys that such people feel old, think that life is passing them by, and seize one last great chance to do something they have always wanted to do. Indeed, one father (William Feather) noted, "Setting a good example for your children takes all the fun out of middle age."

However, a 10-year study of close to 8,000 U.S. adults ages 25 to 74 by the MacArthur Foundation Research Network on Successful Midlife Development revealed that, for most respondents, the middle years brought no crisis at all but a time of good health, productive activity, and community involvement. Less than a quarter (23%) reported a "crisis" in their lives. Those who did experience a crisis were going through a divorce. Two thirds were accepting of getting older; one third did feel some personal turmoil related to the fact that they were aging (Goode, 1999).

Of those who initiated a divorce in midlife, 70% had no regrets and were confident that they did the right thing. This fact is the result of a study of 1,147 respondents ages 40 to 79 who experienced a divorce in their 40s, 50s, or 60s. Indeed, midlife divorcers' levels of happiness or contentment were similar to those of single individuals their own age and those who remarried (Enright, 2004).

Some people embrace middle age. The Red Hat Society (http://www.redhatsociety.com/) is a group of women who have decided to "greet middle age with verve, humor, and élan. We believe silliness is the comedy relief of life [and] share a bond of affection, forged by common life experiences and a genuine enthusiasm for wherever life takes us next." The society traces its beginning to Sue Ellen Cooper's purchase of a bright red hat because of a poem written by Jenny Joseph in 1961 titled "Warning." It says, in part,

> When I am an old woman I shall wear purple With a red hat which doesn't go, and doesn't suit me.

Cooper gave red hats to friends as they turned 50. The group then wore their red hats and purple dresses out to tea, and that's how the society got started. Now there are over 1 million members worldwide.

In the rest of this chapter, we examine how spouses cope with the crisis events of an extramarital affair, unemployment, drug abuse, and death. Each of these events can be viewed either as devastating and the end of meaning in one's life or as an opportunity and challenge to rise above.

12-4d Extramarital Affair (and Successful Recovery)

The term **extramarital affair** refers to a spouse's sexual involvement with someone outside the marriage. Affairs are of different types, which may include the following:

1. **Brief encounter (situationally determined affair).** A spouse hooks up with or meets a stranger at a conference. In this case, the spouse is usually out of town, and alcohol is involved.

extramarital affair a spouse's sexual involvement with someone outside the marriage.

The Red Hat Society is a group of women who have decided to "greet middle age with verve, humor, and élan."

2. **Paid sex.** A spouse seeks sexual variety with a prostitute who will do whatever he wants (e.g., former New York governor Eliot Spitzer). These encounters usually go undetected unless there is an STI, the person confesses, or the prostitute exposes the client.

3. **Instrumental or utilitarian affair.** This is sex in exchange for a job or promotion, to get back at a spouse, to evoke jealousy, or to transition out of a marriage.

4. **Coping mechanism.** Sex can be used to enhance one's self-concept or feeling of sexual inadequacy, compensate for failure in business, cope with the death of a family member, test one's sexual orientation, and so on.

5. **Paraphiliac affairs.** In these encounters, the on-the-side sex partner acts out sexual fantasies or participates in sexual practices that the spouse considers bizarre or abnormal, such as sexual masochism, sexual sadism, or transvestite fetishism.

6. **New love.** A spouse may be in love with the new partner and may plan marriage after divorce (Bagarozzi, 2008).

The computer or Internet affair is another type of affair. Although legally an extramarital affair does not exist unless two people (one being married) have sexual intercourse, an online computer affair can be just as disruptive to a marriage or a couple's relationship. Computer friendships may move to feelings of intimacy, involve secrecy (one's partner does not know the level of emotional involvement), include sexual tension (even though there is no overt sex), and take time, attention, energy, and affection away from one's partner. Schneider (2000) studied 91 women who experienced serious adverse consequences from their partner's cybersex involvement, including loss of interest in relational sex and feeling hurt, betrayed, rejected, abandoned, lonely, jealous, and angry over being constantly lied to. These women noted that the cyber affair was as emotionally painful as an off-line affair and that their partners' cybersex addiction was a major reason for their separation or divorce. Cramer et al. (2008) also noted that women become more upset when their man is emotionally unfaithful with another woman (although men become more upset when their partner is sexually unfaithful with another man). Measure your attitudes toward infidelity by taking the Self-Assessment on the Chapter 12 Assessment Card at the end of the book.

Extradyadic (extrarelational) involvement refers to the sexual involvement of a pair-bonded individual with someone other than the partner. Extradyadic involvements are not uncommon. Of 2,922 undergraduates, 24% agreed with the statement, "I have cheated on a partner I was involved with." Forty-one percent agreed with the statement, "A partner I was involved with cheated on me" (Knox & Hall, 2010).

Men are more upset if their wife has a heterosexual than a homosexual affair while women are equally upset if their spouse has a homosexual or heterosexual affair (Confer & Cloud, 2011). Characteristics associated with spouses who are more likely to have extramarital sex include male gender, a strong interest in sex, permissive sexual values, low subjective satisfaction

extradyadic (extrarelational) involvement emotional or sexual involvement between a member of a pair and someone other than the partner.

"Seldom, or perhaps never, does a **marriage develop** into an **individual relationship** smoothly and without **crises;** there is no coming to **consciousness** without pain."

—CARL JUNG, SWISS PSYCHIATRIST

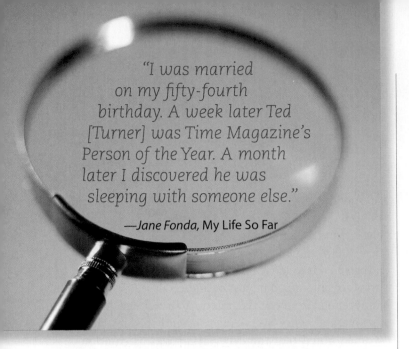

"I was married on my fifty-fourth birthday. A week later Ted [Turner] was Time Magazine's Person of the Year. A month later I discovered he was sleeping with someone else."

—*Jane Fonda,* My Life So Far

© BananaStock/Jupiterimages

in the existing relationship, employment outside the home, low church attendance, greater sexual opportunities, higher social status (power and money), and alcohol abuse (Hall et al., 2008). Elmslie and Tebaldi (2008) noted that a spouse's infidelity behavior is influenced by religiosity (less religious = more likely), city size (urban = more likely), and happiness (less happy = more likely). For more information on extramarital affairs in other countries, see page 229.

Reasons for an Affair Individuals in pair-bonded relationships report a number of reasons why they become involved in a sexual encounter outside their marriage.

1. **Variety, novelty, and excitement.** Extradyadic sexual involvement may be motivated by the desire for variety, novelty, and excitement. One of the characteristics of sex in long-term committed relationships is the tendency for it to become routine. Early in a relationship, the partners cannot seem to have sex often enough. However, with constant availability, partners may achieve a level of satiation, and the attractiveness and excitement of sex with the primary partner seem to wane. A high-end call girl said the following of the Eliot Spitzer affair:

Coolidge effect term used to describe the waning of sexual excitement and the effect of novelty and variety on sexual arousal.

Almost all of my clients are married. I would say easily over 90%. I'm not trying to justify this business, but these are men looking for companionship. They are generally not men that

couldn't have an affair (if they wanted to), but men who want this tryst with no strings attached. They're men who want to keep their lives at home intact. (Kottke, 2008)

The **Coolidge effect** is a term used to describe this waning of sexual excitement and the effect of novelty and variety on sexual arousal:

One day President and Mrs. Coolidge were visiting a government farm. Soon after their arrival, they were taken off on separate tours. When Mrs. Coolidge passed the chicken pens, she paused to ask the man in charge if the rooster copulated more than once each day. "Dozens of times," was the reply. "Please tell that to the President," Mrs. Coolidge requested. When the President passed the pens and was told about the rooster, he asked, "Same hen every time?" "Oh no, Mr. President, a different one each time." The President nodded slowly and then said, "Tell that to Mrs. Coolidge." (Bermant, 1976, pp. 76–77)

Whether or not individuals are biologically wired for monogamy continues to be debated. Monogamy among mammals is rare (from 3 to 10%), and monogamy tends to be the exception more often than the rule (Morell, 1998). Even if such biological wiring for plurality of partners does exist, it is equally debated whether such wiring justifies nonmonogamous behavior—that individuals are responsible for their decisions.

2. **Workplace friendships.** A common place for extramarital involvements to develop is the workplace (Merrill & Knox, 2010). Coworkers share the same world 8 to 10 hours a day and, over a period of time, may develop good feelings for each other that eventually lead to a sexual relationship. Former Democratic presidential candidate John Edwards became involved with a journalist when they were on the campaign trail together. Brad Pitt and Angelina Jolie met on a movie set. Arnold Schwarzenegger's housekeeper was "at work" when she had sex with him that produced their child (and prompted the end of Schwarzenegger's marriage upon disclosure 10 years later).

3. **Relationship dissatisfaction.** It is commonly believed that people who have affairs are not

happy in their marriage. Spouses who feel misunderstood, unloved, and ignored sometimes turn to another who offers understanding, love, and attention. Neuman (2008) confirmed that being emotionally dissatisfied in one's relationship is the primary culprit behind an affair.

One source of relationship dissatisfaction is an unfulfilling sexual relationship. Some spouses engage in extramarital sex because their partner is not interested in sex. Others may go outside the relationship because their partners will not engage in the sexual behaviors they want and enjoy. The unwillingness of the spouse to engage in oral sex, anal intercourse, or a variety of sexual positions sometimes results in the other spouse's looking elsewhere for a more cooperative and willing sexual partner.

4. **Revenge.** Some extramarital sexual involvements are acts of revenge against one's spouse for having an affair. When partners find out that their mate has had or is having an affair, they are often hurt and angry. One response to this hurt and anger is to have an affair to get even with the unfaithful partner.

5. **Homosexual relationship.** Some individuals marry as a front for their homosexuality. Cole Porter, known for such songs as "I've Got You Under My Skin," "Night and Day," and "Every Time We Say Goodbye," was a homosexual who feared no one would buy or publish his music if his sexual orientation were known. He married Linda Lee Porter (alleged to be a lesbian), and their marriage endured for 30 years.

Edw/Shutterstock.com

Other gay individuals marry as a way of denying their homosexuality. These individuals are likely to feel unfulfilled in their marriage and may seek involvement in an extramarital homosexual relationship. Other individuals may marry and then discover later in life that they desire a homosexual relationship. Such individuals may feel that (1) they have been homosexual or bisexual all along, (2) their sexual orientation has changed from heterosexual to homosexual or bisexual, (3) they are unsure of their sexual orientation and want to explore a homosexual relationship, or (4) they are predominantly heterosexual but wish to experience a homosexual relationship for variety. The term **down low** refers to African American married men who have sex with men.

> **down low**
> African-American "heterosexual" man who has sex with men.

EXTRAMARITAL AFFAIRS IN OTHER COUNTRIES

Researcher Pam Druckerman (2007) wrote *Lust in Translation,* in which she reflects on how affairs are viewed throughout the world. First, terms for having an affair vary; for the Dutch, it is called "pinching the cat in the dark"; in Taiwan, it is called "a man standing in two boats"; and in England, "playing off sides." Second, how an affair is regarded differs by culture. In America, the script for discovering a partner's affair involves confronting the partner and ending the marriage. In France, the script does not involve confronting the partner and does not assume that the affair means the end of the marriage; rather, "letting time pass to let a partner go through the experience without pressure or comment is the norm." In America, presidential candidate John Edwards's disclosure of his affair with an office worker ended his political life.

© flas100/Shutterstock

6. **Absence from partner.** One factor that may predispose a spouse to have an affair is prolonged separation from the partner. Some wives whose husbands are away for military service report that the loneliness can become unbearable. Some husbands who are away say that remaining faithful is difficult. Partners in commuter relationships may also be vulnerable to extradyadic sexual relationships.

Finding out about a partner's affair can result from confession, a phone call from the other person, or, when an affair is suspected, snooping. Derby et al. (2012) analyzed snooping behavior in 268 undergraduates and found that almost two thirds (66%) reported that they had engaged in snooping behavior, most often when the partner was taking a shower. Primary motives were "curiosity" and "suspicion" that the partner was cheating. Being female, being jealous, and having cheated were associated with higher frequencies of snooping behavior.

Effects of an Affair Reactions to the knowledge that one's partner or spouse has been unfaithful vary. For most, the revelation is devastating—the partner feels betrayed, cries (women more likely), and drinks more alcohol (men more likely; Barnes et al., 2012). The following is an example of a wife's reaction to her husband's affairs:

My husband began to have affairs within six months of our being married. When I confronted my husband he denied any such affair and said that I was suspicious, jealous and had no faith in him. In effect, I had the problem. He said that I should not listen to what others said because they did not want to see us happy. I was deeply in love with my husband and knew in my heart that he was guilty as sin; I lived in denial so I could continue our marriage.

Of course, my husband continued to have affairs. Some of the effects on me included:

1. I lost the ability to trust my husband and, after my divorce, other men.
2. I developed a negative self-concept—the reason he was having affairs is that something was wrong with me.
3. He robbed me of the innocence and my "VIRGINITY"—clearly he did not value the opportunity to be the only man to have experienced intimacy with me.
4. I developed an intense hatred for my husband.

It took years for me to recover from this crisis. I feel that through faith and religion I have emerged "whole" again. Years after the divorce my husband made a point of apologizing and letting me know that there was nothing wrong with me, that he was just young and stupid and not ready to be serious and committed to the marriage.

But reactions to an affair of the spouse vary. The wife of Eliot Spitzer said that "Sex is the department of the wife and in that regard, I was inadequate." Affairs may also have negative effects on one's health (e.g., heart problems; Fisher et al., 2012) and one's children. The latter may be subjected to witnessing the parents argue and may experience a lack of attention (because the father is not at home) and trauma from the breakup of the marriage (Schneider, 2003).

Successful Recovery From Infidelity When an affair is discovered, a sense of betrayal pervades the nonoffending spouse or partner (Barnes et al., 2012). Keeping the relationship together (with forgiveness and time) is the most frequent outcome. Olson et al. (2002) identified three phases that spouses pass through on their way to successful recovery from the discovery of a partner's affair. The "roller-coaster" phase involves agony at the initial discovery, which elicits an array of feelings including rage or anger, self-blame, the desire to give up, and the desire to work on the marital relationship. The second phase, "moratorium," involves less emotionality and a decision to work the problem out. The partners settle into a focused though tenuous commitment to get beyond the current crisis. The third phase, "trust building," involves taking responsibility for the infidelity, reassurance of commitment, increased communication, and forgiveness. Dean

> One factor that may predispose a spouse to have an affair is prolonged separation from the partner.

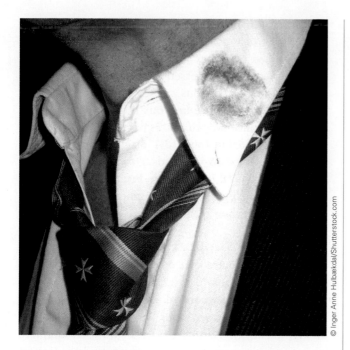
© Inger Anne Hulbækdal/Shutterstock.com

(2011) noted that part of the recovery from an affair is the grieving over what the relationship was and what it represented and still moving forward from the event.

Bagarozzi (2008) defined forgiveness as a conscious decision on the part of the offended spouse to grant a pardon to the offending spouse, to give up feeling angry, and to relinquish the right to retaliate against the offending spouse. In exchange, offending spouses take responsibility for the affair, agree not to repeat the behavior, and grant their partner the right to check up on them to regain trust.

Positive outcomes of having experienced and worked through infidelity include a closer marital relationship, increased assertiveness, placing higher value on each other, and realizing the importance of good marital communication (Linquist & Negy, 2005).

Spouses who remain faithful to their partners have decided to do so. They avoid intimate conversations with members of the other sex and a context (e.g., being alone in a hotel room) that is conducive to physical involvement. The best antidote to an affair is a strong emotional and sexual connection with one's spouse, and avoiding contexts conductive to external involvement.

Prevention of Infidelity Allen et al. (2008) identified the premarital factors predictive of future infidelity. The primary factor for both partners was a negative pattern of interaction. Partners who ended up being unfaithful were in relationships where they did not connect, they argued, and they criticized each other. Hence, spouses least vulnerable are in loving, nurturing, communicative relationships where each affirms the other. Neuman (2008) also noted that avoiding friends who have affairs and establishing close relationships with married couples who value fidelity further insulate individuals from having an affair.

12-4e Unemployment

America continues to downsize and outsource jobs to India and Mexico, resulting in layoffs and insecurity in the lives of American workers. In recent years, unemployment in America has hovered around 9% in the general adult population and even higher among minorities. Forced unemployment is a major stressor for individuals (causing emotional and health problems; Sweet & Moen, 2012), couples, and families, sometimes leading to homelessness (Gould & Williams, 2010). In addition, when spouses or parents lose their jobs as a result of physical illness or disability, a family experiences a double blow—loss of income combined with higher medical bills. Unless an unemployed spouse is covered by the partner's medical insurance, unemployment can also result in loss of health insurance for the family. Insurance for both health care and disability is very important to help protect a family from an economic disaster. One reaction to the husband's unemployment is that the wife becomes employed or seeks work (Mattingly & Smith, 2010).

The effects of unemployment may be more severe for men than for women. Our society expects men to be the primary breadwinners in their families and equates masculine self-worth and identity with job and income. Stress, depression, suicide, alcohol abuse, and lowered self-esteem, as well as increased cigarette smoking (Falba et al., 2005) and child abuse (Euser et al., 2010) are associated with unemployment.

Women tend to adjust more easily to unemployment than do men. Women are not burdened with the cultural expectation of the provider role, and their identity is less tied to their work role. Hence, women may view unemployment as an opportunity to spend more time with their families; many enjoy doing so.

Although unemployment can be stressful, increasing numbers of workers are also experiencing job stress. With the recession, layoffs, and downsizing, workers are given the work of two and told, "We'll hire someone soon." Meanwhile, employers have learned to pay for one and get the work of two. The result is lowered morale, exhaustion, and stress that can escalate into violence.

The effects of unemployment may be more severe for men than for women.

© Brand X Pictures/Jupiterimages

12-4f Substance Abuse

Spouses, parents, and children who abuse drugs contribute to the stress and conflict experienced in their respective marriages and families. Although some individuals abuse drugs to escape from unhappy relationships or the stress of family problems, substance abuse inevitably adds to the individual's marital and family problems when it results in health and medical problems, legal problems, loss of employment, financial ruin, school failure, divorce, and even death (due to accidents or poor health; Dethier et al., 2011). The story titled "Marriage Cancelled" on page 249 reflects the experience of the wife of a man addicted to crack.

Although getting married is associated with significant reductions in cigarette smoking, heavy drinking, and marijuana use for both men and women (Merline et al. 2008), family crises involving alcohol and/or drugs are not unusual. And, it may be the children rather than the parents involved in the drug abuse that can cause the stress in the family. In 2010, Cameron Douglas (son of actor Michael Douglas) was sentenced to five years in prison for heroin possession and dealing cocaine and meth. Having a personality that is characterized by a sense of urgency and sensation-seeking is related to substance abuse (Shin et al., 2012).

Tomlinson and Brown (2012) found that the cause of drinking alcohol may be both depression and a response to anxiety. They studied 400 8th graders who revealed that they used alcohol to self-medicate because they were depressed or because they felt inadequate in social situations and alcohol helped them to cope. Indeed, the more anxious the respondents felt, the more alcohol they drank at parties.

As indicated in Table 12.1, drug use is most prevalent among 18- to 25-year-olds. Drug use among teenagers under age 18 is also high. Because teenage drug use is common, it may compound the challenge parents may have with their teenagers.

Drug Abuse Support Groups The treatments for alcohol abuse are varied—a combination of medications (naltrexone and acamprosate), behavioral interventions (e.g., controlling the social context) and Alcoholics Anonymous (AA; www.alcoholics-anonymous.org). There are over 15,000 AA chapters nationwide; one in your community can be found through the Yellow Pages. The only requirement for membership is the desire to stop drinking.

Former abusers of drugs (other than alcohol) also meet regularly in local chapters of Narcotics Anonymous (NA) to help each other continue to be drug-free. Patterned after Alcoholics Anonymous, the premise of NA is that the best person to help someone stop abusing drugs is someone who once abused drugs. NA members of all ages, social classes, and educational levels provide a sense of support for each other to remain drug-free.

Al-Anon is an organization that provides support for family members and friends of alcohol abusers. Spouses and parents of substance abusers learn how to live with and react to living with a substance abuser.

Parents who abuse drugs may also benefit from the Strengthening Families Program, which provides specific social skills training for both parents and children. After families attend a five-hour retreat, parents and children are involved in face-to-face skills training

Al-Anon an organization that provides support for family members and friends of alcohol abusers.

Table 12.1
Drug Use by Type of Drug and Age Group

Type of Drug Used	Age 12 to 17	Age 18 to 25	Age 26 to 34
Marijuana and hashish	16.5%	50.4%	51.3%
Cocaine	1.9%	14.4%	16.7%
Alcohol	38.3%	85.6%	no data
Cigarettes	22.9%	64.2%	no data

Source: Adapted from Table 207, *Statistical Abstract of the United States* (2012, 131st ed.), Washington, DC: U.S. Census Bureau.

MARRIAGE CANCELLED

Testimonial of a wife married to a man addicted to crack.

Marriage needs a foundation of trust and open communication, but when you are married to an addict, you won't find either. When I first met "Nate," I thought he was everything I wanted in a partner—good looks, a great sense of humor, intelligence, and ambition. It didn't take me long to realize that I was extremely attracted to this man. Since we knew some of the same people, we ended up at several parties together. We both drank and occasionally used cocaine but it wasn't a problem. I loved being with him and whenever I wasn't with him, I was thinking about him.

After only four months of dating, we realized we were falling madly in love with each other and decided to get married. Neither of us had ever been married before but I had a 5-year-old son from a previous relationship. Nate really took to my son and it seemed like we had the perfect marriage—that changed drastically and quickly. I didn't know it but my husband had been using crack cocaine the entire time we dated. I wasn't extremely familiar with this new drug and I knew even less about its addictive power. Of course, I soon found out more about it than I cared to know.

Less than three months after we married, Nate went on his first binge. After leaving work one Friday night, he never came home. I was really worried and was calling all of his friends trying to find out where he was and what was wrong. I talked to the girlfriend of one of his best friends and found out everything. She told me that Nate had been smoking crack off and on for a long time and that he was probably out using again. I didn't know what to do or where to turn. I just stayed home all weekend waiting to hear from my husband.

Finally, on Sunday afternoon, Nate came home. He looked like hell and I was mad as hell. I sent my son to play with his friends next door and as soon as he was out of hearing range, I lost it. I began screaming and crying, asking why he never came home over the weekend. He just hung his head in absolute shame. I found out later that he had spent his entire paycheck, pawned his wedding ring, and had written checks off of a closed checking account. He went through over $1,000 of "our" money in less than three days.

I was devastated and in shock. After I calmed down and my anger subsided, we discussed his addiction. He told me he loved me with all his heart and that he was so sorry for what he had done. He also swore he would never do it again. Well, this is when I became an enabler and I continued to enable my husband for three years. He would stay clean for a while and things would be great between us, then he would go on another binge. It was the roller-coaster ride from hell. Nate was on a downward spiral and he was dragging my son and me down along with him.

By the end of our third year together, we were more like roommates. Our once wonderful sex life was virtually nonexistent and what love I still had for him was quickly fading away. I knew I had to leave before my love turned to hate. It was obvious that my son had been pulling away from Nate emotionally so it was a good time to end the nightmare. I packed our belongings and moved in with my sister.

Less than three months after I left, Nate went into a rehabilitation program. I was happy for him but I knew I could never go back; it was too little, too late. That was nine years ago and the last I heard, he was still struggling with his addiction, living a life of misery. I have no regrets about leaving but I am saddened by what crack cocaine had done to my once wonderful husband— it turned him into a thief and a liar and ended our marriage. Reprinted with permission from the author.

over a four-month period. A 12-month follow-up revealed that parenting skills remained improved and that reported heroin and cocaine use had declined.

12-4g Death of a Family Member

Even more devastating than drug abuse are family crises involving death—of one's child, parent, or loved one. The crisis is particularly acute when the death is a suicide.

Death of One's Child A parent's worst fear is the death of a child. Most people expect the death of their parents but not the death of their children. The feature on Emily below reflects the anguish of parents who experienced the death of their daughter.

Alam et al. (2012) studied parents of children who died of cancer and found that fathers became work focused while mothers focused more on bereaved siblings. Amy Winehouse died at the age of 27 in 2011. Her distraught parents, Mitch and Janis, said that they were "left bereft" at her death. Grief feelings may be particularly acute on the anniversary of the death of an individual with whom one was particularly close.

Mothers and fathers sometimes respond to the death of their child in different ways. When they do, the respective partners may interpret these differences in negative ways, leading to relationship conflict and unhappiness. For example, after the death of their 17-year-old son, one wife accused her husband of not sharing in her grief. The husband explained that, although he was deeply grieved, he poured his grief into working more as a form of distraction. To deal with these differences, spouses might need to be patient and practice tolerance in allowing both to grieve in their own way.

Death of One's Parent Terminally ill parents may be taken care of by their children. Such care over a period of years can be emotionally stressful, financially draining, and exhausting. Hence, by the time the parent dies, a crisis has already occurred.

Reactions to the death of one's parents include depression, loss of concentration, and anger. Whether the death is that of a child or a parent, Burke et al. (1999) noted that grief is not a one-time experience that people adjust to and move on. Rather, for some, there is "chronic sorrow," where grief-related feelings occur periodically throughout the lives of those left behind. The researchers noted that almost all (97%) of the individuals in one study who had experienced the death of a loved one 2 to 20 years earlier met the criteria for chronic sorrow.

12-4h Suicide of a Family Member

Suicide is a devastating crisis event for families, and not that unusual. Annually there are 31,000 suicides (750,000 attempts), and each suicide immediately affects at least six other people in that person's life. These effects include depression (e.g., grief), physical disorders (e.g., shingles due to stress), and social stigma (e.g., the person is viewed as weak and the family as

Emily: Our Daughter's Death at 17

Our lives changed forever when we received a phone call that our daughter was in the hospital and in critical condition. She had been on a trip to the beach with friends and was on the way home when the van flipped, crossed the divided highway, and severely injured her. She died from the head injuries. We discovered that the driver had put the cruise control on 70 and was playing "switch the driver" with his girlfriend, the front passenger. One of them hit the steering wheel inadvertently and our daughter (the only one of the six to die) was gone forever.

Emily was a month shy of her 17th birthday. She was a lovely child. We, as her parents, haven't been the same since her death. Even after all these years, our first thought upon awakening and last thought before we sleep is of our daughter, and how we miss her. At the time of her death, we were not alone in our grief. When we told Emily's grandfather of her death, he was overwhelmed with grief, and two weeks later died of a heart attack. To this day, we wonder if we had been protective enough. The grief goes on.

Also, our marriage has been strained due to the trauma. Since all couples have different grieving trajectories, we found one of us would be battling depression, while the other was halfway level . . . or vice versa. Her death will always remain a nightmare from which we cannot awaken. Time, therapy, support groups, and the like have helped to manage our loss of our child, but we will forever continue to deal with our personal tragedy within our family . . . . We hope for peace . . . but 12 years after her death we know our loss can only be managed.

a failure for not being able to help with the precipitating emotional problems; De Castro & Guterman, 2008). Schum (2007) described family members who experience the suicide of a family member as having "the worst day of their lives." Miers et al. (2012) interviewed six parental units (mother and dad) whose teenager had committed suicide. Critical needs of the couple included that they needed support from another suicide survivor, support when viewing the deceased teen, support in remembering the teen, and support in parents giving back to the community.

People who are between the ages of 15 and 19, homosexual, or male, and those with a family history of suicide or mood disorder, substance abuse, or past history of child abuse and parental sex abuse are more vulnerable to suicide than others (Melhem et al., 2007). As noted earlier, Taliaferro et al. (2008) found that vigorous exercise and involvement in sports are associated with lower rates of suicide among adolescents.

Suicide is viewed as a "rational act" (and can sometimes be predicted; Stefansson et al., 2012). The person feels that suicide is the best option available. Therapists view suicide as a "permanent solution to a temporary problem" and routinely call 911 to have people hospitalized or restrained who threaten suicide or who have been involved in an attempt.

Adjustment to the suicide of a family member takes time. The son of physician T. Schum committed suicide, which set in motion a painful adjustment for Dr. Schum. "You will never get over this, but you can get through it," was a phrase Dr. Schum found helpful (Schum, 2007). Survivors of Suicide is also a helpful support group. Part of the recovery process is accepting that one cannot stop the suicide of those who are adamant about taking their own life and that one is not responsible for the suicide of another. Indeed, family members often harbor the belief that they could have done something to prevent the suicide. Singer Judy Collins lost her son to suicide and began to participate in a support group for people who had lost a loved one to suicide. At a meeting she attended, a group member answered the question of whether there was something she could have done with a resounding *no*:

> I was sitting on his bed saying, "I love you, Jim. Don't do this. How can you do this?" I had my hand on his hand, my cheek on his cheek. He said excuse me, reached his other hand around, took the gun from under the pillow, and blew his head off. My face was inches from his. If somebody wants to kill himself or herself, there is nothing you can do to stop them. (Collins, 1998, p. 210)

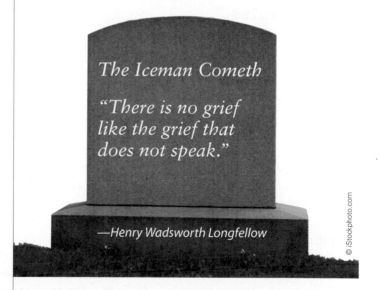

The Iceman Cometh

"There is no grief like the grief that does not speak."

—Henry Wadsworth Longfellow

© iStockphoto.com

STUDY TOOLS 12

Ready to study? In this book, you can:

- ⊃ Rip out the Chapter Review card in the back of the book to study for exams

- ⊃ Take the Self Assessment for this chapter (card in the back of the book) and see where you stand on the vital issues raised in the chapter

Or you can go online to CourseMate at www.cengagebrain.com for these resources:

- ⊃ Complete Practice Quizzes to prepare for tests

- ⊃ Review Key Terms Flash Cards (online or print)

- ⊃ Read about Marriage and Family in the news

- ⊃ Play "Beat the Clock" to master concepts

- ⊃ Check out Personal Applications

Abuse in Relationships

"**Love** does not **dominate,**
it **cultivates.**"

—JOHANN WOLFGANG VON GOETHE, GERMAN PHILOSOPHER

SECTIONS

13-1 Nature of Relationship Abuse

13-2 Explanations for Violence and Abuse in Relationships

13-3 Sexual Abuse in Undergraduate Relationships

13-4 Abuse in Marriage Relationships

13-5 Effects of Abuse

13-6 The Cycle of Abuse

Yeardley Love was a 22-year-old senior and lacrosse player at the University of Virginia who was beaten to death in 2010 by her estranged former boyfriend, George Huguely. He was convicted of second degree murder and sentenced to 26 years in prison. A year earlier, Chris Brown beat girlfriend Rihanna to the point where she called 911. Chris Brown later turned himself in to the police and was released on $50,000 bail. Televised media regularly feature horror stories of women who are beaten up or killed by their partners. Whether the motive is revenge, to be free for another partner, financial gain, or jealousy, the result is the same. What began as an intimate love relationship ends in the death of one's former partner or spouse. In this chapter, we examine the other side of intimacy in one's relationships. We begin by looking at the nature of abuse and defining some terms.

13-1 Nature of Relationship Abuse

Abuse is not uncommon, and there are several types of abuse in relationships. These types include physical and emotional abuse. Both are devastating to the individual and to the relationship.

13-1a Violence

Also referred to as **physical abuse**, **violence** may be defined as the deliberate infliction of physical harm by either partner on the other. Examples of physical violence include pushing, throwing something at the partner, slapping, hitting, and forcing sex on the partner. **Intimate-partner violence (IPV)** is an all-inclusive term that refers to crimes committed against current or former spouses, boyfriends, or girlfriends. John Gottman (2007) identified two types of violence. In one type, conflict escalates over an issue, and one or both partners lose control. The trigger seems to be feeling disrespected and losing one's dignity. The person feels threatened and seeks to defend his or her dignity by posturing and threatening the partner while in a very agitated state. Control is lost, and the partner strikes out. Because both partners may lose control at the same time, this type of violence is symmetrical. Couples can prevent this type of violence by recognizing the sequence of interactions that leads to the violent behavior (Gottman, 2007).

physical abuse (violence) intentional infliction of physical harm by either partner on the other.

intimate-partner violence an all-inclusive term that refers to crimes committed against current or former spouses, boyfriends, or girlfriends.

A second type of violence is based on one partner's controlling the other, creating a clear perpetrator and victim. The best way to avoid the second type of violence is for the victim to leave the relationship.

A syndrome related to violence is **battered-woman syndrome**, which refers to the general pattern of battering that a woman is subjected to and which is defined in terms of the frequency, severity, and injury she experiences. Battering is severe if the person's injuries require medical treatment or the perpetrator could be prosecuted.

Battering may lead to murder. **Uxoricide** is the murder of a woman by a romantic partner; the murder of Yeardley Love by George Huguely is one example. As a prelude to murder, the man may hold the woman hostage (holding one or more people against their will with the actual or implied use of force). A typical example is an estranged spouse who reenters the house of a former spouse and holds his wife and children hostage. These situations are potentially dangerous because the perpetrator often has a weapon and may use it on the victims, himself, or both. A negotiator is usually called in to resolve the situation.

© Nonstock/Jupiterimages

> "Nothing good *ever comes of violence.*"
>
> —*Martin Luther King Jr.*

---

battered-woman syndrome general pattern of battering that a woman is subjected to, defined in terms of the frequency, severity, and injury.

uxoricide the murder of a woman by her romantic partner.

emotional abuse (psychological abuse; verbal abuse; symbolic aggression;) the denigration of an individual with the purpose of reducing the victim's status and increasing the victim's vulnerability so that he or she can be more easily controlled by the abuser.

13-1b Emotional Abuse

In addition to being physically violent, partners may engage in **emotional abuse** (also known as **psychological abuse**, **verbal abuse**, or **symbolic aggression**). Whereas 7.5% of 2,922 undergraduates reported having been involved in a physically abusive relationship, 29.9% reported that they "had been involved in an *emotionally* abusive relationship with a partner" (Knox & Hall, 2010). Although emotional abuse does not involve physical harm, it is designed to make the partner feel bad—to denigrate the partner, reduce the partner's status, and make the partner vulnerable to being controlled by the abuser. Although there is debate about what constitutes psychological abuse, examples of various categories of emotional abuse include the following:

Criticism and ridicule—being called obese, stupid, crazy, ugly, pitiful, and repulsive

Isolation—being prohibited by the partner from spending time with friends, siblings, and parents

Control—being told how to dress and/or accused of dressing like a slut; having one's money controlled

Silent treatment—having a partner who refuses to talk to or to touch the victim

Yelling—being yelled and screamed at by one's partner

Accusation—being told (unjustly) that one is unfaithful

Threats—being threatened by the partner with abandonment or threats of harm to oneself, one's family, or pets

Demeaning behavior—being insulted by the partner in front of others

Torture—being subjected to psychological or physical pain (e.g., a husband puts a gun to his wife's head and threatens to pull the trigger)

Demanding behavior—being required to do as the partner wishes (e.g., have sex)

According to Follingstad and Edmundson (2010), while both partners in a relationship may report that they engage in emotionally abusive behavior, each partner tends to report that he or she engages in less

"**Domestic violence** causes far more **pain** than the visible marks of **bruises** and **scars.** It is devastating to be abused by someone that **you love** and **think loves you** in return."

—DIANNE FEINSTEIN, U.S. SENATOR

frequent and less severe emotionally abusive behavior. In effect, respondents create "a picture of their own use of psychological abuse as limited in scope and not harmful in nature" (p. 506). Each sought to present himself or herself as less abusive.

13-1c Female Abuse of Partner

Although it is assumed that men are more often the perpetrators of abuse, Brownridge (2010) noted that women and men experience abuse with equal frequency (but men report it less often). Some researchers focus on women as abusive partners. Swan et al. (2008) found that women's physical violence is more likely to be motivated by self-defense and fear, whereas men's physical violence is more likely than women's to be driven by control motives. Hence, women tend to be striking back rather than throwing initial blows. An abusive female wrote the following:

When people think of physical abuse in relationships, they get the picture of a man hitting or pushing the woman. Very few people, including myself, think about the "poor man" who is abused both physically and emotionally by his partner—but this is what I do to my boyfriend. We have been together for over a year now and things have gone from really good to terrible.

Abuse of a partner can escalate rapidly from dismissiveness to violence.

The problem is that he lets me get away with taking out my anger on him. No matter what happens during the day, it's almost always his fault. It started out as kind of a joke. He would laugh and say "oh, how did I know that this was going to somehow be my fault." But then it turned into me screaming at him, and not letting him go out and be with his friends as his "punishment."

Other times, I'll be talking to him or trying to understand his feelings about something that we're arguing about and he won't talk. He just shuts off and refuses to say anything but "alright." That makes me furious, so I start verbally and physically attacking him just to get a response.

Or, an argument can start from just the smallest thing, like what we're going to watch on TV, and before you know it, I'm pushing him off the bed and stepping on his stomach as hard as I can and throwing the remote into the toilet. That really gets to him.

13-1d Stalking in Person

Abuse may take the form of stalking. **Stalking** is defined as unwanted following or harassment that induces fear in a target person. Stalking is a crime. A stalker is one who has been rejected by a previous lover or is obsessed with a stranger or acquaintance who fails to return the stalker's romantic overtures. In about 80% of cases, the stalker is a heterosexual male who follows his previous lover. Women who stalk are more likely to target a married male. Stalking is a pathological state which decreases over time (Cattaneo et al., 2011). One in six women in a national study conducted by the Centers for Disease Control and Prevention

stalking unwanted following or harassment that induces fear in a target person.

© iStockphoto.com/Nikolay Mamluke

reported having been stalked (followed and harassed; CDC, 2011). Such stalking is usually designed either to seek revenge or to win a partner back.

13-1e Stalking Online—Cybervictimization

Cybervictimization includes being sent threatening e-mail, unsolicited obscene e-mail, computer viruses, or junk mail (spamming). It may also include flaming (online verbal abuse) and leaving improper messages on message boards. Burke et al. (2011) surveyed 804 undergraduates and found that half of both female and male respondents (females more than males) reported use of communication technology (e.g., cell phones, e-mail, social network sites) to monitor a partner either as the initiator or victim.

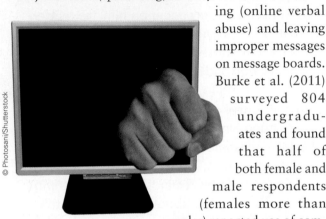

© Photosani/Shutterstock

Less threatening than stalking is **obsessive relational intrusion (ORI)**, the relentless pursuit of intimacy with someone who does not want it. The person becomes a nuisance but does not have the goal of harm, as does the stalker. When ORI and stalking are combined, however, as many as 25% of women and 10% of men can expect to be pursued in unwanted ways (Spitzberg & Cupach, 2007).

People who cross the line in terms of pursuing an ORI relationship or responding to being rejected (stalking) engage in a continuum of eight forms of behavior. Spitzberg and Cupach (2007) have identified these behaviors:

cyber-victimization being sent unwanted e-mail, spam, viruses, or being threatened online.

obsessive relational intrusion the relentless pursuit of intimacy with someone who does not want it.

1. **Hyperintimacy**—telling a person that he or she is beautiful or desirable to the point of making the person uncomfortable.

2. **Relentless electronic contacts**—flooding the person with e-mail messages, cell phone calls, text messages, or faxes.

3. **Interactional contacts**—showing up at the person's workplace or gym. The intrusion may also include joining the same volunteer groups as the pursued.

4. **Surveillance**—monitoring the movements of the pursued (e.g., by following the person or driving past the person's house).

5. **Invasion**—breaking into the person's house and stealing objects that belong to the person; stealing the person's identity; or putting Trojan horses (viruses) in the person's computer. One woman downloaded child pornography on her boyfriend's computer and called the authorities to arrest him. He was sentenced and served time in prison.

6. **Harassment or intimidation**—leaving unwanted notes on the person's desk or a dead animal on the doorstep.

7. **Threat or coercion**—threatening physical violence or harm to the person or to the person's family, friends, or pets.

8. **Aggression or violence**—carrying out a threat by becoming violent (e.g., kidnapping or rape).

13-2 Explanations for Violence and Abuse in Relationships

Research suggests that numerous factors contribute to violence and abuse in intimate relationships. These factors can be found at the cultural, community, and individual and family levels.

13-2a Cultural Factors

In many ways, American culture tolerates and even promotes violence. Violence in the family stems from the acceptance of violence in our society as a legitimate means of enforcing compliance and solving conflicts at interpersonal, familial, national, and international levels. Violence and abuse in the family may be linked to such cultural factors as violence in the media, acceptance of corporal punishment, gender inequality, and the view of women and children as property. The context of stress is also conducive to violence.

Violence in the Media One need only watch the news every night to see the violence in Afghanistan, school shootings, and domestic murders. New films like *The Hunger Games* and TV shows regularly reflect themes of violence. *Dexter* features a serial killer who ties up his victims and stabs them as they are witness to their own execution. This predictable scene is preceded by murder and mayhem perpetrated by the killer Dexter is tracking. Football, which dominates Saturday, Sunday, and Monday television in the fall, is a very violent sport in which some players become injured for life. The New Orleans Saints National Football League team was exposed and penalized for a "bounty system," whereby the players were given money to knock-out or disable a player on the opposing team. Boxing matches continue to draw large, paid TV audiences.

Corporal Punishment of Children Corporal **punishment** is defined as the use of physical force with the intention of causing a child to experience pain, but not injury, for the purposes of correcting or controlling the child's behavior and/or making the child obedient. In the United States, corporal punishment in the form of spanking, hitting, whipping, and paddling or otherwise inflicting punitive pain on the child is legal in all states (as long as the corporal punishment does not meet the individual state's definition of child abuse). Unlike in Sweden, Italy, Germany, and 12 other countries which have banned corporal punishment in the home, violence against children has become a part of America's cultural heritage. Children who are victims of corporal punishment display more antisocial behavior, are more violent, and have an increased incidence of depression as adults. Children who are victims of physical abuse also report lower relationship quality as adult college students (Larsen et al., 2011).

Gender Inequality Domestic violence and abuse may also stem from traditional gender roles. Traditionally, men have also been taught that they are superior to women and that they may use their aggres-

Martina McBride is an advocate against domestic violence reflected in her songs "Independence Day" and "Broken Wing."

sion toward women, believing that women need to be "put in their place." The greater the inequality and the more the woman is dependent on the man, the more likely the abuse.

Some occupations, such as law enforcement and the military, lend themselves to contexts of gender inequality. In military contexts, men notoriously devalue, denigrate, and sexually harass women. In spite of the rhetoric about gender equality in the military, women in the Army, Navy, and Air Force academies continue to be sexually harassed. These male perpetrators may not separate their work roles from their domestic roles. A former student noted that she was the ex-wife of a Navy Seal and that "he knew how to torment someone, and I was his victim."

In cultures where a man's "honor" is threatened if his wife is unfaithful, the husband's violence toward her is tolerated by the legal system. Unmarried women in Jordan who have intercourse (even if is through rape)

corporal punishment the use of physical force with the intention of causing a child to experience pain, but not injury, for the purpose of correction or control of the child's behavior.

"Teenage girls can't tell their parents that their **boyfriend beat them** up. You **don't dare** let your neighbor know that **you fight.** It's one of the things we [women] will hide, because **it's embarrassing."**

—RIHANNA, SINGER

Chapter 13: Abuse in Relationships **241**

are viewed as bringing shame on their parents and siblings and may be killed; this action is referred to as an **honor crime** or **honor killing**. For example, a brother may kill his sister if she has intercourse with a man she is not married to, but may spend no more than a month or two in jail as the penalty.

View of Women and Children as Property Prior to the late 19th century, a married woman was considered the property of her husband. A husband had a legal right and marital obligation to discipline and control his wife through the use of physical force.

Stress Our culture is a context of stress. The stress associated with getting and holding a job, rearing children, staying out of debt, and paying bills may predispose one to lash out at one's partner and/or children. College students are not immune to stress. Gormley and Lopez (2010) found that undergraduate males who reported high stress levels also reported higher rates of psychological abuse of their partners.

13-2b **Community Factors**

Community factors that contribute to violence and abuse in the family include social isolation, poverty, and inaccessible or unaffordable health-care, day-care, elder-care, and respite-care services and facilities.

honor crime (honor killing) killing a daughter because she brought shame to the family by having sex while not married. The killing is typically overlooked by the society. Jordan is a country where honor crimes occur.

Social Isolation Living in social isolation from extended family and community members increases the risk of being abused. Spouses whose parents live nearby are least vulnerable.

Poverty Abuse in adult relationships occurs among all socioeconomic groups. However, poverty and low socioeconomic status are a context of high stress, which is sometimes expressed in violence in interpersonal relationships.

Inaccessible or Unaffordable Community Services Failure to provide medical care to children and elderly family members sometimes results from the lack of accessible or affordable health-care services in the community. Failure to provide supervision for children and adults may result from inaccessible day-care and elder-care services. Without elder-care and respite-care facilities, families living in social isolation may not have any help with the stresses of caring for elderly family members and children.

13-2c **Individual Factors**

Individual factors associated with domestic violence and abuse include psychopathology, personality characteristics, and alcohol or substance abuse.

Personality Factors A number of personality characteristics have been associated with people who are abusive in their intimate relationships. Some of these characteristics follow:

1. **Dependency.** Therapists who work with batterers have observed that they are overly dependent on their partners. Because the thought of being left by their partners induces panic and abandonment anxiety, batterers use physical aggression and threats of suicide to keep their partners with them.

2. **Jealousy.** Along with dependence, batterers exhibit jealousy, possessiveness, and suspicion. An abusive husband may express his possessiveness by isolating his wife from others; he may insist she stay at home, not work, and not socialize with others. His extreme, irrational jealousy may lead him to accuse his wife of infidelity and to beat her for her presumed affair. Undergraduates may also feel jealous when their partner spends a lot of time texting, e-mailing, or talking with an ex.

3. **Need to control.** Abusive partners have an excessive need to exercise power over their partners and to control them. The abusers do not let their partners make independent decisions, and they want to know where their partners are, whom they are with, and what they are doing. Abusers like to be in charge of all aspects of family life, including finances and recreation.

4. **Unhappiness and dissatisfaction.** Abusive partners often report being unhappy and

dissatisfied with their lives, both at home and at work. Many abusers have low self-esteem and high levels of anxiety, depression, and hostility. They may expect their partner to make them happy.

5. **Anger and aggressiveness.** Abusers tend to have a history of interpersonal aggressive behavior. They have poor impulse control and can become instantly enraged and lash out at the partner. Battered women report that episodes of violence are often triggered by minor events, such as a late meal or a shirt that has not been ironed.

6. **Quick involvement.** Because of feelings of insecurity, the potential batterer will move his partner quickly into a committed relationship. If the woman tries to break off the relationship, the man will often try to make her feel guilty for not giving him and the relationship a chance.

7. **Blaming others for problems.** Abusers take little responsibility for their problems and blame everyone else. For example, when they make mistakes, they will blame their partner for upsetting them and keeping them from concentrating on their work. A man may become upset because of what his partner said, hit her because she smirked at him, and kick her in the stomach because she poured him too much (or not enough) alcohol.

8. **Jekyll-and-Hyde personality.** Abusers have sudden mood changes so that a partner is continually confused. One minute an abuser is nice, and the next minute angry and accusatory. Explosiveness and moodiness are the norm.

9. **Isolation.** An abusive person will try to cut off a partner from all family, friends, and activities. Ties with anyone are prohibited. Isolation may reach the point at which an abuser tries to stop the victim from going to school, church, or work.

10. **Alcohol and other drug use.** Whether alcohol reduces one's inhibitions to display violence, allows one to avoid responsibility for being violent, or increases one's aggression, alcohol is associated with violence and abuse (even if the partner is pregnant; Eaton et al., 2012). Younger individuals with more severe drug addictions (e.g., a history of overdose) are more likely to be violent (Fernandez-Montalvo et al., 2012).

11. **Criminal/Psychiatric background.** Eke et al. (2011) examined the characteristics of 146 men who murdered or attempted to murder their intimate partner. Of these, 42% had prior criminal charges, 15% had a psychiatric history, and 18% had both.

13-2d Family Factors

Family factors associated with domestic violence and abuse include being abused as a child, having parents who abused each other, and not having a father in the home.

Child Abuse in Family of Origin Individuals who were abused as children were more likely to be abusive toward their partners as adults.

Oprah Winfrey experienced trauma resulting from child abuse. She shares her personal life experience of rape and sexual molestation at a young age by three different family members, spreading information and hope to others who have experienced abuse in their lives.

© Chris Farina/Corbis

Parents Who Abused Each Other Although most parents are not aggressive toward each other in the presence of their children (Pendry et al., 2011), Busby et al. (2008) found that individuals who observe their parents being violent with each other are more likely to be violent in their adult relationships. In addition, Anderson and Bang (2012) found that children exposed to their parents' violence toward each other (including violence which involved the police coming to the house) reported higher rates of PTSD than did children without such exposure.

13-3 Sexual Abuse in Undergraduate Relationships

Palmer et al. (2010) surveyed 370 college students regarding their past year experiences and found that 34% of women reported unwanted sexual contact. Flack et al. (2008) found no evidence of the **red zone**, the first month of the first year of college when women are most likely to be victims of sexual abuse (the first semester freshman is viewed as an easy sexual target). In a sample of 2,922 undergraduates at a large southeastern university, 33% reported being pressured to have sex by a partner they were dating (Knox & Hall, 2010). Katz and Myhr (2008) noted that some women experience sexual abuse in addition to a larger pattern of physical abuse and that the combination is associated with less general satisfaction, less sexual satisfaction, more conflict, and more psychological abuse from the partner. In effect, women in these co-victimization relationships are miserable.

red zone the first month of the first year of college when women are particularly vulnerable to unwanted sexual advances.

acquaintance rape nonconsensual sex between adults who know each other.

date rape nonconsensual sex between two people who are dating or on a date.

13-3a Acquaintance and Date Rape

The word *rape* often evokes images of a stranger jumping out of the bushes or a dark alley to attack an unsuspecting

victim. However, most rapes (85%) are perpetrated not by strangers but by people who have a relationship with the victim, an acquaintance (12% of victims have been raped by both an acquaintance and a stranger—double victims; Hall & Knox, 2012). This type of rape is known as **acquaintance rape**, which is defined as nonconsensual sex between adults (of the same or other sex) who know each other. The behaviors of sexual coercion occur on a continuum from verbal pressure and threats to the use of physical force to obtain sexual acts, such as kissing, petting, or intercourse.

Men are also raped. Chapleau et al. (2008) found that about 13% of men reported being raped. Rape myths abound as it is assumed that "most rapes are by strangers at night" and that "false accusations are prevalent" (McGee et al., 2011). Not only is there a double standard evident in the perception that only women can be raped, but there is also a double standard operative in the perception of the gender of the person engaging in sexual coercion. Men who rape are aggressive; women who rape are promiscuous (Oswald & Russell, 2006).

One type of acquaintance rape is **date rape**, which refers to nonconsensual sex between people who are dating or on a date. The woman is least suspecting of this rape; she may go out to dinner, have a pleasant evening, and be walked back to her apartment by her partner, who then rapes her.

Women are also vulnerable to repeated sexual force. Daigle et al. (2008) noted that 14 to 25% of college women experience repeat sexual victimization during the same academic year—they are victims of rape or other unwanted sexual force more than once, often in the same month. The primary reason is that they do not change the context; they stay in the relationship with the same person and may continue to use alcohol or other drugs.

Although acts of sexual aggression by women on men do occur, they are less frequent than women being raped. The sidebar "I Was Raped by My Boyfriend" details an experience provided by a former student.

I Was Raped by My Boyfriend

I was 13, a freshman in high school. He was 16, a junior. We were both in band, that's where we met. We started dating and I fell head over heels in love. He was my first real boyfriend. I thought I was so cool because I was a freshman dating a junior. Everything was great for about 6 maybe 8 months and that's when it all started. He began to verbally and emotionally abuse me. He would tell me things like I was fat and ugly and stupid. It hurt me but I was young and was trying so hard to fit in that I didn't really do anything about it.

When he began to drink and smoke marijuana excessively everything got worse. I would try to talk to him about it and he would only yell at me and tell me I was stupid. He began to get physical with the abuse about this time too. He would push me around, grab me really hard, and bruise me up, things like that.

I tried to break up with him and he would tell me that if I broke up with him I would be alone because I was so ugly and fat that no one else would want me. He would say that I was lucky to have him. He began to pressure me for sex. I wasn't ready for sex. I was only 14 at that time. He told everyone that we were having sex and I never said anything different. I was scared to say anything different. It got to the point that I was trying my best to avoid being around him alone. I was afraid of him; I never knew what he would do to me.

Our school band had a competition in Myrtle Beach, South Carolina. When we were there we stayed overnight in a hotel. At the hotel he asked me to come by his room so we could go to dinner together. When I went by his room he pulled me into the room. He locked the door behind me. He pushed me into the bathroom and shut and locked the door behind us. He had his hand over my mouth and told me not to make a sound or he would hurt me. He said that it was time for him to get what everyone already thought he had.

He pushed me up against the counter where the sink was and pulled my clothes off. He put a condom on and began to rape me. I closed my eyes, tears streamed down my face. I had no idea what to do. When he was finished he cleaned himself up and turned on the shower. He pushed me into the shower and told me to take a shower and get ready for dinner. Before he left the room he told me that if I had given in to him and had sex earlier he wouldn't have had to force me. He also told me not to tell anyone or he'd hurt me plus no one would believe me because everyone already thought we had been having sex.

I sat in the shower and cried for a while. I thought it was my fault. I didn't tell anyone about it for 2 years. I decided not to press charges; I didn't want to go through it again. I regret that decision still to this day. I was an emotional wreck after I was raped. I didn't know how to act. He acted normal, like nothing had happened. A few months later he broke up with me for another girl. It was a relief to me and it made me angry at the same time. It angered me that he would rape me and then break up with me.

I didn't know what to do with myself. I started working out all the time, several hours a day, every day. After high school I didn't have the time to work out so I developed an eating disorder. This went on for almost 6 years. I am just now dealing with moving past the rape and rebuilding my self-esteem, ability to trust men, and ability to enjoy sex in a new relationship. I encourage anyone who has been raped to speak up about it. Do not let the person get away with it. I also encourage anyone who has been raped to get immediate professional help, don't try to get past it on your own.

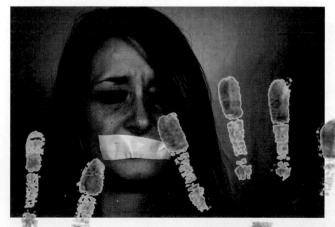

© iStockphoto.com/© Peter Finnie/iStockphoto.com

13-3b Rophypnol— The Date Rape Drug

Of those occasions in which a woman is incapacitated due to alcohol or drugs, 85% are voluntary (the woman is willingly drinking or doing drugs). In 15% of the cases, however, drugs are used against her will (Lawyer et al., 2010). **Rophypnol**—also known as the date rape drug, rope, roofies, Mexican Valium, or the "forget (me) pill"—causes profound, prolonged sedation and short-term memory loss. Similar to Valium but 10 times as strong, Rophypnol is a prescription drug in Europe and is used as a potent sedative. It is sold in the United States for about $5, is dropped in a drink (where it is tasteless and odorless), and causes victims to lose their memory for 8 to 10 hours. During this time, victims may be raped yet be unaware until they notice signs of it (e.g., blood in panties).

The Drug-Induced Rape Prevention and Punishment Act of 1996 makes it a crime to give a controlled substance to anyone without his or her knowledge and with the intent of committing a violent crime (such as rape). Violation of this law is punishable by up to 20 years in prison and a fine of $250,000.

Women are defenseless when drugged. When an assault begins when the woman is not drugged, her responses may vary from pleading to resisting, including "turning cold" and "running away." Gidycz et al. (2008) discussed the various responses in terms of background. Women who had been victimized as children were more likely to "freeze and turn cold."

The effects of rape include loss of self-esteem, loss of trust, and the inability to be sexual. Zinzow et al. (2010) studied a national sample of women and found that forcible rape was more likely to be associated with PTSD and with major depression than was drug- or alcohol-facilitated or incapacitated rape. (Sexually abused children also report drug abuse as adults; Robboy & Anderson, 2011.)

rophypnol date rape drug that causes profound, prolonged sedation and short-term memory loss.

marital rape forcible rape by one's spouse.

© corepics/Shutterstock.com

13-4 Abuse in Marriage Relationships

The chance of abuse in a relationship increases with marriage. Indeed, the longer individuals know each other and the more intimate the relationship, the greater the abuse.

13-4a General Abuse in Marriage

Abuse in marriage differs from unmarried abuse in that the husband may feel "ownership" of the wife and feel the need to "control her." The feature "Twenty-Three Years in an Abusive Marriage," on page 247, reveals the experience of horrific marital abuse by a former student who survived it.

13-4b Rape in Marriage

Marital rape, now recognized in all states as a crime, is forcible rape by one's spouse. The forced sex may take the form of sexual intercourse, fellatio, or anal intercourse. Sexual violence against women in an intimate relationship is often repeated.

13-5 Effects of Abuse

Abuse affects the physical and psychological well-being of victims. Abuse between parents also affects the children.

23 YEARS
IN AN ABUSIVE MARRIAGE

My name is Jane and I am a survivor (though it took me forever to get out) of a physically and mentally abusive marriage. I met my husband-to-be in high school when I was 15. He was the most charming person that I had ever met. He would see me in the halls and always made a point of coming over to speak. I remember thinking how cool it was that an older guy would be interested in a little freshman like me. He asked me out but my parents would not let me go out with him until I turned 16 so we met at the skating rink every weekend until I was old enough to go out on a date with him. Needless to say, I was in love and we didn't date other people.

When we married I remember the wedding day very well as I was on top of the world because I was marrying the man of my dreams whom I loved very much. My family (they did not like him) tried for the longest time to talk me out of marrying him. My dad always told me that something was not right with him but I did not listen. I wish to God that I had listened to my dad and walked out of the church before I said "I do."

I remember the first slap, which came six months after we had been married. I had burned the toast for his dinner. I was shocked when he slapped me, cried, and when he saw the blood coming from my lip he started crying and telling me he was sorry and that he would never do that again. But the beatings never stopped for twenty-three years. His most famous way of torturing me was to put a gun to my head while I was sleeping and wake me up and pull the trigger and say next time you might not be so lucky.

I left this man a total of seventeen times but I always went back after I got well from my beatings. At one point during this time every bone in my body had been broken with the exception of my neck and back. Some bones more than once and sometimes more than one at a time. I was pushed down a flight of stairs and still suffer from the effects of it to this day. I was stabbed several times and I had to be hospitalized a total of six times.

After our children were born I thought the beatings would stop but they didn't. When our children got older he started on them so I took many beatings for my children just so he would leave them alone. Three years ago he beat me so badly that I almost died. I remember being in the hospital wanting to die because if I did I would not hurt anymore and I would finally be safe. At that time I was thinking that death would be a blessing since I would not have to go back to live with this monster.

While I was in the hospital a therapist came to see me and told me that I had two choices— go home and be killed or fight for my life and live. Instead of going back to the monster, I moved my children in with my mother. It took me three years to get completely clear of him, and even today he harasses me but I am free. I had to learn how to think for myself and how to love again. I am now remarried to a wonderful man.

If you are in an abusive relationship there is help out there and please don't be afraid to ask for it. I can tell you this much—the first step (deciding to leave) is the hardest to take. If and when you do leave I promise your life will be better—it may take awhile but you will get there.

13-5a Effects of Partner Abuse on Victims

Becker et al. (2010) confirmed that being a victim of intimate partner violence is associated with symptoms of PTSD—loss of interest in activities and life in general, a feeling of detachment from others, inability to sleep, irritability, and so on. Rhatigan and Nathanson (2010) found that women took into account their own behavior in evaluating their boyfriends' abusive behavior. If they felt they had "set him up" by their own behavior, they were more understanding of his aggressiveness. Additionally, those with low self-esteem were more likely to take some responsibility for their boyfriends' abusive behavior. In effect, they excused the rape as their fault.

13-5b Effects of Partner Abuse on Children

In the most dramatic situation, some women are abused during their pregnancy, resulting in higher rates of miscarriage and birth defects, affecting the child directly (Stockl et al., 2012). Negative effects may also accrue to children who witness domestic abuse. Twenty percent of undergraduates report having observed their parents engage in abusive behavior toward each other (DiLillo et al., 2010). Russell et al. (2010) found that children who observed parental domestic violence were more likely to be depressed as adults. Hence, a child need not be a direct target of abuse but merely a witness to incur negative effects.

13-6 The Cycle of Abuse

The cycle of abuse exemplified in "I Got Flowers Today" begins when a person is abused and the perpetrator feels regret, asks for forgiveness, and starts acting nice (e.g., giving flowers). The victim, who perceives few options and feels guilty about terminating the relationship with a partner who asks for forgiveness, feels hope for the relationship because of the contriteness of the abuser and does not call the police or file charges. Forgiving the partner and taking him back usually occurs *seven* times before the partner leaves for good. Shakespeare (in *As You Like It*) said of such forgiveness, "Thou prun'st a rotten tree."

I got flowers today. It wasn't my birthday or any other special day. We had our first argument last night, and he said a lot of cruel things that really hurt me. I know he is sorry and didn't mean the things he said, because he sent me flowers today.

I got flowers today. It wasn't our anniversary or any other special day. Last night, he threw me into a wall and started to choke me. It seemed like a nightmare. I couldn't believe it was real. I woke up this morning sore and bruised all over. I know he must be sorry, because he sent me flowers today.

Last night, he beat me up again. And it was much worse than all the other times. If I leave him, what will I do? How will I take care of my kids? What about money? I'm afraid of him and scared to leave. But I know he must be sorry, because he sent me flowers today.

I got flowers today. Today was a very special day. It was the day of my funeral. Last night, he finally killed me. He beat me to death.

If only I had gathered enough courage and strength to leave him, I would not have gotten flowers today.

Anonymous

I GOT FLOWERS TODAY

After the forgiveness, couples usually experience a period of making up or honeymooning, during which the victim feels good again about the partner and is hopeful for a nonabusive future. However, stress, anxiety, and tension mount again in the relationship and are relieved by violence toward the victim. Such violence is followed by the familiar sense of regret and pleadings for forgiveness, accompanied by being nice (a new bouquet of flowers, and so on).

As the cycle of abuse reveals, some victims do not prosecute their partners who abuse them. To deal with this problem, Los Angeles has adopted a "zero tolerance" policy toward domestic violence. Under the law, an arrested person is required to stand trial and his victim required to testify against the perpetrator. The sentence in Los Angeles County for partner abuse is up to six months in jail and a fine of $1,000.

Figure 13.1 illustrates this cycle, which occurs in clockwise fashion. In the rest of this section, we discuss reasons why people stay in abusive relationships and how to get out of such relationships.

13-6a Why People Stay in Abusive Relationships

One of the most frequently asked questions of people who remain in abusive relationships is, "Why do you stay?" Few and Rosen (2005) interviewed 25 women who had been involved in abusive dating relationships from three months to nine years (the average was 2.4 years) to find out why they stayed. The researchers conceptualized the women as **entrapped**—stuck in an abusive relationship and unable to extricate themselves from the abusive partner. Indeed, these women escalated their commitment to stay in hopes that doing so would eventually pay off. Among the factors of their perceived investment were the sharing of their emotional self with their partner (they were in love), the time they had already spent in the relationship, and the relationships to which they were connected because of the partner. In effect, they had invested time with a partner they were in love with and wanted to turn the relationship around into a safe, nonabusive one. The following are some of the factors explaining how abused women become entrapped:

- Fear of loneliness ("I'd rather be with someone who abuses me than alone")
- Love ("I love him")
- Emotional dependency ("I need him")

Figure 13.1
The Cycle of Abuse

- Commitment to the relationship ("I took a vow 'for better or for worse'")
- Hope ("He will stop")
- A view of violence as legitimate ("All relationships include some abuse")
- Guilt ("I can't leave a sick man")
- Fear ("He'll kill me if I leave him")
- Economic dependence ("I have no place to go")
- Isolation ("I don't know anyone who can help me" "No one will believe me")

Battered women also stay in abusive relationships because they rarely have escape routes related to educational or employment opportunities, their relatives are critical of plans to leave the partner, they do not want to disrupt the lives of their children, and they may be so emotionally devastated by the abuse (anxious, depressed, or suffering from low self-esteem) that they feel incapable of planning and executing their departure.

13-6b How to Leave an Abusive Relationship

Leaving an abusive partner begins with the decision to do so. Such a choice often follows

entrapped stuck in an abusive relationship and unable to extricate one's self from the abusive partner.

Leaving an abusive partner begins with the decision to do so. Unfortunately, returning is common and typically happens seven times.

the belief that one will end up being seriously hurt or one's children will be harmed by staying. The person makes a plan and acts on the plan—packs clothes and belongings, moves in with a sister, mother, or friend, or goes to a homeless shelter. If the new context is better than being in the abusive context, the person will stay away. Otherwise, the person may go back and start the cycle all over. As noted previously, this leaving and returning typically happens seven times.

Sometimes the woman does not just disappear while the abuser is away but calls the police and has the man arrested for violence and abuse. While the abuser is in jail, she may move out and leave town. In either case, disengagement from the abusive relationship takes a great deal of courage. Calling the National Domestic Violence Hotline (800-799-7233 [SAFE]), available 24 hours, is a point of beginning. Earlier, we discussed the need to be cautious in leaving a partner because this is the time the woman is most vulnerable to being murdered. Moving quickly to a safe context (e.g., a parent's home) is important.

Kress et al. (2008) noted that involvement with an intimate partner who is violent may be life-threatening. Particularly if the individual decides to leave the violent partner, the abuser may react with more violence and murder the person who has left. Indeed, a third of murders that occur in domestic violence cases occur shortly after a breakup. Specific actions that could be precursors to murder of an intimate partner are stalking, strangulation, forced sex, physical abuse, perpetrator suicidality, gun ownership, and drug or alcohol use on the part of the violent partner.

Individuals in such relationships should be cautious about how they react and develop a safe plan of withdrawal. Safety plans will vary but include the following:

- Identifying a safe place the person can go to. Arrangements to go to a one's parents, a friend's home, or to a women's shelter must be set up in advance. The victim needs to stay in a protected context.

- Telling friends or neighbors about the violence and requesting that they call the police if they hear suspicious noises or witness suspicious events.

- Storing an escape kit (e.g., keys, money, checks, important phone numbers, medications, Social Security cards, bank documents, birth certificates, change of clothes, and so on) somewhere safe (and usually not in the house; Kress et al., 2008).

Above all, individuals should do what they can to de-escalate the situation, develop a plan of escape, and execute the plan.

13-6c Strategies to Prevent Domestic Abuse

Family violence and abuse prevention strategies are focused at three levels: the general population, specific groups thought to be at high risk for abuse, and families who have already experienced abuse. Public education and media campaigns aimed at the general population convey the criminal nature of domestic assault, suggest ways abusers might learn to prevent abuse (seek therapy for anger,

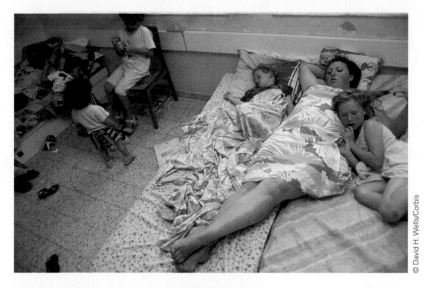

© David H. Wells/Corbis

Another important cultural change is to reduce violence-provoking stress by reducing poverty and unemployment and by providing adequate housing, nutrition, medical care, and educational opportunities for everyone. Integrating families into networks of community and kin would also enhance family well-being and provide support for families under stress.

13-6d Treatment of Partner Abusers

LeCouteur and Oxlad (2011) interviewed men who talked of why they abused their partners and found that they constructed a justification that their partner had "breached the normative moral order." Successful therapy required the men to acknowledge the wrongness of their violent and abusive actions and to move forward. Shamai and Buchbinder (2010) surveyed men who had participated in a treatment program for partner-violent men. Most experienced therapy as positive and meaningful and underwent personal changes, especially the acquisition of self-control. Taking a "time out," counting to 10, and reassessing the situation before reacting were particularly helpful techniques. While gains were made, the men still tended to create dominant relationships with women, which left open the potential to explode again.

jealousy, or dependency), and identify where abuse victims and perpetrators can get help. Rothman and Silverman (2007) noted that preventive abuse programs for college students were effective in reducing violence for both men and women. However, people with a prior history of assault; who were gay, lesbian, or bisexual; or who were binge drinkers did not benefit.

Preventing or reducing family violence through education necessarily involves altering aspects of American culture that contribute to such violence. For example, violence in the media (including violent video games) must be reduced and egalitarian gender roles must become the relationship norm.

Divorce and Remarriage

"**I don't care** if
I'm your **first love;**
I just want to be **your last.**"

—GRETCHEN WILSON

SECTIONS

14-1 Macro Factors
Contributing to Divorce

14-2 Micro Factors
Contributing to Divorce

14-3 Ending an Unsatisfactory
Relationship

14-4 Gender Differences
in Filing for Divorce

14-5 Consequences for
Spouses Who Divorce

14-6 Effects of Divorce
on Children

14-7 Prerequisites for Having
a "Successful" Divorce

14-8 Divorce Prevention

14-9 Remarriage

14-10 Stepfamilies

14-11 Strengths of Stepfamilies

14-12 Developmental Tasks
for Stepfamilies

Just as weddings are a time of celebration and joy, divorce is a time of dismay and sadness. **Divorce** is the termination of a valid marriage contract. Whether a divorce is the beginning of the end of one's life (some never recover) or the beginning of a new life depends on the person. College students are not unacquainted with divorce. In a sample of 2,922 undergraduates, 26% reported that their parents were divorced (Knox & Hall, 2010). The United States has one of the highest divorce rates in the world, higher even than vanguard countries such as Sweden (Cherlin, 2009).

14-1 Macro Factors Contributing to Divorce

divorce the termination of a valid marriage contract

Sociologists emphasize that social context creates outcomes. This concept is best illustrated by the fact that the Puritans in Massachusetts, from 1639 to 1760, averaged only one divorce per year (Morgan, 1944). The social context of that era involved strong pro-family values and strict divorce laws, with the

© Britt Erlanson/cultura/Corbis

result that divorce was almost nonexistent. In contrast, divorce occurs more frequently today as a result of various structural and cultural factors, also known as macro factors.

14-1a Increased Economic Independence of Women

In the past, an unemployed wife was dependent on her husband for food and shelter. No matter how unhappy her marriage was, she stayed married because she was economically dependent on her husband. Her husband literally represented her lifeline. Finding gainful employment outside the home made it possible for a wife to afford to leave her husband if she wanted to. Now that about three fourths of wives are employed, fewer wives are economically trapped in unhappy marriage relationships. A wife's employment does not increase the risk of divorce in a happy marriage. However, it does provide an avenue of escape for women in unhappy or abusive marriages (Kesselring & Bremmer, 2006).

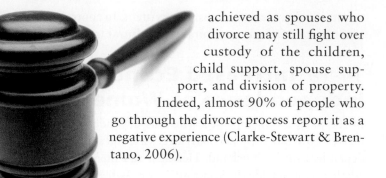

Employed wives are also more likely to require an egalitarian relationship; although some husbands prefer this role relationship, others are unsettled by it. Another effect of a wife's employment is that she may meet someone new in the workplace and become aware of an alternative to her current partner. Finally, unhappy husbands may be more likely to divorce if their wives are employed and able to be financially independent (requiring less alimony and child support).

14-1b Changing Family Functions and Structure

Many of the protective, religious, educational, and recreational functions of the family have been largely taken over by outside agencies. Family members may now look to the police for protection, the church or synagogue for meaning, the school for education, and commercial recreational facilities for fun rather than to others within the family to fulfill these needs. The result is that, although meeting emotional needs remains a primary function of the family, fewer reasons exist to keep a family together.

In addition to the change in functions of the family brought on by the Industrial Revolution, family structure has changed from that of the larger extended family in a rural community to a smaller nuclear family in an urban community. In the former, individuals could turn to a lot of people in times of stress; in the latter, more stress necessarily falls on fewer shoulders. Also, with marriages more isolated and scattered, kin may not live close enough to express their disapproval for the breakup of a marriage. With fewer social consequences for divorce, Americans escape unhappy unions.

14-1c Liberal Divorce Laws

All states recognize some form of **no-fault divorce** in which neither party is identified as the guilty party or the cause of the divorce (e.g., committing adultery). In effect, divorce is granted after a period of separation (typically 12 months). The goal of no-fault divorce is to make divorce less acrimonious. However, this objective has not always been

no-fault divorce
a divorce in which neither party is identified as the guilty party or the cause of the divorce.

achieved as spouses who divorce may still fight over custody of the children, child support, spouse support, and division of property. Indeed, almost 90% of people who go through the divorce process report it as a negative experience (Clarke-Stewart & Brentano, 2006).

14-1d Prenuptial Agreements and the Internet

New York family law attorney Nancy Chemtob notes that those who have prenuptial agreements are more likely to divorce, since one can cash out without economic devastation. Additionally, she suggested that the Internet contributes to divorce since a bored spouse can go online to various dating sites and see what alternatives are out there. Starting up a new relationship online before dumping the spouse of many years is not uncommon. Brown and Lin (2012) documented that divorce is becoming more common (1 in 4) in adults age 50 and older.

14-1e Fewer Moral and Religious Sanctions

Many priests and clergy recognize that divorce may be the best alternative in particular marital relationships and attempt to minimize the guilt that congregation members may feel at the failure of their marriage. Churches increasingly embrace single and divorced or separated individuals, as evidenced by "divorce adjustment groups."

14-1f More Divorce Models

As the number of divorced individuals in our society increases, the probability increases that a person's friends, parents, siblings, or children will be divorced. The more divorced people a person knows, the more normal divorce will seem to that person. The less deviant the person perceives divorce to be, the greater the probability the person will divorce if that person's own marriage becomes strained. Divorce has become so common that numerous websites exclusively for divorced individuals are now available (e.g., www .heartchoice.com/divorce).

14-1g Mobility and Anonymity

When individuals are highly mobile, they have fewer roots in a community and greater anonymity. Spouses who move away from their respective families and friends often discover that they are surrounded by strangers who don't care if they stay married or not. Divorce thrives when pro-marriage social expectations are not operative. In addition, the factors of mobility and anonymity also result in the removal of a consistent support system to help spouses deal with the difficulties they may encounter in marriage.

14-1h Social Class, Ethnicity, and Culture

Charles Murray (2012) argues that social class influences who stays married, and points out that educated, white Americans are the least likely to divorce. Indeed, the less educated with less income are more likely to divorce.

Asian Americans and Mexican Americans have lower divorce rates than European Americans or African Americans because they consider the family unit to be of greater value (familism) than their individual interests (individualism). Unlike familistic values in Asian cultures, individualistic values in American culture emphasize the goal of personal happiness in marriage. When spouses stop having fun (when individualistic goals are no longer met), they sometimes feel no reason to stay married. Of 2,922 undergraduates, only 8% agreed with the statement, "I would not divorce my spouse for any reason" (Knox & Hall, 2010). In two national samples, 36% agreed that "[t]he personal happiness of an individual is more important than putting up with a bad marriage" (Amato et al., 2007). Reflecting an individualistic philosophy, Geraldo Rivera asked, "Who cares if I've been married five times?"

Larry King has been married seven times to six women. Journalist Belinda Luscombe (2010) of *Time Magazine* suggested that in light of King's relentless divorces, maybe our society should create a "Revoking the Marriage License" policy. She noted that in no other area of life are people allowed to fail and then do the same thing again and again.

14-2 Micro Factors Contributing to Divorce

Although macro factors may make divorce a viable cultural alternative to marital unhappiness, they are not sufficient to "cause" a divorce. Micro factors are individual factors which are predictive of divorce. Relationships may break for numerous reasons—affair, abuse, in-laws, etc (Stork-Hestad et al., 2011). A partner whose behavior does not meet one's expectations (fidelity, no abuse, value spouse over parents) spells marital dissatisfaction (Dixon et al., 2012).

14-2a Falling Out of Love

Fish et al. (2012) found in a sample of 478 undergraduates that the most frequent reason for breaking up was "feeling unhappy or bored with the relationships" (22.6%). Benjamin et al (2010) noted that the absence of love in a relationship is associated with an increased chance of divorce. Indeed, almost half (48%) of 2,922 undergraduates reported that they would divorce a spouse they no longer loved (Knox & Hall, 2010). Even having dissimilar depths of love feelings between the partners in courtship is predictive of divorce (Wilson & Huston, 2011). No couple, even those reporting satisfaction over several years, is immune to divorce (Lavner & Bradbury, 2010).

© iStockphoto.com

14-2b Spending Limited Time Together

Some spouses do not make time to be together. Their careers are so demanding that they rarely see each other. Barbara Walters reported that this happened with her husband, Lee, who was busy producing Broadway plays. She was career-focused during the weekdays and he was career-focused at night and on weekends, so they drifted apart (Walters, 2008, pp. 213–214).

14-2c Decreasing Positive Behaviors

People marry because they anticipate greater rewards from being married than from being single. During courtship, each partner engages in a high frequency of positive verbal (compliments) and nonverbal (eye contact, physical affection) behaviors toward the other. The good feelings the partners experience as a result of these positive behaviors encourage them to marry to "lock in" these feelings across time. Mitchell (2010) interviewed 390 married couples and found that intimacy and doing things together were associated with marital happiness. Just as love feelings are based on positive behaviors from a partner, negative feelings are created when these behaviors stop and the partner engages in a high frequency of negative behaviors. Thoughts of divorce then begin (to escape the negative behaviors).

14-2d Having an Affair

Some extramarital affairs result in a divorce. For the spouse having the affair, the love and sex with a new partner (in contrast to what may be absence of love and sex with the spouse at home) can speed the movement toward divorce. Alternatively, the spouse at home may become indignant and demand that the partner leave. In an Oprah.com survey of 6,069 adults, an affair was the reason most respondents said they would seek a divorce (Healy & Salazar, 2010). Thirty-three percent said an affair would be the deal breaker. Other top responses were chronic fighting (28%), no longer being in love (24%), boredom (8%) and sexual incompatibility (4%). Of 2,922 undergraduates, over half (54%) agreed with the statement, "I would divorce a spouse who had an affair" (Knox & Hall, 2010). However, most couples do not end a marriage because of an affair. Although they may consider divorce, they may recover and regard their marriage as a "great" marriage (Tulane et al., 2011).

satiation the state in which a stimulus loses its value with repeated exposure.

14-2e Having Poor Conflict Resolution Skills

Managing differences and conflict in a relationship helps to reduce the negative feelings that develop in a relationship. Some partners respond to conflict by withdrawing emotionally from their relationship; others respond by attacking, blaming, and failing to listen to their partner's point of view.

14-2f Changing Values

Both spouses change throughout the marriage. "He's not the same person I married" is a frequent observation of people contemplating divorce. People may undergo radical value changes after marriage. One minister married and decided seven years later that he did not like the confines of the marriage role. He left the ministry, earned a PhD, and began to drink and have affairs. His wife, who had married him when he was a minister, now found herself married to a clinical psychologist who spent his evenings at bars with other women. The couple divorced.

Because people change throughout their lives, the person that one selects at one point in life may not be the same partner one would select at another point. Margaret Mead, the famous anthropologist, noted that her first marriage was a student marriage; her second, a professional partnership; and her third, an intellectual marriage to her soul mate, with whom she had her only child. At each of several stages in her life, she experienced a different set of needs and selected a mate in reference to those needs.

14-2g Onset of Satiation

Satiation, also referred to as *habituation*, refers to the state in which a stimulus loses its value with repeated exposure. Spouses may tire of each other. Their stories are no longer new, their sex is repetitive, and their presence no longer stimulates excitement as it did in courtship. Some people, who feel trapped by the boredom of constancy, divorce and seek what they believe to be more excitement by returning to singlehood and new partners. A developmental task of marriage is for couples to enjoy being together and not demand a constant state of excitement (which is

What Divorcing Parents Might Tell Their Children: An Example

© Blend Images/JupiterImages/© John Solie/iStockphoto.com

The following are the words of a mother of two children (ages 8 and 12) as she tells them of the pending divorce. The approach assumes that both parents take some responsibility for the divorce and are willing to provide a united front for the children. The script should be adapted for one's own unique situation.

Daddy and I want to talk to you about a big decision that we have made. A while back we told you that we were having a really hard time getting along, and that we were having meetings with someone called a therapist who has been helping us talk about our feelings, and deciding what to do about them.

We also told you that the trouble we are having is not about either of you. Our trouble getting along is about our grown-up relationship with each other. That is still true. We both love you very much, and love being your parents. We want to be the best parents we can be.

Daddy and I have realized that we don't get along so much, and disagree about so many things all the time, that we want to live separately, and not be married to each other anymore. This is called getting divorced. Daddy and I care about each other but we don't love each other in the way that happily married people do. We are sad about that. We want to be happy, and want each other to be happy. So to be happy we have to be true to our feelings.

It is not your fault that we are going to get divorced. And it's not our fault. We tried for a very long time to get along living together but it just got too hard for both of us.

We are a family and will always be your family. Many things in your life will stay the same. Mommy will stay living at our house here, and Daddy will move to an apartment close by. You both will continue to live with mommy and daddy but in two different places. You will keep your same rooms here, and will have a room at Daddy's apartment. You will be with one of us every day, and sometimes we will all be together, like to celebrate somebody's birthday, special events at school, or scouts. You will still go to your same school, have the same friends, go to soccer, baseball, and so on. You will still be part of the same family and will see your aunts, uncles, and cousins.

The most important things we want you both to know are that we love you, and we will always be your mom and dad . . . nothing will change that. It's hard to understand sometimes why some people stop getting along and decide not to be friends anymore, or if they are married decide to get divorced. You will probably have lots of different feelings about this. While you can't do anything to change the decision that daddy and I have made, we both care very much about your feelings. Your feelings may change a lot. Sometimes you might feel happy and relieved that you don't have to see and feel daddy and me not getting along. Then sometimes you might feel sad, scared, or angry. Whatever you are feeling at any time is okay. Daddy and I hope you will tell us about your feelings, and it's okay to ask us about ours. This is going to take some time to get used to. You will have lots of questions in the days to come. You may have some right now. Please ask any question at any time.

Amato, P. R., J. B. Kane and S. James. 2011. Reconsidering the 'Good Divorce' *Family Relations* 60: 511-524. Reprinted with permission from Pam Lewis, the author.

> **"When two people** decide to **get a divorce,**
> it isn't a sign that they **"don't understand"**
> one another, but **a sign** that they have,
> at last, **begun to."**
>
> —HELEN ROWLAND, JOURNALIST

divorced, nonresident parent (usually the father) and found that overnight stays with the nonresident parent were associated with reported greater closeness and better quality relationships than was true if the parent and child had only daytime contact.

5. **Attention from the noncustodial parent.** Children benefit when they receive frequent and consistent attention from noncustodial parents, usually the fathers. Noncustodial parents who do not show up at regular intervals exacerbate their children's emotional insecurity by teaching them, once again, that parents cannot be depended on. Parents who show up often and consistently teach their children to feel loved and secure. Sometimes joint custody solves the problem of children's access to their parents.

6. **Assertion of parental authority.** Children benefit when both parents continue to assert their parental authority and continue to support the discipline practices of each other.

7. **Regular and consistent child support payments.** Support payments (usually from the father to the mother) are associated with economic stability for the child.

8. **Stability.** Not moving and keeping children in the same school system is beneficial to children. Some parents—called *latchkey parents*—spend every other week with the children in the family home so the children do not have to alternate between residences (Luscombe, 2011).

9. **Children in a new marriage.** Manning and Smock (2000) found that divorced, noncustodial fathers who remarried and who had children in the new marriages were more likely to shift their emotional and economic resources to the new family unit than were fathers who did not have new biological children. Fathers might be alert to this potential and consider each child, regardless of when or with whom the child was born, as worthy of continued love, time, and support.

10. **Age and reflection on the part of children of divorce.** Sometimes children whose parents are divorced benefit from growing older and reflecting on their parents' divorce as adults rather than as children. Nielsen (2004) emphasized that daughters who feel distant from their fathers can benefit from examining the divorce from the viewpoint of the father (was he alienated by the mother?), the cultural bias against fathers (they are maligned as "deadbeat dads" who "abandon their families for a younger woman"), and the facts about divorced dads (they are more likely than mothers to be depressed and suicidal following divorce). Adolescents also bear some responsibility for the post-divorce relationships with their parents. Menning (2008) emphasized that adolescents are not passive recipients of their parents' divorce but may actively accelerate or decelerate a positive or negative relationship with their parents by using a variety of relationship management strategies. For example, by deciding to shut down and disclose nothing to their parents about their lives, adolescents increase the emotional distance between themselves and their parents.

11. **Divorce education program for children.** Communities also make divorce education programs available for children. These may be effective in helping children to talk about their feelings and to know that other children going through divorce have similar feelings.

14-7 Prerequisites for Having a "Successful" Divorce

Although the concept of a "successful" divorce has been questioned (Amato et al., 2011), some divorces may cause limited damage to the partners and their children. Indeed, most people are resilient and "are able to adapt constructively to their new life situation within two to three years following divorce, a minority being defeated by the marital breakup, and a substantial group of women being enhanced" (Hetherington, 2003, p. 318). The following are some of the behaviors spouses can engage in to achieve a successful divorce:

1. **Mediate rather than litigate the divorce.** Divorce mediators encourage a civil, cooperative, compromising relationship while moving the couple toward an agreement on the division of property, custody, and child support. By contrast, attorneys make their money by encouraging hostility so that spouses will prolong the conflict, thus running up higher legal bills. In addition, the couple cannot divide money spent on divorce attorneys (the average cost is $15,000 for *each side* so a litigated divorce will start at $30,000). Benton (2008) noted that the worst thing divorcing spouses can do is to each hire the "meanest, nastiest, most expensive yard dog lawyer in town" because doing so will only result in a protracted, expensive divorce in which neither spouse will "win." Because the greatest damage to children from a divorce is caused by a continuing hostile and bitter relationship between their parents, some states require **divorce mediation** as a mechanism to encourage civility in working out differences and to clear the court calendar of protracted legal battles (Amato, 2010).

2. **Coparent with your ex-spouse.** Setting aside negative feelings about your ex-spouse so as to cooperatively coparent not only facilitates parental adjustment but also takes children out of the line of fire. Such coparenting translates into being cooperative when one parent needs to change a child-care schedule, sitting together during a performance by the children, and showing appreciation for the other parent's skill in responding to a crisis with the children.

3. **Take some responsibility for the divorce.** Because marriage is an interaction between spouses, one person is seldom totally to blame for a divorce. Rather, both spouses share reasons for the demise of the relationship. Take some responsibility for what went wrong.

4. **Learn from the divorce.** View the divorce as an opportunity to improve yourself for future relationships. What did you do that you might consider doing differently in the next relationship?

5. **Create positive thoughts.** Divorced people are susceptible to feeling as though they are failures. They see themselves as Divorced people with a capital D, a situation sometimes referred to as "hardening of the categories" disease. Improving self-esteem is important for divorced people. They can do this by systematically thinking positive thoughts about themselves. One technique is to write down 21 positive statements about yourself ("I am honest," "I have strong family values," "I am a good parent," and so on) and transfer them to three-by-five cards, each containing three statements. Take one of the cards with you each day and read the thoughts at three regularly spaced intervals (e.g., 7:00 a.m., 1:00 p.m., and 7:00 p.m.). This ensures that you are thinking positive thoughts about yourself and are not allowing yourself to drift into a negative set of thoughts ("I am a failure" or "no one wants to be with me.")

6. **Avoid alcohol and other drugs.** The stress and despair that some people feel following a divorce make them particularly vulnerable to the use of alcohol or other drugs. These should be avoided because they produce an endless negative cycle. For example, stress is relieved by alcohol; alcohol produces a hangover and negative feelings; the negative feelings are relieved by more alcohol, producing more negative feelings, and so on.

7. **Relax without drugs.** Deep muscle relaxation can be achieved by systematically tensing and relaxing each of the major muscle groups in the body. Alternatively, yoga, transcendental meditation, and massage can induce a state of relaxation in some people. Whatever the form, it is important to schedule a time each day for relaxation.

divorce mediation process in which divorcing parties make agreements with a third party (mediator) about custody, visitation, child support, property settlement, and spousal support.

8. **Engage in aerobic exercise.** Exercise helps one to not only counteract stress but also avoid it. Jogging, swimming, riding an exercise bike, or engaging in other similar exercise for 30 minutes every day increases the oxygen to the brain and helps facilitate clear thinking. In addition, aerobic exercise produces endorphins in the brain, which create a sense of euphoria ("runner's high").

9. **Engage in fun activities.** Some divorced people sit at home and brood over their "failed" relationship. This ruminating only compounds their depression. Doing what they have previously found enjoyable—swimming, horseback riding, skiing, attending sporting events with friends—provides an alternative to sitting on the couch alone.

10. **Continue interpersonal connections.** Adjustment to divorce is facilitated by continuing relationships with friends and family. These individuals provide emotional support and help buffer the feeling of isolation and aloneness. First Wives World (www.firstwivesworld.com) is a new interactive website that provides an Internet social network for women transitioning through divorce.

11. **Let go of the anger for your ex-partner.** Former spouses who stay negatively attached to an ex by harboring resentment and trying to get back at the ex prolong their adjustment to divorce. The old adage that you can't get ahead by getting even is relevant to divorce adjustment.

12. **Allow time to heal.** Because self-esteem usually drops after divorce, a person is often vulnerable to making commitments before working through feelings about the divorce. Most people need between 12 and 18 months to adjust to the legal end of a marriage. Although being available to others may help to repair one's self-esteem, getting remarried during this time should be considered cautiously. Two years between marriages is recommended.

divorcism the belief that divorce is a disaster.

13. **Progress through the psychological stages of divorce.** When getting a divorce, individuals go through a variety of psychological stages (Wiseman, 1975). These stages include the following:

   a. **Denial.** Marital problems are ignored or attributed to an external cause. "We are fine" or "this is normal" are ways of coping with the emotional and physical distance from the partner.

   b. **Depression.** Spouses confront the reality that their marriage is in trouble, that they will divorce and lose their once-intimate relationship permanently. Depression sets in.

   c. **Anger or ambivalence.** Spouses turn their anger toward each other and become critical, vindictive, and even violent. Strong negative emotions ensue during the marital separation, which is a normal response to the loss of an important attachment figure (e.g., a spouse). Some spouses are not capable of detaching and maintain a negative attachment; by staying bitter and resentful, they remain attached (some until their deaths). They may also want to hold on to the dying relationship because it feels safe.

   d. **New lifestyle and identity.** Ex-spouses begin a new life as single adults and detach from marital identity. Men typically drink more alcohol. Women typically turn to girlfriends for support. Both may enter a period of being sexually indiscriminate.

   e. **Acceptance and integration.** Individuals recognize their new status, loss of anger, and acceptance, and move on to new relationships. Some never reach this stage but harbor resentments or blame the ex for the end of the marriage.

14-8 Divorce Prevention

Divorce remains stigmatized in our society, as evidenced by the term **divorcism**—the belief that divorce is a disaster. In view of this cultural attitude, a number of attempts have been made to reduce it. Marriage education workshops provide an opportunity for couples to meet with other couples and

© iStockphoto.com/Nic Taylor

a leader who provides instruction in communication, conflict resolution, and parenting skills. Lucier-Greer et al. (2012) found positive outcomes for remarried couples with children who were involved in couple and relationship education (CRE) that focused on stepfamily issues.

Another attempt at divorce prevention is **covenant marriage** (now available in Arizona, Arkansas, and Louisiana), which emphasizes the importance of staying married. In these states, couples agree to the following when they marry: (1) marriage preparation (meeting with a counselor who discusses marriage and their relationship); (2) full disclosure of all information that could reasonably affect a partner's decision to marry (e.g., previous marriages, children, STIs, one's homosexuality); (3) an oath that their marriage is a lifelong commitment; (4) an agreement to consider divorce only for "serious" reasons such as abuse, adultery, and imprisonment for a felony or separation of more than two years; (5) an agreement to see a marriage counselor if problems threaten the marriage; and (6) not to divorce until after a two-year "cooling off" period (Economist, 2005).

Although most of a sample of 1,324 adults in a telephone survey in Arizona, Louisiana, and Minnesota were positive about covenant marriage (Hawkins et al., 2002), fewer than 3% of marrying couples elected covenant marriages when given the opportunity to do so (Licata, 2002). Although already married couples can convert their standard marriages to covenant marriages, there are no data on how many have done so (Hawkins et al., 2002).

14-9 Remarriage

Divorced spouses are not sour on marriage. Although they may want to escape from the current spouse, they are open to having a new spouse. Two thirds of divorced females and three fourths of divorced males remarry (Sweeney,

2010). And these new unions typically occur within two years. When comparing divorced individuals who have remarried and divorced individuals who have not remarried, the remarried individuals report greater personal and relationship happiness.

14-9a Remarriage for Divorced Individuals

Ninety percent of remarriages are of people who are divorced rather than widowed. The principle of **homogamy** is illustrated in the selection of a new

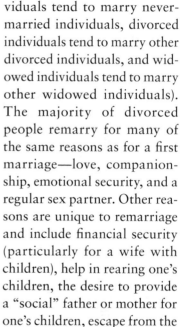

spouse (never-married individuals tend to marry never-married individuals, divorced individuals tend to marry other divorced individuals, and widowed individuals tend to marry other widowed individuals). The majority of divorced people remarry for many of the same reasons as for a first marriage—love, companionship, emotional security, and a regular sex partner. Other reasons are unique to remarriage and include financial security (particularly for a wife with children), help in rearing one's children, the desire to provide a "social" father or mother for one's children, escape from the stigma associated with the label "divorced person," and legal threats regarding the custody of one's children. With regard to the latter, the courts view a parent seeking custody of a child more favorably if the parent is remarried.

Regardless of the reason for remarriage, it is best to proceed slowly into a remarriage. Some may benefit from involvement in a divorce adjustment program. Vukalovich and Caltabiano (2008) assessed the effectiveness of an adjustment separation or divorce program. Twenty females and 10 males completed a pre- and post-program questionnaire. Overall, the results indicated that participants made significant adjustment gains following participation in the program.

covenant marriage
type of marriage that permits divorce only under specific conditions.

homogamy
tendency to select someone with similar characteristics to marry.

For others, grieving over the loss of a first spouse through divorce or death and developing a relationship with a new partner take time. At least two years is the recommended interval between the end of a marriage and a remarriage (Marano, 2000). Older divorced women (over age 40) are less likely than younger divorced women to remarry. Not only are fewer men available, but also the **mating gradient** whereby men tend to marry women younger than themselves is operative. In addition, older women are more likely to be economically independent, to enjoy living alone, to value the freedom of singlehood, and to want to avoid the restrictions of marriage. Divorced people getting remarried are usually about 10 years older than those marrying for the first time. People in their mid-30s who are considering remarriage are more likely to have finished school and to be established in a job or career than individuals in their first marriages.

Courtship for previously married individuals is usually short and takes into account the individuals' respective work schedules and career commitments. Because each partner may have children, much of the couple's time together includes their children. In subsequent marriages, eating pizza at home and renting a DVD to watch with the kids often replaces the practice of going out alone on an expensive dinner date during courtship before a first marriage.

© Food Pix/Jupiterimages

14-9b Preparation for Remarriage

Higginbotham and Skogrand (2010) studied 356 adults who attended a 12-hour stepfamily relationship education course. These participants were either remarried, cohabiting, or seriously dating someone who had children from a previous relationship. Results revealed that regardless of their race or marital status, the individuals benefited from the stepfamily relationship education—commitment to the relationship, agreement on finances, relationships with ex-partners, and parenting all improved over time.

Children also benefit when parents participate in stepfamily education programs. Higginbotham et al. (2010) interviewed 40 parents and 20 facilitators who were part of a stepfamily education program. They found that children were perceived to benefit by their parents' increased empathy, engagement in family time, and enhanced relationship skills.

Self-help books on remarriage and stepfamily issues are lacking. One researcher (Shafer, 2010) noted that such books tend to lack empirical research and are unable to provide practical solutions for blended families and parents.

14-9c Issues of Remarriage

Several issues challenge people who remarry (Ganong & Coleman, 1994; Goetting, 1982; Kim, 2011).

Boundary Maintenance Movement from divorce to remarriage is not a static event that is over after a brief ceremony. Rather, ghosts of the first marriage—in terms of the ex-spouse and, possibly, the children—must be dealt with. A parent must decide how to relate to an ex-spouse to maintain a good parenting relationship for the biological children while keeping an emotional distance to prevent problems from developing with a new partner. Some spouses continue to be emotionally attached to and have difficulty breaking away from an ex-spouse. These former spouses have what Masheter (1999) terms a **negative commitment**. Masheter says such individuals "have decided to remain [emotionally] in this relationship

mating gradient
the tendency for husbands to be more advanced than their wives with regard to age, education, and occupational success.

negative commitment also known as negative attachment, individuals who remain emotionally invested in their relationship with their former spouse, despite remarriage.

and to invest considerable amounts of time, money, and effort in it ... [T]hese individuals do not take responsibility for their own feelings and actions, and often remain 'stuck,' unable to move forward in their lives" (p. 297).

One former spouse recalled a conversation with his ex from whom he had been divorced for more than 17 years. "Her anger coming through the phone seemed like she was still feeling the divorce as though it had happened that morning," he said.

© Medioimages/Jupiterimages

Emotional Remarriage Remarriage involves beginning to trust and love another person in a new relationship. Such feelings may come slowly as a result of negative experiences in a previous marriage.

Psychic Remarriage Divorced individuals considering remarriage may find it difficult to give up the freedom and autonomy of being single and to develop a mental set conducive to pairing. This transition may be particularly difficult for people who sought a divorce as a means to personal growth and autonomy. These individuals may fear that getting remarried will end their freedom.

Community Remarriage This aspect involves a change in focus from single friends to a new mate and other couples with whom the new pair will interact. The bonds of friendship established during the divorce period may be particularly valuable because they have given support at a time of personal crisis. Care should be taken not to drop these friendships.

Parental Remarriage Because most remarriages involve children, people must work out the nuances of living with someone else's children. Mothers are usually awarded primary physical custody, and this circumstance translates into a new stepfather adjusting to the mother's children and vice versa. For individuals with children from a previous marriage who do not live primarily with them, a new spouse must adjust to these children on weekends, holidays, and vacations or at other visitation times.

Economic and Legal Remarriage A second marriage may begin with economic responsibilities to a first marriage. Alimony and child support often threaten the harmony and sometimes even the economic survival of second marriages. Although the income of a new wife is not used legally to decide the amount her new husband is required to pay in child support for his children of a former marriage, his ex-wife may petition the court for more child support. The ex-wife may do so, however, on the premise that his living expenses are reduced with a new wife and that, therefore, he should be able to afford to pay more child support. Although an ex-wife is not likely to win, she can force the new wife to go to court and disclose her income (all with considerable investment of time and legal fees for a newly remarried couple).

Economic issues in a remarriage may become evident in another way. A remarried woman who receives inadequate child support from an ex-spouse and needs money for her child's braces, for instance, might wrestle with how much money to ask her new husband for.

14-9d Remarriage for the Widowed

About 10% of remarriages involve widows or widowers. Unlike divorced individuals, widowed individuals are usually much older and their children are grown. Marriages in which one spouse is considerably older than the other are referred to as May-December marriages. Here we will discuss only **December marriages**, in which both spouses are elderly.

A study of 24 elderly couples found that the primary motivation for remarriage was the need to escape loneliness or the need for companionship (Vinick, 1978). Men reported a greater need to remarry than did the women. Most of the spouses (75%) met through a mutual friend or relative and married less than a year after their partner's death (63%) Increasingly, elderly individuals are meeting online. Some sites cater to older individuals seeking partners, including senior friendfinder.com and thirdage.com. The children of the couples in Vinick's study had mixed reactions to their parent's remarriage. Most of the children were happy that their parent was happy and felt relieved that someone would now meet the companionship needs of their elderly parent on a more regular basis. Brimhall and Engblom-Deglmann (2011) studied remarriages in which at

December marriage a new marriage in which both spouses are elderly.

> "When one **door of happiness** closes, another opens; but often we look so long at the **closed door** that we do not **see** the one which has **been opened for us.**"
>
> —HELEN KELLER, BLIND AND DEAF AMERICAN AUTHOR, POLITICAL ACTIVIST, AND LECTURER

least one of the partners was a widow. They interviewed 24 remarried individuals about the death of a previous spouse, either theirs or their partner's, and how this was affecting their marriage. Participants were interviewed individually and as a couple. Several themes emerged including putting the past spouse on a pedestal, comparing the current and past spouses, insecurity of the current spouse, curiosity about the past spouse and relationship, the new partner's response to this curiosity, and its impact on the current relationship. The best new relationship outcomes seemed to happen when the spouse of a deceased partner talked openly about the past relationship and reassured the current partner of his or her love for the new partner and the new relationship.

14-9e Stages of Involvement With a New Partner

After a legal separation or divorce (or being widowed), a parent who becomes involved in a new relationship passes through various transitions. These include deciding on the level of commitment to the new partner, introducing one's children to the new partner, and allowing time for the children to adjust to the new partner. Progressing through the stages may not be easy.

Stability of Remarriages National data reflect that remarriages are more likely than first marriages to end in divorce in the early years of remarriage (Sweeney, 2010). Remarriages most vulnerable to divorce are those that involve a woman bringing a child into the new marriage. Teachman (2008) analyzed data on women (N = 655) from the National Survey of Family Growth to examine the correlates of second marital dissolution. He found that women who brought stepchildren into their second marriage experienced an elevated risk of marital disruption. Premarital

blended family
a family in which new spouses both have children from previous marriages.

cohabitation or having a birth while cohabiting with a second husband did not raise the risk of marital dissolution, however. In addition, marrying a man who brought a child to the marriage did not increase the risk of marital disruption. One possible explanation for why a woman's bringing a child into a second marriage is related to greater instability is that she may be less attentive to the new husband and more of a mother than a wife.

Second marriages, in general, are more susceptible to divorce than first marriages because divorced individuals are less fearful of divorce than individuals who have never divorced. So, rather than stay in an unhappy second marriage, the spouses leave.

Though remarried people are more vulnerable to divorce in the early years of their subsequent marriage, they are less likely to divorce after 15 years of staying in the second marriage than are those in first marriages (Clarke & Wilson, 1994). Hence, these spouses are likely to remain married because they want to, not because they fear divorce. McCarthy and Ginsberg (2007) also noted that couples in functional and stable second marriages take greater pride and report higher satisfaction in their marriage than do couples in their first marriage. Not only are they relieved at not having to live in their former relationships, but they are older and more skilled in working out the nuances of a life together.

14-10 Stepfamilies

Stepfamilies, also known as blended, binuclear, remarried, or reconstituted families, represent the fastest-growing type of family in the United States. Of 2,691 adults, 42% said they have at least one step relative. Three in 10 have a step or half sibling, 18% have a living stepparent, and 13% have at least one stepchild (Parker, 2011). A **blended family** is one in which spouses in a new marriage relationship blend their children from at least one other spouse from a

previous marriage. The term **binuclear** refers to a family that spans two households; when a married couple with children divorce, their family unit typically spreads into two households.

There is a movement away from the use of the term *blended* because stepfamilies really do not blend. The term **stepfamily** (sometimes referred to as **step relationships**) is currently in vogue. This section examines how stepfamilies differ from nuclear families and explores the various developmental tasks of stepfamilies.

14-10a Definition and Types of Stepfamilies

Although there are various types of stepfamilies (Sweeney, 2010), the most common is a family in which the partners bring children from previous relationships into the new home, where they may also have a child of their own (Sweeney, 2010). The couple may be married or living together, heterosexual or homosexual, and of any race. Although a stepfamily can be created when an individual who has never married or a widowed parent with children marries a person with or without children, most stepfamilies today are composed of spouses who are divorced and who bring children into a new marriage. This arrangement is different from stepfamilies characteristic of the early 20th century, which more often were composed of spouses who had been widowed.

As noted, stepfamilies may be both heterosexual and homosexual. Lesbian stepfamilies model gender flexibility in that a biological mother and a stepmother tend to share parenting (in contrast to a traditional family, in which the mother may take primary responsibility for parenting and the father is less involved). This sharing allows a biological mother some freedom from motherhood as well as support in it. In gay male stepfamilies, the gay men may also share equally in the work of parenting.

14-10b Myths of Stepfamilies

Various myths abound regarding stepfamilies, including that new family members will instantly bond emotionally, that children in stepfamilies are damaged and do not recover, that stepmothers are "wicked" and "home-wreckers," that stepfathers are uninvolved with their stepchildren, and that stepfamilies are not "real" families. Regarding the latter, Sweeney (2010) reported that almost half of adult children chose a middle-range category—"quite a bit" or "a little" rather than "fully" or "not at all"—when asked to identify the extent to which they perceived a stepparent as being a family member.

14-10c Unique Aspects of Stepfamilies

Stepfamilies differ from nuclear families in a number of ways. To begin with, children in nuclear families are biologically related to both parents, whereas children in stepfamilies are biologically related to only one parent. Also, in nuclear families, both biological parents live with their children, whereas only one biological parent in stepfamilies lives with the children. In some cases, the children alternate living with each parent.

Though nuclear families are not immune to loss, everyone in a stepfamily has experienced the loss of a love partner, which results in grief. About 70% of children whose parents have divorced are living without their biological father (who some children desperately hope will reappear and reunite the family). The respective spouses may also have experienced emotional disengagement and physical separation from a once loved partner. Stepfamily members may also experience losses because of having moved away from the house in which they lived, their familiar neighborhood, and their circle of friends.

Stepfamily members are also connected psychologically to others outside their unit. Bray and Kelly (1998) referred to these relationships as the "ghosts at the table":

> Children are bound to absent parents, adults to past lives

binuclear a family that spans two households, often because of divorce.

stepfamily (step relationships) a family in which partners bring children from previous marriages into the new home, where the partners may also have a child of their own.

and past marriages. These invisible psychological bonds are the ghosts at the table and because they play on the most elemental emotions—emotions like love and loyalty and guilt and fear—they have the power to tear a marriage and stepfamily apart. (p. 4)

Children in nuclear families have also been exposed to a relatively consistent set of beliefs, values, and behavior patterns. When children enter a stepfamily, they "inherit" a new parent, who may bring into the family unit a new set of values and beliefs and a new way of living. Likewise, a stepparent may want to rear a child differently than the biological parents reared the child. "My new spouse's kids had never been to church and didn't have good manners," reported one frustrated stepparent.

Another unique aspect of stepfamilies is that the relationship between the biological parent and the children has existed longer than the relationship between the adults in the remarriage. Jane and her twin children have a nine-year relationship and are emotionally bonded to each other. However, Jane has known her new partner only a year, and although her children like their new stepfather, they hardly know him.

In addition, the relationship between biological parents and their children is of longer duration than that of stepparents and stepchildren. The short history of the relationship between children and stepparents is one factor that may contribute to increased conflict between these two during children's adolescence. Children may also become confused and wonder whether they are disloyal to their biological parent if they become friends with a stepparent.

Stepfamilies are also unique in that stepchildren have two homes they may regard as theirs, unlike children in a nuclear family, who have one home they regard as theirs. In some cases of joint custody, children spend part of each week with one parent and part with the other; they live with two sets of adult parents in two separate homes.

Stepfamilies typically have economic problems in the form of reduced disposable income (Adler-Baeder et al., 2010). Money, or lack of it, from an ex-spouse may also be a source of conflict. In some stepfamilies, an ex-spouse is expected to send child support payments to the parent who has custody of the children. Fewer than one half of these fathers send any money. Fathers who

stepism the assumption that stepfamilies are inferior to biological families.

> Stepfamilies typically have economic problems in the form of reduced disposable income.

pay regular child support tend to have higher incomes, to have remarried, to live close to their children, and to visit them regularly. They are also more likely to have legal shared or joint custody, which helps to ensure that they will have access to their children.

Fathers who do not voluntarily pay child support and are delinquent by more than one month may have their wages garnisheed by the state. Some fathers change jobs frequently and move around to make it difficult for the government to keep up with them. Such dodging of the law is frustrating to custodial mothers who need the child support money. Added to the frustration for some mothers is the fact that fathers are legally entitled to see their children even when they do not pay court-ordered child support. This requirement angers mothers who must give up their children on weekends and holidays to a man who is not supporting his children financially. Such distress on the part of mothers may be conveyed to the children.

New relationships in stepfamilies experience almost constant flux. Each member of a new stepfamily has many adjustments to make. Issues that must be dealt with include how the mate feels about and interacts with the partner's children from a former marriage and how the children feel about and interact with the new stepparent. Saint-Jacques et al. (2011) noted the complexity of stepfamily living and the numerous issues new spouses with children must navigate. In general, families in the Bray and Kelly (1998) study did not begin to think and act like a family until the end of the second or third year. These early years are the most vulnerable, with a quarter of stepfamilies ending during that period.

Stepfamilies are also stigmatized. **Stepism** is the assumption that stepfamilies are inferior to biological families. Stepism, like racism, heterosexism, sexism, and ageism, involves prejudice and discrimination. Social changes need to be made to give support to stepfamilies. For example, if there is a graduation banquet, is this an opportunity for children to invite all four of their parents? More often, children are forced to choose, which usually results in selecting the biological parents and ignoring the stepparents. The more adults children have who love and support them (e.g., both biological parents and stepparents), the better for the children, and our society should support this arrangement.

Stepparents also have no child-free period. Unlike newly married couples in nuclear families, who typically have their first child about two and one-half years after their wedding, many remarried couples begin their marriage with children in the house.

Profound legal differences exist between nuclear and blended families. Whereas biological parents in nuclear families are required in all states to support their children, only five states require stepparents to provide financial support for their stepchildren. Thus, when stepparents divorce, this and other discretionary types of economic support usually stop.

Other legal matters with regard to nuclear families versus stepfamilies involve inheritance rights and child custody. Stepchildren do not automatically inherit from their stepparents, and courts have been reluctant to give stepparents legal access to stepchildren in the event of a divorce. In general, U.S. law does not consistently recognize stepparents' roles, rights, and obligations regarding their stepchildren. Without legal support to ensure such access, these relationships tend to become more distant and nonfunctional (Sweeney, 2010).

Finally, extended family networks in nuclear families are smooth and comfortable, whereas those in stepfamilies often become complex and strained. Table 14.2 summarizes the differences between nuclear families and stepfamilies.

Table 14.2
Differences Between Nuclear Families and Stepfamilies

Nuclear Families	Stepfamilies
1. Children are (usually) biologically related to both parents.	1. Children are biologically related to only one parent.
2. Both biological parents live together with children.	2. As a result of divorce or death, one biological parent does not live with the children. In the case of joint physical custody, children may live with both parents, alternating between them.
3. Beliefs and values of members tend to be similar.	3. Beliefs and values of members are more likely to be different because of different backgrounds.
4. The relationship between the adults has existed longer than the relationship between children and parents.	4. The relationship between children and parents has existed longer than the relationship between adults.
5. Children have one home they regard as theirs.	5. Children may have two homes they regard as theirs.
6. The family's economic resources come from within the family unit.	6. Some economic resources may come from an ex-spouse.
7. All money generated stays in the family.	7. Some money generated may leave the family in the form of alimony or child support.
8. Relationships are relatively stable.	8. Relationships are in flux: new adults adjusting to each other; children adjusting to a stepparent; a stepparent adjusting to stepchildren; stepchildren adjusting to each other.
9. No stigma is attached to the nuclear family.	9. Stepfamilies are stigmatized.
10. Spouses had a child-free period.	10. Spouses had no child-free period.
11. Inheritance rights are automatic.	11. Stepchildren do not automatically inherit from stepparents.
12. Rights to custody of children are assumed if divorce occurs.	12. Rights to custody of stepchildren are usually not available.
13. Extended family networks are smooth and comfortable.	13. Extended family networks become complex and strained.
14. Nuclear family may not have experienced loss.	14. Stepfamily has experienced loss.

14-10d Stepfamilies in Theoretical Perspective

Structural functionalists, conflict theorists, and symbolic interactionists view stepfamilies from the following different points of view:

1. **Structural-functional perspective.** To the structural functionalist, integration or stability of the system is highly valued. The very structure of the stepfamily system can be a threat to the integration and stability of a family system. The social structure of stepfamilies consists of a stepparent, a biological parent, biological children, and stepchildren. Functionalists view the stepfamily system as vulnerable to an alliance between the biological parent and the biological children who have a history together.

   In some cases, the mother and her biological children may create an alliance. The stepfather, as an outsider, may view this alliance between the mother and her children as the mother's giving the children too much status or power in the family. Whereas a mother and her children may relate as equals, a stepfather may relate to the children as unequals whom he attempts to discipline. The result is a fragmented parental subsystem whereby the stepfather accuses the mother of being too soft and she accuses him of being too harsh.

   Structural family therapists suggest that parents should have more power than children and that they should align themselves with each other. Not to do so is to give children family power, which they may use to splinter the parents off from each other and create another divorce.

2. **Conflict perspective.** Conflict theorists view conflict as normal, natural, and inevitable as well as functional in that it leads to change. Conflict in a stepfamily system is seen as desirable in that it leads to equality and individual autonomy.

   Conflict is a normal part of stepfamily living. The spouses, parents, children, and stepchildren are constantly in conflict for the limited resources of space, time, and money. *Space* refers to territory (rooms) or property (television,

speakers, or electronic games) in the house that the stepchildren may fight over. *Time* refers to the amount of time that the parents will spend with each other, with their biological children, and with their stepchildren. Money must be allocated in a reasonably equitable way so that each member of the family has a sense of being treated fairly.

Problems arise when space, time, and money are limited. Two new spouses who each bring a child from a former marriage into the house have a situation fraught with potential conflict. Who sleeps in which room? Who gets to watch which channel on television?

To further complicate the situation, suppose the newly remarried couple has a baby. Where does the baby sleep? Because both parents may have full-time jobs, the time they have for the three children is scarce, not to mention that a baby will require a major portion of their available time. As for money, the cost of the baby's needs, such as formula and disposable diapers, will compete with the economic needs of the older children. Meanwhile, the spouses may need to spend time alone and may want to spend money as they wish. All these conflicts are functional, because their resolution will increase the chance that a greater range of needs will be met within the stepfamily.

3. **Interactionist perspective.** Symbolic interactionists emphasize the meanings and interpretations that members of a stepfamily develop for events and interactions in the family. Children may blame themselves for their parents' divorce and feel that they and their stepfamily are stigmatized; parents may view stepchildren as spoiled.

   Stepfamily members also nurture certain myths. Stepchildren sometimes hope that their parents will reconcile and that their nightmare of divorce and stepfamily living will end. This is the myth of reconciliation. Another is the myth of instant love, usually held by stepparents, who hope that the new partner's children will instantly love them. Although this does happen, particularly if the child is young and has no negative influences from the other parent, it is unlikely.

14-10e Stages in Becoming a Stepfamily

Just as a person must pass through various developmental stages in becoming an adult, a stepfamily goes through a number of stages as it overcomes various obstacles. Researchers such as Bray and Kelly (1998) and Papernow (1988) have identified various stages of development in stepfamilies. These stages include the following:

Stage 1: Fantasy. Both spouses and children bring rich fantasies into a new marriage. Spouses fantasize that their new marriage will be better than the previous one. If the new spouse has adult children, the couple assumes that these children will be open to a rewarding relationship with them. Young children have their own fantasy—they hope that their biological parents will somehow get back together and that the stepfamily will be temporary.

Stage 2: Reality. Instead of realizing their fantasies, new spouses may find that stepchildren ignore or are rude to them. Shapiro and Stewart (2011) compared stepmothers with biological mothers and found the former reported more stress and depression.

Stage 3: Being Assertive. Initially a stepparent assumes a passive role and accepts the frustrations and tensions of stepfamily life. Eventually, however, resentment can reach a level where the stepparent is driven to make changes. The stepparent may make the partner aware of the frustrations and suggest that the marital relationship should have priority some of the time. The stepparent may also make specific requests, such as reducing the number of conversations the partner has with the ex-spouse, not allowing the dog on the furniture, or requiring the stepchildren to use better table manners. This stage is successful to the degree that the partner supports the recommendations for change. A crisis may ensue.

Stage 4: Strengthening Pair Ties. During this stage, the remarried couple solidifies their relationship by making it a priority. At the same time, the biological parent must back away somewhat from the parent–child relationship so that the new partner can have the opportunity to establish a relationship with the stepchildren.

This relationship is the product of small units of interaction and develops slowly. Many day-to-day activities, such as watching television, eating meals, and riding in the car together, provide opportunities for the stepparent–stepchild relationship to develop. It is important that the stepparent not attempt to replace the relationship that the stepchildren have with their biological parents.

Stage 5: Recurring Change. A hallmark of all families is change, but this is even more true of stepfamilies. Bray and Kelly (1998) note that, even though a stepfamily may function well when the children are preadolescent, a new era can begin when the children become teenagers and begin to question how the family is organized and run. Stepfamily couples who survive have an intense desire to resolve issues and employ numerous strategies (Saint-Jacques et al., 2011).

The relationship a parent has with the former spouse impacts the relationship the biological child has with the new stepparent (Ganong et al., 2011). "If my dad had a better relationship with my mom I would have felt more comfortable being closer with him and his new wife."

14-11 Strengths of Stepfamilies

Stepfamilies have both strengths and weaknesses. Strengths include children's exposure to a variety of behavior patterns, their observation of a happy remarriage, adaptation to stepsibling relationships inside the family unit, and greater objectivity on the part of the stepparent.

14-11a Exposure to a Variety of Behavior Patterns

Children in stepfamilies experience a variety of behaviors, values, and lifestyles. They have the advantage of living on the inside of two families. Although this may be confusing and challenging for children or adolescents, they are learning early how different family patterns can be. For example, one 12-year-old had never seen a couple pray at the dinner table until his mom remarried a man who was accustomed to a "blessing" before meals.

14-11b Happier Parents

Although children may want their parents to get back together, they often come to understand that their parents are happier divorced, alone or remarried than married to their other parent. If a parent remarries, the children may see the parent in a new and happier relationship. Such happiness may spill over into the context of family living so that there is less tension and conflict.

14-11c Opportunity for a New Relationship With Stepsiblings

Though some children reject their new stepsiblings, others are enriched by the opportunity to live with a new person to whom they are now "related." One 14-year-old remarked, "I have never had an older brother to do things with. We both like to do the same things and I couldn't be happier about the new situation." Some stepsibling relationships are maintained throughout adulthood.

Nevertheless, there is stress associated with having stepsiblings. Tillman (2008) studied sibling relationships in stepfamilies (referring to them as "nontraditional siblings") and noted that adolescents with stepsiblings living in the same house made lower grades and had more school-related behavior problems. The researcher noted that these outcomes are related to the complexity and stress of living with stepsiblings.

14-11d More Objective Stepparents

Because of the emotional tie between a parent and a child, some parents have difficulty discussing certain issues or topics. A stepparent often has the advantage of being less emotionally involved and can relate to a child at a different level. One 13-year-old said of the relationship with his new

developmental task a skill that allows a family to grow as a cohesive unit.

© PhotoAlto sas/Alamy

stepmom, "She went through her own parents' divorce and knew what it was like for me to be going through my dad's divorce. She was the only one I could really talk to about this issue. Both my dad and mom were too angry about the subject to be able to talk about it."

14-12
Developmental Tasks for Stepfamilies

A developmental task is a skill that, if mastered, allows a family to grow as a cohesive unit. Developmental tasks that are not mastered will edge the family closer to the point of disintegration. Some of the more important developmental tasks for stepfamilies are discussed in this section.

14-12a Acknowledge Losses and Changes

As noted previously, each stepfamily member has experienced the loss of a spouse or a biological parent in the home. Indeed, children have experienced the loss of an adult relationship. These losses are sometimes compounded by home, school, neighborhood, and job changes. Feelings about these losses and changes should be acknowledged as important and consequential. In addition, children should not be required to love their new stepparent or stepsiblings (and vice versa). Such feelings will develop only as a consequence of positive interaction over an extended period of time. Resiliency in stepfamilies is also facilitated by having a strong marriage, support from family and friends, and good communication (Greeff & Du Toit, 2009).

Preserving original relationships is helpful in reducing a child's grief over loss. It is sometimes helpful for the biological parent and child to take time to nurture their relationship apart from stepfamily activities. This special consideration will reduce the child's sense of loss and any feelings of jealousy toward new stepsiblings.

14-12b Nurture the New Marriage Relationship

It is critical to the healthy functioning of a new stepfamily that the new spouses nurture each other and form a strong marital or relationship unit (Kim, 2011). Indeed, the adult dyad is vulnerable because the couple has had no child-free time upon which to build a common relationship base. Once a couple develops a core relationship, they can communicate, cooperate, and compromise with regard to the various issues in their new blended family. Too often spouses become child-focused and neglect the relationship on which the rest of the family depends. Such nurturing translates into spending time alone with each other, sharing each other's lives, and having fun with each other. One remarried couple goes out to dinner at least once a week without the children. "If you don't spend time alone with your spouse, you won't have one," says one stepparent. Failure to develop a close emotional bond between the partners is to make their relationship vulnerable to "splitting off" and siding with one or more of the children (with negative outcomes for the family).

14-12c Integrate the Stepfather Into the Child's Life

Stepfathers who become interested in what their stepchild does and who spend time alone with the stepchild report greater integration into the life of the stepchild and the stepfamily. The mutual benefits (stepfather/stepchild) are enormous. In addition, the mother of the child feels closer to her new husband if he has bonded with her offspring. Some stepfathers benefit from specific stepfather instruction (*Higginbotham* et al., 2012).

14-12d Allow Time for the Relationship Between the New Partner and One's Children to Develop

In an effort to escape single parenthood and to live with one's beloved, some individuals rush into remarriage without getting to know each other. Not only do they have limited information about each other, but their respective children may also have spent little or no time with their future stepparent. One stepdaughter remarked, "I came home one afternoon to find a bunch of plastic bags in the living room with my soon-to-be stepdad's clothes in them. I had no idea he was moving in. It hasn't been easy." Both adults and children should have had meals together and spent some time in the same house before becoming bonded by marriage as a family. Schrodt et al. (2008) reported the positive effects of daily communication between stepparents and stepchildren. Both reported higher satisfaction when they had frequent daily communication. As noted previously, stepparents are encouraged to discipline their own children because often the stepparent and stepchild have had insufficient time to experience a relationship that allows for discipline from the stepparent.

© bilderlounge/Jupiterimages

14-12e Have Realistic Expectations

Because of the complexity of meshing the numerous relationships involved in a stepfamily, it is important to be realistic. Dreams of one big happy family often set up stepparents for disappointment, bitterness, jealousy, and guilt. As noted previously, stepfamily members often do not begin to feel comfortable with each other until the third year (Bray & Kelly, 1998). Just as nuclear and single-parent families do not always run smoothly, neither do stepfamilies.

© Jaren Jai Wicklund/Shutterstock

14-12f Accept Your Stepchildren

Rather than wishing your stepchildren were different, accepting them is more productive. All children have positive qualities; find them and make them the focus of your thinking. Stepparents may communicate acceptance of their stepchildren through verbal praise and positive or affectionate statements and gestures. In addition, stepparents may communicate acceptance by engaging in pleasurable activities with their stepchildren and participating in daily activities such as homework, bedtime preparation, and transportation to after-school activities.

Acceptance continues beyond childhood. After graduating from high school the stepchild can benefit from emotional and financial support into the next phase of life. This support may include the stepparent taking the child to visit colleges so that he or she can assess their "fit."

14-12g Establish Your Own Family Rituals

Rituals are one of the bonding elements of nuclear families. Stepfamilies may integrate the various family members by establishing common rituals, such as summer vacations, visits to and from extended kin, and religious celebrations. These rituals are most effective if they are new and unique, not mirrors of rituals in the previous marriages and families.

14-12h Decide About Money

Money is an issue of potential conflict in stepfamilies because it can be a scarce resource, and several people may want to use it for their respective needs. For example, a father may want a new computer; the mother may want a new car; the mother's children may want bunk beds, dance lessons, and a satellite dish; the father's children may want a larger room, clothes, and a cell phone. How do the newly married couple and their children decide how money should be spent?

Some stepfamilies put all their resources into one bank and draw out money as necessary without regard for whose money it is or for whose child the money is being spent. Others keep their money separate; the parents have separate incomes and spend them on their respective biological children. Although no one pattern is superior to another, it is important for remarried spouses to agree on whatever financial arrangements they live by.

In addition to deciding how to allocate resources fairly in a stepfamily, remarried couples may face decisions regarding sending the children and stepchildren to college. Remarried couples may also make a will that is fair to all family members.

14-12i Give Parental Authority to Your Spouse/Coparent

Parents about to remarry should discuss the degree to which the stepparent will have authority over the biological children of each parent. Will each support the other? If Fred tells Mary's child to keep elbows off the dinner table, will Mary back Fred up? Schrodt and Braithwaite (2011) found that stepparents who coparent their children benefit from providing a united front for their children and, in turn, the marital relationship benefits.

14-12j Support the Children's Relationship With Their Absent Parent

A continued relationship with both biological parents is critical to the emotional well-being of children. Ex-spouses and stepparents should encourage children to have a positive relationship with both biological parents. Respect should also be shown for the biological parent's values. Bray and Kelly (1998) note that this consideration is "particularly difficult" to exercise. "But asking a child about an absent parent's policy on movies or curfews shows the child that, despite their differences, his/her mother and father still respect each other" (p. 92).

14-12k Cooperate With the Children's Biological Parent and Coparent

A cooperative, supportive, and amicable coparenting relationship between the biological parents and stepparents is a win-win situation for the children and parents. Otherwise, children are continually caught in the cross fire of the conflicted parental sets.

14-12l Support the Children's Relationship With Grandparents

It is important to support children's continued relationships with their natural grandparents on both sides of the family. This is one of the more stable relationships in the child's changing world of adult relationships.

Regardless of how ex-spouses feel about their former in-laws, they should encourage their children to have positive feelings for their grandparents. One mother said, "Although I am uncomfortable around my ex-in-laws, I know they are good to my children, so I encourage and support my children spending time with them."

14-12m Participate in a Stepfamily Education Web-based Program

Gelatt et al. (2010) studied the effect of a self-administered, interactive, and web-based stepfamily parent education program (*Parenting Toolkit: Skills for Stepfamilies*, http://stepfamily.orcasinc.com) on a sample of 300 parents and stepparents of children ages 11 to 15 who were randomized into either treatment or delayed-access control groups. Results revealed that the stepfamily education program positively influenced several key areas of parenting and family functioning at post-program and follow-up. Examples of positive change include reports of greater adjustment, harmony, and life satisfaction, and a reduction of parent–child conflict. Few online resources that can directly benefit stepfamilies are available. Parents might consider the potential benefits of this one.

STUDY TOOLS **14**

Ready to study? In this book, you can:

- Rip out the Chapter Review card in the back of the book to study for exams

- Take the Self Assessment for this chapter (card in the back of the book) and see where you stand on the vital issues raised in the chapter

Or you can go online to CourseMate at www.cengagebrain.com for these resources:

- Complete Practice Quizzes to prepare for tests

- Review Key Terms Flash Cards (online or print)

- Read about Marriage and Family in the news

- Play "Beat the Clock" to master concepts

- Check out Personal Applications

Relationships in the Later Years

"The **years teach** much which the **days never knew.**"

—RALPH WALDO EMERSON

SECTIONS

15-1 Age and Ageism

15-2 Caring for the Frail Elderly—the "Sandwich Generation"

15-3 Issues Confronting the Elderly

15-4 Successful Aging

15-5 Relationships and the Elderly

15-6 The End of Life

Our society is ambivalent in its attitude toward growing old and becoming the elderly. Although we all want to live a long and happy life, no one wants to be old and infirm. Fitness guru Jack LaLanne died at age 96. He had told his agent, "Dying will be bad for my image." LaLanne was not alone as an elderly member of society. Of the over 330 million people in the United States, around 40 million are adults over the age of 65 (*Statistical Abstract of the United States*, 2012, Table 34). By 2030, the elderly will represent 20% of our population. By 2050, almost 90 million will be over the age of 65 (Potter, 2010). Politicians are paying increased attention to the elderly as they are a considerable voting block.

The end of life comes for us all, and we face challenges as we age. These challenges include adapting to the death of loved ones, adapting to our bodies as they deteriorate over time, adjusting to new roles as those of parenting and work subside, and maintaining joy in our relationships. Although most young couples marrying today rarely give a thought to their own aging, the care of their aging parents is an issue that looms ahead.

In this chapter, we focus on the factors that confront individuals and couples as they age and the dilemma of how to care for aging parents. We begin by looking at the concept of age and how it is conceptualized in different ways.

15-1 Age and Ageism

All societies have a way to categorize their members by age. And all societies provide social definitions for particular ages.

15-1a The Concept of Age

A person's **age** may be defined chronologically, physiologically, psychologically, sociologically, and culturally. Chronologically, an "old" person is defined as one who has lived a certain number of years. The concept has obvious practical significance in everyday life. Bureaucratic organizations and social

age term defined chronologically, physiologically, sociologically, and culturally.

programs identify chronological age as a criterion of certain social rights and responsibilities. One's age determines the right to drive, vote, buy alcohol or cigarettes, and receive Social Security and Medicare benefits.

Age has meaning in reference to the society and culture of the individual. In ancient Greece and Rome, where the average life expectancy was 20 years, one was old at 18; similarly, one was old at 30 in medieval Europe and at 40 in the United States in 1850. In the United States today, however, people are usually not considered old until they reach age 65. However, our society is moving toward new chronological definitions of "old." Three groups of the elderly are the "young-old," the "middle-old," and the "old-old." The young-old are typically between the ages of 65 and 74; the middle-old, 75 to 84; and the old-old, 85 and beyond. Current life expectancy is shown in Table 15.1.

A person's age influences her or his view of what "old" means. Individuals ages 18 to 35 identify 50 as the age when the average man or woman becomes "old." However, those between the ages of 65 and 74 define "old" as 80 (Cutler, 2002). Hostetler (2011) noted that some senior citizens may avoid senior centers because of an "image" problem—they don't want to be labeled as old folks. Dr. Aubrey de Grey of the Department of Genetics, University of Cambridge, predicted that continued research will make adding *hundreds* of years to one's life possible. Maher and Mercer (2009) noted that according to the most optimistic predictions, we are two or three decades away from significant breakthroughs (Ray Kurzweil says we are 49 years away). Maher and Mercer (2009) also noted that the paths to longer life include genetic engineering, tissue or organ replacement, and the merging of computer technology with human biology. Should the human lifespan be extended hundreds of years, imagine the impact on marriage—"till death do us part."

Many individuals remain active in their later years. Clint Eastwood, now in his 80s, continues to make Academy Award–winning movies. Dick Van Dyke is in his mid-80s. His theme is "Keep moving." In her nineties, actress Betty White continues to host *Saturday Night Live,* make movies, and appear in commercials. Fitness guru Jack LaLanne was active until his death at age 96. He exercised two hours a day and did so the day before he died.

Table 15.1
Life Expectancy

Year	White Males	Black Males	White Females	Black Females
2010	76.5	70.2	81.3	77.2
2015	77.1	71.4	81.8	78.2
2020	77.7	72.6	82.4	79.2

Source: *Statistical Abstract of the United States,* 2012, 131st ed. (Washington, DC: U.S. Census Bureau).

Physiologically, people are old when their auditory, visual, respiratory, and cognitive capabilities decline significantly. At age 80 and over, 45% are hearing impaired and 25% have visual impairment (Dillion et al., 2010). Becoming "disabled" is associated with being "old." Individuals see themselves as disabled when their driver's licenses are taken away and when health-care workers come to their home to care for them (Kelley-Moore et al., 2006). Sleep changes occur for the elderly, including going to bed earlier, waking up during the night, and waking up earlier in the morning, along with such disorders as restless legs syndrome, snoring, and obstructive sleep apnea (Wolkove et al., 2007).

People who need full-time nursing care for eating, bathing, and taking medication properly and who are placed in nursing homes are thought of as being old. Failing health is the criterion the elderly use to define themselves as old (O'Reilly, 1997), and successful aging is typically defined as maintaining one's health, independence, and cognitive ability. Jorm et al. (1998) observed that the prevalence of successful aging declines steeply from age 70 to age 80.

People who have certain diseases are also regarded as old. Although younger individuals may suffer from Alzheimer's, arthritis, and heart problems, these ailments are more often associated with aging. As medical science conquers more diseases, the physiological definition of aging changes so that it takes longer for people to be defined as "old."

Psychologically, a person's self-concept is important in defining how old that person is. As individuals begin to fulfill the roles associated with the elderly—retiree, grandparent, nursing home resident—they begin to see themselves as aging. Sociologically, once they occupy these roles, others begin to see them as "old."

Culturally, the society in which an individual lives defines when and if a

© Mikael Damkier/iStockphoto.com

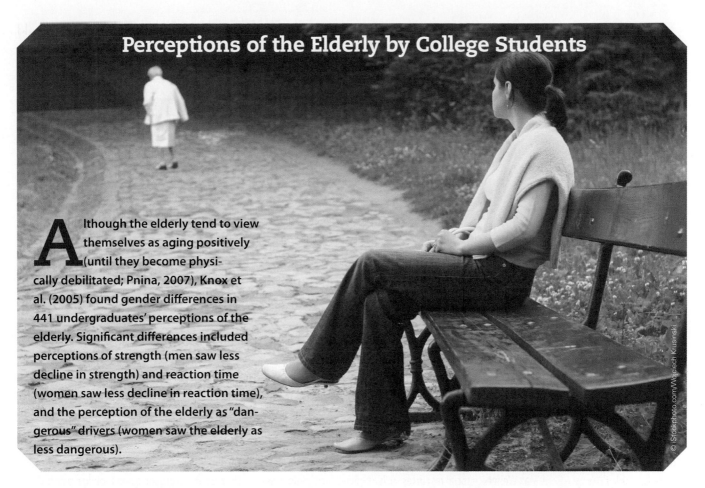

Perceptions of the Elderly by College Students

Although the elderly tend to view themselves as aging positively (until they become physically debilitated; Pnina, 2007), Knox et al. (2005) found gender differences in 441 undergraduates' perceptions of the elderly. Significant differences included perceptions of strength (men saw less decline in strength) and reaction time (women saw less decline in reaction time), and the perception of the elderly as "dangerous" drivers (women saw the elderly as less dangerous).

person becomes old and what being old means. In U.S. society, the period from ages 18 through 64 is generally subdivided into young adulthood, adulthood, and middle age. Cultures also differ in terms of how they view and take care of their elderly. Spain is particularly noteworthy in terms of care for the elderly, with 8 of 10 elderly people receiving care from family members and other relatives.

The elderly often view themselves negatively. In a study of 1,422 individuals ages 65 to 70, those who were of low economic status, were living alone, had multiple chronic medical conditions, and were depressed were most likely to have a negative self-perception of aging (Moser et al., 2011). Similarly, those who were healthy, lived with a partner in a fulfilling relationship, and had economic resources were more likely to view their being elderly positively. Researchers Henry et al. (2011) found that nursing and nutrition students (127) who played the simulated Aging Game increased their empathy and understanding of the elderly.

15.1b Ageism

Most societies have some form of **ageism**—the systematic persecution and degradation of people because

they are old. Ageism is similar to sexism, racism, and heterosexism. The elderly are shunned, discriminated against in employment, and sometimes victims of abuse. Popham et al. (2011) found that having an ageist attitude was associated with risk-taking behaviors (e.g., sex, alcohol, drugs, cigarettes) as people attempt to distance themselves from old age.

Negative stereotypes and media images (Haboush et al., 2012) of the elderly engender **gerontophobia**—a shared fear or dread of the elderly, which may create a self-fulfilling prophecy. For example, an elderly person forgets something and attributes the behavior to age. A younger person, however, engaging in the same behavior, is unlikely to attribute forgetfulness to age, given cultural definitions surrounding the age of the onset of senility. Individuals are also thought to become more inflexible and more conservative as they age. Analysis of national data suggests this is not true. In fact, the elderly become more tolerant (Danigelis et al., 2007).

The negative meanings associated with aging underlie

ageism systematic persecution and degradation of people because they are old.

gerontophobia fear or dread of the elderly.

the obsession of many Americans to conceal their age by altering their appearance. With the hope of holding on to youth a little bit longer, aging Americans spend billions of dollars each year on plastic surgery, exercise equipment, hair products, facial creams, and Botox injections.

The latest attempt to reset the aging clock is to have regular injections of human growth hormone (HGH), which promises to lower blood pressure, build muscles without extra exercise, increase the skin's elasticity, thicken hair, and heighten sexual potency. It is part of the regimen of clinics such as Lifespan (in Beverly Hills), which costs $1,000 a month after an initial workup costing $5,000. In the absence of long-term data, many physicians remain skeptical—there is no fountain of youth.

Table 15.2 **Theories of Aging**

Name of Theory	Level of Theory	Theorists	Basic Assumptions	Criticisms
Disengagement	Macro	Elaine Cumming, William Henry	The gradual and mutual withdrawal of the elderly and society from each other is a natural process. It is also necessary and functional for society that the elderly disengage so that new people can be phased in to replace them in an orderly transition.	Not all people want to disengage; some want to stay active and involved. Disengagement does not specify what happens when the elderly stay involved.
Activity	Macro	Robert Havighurst	People continue the level of activity they had in middle age into their later years. Though high levels of activity are unrelated to living longer, they are related to reporting high levels of life satisfaction.	Ill health may force people to curtail their level of activity. The older a person, the more likely the person is to curtail activity.
Conflict	Macro	Karl Marx, Max Weber	The elderly compete with youth for jobs and social resources such as government programs (Medicare).	The elderly are presented as disadvantaged. Their power to organize and mobilize political resources such as the American Association of Retired Persons is underestimated.
Age stratification	Macro	M. W. Riley	The elderly represent a powerful cohort of individuals passing through the social system that both affect and are affected by social change.	Too much emphasis is put on age, and little recognition is given to other variables within a cohort such as gender, race, and socioeconomic differences.
Modernization	Macro	Donald Cowgill	The status of the elderly is in reference to the evolution of the society toward modernization. The elderly in premodern societies have more status because what they have to offer in the form of cultural wisdom is more valued. The elderly in modern technologically advanced societies have low status because they have little to offer.	Cultural values for the elderly, not level of modernization, dictate the status of the elderly. Japan has high respect for the elderly and yet is highly technological and modernized.
Symbolic	Micro	Arlie Hochschild	The elderly socially construct meaning in their interactions with others and society. Developing social bonds with other elderly can ward off being isolated and abandoned. Meaning is in the interpretation, not in the event.	The power of the larger social system and larger social structures to affect the lives of the elderly is minimized.
Continuity	Micro	Bernice Neugarten	The earlier habit patterns, values, and attitudes of the individual are carried forward as a person ages. The only personality change that occurs with aging is the tendency to turn one's attention and interest on the self.	Other factors than one's personality affect aging outcomes. The social structure influences the life of the elderly rather than vice versa.

15.1c Theories of Aging

Gerontology is the study of aging. Table 15.2 identifies several theories, the level (macro or micro) of the theory, the theorists typically associated with the theory, assumptions, and criticisms. As noted, there are diverse ways of conceptualizing the elderly. Currently popular in sociology is the life-course perspective (Willson, 2007). This approach examines differences in aging across cohorts by emphasizing that "individual biography is situated within the context of social structure and historical circumstance" (p. 150).

15-2 Caring for the Frail Elderly— the "Sandwich Generation"

Elderly people are defined as **frail** if they have difficulty with at least one personal care activity or other activity related to independent living; the severely disabled are unable to complete three or more personal care activities. These personal care activities include bathing, dressing, getting in and out of bed, shopping for groceries, and taking medications. About 6.1% of the U.S. adult population over the age of 65 are defined as being severely disabled and are not living in nursing homes.

Only 6.8% of the frail elderly have long-term health-care insurance (Johnson & Wiener, 2006). Although older adults may have other resources to pay for care, most children choose to take care of their parents. The term *children* typically means female adult children as women account for about two thirds of unpaid caregivers (Haley, 2011). About 40% of adults provide **family caregiving** to their elderly parents. This group is known as the "**sandwich generation**" because they take care of their parents and their children simultaneously (Aumann et al., 2010).

Caregiving for an elderly parent has two meanings. One form of caregiving refers to providing personal help with the basics of daily living, such as getting in and out of bed, bathing, toileting, and eating. A second form of caregiving refers to performing instrumental activities, such as shopping for groceries, managing money (including paying bills), and driving the parent to the doctor.

The typical parental caregiver is a middle-aged, married woman who works outside the home. High levels of stress and fatigue may accompany caring for one's elders. Lee et al. (2010) studied family members (average age = 46) providing care for their elderly parents as well as their own children and found that the respective responsibilities resulted in not having enough time for both and in an increase in problems associated with stress. When the two roles were compared, taking care of the parents often translated into missing more workdays. The number of individuals in the sandwich generation will increase for the following reasons:

1. **Longevity.** The over-85 age group, the segment of the population most in need of care, is the fastest-growing segment of our population.

2. **Chronic disease.** In the past, diseases took the elderly quickly. Today, diseases such as arthritis and Alzheimer's are associated not with an immediate death sentence but with a lifetime of managing the illness and being cared for by others. Family caregivers of individuals with Alzheimer's disease note the difficulty of the role: "He's not the man I married," lamented one wife.

3. **Fewer siblings to help.** The current generation of elderly had fewer children than did the elderly in previous generations. Hence, the number of adult siblings available to help look after parents is more limited. Children without siblings are more likely to feel the weight of caring for elderly parents alone.

4. **Commitment to parental care.** Contrary to the myth that adult children in the United States abrogate responsibility for taking care of their elderly parents, most children institutionalize their parents only as a last resort. Indeed, most adult children want to take care of their aging parents either in the parents' own home or in the adult child's home. When parents can no longer be left alone and can no longer cook for themselves, full-time nursing care is sought. Parents are usually resistant but become resigned to such care when there is no other option.

gerontology the study of aging.

frail elderly person who has difficulty with at least one personal care activity or other activities related to independent living.

family caregiving care provided to the elderly by family members.

sandwich generation individuals who attempt to meet the needs of their children and elderly parents at the same time.

> Some adult children reduce the strain of caring for an elderly parent by arranging for home health care.

5. **Lack of support for the caregiver.** Caring for a dependent, aging parent requires effort, sacrifice, and decision making on the part of U.S. adults who are challenged with this situation. The emotional toll on the caregiver may be heavy. Guilt (over not doing enough), resentment (over feeling burdened), and exhaustion (over the relentless care demands) are common feelings. One caregiver adult child said, "I must be an awful person to begrudge taking my mother supper, but I feel that my life is consumed by the demands she makes on me, and I have no time for myself, my children, or my husband." Sheehy (2010) cared for her husband, who had terminal cancer. She noted the importance of *not* trying to be the sole caretaker but of getting help. But professional caregiving (e.g., in a nursing home) is expensive—$70,000 is the average annual cost (Jackson, 2011).

© Thomas Flügge/iStockphoto.com

Some adult children reduce the strain of caring for an elderly parent by arranging for home health care. This involves having a nurse go to the home of a parent and provide such services as bathing and giving medication. Other services include taking meals to the elderly (e.g., through Meals on Wheels). The National Family Caregiver Support Program provides support services for individuals (including grandparents) who provide family caregiving services. Such services include elder-care resource and referral services, caregiver support groups, and classes on how to care for an

age discrimination discriminating against a person because of age.

aging parent. In addition, states are increasingly providing family caregivers a tax credit or deduction.

Offspring who have no help may become overwhelmed and frustrated. Elder abuse, an expression of such frustration, is not unheard of. Many wrestle with the decision to put their parents in a nursing home or other long-term care facility.

15.3 Issues Confronting the Elderly

Numerous issues become concerns as people age. In middle age, the issues are early retirement (sometimes forced), job layoffs (recession-related cutbacks), **age discrimination** (older people are often not hired and younger workers are hired to take their place), separation or divorce from a spouse, and adjustment to children leaving home. For some in middle age, grandparenting is an issue if they become the primary caregiver for their grandchildren. As a couple moves from the middle to the later years, the issues become more focused on income, housing, health, retirement, and sexuality.

15-3a Income

For most individuals, the end of life is characterized by reduced income. Social Security and pension benefits, when they exist, are rarely equal to the income a retired person formerly earned.

Financial planning to provide end-of-life income is important. The event which triggers end-of-life planning varies—a peer mentions it, one's children ask about it, an advertisement or one's employer offers a retirement seminar, health changes for the worse, a spouse dies, a divorce or remarriage happens. Some adults buy long-term health-care insurance. Such insurance can be costly, is frequently unaffordable, and does not always cover needed expenses.

To be caught at the end of life without adequate resources is not unusual. And surviving a major health crisis if there is no insurance can be a catastrophe (Cook et al., 2010). Women are particularly disadvantaged because their out-of-home employment has often been discontinuous, part-time, and low-paying. Social Security and private pension plans favor those with continuous, full-time work histories. Even so, Social Security benefits amount to only 42% of the worker's preretirement wage (Potter, 2010).

15-3b Housing

Of those over the age of 75, about 80% live in and own their home—hence, most do not live in a nursing home (*Statistical Abstract*, 2012). Elderly people typically have lived in the same neighborhood for many years. Even those who do not live in their own home are likely to live in a family setting. The elderly typically do not go to a residential facility until their functional status declines to the point that there is no alternative (Caro et al., 2012).

For the most part, the physical housing of the elderly is adequate. Indeed, only 6% of the housing units inhabited by the elderly are inadequate. Where deficiencies exist, the most common are inadequate plumbing and heating. However, as individuals age, they find themselves living in homes that are not "elder-friendly." Such elder-friendly homes feature bathroom doors wide enough for a wheelchair, grab bars in the bathrooms, and the absence of stairs. The most recent trend in housing for the elderly is home health care (mentioned previously) as an alternative to nursing home care. In this situation, the elderly person lives in a single-family dwelling, and other people are hired to come in regularly to help with various needs.

Other elderly individuals live in group living or shared housing arrangements, referred to as **cohousing**.

Only 6 percent of the housing units inhabited by the elderly are inadequate.

© Gualtiero Boffi/iStockphoto.com

Residents essentially plan the communities in which they own a unit and have a common area where they meet for meals several times a week (they typically rotate responsibilities of cooking these meals). The upside of cohousing is companionship with others. The downside is the need to make decisions by consensus on what to do when members can no longer take care of themselves. Silver Sage Village (www.silversagevillage .com) and ElderSpirit Community (www.elderspirit .net) are two such cohousing arrangements.

Some elderly singles choose to maintain their own home even though they are in a partnered relationship. The arrangement is living apart together discussed earlier in the text. This arrangement allows for using one's own house as one wants without being a problem to the partner. For example, one widow insisted on keeping her house and living apart from her new boyfriend so that she could have her children visit when she wanted and for as long as she wanted. "If I lived with him, he would get upset every time my grandchildren came around. I wouldn't like that," she said.

15-3c Physical Health

Good physical health is the single most important determinant of an elderly person's reported happiness. Franks et al. (2010) noted that diabetes on the part of one elderly spouse affects both in terms of depressive symptoms. Weight also has an effect on health as one ages. Gadalla (2010) studied the characteristics of those over 65 who had limitations in IADL (instrumental activities of daily living) due to body weight and found that older women and those not in a relationship were more vulnerable. Weight may also affect mental health. Carroll et al. (2010) found that (for women) becoming obese was associated with depressive symptoms. Health status varies by race and ethnicity. Liang et al. (2010) compared the self-rated health of 18,486 Americans ages 50 and above and found that White Americans rated their health most positively, followed by Black Americans, with Hispanics rating their health least positively. Despite their ethnic background, most elderly individuals, even those of advanced years, continue to define themselves as being in good health. Felding et al. (2011) found that inability of the elderly to walk 400 meters was associated with negative outcomes such as higher morbidity, mortality, hospitalizations, and a poorer quality of life. Walking at

cohousing an arrangement where elderly individuals live in group housing or shared living arrangements.

least 20 minutes a day results in physical and cognitive benefits for the elderly (Aoyagi & Shephard, 2011). Drinking tea or coffee is associated with slowing cognitive decline in women (Arab et al., 2012).

Even individuals who have a chronic, debilitating illness maintain a perception of good health as long as they are able to function relatively well. However, when elders' vision, hearing, physical mobility, and strength are markedly diminished, their sense of well-being is often significantly negatively impacted (Smith et al., 2002). For many, the ability to experience the positive side of life seems to become compromised after age 80 (Smith et al., 2002). Driving accidents also increase with aging.

Some elderly question the quality of their life, and this concern may lead to consideration of suicide. Paraschakis et al. (2012) studied the elderly in Greece and noted that the elderly had the highest suicide rate of any age group (elderly suicide victims represented 35% of the total suicides). Predictors of suicide for the elderly included being over age 75, having a history of psychiatric treatment, being depressed, and experiencing physical illness. For the majority (82%) of the elderly suicide victims, their first suicide attempt was, unfortunately, successful.

15-3d Mental Health

Aging also affects mental processes. Elderly people (particularly those 85 and older) more often have a reduced capacity for processing information quickly, for cognitive attention to a specific task, for retention, and for motivation to focus on a task. However, judgment may not be affected, and experience and perspective are benefits to decision making. Silverstein et al. (2006) emphasized that the elderly in rural China who live with their children and grandchildren report greater psychological well-being than the elderly who live with just their children. Hispanic elderly are also likely to live with their children, and some elderly grandparents take care of the grandchildren while the parents work outside the home.

Mental health may worsen for some elderly. Mood disorders, with depression being the most frequent, are more common among the elderly. Ryan et al. (2008) analyzed detailed reproductive histories of 1,013 women ages 65 years and over and found that the prevalence of depressive

dementia the mental disorder most associated with aging, whereby the normal cognitive functions are slowly lost.

symptoms was 17%. Wu et al. (2012) noted that having medical problems over the age of 65 is associated with depression. Avoiding medical problems may also be related to one's mind set. Peterson et al. (2012) emphasized that having a positive outlook is associated with good health and a long life.

Regrets over not having had children may be related to depression. Elderly women who have not had children and who have not accepted their childlessness report more mental distress than those who had children or those who have accepted their child-free status (Wu & Hart, 2002). The mental health of elderly men seems unaffected by their parental status.

Regardless of the source of depression, it has a negative impact on an elderly couple's relationship. Sandberg et al. (2002) studied depression in 26 elderly couples and concluded, "The most striking finding was the frequent mention of marital conflict and confrontation among the depressed couples and the almost complete absence of it among the nondepressed couples" (p. 261).

Dementia, which includes Alzheimer's disease, is the mental disorder most associated with aging. In spite of the association, only 3% of the aged population experience severe cognitive impairment—the most common symptom is loss of memory. It can be devastating to an individual and the partner. An 87-year-old woman, who was caring for her 97-year-old demented husband, said, "After 56 years of marriage, I am waiting for him to die, so I can follow him. At this point, I feel like he'd be better off dead. I can't go before him and abandon him" (Johnson & Barer, 1997, p. 47).

15-3e Retirement

Retirement represents a rite of passage through which most elderly pass. In 1983, Congress increased the retirement age at which individuals can receive full Social Security benefits, from 65 (for those born before 1938) to 67 (for those born after 1960). People can take early retirement at age 62, with reduced benefits. Retirement affects an individual's status, income, privileges, power, and prestige. Fabian (2007) noted that, for most of our history, the concept of retirement did not exist—older individuals were viewed as a source of wisdom, and they continued to work. In the 20th century, retirement was developed in reference to the economy, which was faced with an aging population and surplus labor (Willson, 2007). Indeed, retirement is a socially programmed stage of life that is being reevaluated.

> "To be **young** is all there is in the world . . . [Adults] talk
> so **beautifully** about work and **having a family** . . .
> telling people **how nice it is,** when, **in reality,**
> you would rather **give all** of your last
> **thirty years** for one hour of your first thirty.
> Old people are **tremendous frauds."**
>
> —WALTER STEVENS

People least likely to retire are unmarried, widowed, single-parent women who need to continue working because they have no pension or even Social Security benefits—if they don't work or continue to work, they will have no income, so retirement is not an option. Some workers experience what is called **blurred retirement** rather than a clear-cut one. A blurred retirement means the individual works part-time before completely retiring or takes a "bridge job" that provides a transition between a lifelong career and full retirement. Increasingly, individuals are delaying retirement since they need to continue working to pay their bills.

Individuals who have a positive attitude toward retirement are those who have a pension waiting for them, are married (and thus have social support for the transition), have planned for retirement, are in good health, and have high self-esteem. Those who regard retirement negatively have no pension waiting for them, have no spouse, gave no thought to retirement, have bad health, and have negative self-esteem (Mutran et al., 1997). Nordenmark and Stattin (2009) analyzed national data in Sweden and found that better psychosocial retirement adjustment outcomes resulted from voluntary retirement than from retirement forced by heath problems. Also, men whose skills were no longer needed reported a more difficult adjustment than did women.

Largier (2010) identified several factors related to a happy retirement. These include good health (the single most important factor), involvement with a significant other (and even better if that person is also retired), friends, lots of interests and activities (not TV), reading (or other brain activity), enough money not to worry, and a willingness to drop out of the achievement value ("I must be the best at whatever") promoted by society. Cozijnsen et al. (2010) noted that once individuals retire, they are likely to maintain at least one relationship with someone they knew at work. Hence, some social ties are maintained. Paul Yelsma is a retired university professor. For a successful retirement, he recommends the following:

- Just like going to high school, the military, college, or your first job, know that the learning curve is very sharp. Don't think retirement is a continuation of what you did a few years earlier. Learn fast or you will be bored.

- For every 10 phone calls you make, expect about 1 back; your status has dwindled, and you have less to offer those who want something from you.

- Have at least three to five hobbies that you can do almost any time and any place. Feed your hobbies or they will die.

- Having several things to look forward to is so much more exciting than looking backward.

- Develop a new physical exercise program that you want to do. Don't expect others to support your activities. The "pay off" in retirement is different from what colleagues expected of you.

blurred retirement the process of retiring gradually so that the individual works part-time before completely retiring or takes a bridge job that provides a transition between a lifelong career and full retirement.

15-3f Sexuality

Levitra, Cialis, and Viagra (prescription drugs that help a man obtain and maintain an erection) have given cultural visibility to the issue of sexuality among the elderly. Though the elderly (both men and women) experience physiological changes that impact sexuality (e.g., testosterone levels decrease 1% a year after age 40 in men; Powell, 2011), older adults often continue their interest in sexual activity. Chao et al. (2011) found that although sexual intercourse decreased in the elderly (to once a month), the respondents rated themselves more interested in sex than others their age.

Table 15.3 describes the physiological changes that elderly men experience during the sexual response cycle. Table 15.4 describes the physiological changes elderly women experience during the sexual response cycle.

The National Institute on Aging surveyed 3,005 men and women ages 57 to 85 and found that sexual activity decreases with age (Lindau, 2007). Almost three fourths (73%) of those ages 57 to 64 reported being sexually active in the last 12 months. This percentage declined to about half (53%) for those ages 65 to 74 and to about a fourth (26%) for those ages 75 to 80. An easy way to remember these percentages is that three fourths of those around 60, a half of those about 70, and a fourth of those around 80 report being sexually active. In general, women reported less sexual activity due to the absence of a sexual partner. Those most sexually active were also in good health. Diabetes and hypertension were major causes of sexual dysfunction. Only one out of seven reported using Viagra or other such medications. The most frequent sexual problem for men was erectile dysfunction; for women, the problems were low sexual desire (43% of those reporting), less vaginal lubrication (39%), and inability to climax (34%). Alford-Cooper (2006) studied the sexuality of couples married for more than 50 years and found that sexual interest, activity, and capacity declined with age but that marital satisfaction did not decrease with these changes.

Although sex may occur less often, it remains satisfying for most elderly. Winterich (2003) found that, in spite of vaginal, libido, and orgasm changes past menopause, women of both sexual orientations reported that they continued to enjoy active, pleasurable sex lives due to open communication with their partners.

The debate continues about whether menopausal women should become involved in estrogen replacement therapy and estrogen-progestin replacement therapy (collectively referred to as HRT—hormone replacement therapy). Schairer et al. (2000) studied 2,082 cases of breast cancer and concluded that the estrogen-progestin regimen increased the risk of breast cancer beyond that associated with estrogen alone. Beginning in 2003, women were no longer routinely encouraged to take HRT, and the effects of their not doing so on their physical and emotional health were minimal. A classic study (Hays et al., 2003) of more than 16,608

Table 15.3
Physiological Sexual Changes in Elderly Men

Phases of Sexual Response	Changes in Men
Excitement phase	As men age, getting an erection can take them longer. Although a young man may get an erection within 10 seconds, elderly men may take several minutes (10 to 30). During this time, they usually need intense stimulation (manual or oral). Unaware that the greater delay in getting erect is a normal consequence of aging, men who experience this for the first time may panic and have erectile dysfunction.
Plateau phase	The erection may be less rigid than when the man was younger, and there is usually a longer delay before ejaculation. This latter change is usually regarded as an advantage by both the man and his partner.
Orgasm phase	Orgasm in the elderly male is usually less intense, with fewer contractions and less fluid. However, orgasm remains an enjoyable experience, as over 70% of older men in one study reported that having a climax was very important when having a sexual experience.
Resolution phase	The elderly man loses his erection rather quickly after ejaculation. In some cases, the erection will be lost while the penis is still in the woman's vagina and she is thrusting to cause her orgasm. The refractory period is also increased. Whereas a young male needs only a short time after ejaculation to get another erection, the elderly man may need considerably longer.

Source: Adapted from W. Boskin, G. Graf, and V. Kreisworth. *Health Dynamics: Attitudes and Behaviors*, 1e, p. 209. © 1990, Brooks/Cole, a part of Cengage Learning, Inc. Reproduced by permission. www.cengage.com/permissions

Table 15.4
Physiological Sexual Changes in Elderly Women

Phases of Sexual Response	Changes in Women
Excitement phase	Vaginal lubrication takes several minutes or longer, as opposed to 10 to 30 seconds when younger. Both the length and the width of the vagina decrease. Considerable decreased lubrication and vaginal size are associated with pain during intercourse. Some women report decreased sexual desire and unusual sensitivity of the clitoris.
Plateau phase	Little change occurs as the woman ages. During this phase, the vaginal orgasmic platform is formed and the uterus elevates.
Orgasm phase	Elderly women continue to experience and enjoy orgasm. Of women ages 60 to 91, almost 70% reported that having an orgasm made for a good sexual experience. With regard to their frequency of orgasm now as opposed to when they were younger, 65% said "unchanged," 20% "increased," and 14% "decreased."
Resolution phase	Defined as a return to the preexcitement state, the resolution phase of the sexual response cycle happens more quickly in elderly than in younger women. Clitoral retraction and orgasmic platform disappear quickly after orgasm. This is most likely a result of less pelvic vasocongestion to begin with during the arousal phase.

Source: Adapted from W. Boskin, G. Graf, and V. Kreisworth. *Health Dynamics: Attitudes and Behaviors*, 1e, p. 209.
© 1990, Brooks/Cole, a part of Cengage Learning, Inc. Reproduced by permission. www.cengage.com/permissions

postmenopausal women ages 50 to 79 found no significant benefits from HRT in terms of quality of life. Those with severe symptoms (e.g., hot flashes, sleep disturbances, irritability) do seem to benefit without negative outcomes. Indeed, one woman reported, "I'd rather be on estrogen so my husband can stand me (and I can stand myself)." New data suggest that women who have had a hysterectomy can benefit from estrogen-alone therapy without raising their breast cancer risk.

15-4 Successful Aging

Researchers who worked on the Landmark Harvard Study of Adult Development (Vaillant, 2002) followed 824 men and women from their teens into their 80s and identified those factors associated with successful aging. These include not smoking (or quitting early), developing a positive view of life and life's crises, avoiding alcohol and substance abuse, maintaining healthy weight, exercising daily, continuing to educate oneself, and having a happy marriage. Indeed, those who were identified as "happy and well" were six times more likely to be in a good marriage than those who were identified as "sad and sick."

Not smoking is "probably the single most significant factor in terms of health" according to Vaillant (2002), of the Landmark Harvard Study of Adult Development. Smokers who quit before age 50 were as healthy at 70 as those who had never smoked.

Exercise is one of the most beneficial activities the elderly can engage in to help them maintain good health. Yet participation in exercise activities significantly decreases as a person ages. Wrosch et al. (2007) studied the exercise behavior of 172 elderly adults and found that the greatest predictor of whether elderly people exercised was their having done so previously. Hence, people who exercised at year 2 of the study were likely to still be exercising at year 5 of the study. Jack LaLanne, the health fitness guru, remarked at age 90, "I hate to exercise; I love the results." At his 90th birthday party, he challenged his well-wishers to be at his 115th birthday party. He died at 96.

Agahi (2008) also noted a strong positive effect of involvement in social leisure activities and successful aging among a representative sample of 1,246 men and women ages 65 to 95. Participating in only a few activities doubled mortality risk compared to those with the highest participation levels, even after controlling for age, education, walking ability, and other health indicators. Strongest benefits were found for engagement in organizational activities and study circles among women and hobby activities and gardening among men.

Although genetics plays an important role in healthy aging, "maintaining a healthy environment (including dietary intake, social influences, and exercise) contributes greatly to healthy aging and slowing age-related disease incidence and progression" (Brown-Borg et al., 2012). Hence, individuals can take control of their health and longevity by making sensible choices in their eating, exercise, and lifestyle.

15-5 Relationships and the Elderly

Relationships continue into old age. Here we examine relationships with one's spouse, siblings, and children.

15-5a Relationships Between Elderly Spouses

Marriages that survive into late life are characterized by little conflict, considerable companionship, and mutual supportiveness. All but one of the 31 spouses over age 85 in the Johnson and Barer (1997) study reported "high expressive rewards" from their mate. Walker and Luszcz (2009) reviewed the literature on elderly couples and found marital satisfaction related to equality of roles and marital communication. Health may be both improved by positive relationships and decreased by negative relationships.

Field and Weishaus (1992) reported interview data on 17 couples who had been married an average of 59 years and found that the husbands and wives viewed their marriages very differently. Men tended to report more marital satisfaction, more pleasure in the way their relationships had been across time, more pleasure in shared activities, and closer affectional ties.

© Chris Schmidt/iStockphoto.com

Even though club activities, including church attendance, were related to marital satisfaction, this study found that financial stability, amount of education, health, and intelligence were not. Sex was also more important to the husbands. Every man in the study reported that sex was always an important part of the relationship with his wife, but only 4 of the 17 wives reported the same.

The wives, according to the researchers, presented a much more realistic view of their long-term marriages. For this generation of women, "it was never as important for them to put the best face on things" (Field & Weishaus, 1992, p. 273). Hence, many of these wives were not unhappy; they were just more willing to report disagreements and changes in their marriages across time.

Only a small percentage (8%) of individuals older than 100 are married. Most married centenarians are men in their second or third marriage. Many have outlived some of their children. Marital satisfaction in these elderly marriages is related to a high frequency of expressing love feelings to one's partner. Though it is assumed that spouses who have been married for a long time should know how their partners feel, this is often not the case. Telling each other "I love you" is very important to these elderly spouses.

15-5b Renewing an Old Love Relationship

Some widowed or divorced elderly try to find and renew an earlier love relationship. Researcher Nancy Kalish (1997) surveyed 1,001 individuals who reported that they had renewed an old love relationship. Two thirds of those who had contacted a previous love (mostly by phone or letter) were female; one third were male. Almost two thirds (62%) reported that the person they made contact with was their first love, and 72% reported that they were (at the time of the survey) still in love with and together with the person with whom they had renewed their relationship.

Weintraub (2006) suggested some cautions when renewing an old love relationship. These include (1) discussing the original breakup or what went wrong—if one partner was hurt badly, the other must take responsibility or make amends; and (2) going slow in reviving the relationship. She also identified some successful couples who reunited after a long interval. Harry Kullijian contacted Carol Channing 70 years after they left high school. Carol's mother had broken up the relationship. Kullijian and Channing married in 2003.

Married 84 Years, and Still Loving

Herbert and Zelmyra Fisher of the Brownsville community in North Carolina have been married for more than 84 years. They have the world record of the longest marriage for a living couple (they have a certificate from the *Guinness Book of Records*).

Herbert was born June 10, 1905. His hearing is going but his mind is sharp. Zelmyra was born December 10, 1907. She uses a walker to get around the house and yard. The two of them can still give their reasons for marrying on May 13, 1924. "He was not mean; he was not a fighter," Zelmyra said. "He was quiet and kind. He was not much to look at but he was sweet." Herbert said Zelmyra never gave him any trouble. "No, no trouble at all. We never argued, but we might have disagreed," he said.

Norma Godette, one of the couple's five children, said her parents have gotten along well through the years. "One time, mama wanted to work. Daddy told her she could not work, that he could take care of the family. She slipped down to Cherry Point and got a job as a caretaker there," Godette said. "Well, it was done; she got the job. I had to let it be," Herbert said.

Different religions did not tear the two apart. He is a member of Pilgrim Chapel Missionary Baptist Church. She is a member of Jones Chapel African Methodist Episcopal Zion Church. The churches are in James City, where they both grew up. For all of their married life, they have attended their own churches. They go their own ways on Sunday morning. She reads the Bible daily.

The two watch television together. "We separate when the baseball comes on," Zelmyra said. Herbert loves baseball, especially the Atlanta Braves. He also enjoys golf, because one of his sons-in-law plays the game.

They have no secret or sage advice as to why their marriage has lasted so long. "I didn't know I would be married this long," Herbert said. "But I lived a nice holy life and go to church every Sunday. Yes sir, anything for her."

Zelmyra said Herbert was the only boyfriend she ever had. "We got along good," she said. "There was no trouble." She said she is not tired of seeing him. "I didn't think I'd be married this long. He is quiet," she said. Zelmyra said her husband had no annoying habits. They both said they shared the title of "boss." They worked hard and put all five of the children through college.

The two sit on the porch, and as a train goes by, they count the cars. They also watch the neighbors who walk by. Herbert makes his bed each day and sweeps the floor. He also checks on his wife as she rests. Between the rests, they enjoy their children, ten grandchildren, nine great-grandchildren, and nieces and nephews. Both say that if they had it to do over, they would not change their life.

Courtesy of Liz Bowles/*New Bern Sun Journal* © Nicholas Belton/iStockphoto.com Mr. Fisher died in 2011 after 86 years of marriage.
Source: Adapted from an article in the *New Bern (North Carolina) Sun Journal* by F. Sawyer (September 13, 2008). Used by permission.

iStockphoto.com/Eliza

15-5c Use of Technology to Maintain Relationships

Youth regularly text friends and romantic partners throughout the day. Madden (2010) noted that older adults and senior citizens are increasing their use of technology to stay connected as well. Over 40% of adults over the age of 50 use e-mail. And almost half (47%) of Internet users ages 50 to 64 and 25% of users ages 65 and older use social networking sites such as Facebook. Indeed, these older adults and seniors view themselves as being among the Facebooking and LinkedIn masses. These figures have doubled since 2009.

Use of Twitter has not caught on among older adults and senior citizens, with only 11% of online adults ages 50 to 64 reporting such use. E-mailing and reading online news are the most frequent Internet behaviors of this age group.

15-5d Relationships With Siblings at Age 85 and Beyond

Relationships between the elderly and their siblings are primarily emotional (enjoying time together) rather than functional (providing money or services). Earlier in the text, we noted that sibling relationships, particularly those between sisters, are the most enduring of all relationships.

15-5e Relationships With Children at Age 85 and Beyond

In regard to the relationships of the elderly with their children, emotional and expressive rewards are high. Actual caregiving by one's children is rare. Only 12% of the Johnson and Barer (1997) sample of adults older than 85 lived with their children. Most preferred to be independent and to live in their own residence. "This independent stance is carried over to social supports; many prefer to hire help rather than bother their children. When hired help is used, children function more as mediators than regular helpers, but most are very attentive in filling the gaps in the service network" (Johnson & Barer, 1997, p. 86).

Relationships among multiple generations will increase.

thanatology
examination of the social dimensions of death, dying, and bereavement.

Whereas three-generation families have been the norm, four- and five-generation families will increasingly become the norm. These changes have already become visible.

15-6 The End of Life

Thanatology is the examination of the social dimensions of death, dying, and bereavement (Bryant, 2007). The end of life sometimes involves the death of one's spouse.

15-6a Death of a Spouse

The death of a spouse is one of the most stressful life events a person ever experiences. Compared to both the married and divorced, the widowed are the most lonely and have the lowest life satisfaction (Ben-Zur, 2012; Hensley, 2012). Because women tend to live longer than men, they are more likely to experience the role of widow.

Although individual reactions and coping mechanisms for dealing with the death of a loved one vary, several reactions to death are common. These include shock, disbelief and denial,

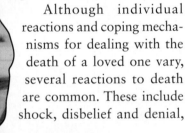

Özgür Donmaz/iStockphoto.com/© iStockphoto.com/Kevin Russ

"I am not looking forward to **death** but I don't fear it.
I would rather **stay alive** as long as I can.
I have an **adventurous life.**"

—JIMMY CARTER, 39TH PRESIDENT

confusion and disorientation, grief and sadness, anger, numbness, physiological symptoms such as insomnia or lack of appetite, withdrawal from activities, immersion in activities, depression, and guilt. Eventually, surviving the death of a loved one involves the recognition that life must go on, the need to make sense out of the loss, and the establishment of a new identity.

Women and men tend to have different ways of reacting to and coping with the death of a loved one. Women are more likely than men to express and share feelings with family and friends and are also more likely to seek and accept help, such as attending support groups of other grievers. Initial responses of men are often cognitive rather than emotional. From early childhood, males are taught to be in control, to be strong and courageous under adversity, and to be able to take charge and fix things. Showing emotions is labeled as weak.

Men sometimes respond to the death of their spouse in behavioral rather than emotional ways. Sometimes they immerse themselves in work or become involved in physical action in response to the loss. For example, a widower immersed himself in repairing a beach cottage he and his wife had recently bought. Later, he described this activity as crucial to getting him through those first two months. Another coping mechanism for men is the increased use of alcohol and other drugs.

Women's response to the death of their husbands may necessarily involve practical considerations. Johnson and Barer (1997) identified two major problems of widows—the economic effects of losing a spouse and the practical problems of maintaining a home alone. The latter involves such practical issues as cleaning the gutters, painting the house, and changing the filters in the furnace.

Whether a spouse dies suddenly or after a prolonged illness has an impact on the reaction of the remaining spouse. The sudden rather than prolonged death of one's spouse is associated with being less at peace with death and being more angry. The suddenness of the death provides no cognitive preparation time.

The age at which one experiences the death of a spouse is also a factor in one's adjustment. People in their 80s may be so consumed with their own health

and disability concerns that they have little emotional energy left to grieve. But even after the spouse's death, the emotional relationship with the deceased may continue. Some widows and widowers report a feeling that their spouses are with them and are watching out for them a year after the death of their beloved. Some may also dream of their deceased spouses, talk to their photographs, and remain interested in carrying out their wishes. Such continuation of the relationship may be adaptive by providing meaning and purpose for the living or maladaptive in that it may prevent the surviving spouse from establishing new relationships. Research is not consistent on the degree to which individuals are best served by continuing or relinquishing the emotional bonds with the deceased (Stroebe & Schut, 2005).

15-6b Involvement With New Partners at Age 80 and Beyond

Most women who live to age 80 have lost their husbands. At age 80, only 53 men are available for every 100 women. Patterns women use to adjust to this lopsided man–woman ratio include dating younger men, engaging in romance without marriage, and forming

These individuals were both widowed and met on the Internet.

David Knox/© Comstock Images/Jupiterimages

"share-a-man" relationships. For example, in elderly retirement communities such as Palm Beach, Florida, women count themselves lucky to have a man who will come for lunch, take them to a movie, or be an escort to a dance. They accept the fact that the man may also have lunch, go to a movie, or go dancing with other women. Some elderly women (and men) use the Internet to find a new partner (http://www.ourtime.com/).

Finding a companion to share life with (not a new spouse) is more often the goal of the widowed elderly. Women in their later years become more accepting of the idea that they can enjoy the romance of a relationship without the obligations of a marriage. Many enjoy their economic independence, their control over their life space, and their freedom not to be a nurse to an aging partner.

To avoid marriage, some elderly couples live together. Parenthood is no longer a goal, and many do not want to entangle their assets. For some, marriage would mean the end of their Social Security benefits or other pension moneys.

15-6c **Preparing for Death**

What is it like for those near the end of life to think about death? To what degree do they go about actually "preparing" for death? Johnson and Barer (1997) interviewed 48 individuals with an average age of 93 to find out their perspectives on death. Most interviewees were women

filial piety respect for one's parents

(77%); of those, 56% lived alone, but 73% had some sort of support from their children or from one or more social support services. The following findings are specific to those who died within a year after the interview.

Thoughts in the Last Year of Life Most had thought about death and saw their life as one that would soon end. Most did so without remorse or anxiety. With their spouses and friends dead and their health failing, they accepted death as the next stage in life. Some comments follow:

> If I die tomorrow, it would be all right. I've had a beautiful life, but I'm ready to go.
>
> My husband is gone, my children are gone, and my friends are gone.
>
> That's what is so wonderful about living to be so old. You know death is near and you don't even care.
>
> I've just been diagnosed with cancer, but it's no big deal. At my age, I have to die of something. (Johnson and Barer 1997, 205)

The major fear these respondents expressed was not the fear of death but of the dying process. Dying in a nursing home after a long illness is a dreaded fear. Sadly, almost 60% of the respondents died after a long, progressive illness. They had become frail, fatigued, and burdened by living. They identified dying in their sleep as the ideal way to die. Some hastened their death by no longer taking their medications; others wished they could terminate their own life. "I'm feeling kind of

Diversity in Elder Care

Whereas female children in the United States have the most frequent contact with and are more involved in the caregiving of their elderly parents than male children, the daughters-in-law in Japan offer the most help to elderly individuals. The female child who is married gives her attention to the parents of her husband (Ikegami 1998).

Eastern cultures emphasize **filial piety**, which is love and respect toward their parents. Filial piety involves respecting parents, bringing no dishonor to parents, and taking good care of parents (Jang and Detzner 1998).

Part of filial piety involves financial support, in evidence in Chinese culture. There, loss of income is minimized among the elderly Chinese, who receive money from their children. Logan and Bian (2003) noted that money from "children accounts for nearly a third of parents' incomes" (p. 85).

useless. I don't enjoy anything anymore.... What the heck am I living for? I'm ready to go anytime—straight to hell. I'd take lots of sleeping pills if I could get them" (Johnson & Barer, 1997, p. 204). Legally, competent adults have the right to refuse or discontinue medical interventions. For incompetent individuals, decisions are made by a surrogate—typically a spouse or child (McGowan, 2011).

Behaviors in the Last Year of Life Aware that they are going to die, most simplify their life, disengage from social relationships, and leave final instructions. In simplifying their life, they sell their home and belongings and move to smaller quarters.

One 81-year-old woman sold her home, gave her car away to a friend, and moved into a nursing home. The extent of her belongings became a chair, lamp, and TV.

Disengaging from social relationships is selective. Some maintain close relationships with children and friends, but others "let go." They may no longer send letters and Christmas cards, and phone calls become the primary way to maintain social connections. Some leave final instructions in the form of a will or handwritten note expressing wishes about where to be buried, how to handle costs associated with disposal of the body, and what to do about pets. One of Johnson and Barer's respondents left $30,000 to specific caregivers to take care of each of several pets (1997, p. 204).

Elderly individuals who have counted on children to take care of them may be disappointed. Unlike Asian cultures which emphasize familism, the American culture of individualism is more conducive to children who live elsewhere, maintain less contact with parents, and may not consider elderly parents their responsibility. The result is that the elderly may have to fend for themselves.

One of the last legal acts of the elderly is to make a will. Stone (2008) emphasized how wills may stir up sibling rivalry (e.g., one sibling may be left more than another), be used as a weapon against a second spouse (e.g., by leaving all one's possessions to one's children), or reveal a toxic secret (e.g., by leaving money to a mistress and any children from the relationship).

STUDY TOOLS ➡ **15**

Ready to study? In this book, you can:

➲ Rip out the Chapter Review card in the back of the book to study for exams

➲ Take the Self Assessment for this chapter (card in the back of the book) and see where you stand on the vital issues raised in the chapter

Or you can go online to CourseMate at www.cengagebrain.com for these resources:

➲ Complete Practice Quizzes to prepare for tests

➲ Review Key Terms Flash Cards (online or print)

➲ Read about Marriage and Family in the news

➲ Play "Beat the Clock" to master concepts

➲ Check out Personal Applications

REFERENCES

Chapter 1

Allen, K. R., E. K. Husser, D. J. Stone, & C. E. Jordal. (2008). Agency and error in young adults' stories of sexual decision making. *Family Relations, 57*, 517–529.

Amato, P. R., A. Booth, D. R. Johnson, & S. F. Rogers. (2007). *Alone together: How marriage in America is changing.* Cambridge, MA: Harvard University Press.

Anderson, S. (2010, February). The polygamists: An exclusive look inside the FLDS. *National Geographic,* 38–51.

Aulette, J. R. (2010). *Changing American families.* Boston, MA: Allyn and Bacon.

Blumer, H. G. (1969). The methodological position of symbolic interaction. In *Symbolic interactionism: Perspective and method.* Englewood Cliffs, NJ: Prentice-Hall.

Brown, G. L., S. C. Mangelsdorf, C. Neff, S. J. Schoppe-Sullivan, & C. A. Frosch. (2009). Young children's self-concepts: Associations with child temperament, mothers' and fathers' parenting, and triadic family interaction. *Merrill-Palmer Quarterly, 55,* 207–221.

Cherlin, A. J. (2009). *The marriage-go-round: The state of marriage and the family in America today.* New York, NY: Knopf.

Cherlin, A. J. (2010). Demographic trends in the United States: A review of research in the 2000s. *Journal of Marriage and Family, 72,* 403–419.

Chiu, H. and D. Busby. (2010). Parental influence in adult children's marital relationship. Poster, National Council on Family Relations annual meeting, November 3-5. Minneapolis, MN.

Cooley, C. H. (1964). *Human nature and the social order.* New York, NY: Schocken.

Couric, K. (2011). *The best advice I ever got: Lessons from extraordinary lives.* New York, NY: Random House.

Curran, M. A., E. A. Utley, & J. A. Muraco. (2010). An exploratory study of the meaning of marriage for African Americans. *Marriage and Family Review, 46,* 346–365.

Darnton, K. (2012). Deception at Duke. *Sixty Minutes*/CBS Television Feb 12.

Fish, J., T. Pavkov, J. Wetchler and J. Bercik. (2011). The role of adult attachment and differentiation in extradyadic experiences. Poster, National Council on Family Relations annual meeting, Orlando, Florida. November 18.

Girgis, S., R. P. George, & R. Anderson. (2011). What is marriage? *Harvard Journal of Law and Public Policy, 34,* 245–287.

Gregory, J. D. (2010). Pet custody: Distorting language and the law. *Family Law Quarterly, 44,* 35–64.

James, S. D. (2008, May 7). Wild child speechless after tortured life. *ABC News.*

Kaiser Family Foundation. (2010). Media use among teens. Retrieved February 7, 2010, from http://www.kff.org/entmedia/entmedia 012010nr.cfm.

Kefalas, M., F. F. Furstenberg, P. J. Carr, & L. Napolitano. (2011). Marriage is more than being together: The meaning of marriage for young adults. *Journal of Family Issues, 32,* 845–875.

Klinenberg, E. (2012). *Going solo: The extraordinary rise and surprising appeal of living alone.* New York: Penguin.

Knox, D., & S. Hall. (2010). Relationship and sexual behaviors of a sample of 2,922 university students. Unpublished data. Department of Sociology, East Carolina University, and Department of Family and Consumer Sciences, Ball State University.

Lorber, J. (1998). *Gender inequality: Feminist theories and politics.* Los Angeles, CA: Roxbury.

Malinen, K., U. Kinnunen, A. Tolvanen, A. Rönkä, H. Wierda-Boer, & J. Gerris (2010). Happy spouses, happy parents? Family relationships among Finnish and Dutch dual earners. *Journal of Marriage and Family, 72,* 293–306.

Mead, G. H. (1934). *Mind, self, and society.* Chicago: University of Chicago Press.

Meinhold, J. L., A. Acock, & A. Walker. (2006). The influence of life transition statuses on sibling intimacy and contact in early adulthood. Paper presented at the annual meeting of the National Council on Family Relations, Orlando, FL.

Murdock, G. P. (1949). *Social structure.* New York, NY: Free Press.

Pew Research Center. (2010). Social and demographic trends: The decline of marriage and rise of new families. Retrieved from http://pewresearch.org/pubs/1802/decline-marriage-rise-new-families.

Pryor, J. H., L. DeAngelo, L. Paluki Blake, S. Hurtado & S. Tran. (2011). *The American freshman: National norms Fall 2011.* Los Angeles, CA: Higher Education Research Institute, UCLA Graduate School of Education and Information Studies.

Randall, B. (2008). *Songman: The story of an Aboriginal elder of Uluru.* Sydney, Australia: ABC Books.

Sayare, S., & M. De La Baume. (2010, December 15). In France, civil unions gain favor over marriage. *New York Times.*

Silverstein, L. B., and C. F. Auerbach. 2005. (Post) modern families. In *Families in global perspective,* ed. Jaipaul L. Roopnarine and U. P. Gielen, 33P48. Boston, MA. Pearson Education.

Starnes, T. (2011, June 15). School surveys 7th graders on oral sex. *Fox News.*

Statistical Abstract of the United States. (2012, 131st ed.). Washington, DC: U.S. Census Bureau.

Veenhoven, R. (2007). Quality-of-life-research. In C. D. Bryant & D. L. Peck (Eds.), *21st century sociology: A reference handbook* (pp. 54–62). Thousand Oaks, CA: Sage.

White, J. M., & D. M. Klein. (2002). *Family theories* (2nd ed.). Thousand Oaks, CA: Sage.

Zeitzen, M. K. (2008). *Polygamy: A cross-cultural analysis.* Oxford: Berg.

Chapter 2

Abowitz, D., D. Knox, & K. Berner. (2011). Traditional and non-traditional husband preference among college women. Annual Meeting Eastern Sociological Society, Philadelphia, Pennsylvania.

Abowitz, D. A., D. Knox, M. Zusman, & A. McNeely. (2009). Beliefs about romantic relationships: Gender differences among undergraduates. *College Student Journal, 43,* 276–284.

Barnes, H., D. Knox, & J. Brinkley. (2012, March 23). CHEATING: Gender differences in reactions to discovery of a partner's cheating. Poster, Southern Sociological Society, New Orleans.

Bem, S. L. (1983). Gender schema theory and its implications for child development: Raising gender-aschematic children in a gender-schematic society. *Signs, 8,* 596–616.

Bloch, K. and T. Taylor. 2012. Overworked or underworked? Hour mismatches for women and men in the United States. *Sociological Spectrum, 32,* 37–60.

Boehnke, M. (2011). Gender role attitudes around the globe: Egalitarian vs. traditional views. *Asian Journal of Social Science, 39,* 57–74.

Bono, Chaz. (2011). *The story of how I became a man.* New York, NY: E. P. Dutton.

Bulanda, J. R. (2011). Gender, marital power, and marital quality in later life. *Journal of Women and Aging, 23,* 3–22.

Cheng, C. (2005). Processes underlying gender-role flexibility: Do androgynous individuals know more or know how to cope? *Journal of Personality, 73,* 645–674.

Cohen-Kettenis, P. T. (2005). Gender change in 46, XY persons with 5[alpha]-reductase-2 deficiency and 17[beta]-hydroxysteroid dehydrogenase-3 deficiency. *Archives of Sexual Behavior, 34,* 399–411.

Colapinto, J. (2000). *As nature made him: The boy who was raised as a girl.* New York, NY: HarperCollins.

Cordova, J. V., C. B. Gee, & L. Z. Warren. (2005). Emotional skillfulness in marriage: Intimacy as a mediator of the relationship between emotional skillfulness and marital satisfaction. *Journal of Social and Clinical Psychology, 24*, 218–235.

Corra, M., S. Carter, J. Scott Carter, & D. Knox. (2009). Trends in marital happiness by sex and race, 1973–2006. *Journal of Family Issues, 30*, 1379–1404.

Craig, L., & K. Mullan. (2010). Parenthood, gender and work-family time in the United States, Australia, Italy, France, and Denmark. *Journal of Marriage and Family, 72*, 1344–1361.

Crawley, S. L., L. J. Foley, & C. L. Shehan. (2008). *Gendering bodies.* Boston, MA: Rowman and Littlefield.

Dabbous, Y., & A. Ladley. (2010). A spine of steel and a heart of gold: Newspaper coverage of the first female Speaker of the House. *Journal of Gender Studies, 19*, 181-194.

Dotson-Blake, K., D. Knox, and A. Holman. (2008). College student attitudes toward marriage, family, and sex therapy. Unpublished data from 288 undergraduate/graduate students. East Carolina University, Greenville, NC.

Dysart-Gale, D. (2010). Social justice and social determinants of health: Lesbian, gay, bisexual, transgendered, intersexed, and queer youth in Canada. *Journal of Child and Adolescent Psychiatric Nursing, 23*, 23–28.

East, L., D. Jackson, L. O'Brien, & K. Peters. (2011). Condom negotiation: Experiences of sexually active young women. *Journal of Advanced Nursing, 67*, 77–85.

England, P. (2010). The gender revolution: Uneven and stalled. *Gender and Society, 24*, 149–166.

Eubanks Fleming, C. J. and J. V. Córdova. (2012), Predicting relationship help seeking prior to a Marriage Checkup. *Family Relations, 61*, 90–100

Garfield, R. (2010). Male emotional intimacy: How therapeutic men's groups can enhance couples therapy. *Family Process, 49*, 109–122.

Grief, G. L. 2006. Male friendships: Implications from research for family therapy. *Family Therapy, 33*, 1–15.

Grogan, S. (2010). Promoting positive body image in males and females: Contemporary issues and future directions. *Sex Roles, 63*, 757–765.

Heller, N. (2008). Will the transgender dad be a father? What goes on the birth certificate? Retrieved June 13, 2008, from http://www.slate.com/id/2193475/

Hepp, U., A. Spindler, & G. Milos. (2005). Eating disorder symptomatology and gender role orientation. *International Journal of Eating Disorders, 37*, 227–233.

Hogue, M., C. L. Z. Dubois, & L. Fox-Cardamone. (2010). Gender differences in pay expectations: The roles of job intention and self-view. *Psychology of Women Quarterly, 34*, 215–227.

Irvolino, A. C., M. Hines, S. E. Golombok, J. Rust, & R. Plomin. (2005). Genetic and environmental influences on sex-typed behavior during the preschool years. *Child Development, 76*, 826–840.

Kim, J. L., C. L. Sorsoli, K. Collins, B. A. Zylbergold, D. Schooler, & D. L. Tolman. (2007). From sex to sexuality: Exposing the heterosexual script on primetime network television. *Journal of Sex Research, 44*, 145–157.

Kimmel, M. S. (2001). Masculinity as homophobia: Fear, shame, and silence in the construction of gender identity. In T. F. Cohen (Ed.), *Men and masculinity: A text reader* (pp. 29–41). Belmont, CA: Wadsworth.

Knox, D., & M. Zusman. (2007). Traditional wife? Characteristics of college men who want one. *Journal of Indiana Academy of Social Sciences, 11*, 27–32.

Knox, D., M. E. Zusman, & H. R. Thompson. (2004). Emotional perceptions of self and others: Stereotypes and data. *College Student Journal, 38*, 130–142.

Kohlberg, L. (1966). A cognitive-developmental analysis of children's sex-role concepts and attitudes. In E. E. Macoby (Ed.), *The development of sex differences*. Stanford, CA: Stanford University Press.

Kohlberg, L. (1969). State and sequence: The cognitive developmental approach to socialization. In D. A. Goslin (Ed.), *Handbook of socialization theory and research* (pp. 347–480). Chicago, IL: Rand McNally.

Lareau, A., & E. B. Weininger. (2008). Time, work, and family life: Reconceptualizing gendered time patterns through the case of children's organized activities. *Sociological Forum, 23*, 419–454.

Lease, S. H., A. B. Hampton, K. M. Fleming, L. R. Baggett, S. H. Montes, & R. J. Sawyer. (2010). Masculinity and interpersonal competencies: Contrasting White and African American men. *Psychology of Men and Masculinity, 11*, 195–207.

Lewis, R. (2007). *Raising a modern-day knight.* Wheaton, IL: Tyndale House Publishers.

Ma, M. K. (2005). The relation of gender-role classifications to the prosocial and antisocial behavior of Chinese adolescents. *Journal of Genetic Psychology, 166*, 189–201.

Maltby, L. E., M. E. L. Hall, T. L. Anderson & K. Edwards. (2010). Religion and sexism: The moderating role of participant gender. *Sex Roles, 62*, 615–622.

Mandara, J., F. Varner, & S. Richman. (2010). Do African American mothers really "love" their sons and "raise" their daughters? *Journal of Family Psychology, 24*, 41–50.

Maume, D. J. (2006). Gender differences in taking vacation time. *Work and Occupations, 33*, 161–190.

McPherson, M., L. Smith-Lovin, & M. E. Brashears. (2006). Social isolation in America, 1985–2004. *American Sociological Review, 71*, 353–375.

Mead, M. (1935). *Sex and temperament in three primitive societies.* New York, NY: William Morrow.

Meinhold, J. L., A. Acock, & A. Walker. (2006), November). The influence of life transition statuses on sibling intimacy and contact in early adulthood. Paper presented at the annual meeting of the National Council on Family Relations, Orlando, FL.

Mellor, D., L. A. Ricciardelli, M. P. McCabe, J. Yeow, N. Hidayah bt Mamat, & N. Fizlee bt Mohd Hapidzal. (2010). Psychosocial correlates of body image and body change behaviors among Malaysian adolescent boys and girls. *Sex Roles, 63*, 386–398.

Minnottea, K. L., D. E. Pedersena, S. E. Mannonb & G. Kiger. (2010). Tending to the emotions of children: Predicting parental performance of emotion work with children. *Marriage and Family Review, 46*, 224–241.

Mirchandani, R. (2005). Postmodernism and sociology: From the epistemological to the empirical. *Sociological Theory, 23*, 86–115.

Monro, S. (2000). Theorizing transgender diversity: Towards a social model of health. *Sexual and Relationship Therapy, 15*, 33–42.

Moore, O., S. Kreitler, M. Ehrenfeld, & N. Giladi. (2005). Quality of life and gender identity in Parkinson's disease. *Journal of Neural Transmission, 112*, 1511–1522.

Nelms, Bobbie Jo, D. Knox and B. Easterling. 2012. THE RELATIONSHIP TALK: Assessing partner commitment. *College Student Journal, 46*, 178–182

Pew Research Center. (2010). Religion among the Millennials. Pew Forum on Religion and Public Life. Retrieved from http://pewforum.org/Age/Religion-Among-the-Millennials.aspx

Peoples, J. G. (2001). The cultural construction of gender and manhood. In T. F. Cohen (Ed.), *Men and masculinity: A text reader* (pp. 9–18). Belmont, CA: Wadsworth.

Pollack, W. S. (with T. Shuster). (2001). *Real boys' voices.* New York, NY: Penguin Books.

Read, S., & E. Grundy. (2011). Mental health among older married couples: The role of gender and family life. *Social Psychiatry and Psychiatric Epidemiology, 46*, 331–341.

Rees, C., & G. Pogarsky. (2011). One bad apple may not spoil the whole bunch: Best friends and adolescent delinquency. *Journal of Quantitative Criminology, 27*, 197–223.

Rivadeneyraa, R., & M. J. Lebob. (2008). The association between television-viewing behaviors and adolescent dating role attitudes and behaviors. *Journal of Adolescence, 31*, 291–305.

Robnett, R. D., & J. E. Susskind. (2010). Who cares about being gentle? The impact of social identity and the gender of one's friends on children's display of same-gender favoritism. *Sex Roles, 63,* 820–832.

Ross, C., D. Knox, & M. Zusman. (2008). "Hey big boy": Characteristics of university women who initiate relationships with men. Poster presented at the meeting of the Southern Sociology Society, Richmond, VA.

Shin, K., H. Shin, J. A. Yang, & C. Edwards. (2010). Gender role identity among Korean and American college students: Links to gender and academic achievement. *Social Behavior and Personality: An International Journal, 38,* 267–272.

Simon, R. W., & K. Lively. (2010). Sex, anger, and depression. *Social Forces, 88,* 1543–1568.

Statistical Abstract of the United States. (2012, 131th ed.). Washington, DC: U.S. Census Bureau.

Taylor, R. L. (2002). Black American families. In R. L. Taylor (Ed.), *Minority families in the United States: A multicultural perspective* (pp. 19–47). Upper Saddle River, NJ: Prentice Hall.

Ter Gogt, T. F. M., R. C. M. E. Engles, S. Bogers, & M. Kloosterman. (2010). "Shake it baby, shake it": Media preferences, sexual attitudes and gender stereotypes among adolescents. *Sex Roles, 63,* 844–859.

Thompson, M. (2008, July 16). America's medicated army. *Time,* 38–42.

Trucco, E. M., C. R. Colder, & W. F. Wieczorek. (2011). Vulnerability to peer influence: A moderated mediation study of early adolescent alcohol use initiation. *Addictive Behaviors, 36,* 729–736.

Vail-Smith, K., D. Knox, & M. Zusman. (2007). The lonely college male. *International Journal of Men's Health, 6,* 273–279.

Walker, R. B., & M. A. Luszcz. (2009). The health and relationship dynamics of late-life couples: A systematic review of the literature. *Ageing and Society, 29,* 455–481.

Wallis, C. (2011). Performing gender: A content analysis of gender display in music videos. *Sex Roles, 64,* 160–172.

Walzer, S. (2008). Redoing gender through divorce. *Journal of Social and Personal Relationships, 25,* 5–21.

Weisgram, E. S., R. S. Bigler, & L. S. Liben. (2010). Gender, values, and occupational interests among children, adolescents, and adults. *Child Development, 81,* 778–796.

Wells, R. S., T. A. Seifert, R. D. Padgett, S. Park, & P. Umbach. (2011). Why do more women than men want to earn a four-year degree? Exploring the effects of gender, social origin, and social capital on educational expectations. *Journal of Higher Education, 82,* 1–32.

Wilcox, W. B., & S. L. Nock. (2006). What's love got to do with it? Equality, equity, commitment and marital quality. *Social Forces, 84,* 1321–1345.

Woodhill, B. M., & C. A. Samuels. (2003). Positive and negative androgyny and their relationship with psychological health and well-being. *Sex Roles, 48,* 555–565.

Yu, L., & D. Zie. (2010). Multidimensional gender identity and psychological adjustment in middle childhood: A study in China. *Journal of Sex Roles, 62,* 100–113.

Chapter 3

Bauerlein, M. (2010). Literary learning in the hyperdigital age. *Futurist, 44,* 24–25.

Behringer, A. M. (2005). Bridging the gap between Mars and Venus: A study of communication meanings in marriage. *Dissertation Abstracts International, A: The Humanities and Social Sciences, 65,* 4007A–8A.

Berle, W. (with B. Lewis). (1999). *My father uncle Miltie.* New York, NY: Barricade Books.

Braun, M., K. Mura, M. Peter-Wright, R. Hornhung, & U. Scholz. (2010). Toward a better understanding of psychological well-being in dementia caregivers: The link between marital communication and depression. *Family Process, 49,* 185–203.

Carey, A., & V. Salazar. (2011, February 1). Women talk and text more. *USA Today,* p. 1.

Carey, A. R., & P. Trap. (2010, January 3). How honest are you on your social networking sites? *USA Today,* p. A1.

Caughlin, J. P., & M. B. Ramey. (2005). The demand/withdraw pattern of communication in parent-adolescent dyads. *Personal Relationships, 12,* 337–355.

Coyne, S. M., L. Stockdale, D. Busby, B. Iverson, & D. M. Grant. (2011). "I luv u :)!": A descriptive study of the media use of individuals in romantic relationships. *Family Relations, 60,* 150–162.

Coyne, S. M., L. M. Padilla-Walker, L. Stockdale and A. Fraser. (2011b). Media and the family: Associations between family media use and family connections. Poster, presented at the annual meeting of the National Council on Family Relations, Orlando, November 18, 2011.

Easterling, B., D. Knox, and A. Brackett. (2012). Secrets in romantic relationships: Does sexual orientation matter? *Journal of GLBT Family Studies, 8,* 198–210

Faircloth, M., D. Knox, and J. Brinkley. (2012). The good, the bad, and technology mediated communication in romantic relationships. Paper, presented at the annual meeting of the Southern Sociological Society, New Orleans, March 23, 2012.

Gallmeier, C. P., M. E. Zusman, D. Knox, & L. Gibson. (1997). Can we talk? Gender differences in disclosure patterns and expectations. *Free Inquiry in Creative Sociology, 25,* 129–225.

Ganong, L. H., M. Coleman, R. Feistman, T. Jamison and M. S. Markham. (2011). Communiction technology and post-divorce coparenting. Poster, presented at the annual meeting of the National Council on Family Relations, Orlando, November 18, 2011.

Gershon, I. (2010). *The breakup 2.0.* New York: Cornell University Press.

Gottman, J. (1994). *Why marriages succeed or fail.* New York: Simon and Schuster.

Greeff, A. P., & T. De Bruyne. (2000). Conflict management style and marital satisfaction. *Journal of Sex and Marital Satisfaction, 26,* 321–334.

Hertenstein, M. J., J. M. Verkamp, A. M. Kerestes, & R. M. Holmes. (2007). The communicative functions of touch in humans, nonhuman primates, and rats: A review and synthesis of the empirical research. *Genetic, Social, and General Psychology Monographs, 132,* 5–94.

Hill, E. W. (2010). Discovering forgiveness through empathy: Implications for couple and family therapy. *Journal of Family Therapy, 32,* 169–185.

Kalush, W., & L. Sloman. (2006). *The secret life of Houdini.* New York, NY: Altria Books.

Knox, D., & S. Hall. (2010). Relationship and sexual behaviors of a sample of 2,922 university students. Unpublished data. Department of Sociology, East Carolina University, and Department of Family and Consumer Sciences, Ball State University.

Knox, D., C. Schacht, J. Turner, & P. Norris. (1995). College students' preference for win-win relationships. *College Student Journal, 29,* 44–46.

Kurdek, L. A. (1995). Predicting change in marital satisfaction from husbands' and wives' conflict resolution styles. *Journal of Marriage and Family, 57,* 153–164.

Lavner, J. A., & T. N. Bradbury. (2010). Patterns of change in marital satisfaction over the newlywed years. *Journal of Marriage and Family, 72,* 1171–1187.

Looi, C., P. Seow, B. Zhang, H. So, W. Chen, & L. Wong. (2010). Leveraging mobile technology for sustainable seamless learning: A research agenda. *British Journal of Educational Technology, 41,* 154–169.

Markman, H. J., G. K. Rhoades, S. M. Stanley, E. P. Ragan, & S. W. Whitton. (2010). The premarital communication roots of marital distress and divorce: The first five years of marriage. *Journal of Family Psychology, 24,* 289–298.

Merolla, A. J., & S. Zhang. (2011). In the wake of transgressions: Examining forgiveness communication in personal relationships. *Personal Relationships, 18,* 79–95.

Moore, M. M. (2010). Human nonverbal courtship behavior—A brief historical review. *Journal of Sex Research, 47,* 171–180.

Oyamot, C. M., P. T. Fuglestad, & M. Snyder. (2010). Balance of power and influence in relationships: The role of self-monitoring. *Journal of Social and Personal Relationships, 27,* 23–46.

Perry, M. S. and R. J. Werner-Wilson. (2011). Couples and computer-mediated communication: A closer look at the affordances and use of the channel. *Family & Consumer Sciences Research Journal, 40,* 120–134

Punyanunt-Carter, N. N. (2006). An analysis of college students' self-disclosure behaviors. *College Student Journal, 40,* 329–331.

Robbins, C. A. (2005). ADHD couple and family relationships: Enhancing communication and understanding through Imago Relationship Therapy. *Journal of Clinical Psychology, 61,* 565–578.

Sanford, K. (2007). Hard and soft emotion during conflict: Investigating married couples and other relationships. *Personal Relationships, 14,* 65–90.

Sharpe, A. (2012). Transgender marriage and the legal obligation to disclose gender history. *Modern Law Review, 75,* 33–53.

Shinn, L. K., & M. O'Brien. (2008). Parent-child conversational styles in middle childhood: Gender and social class differences. *Sex Roles, 59,* 61–69.

Strickler, B. L., & J. D. Hans. (2010, November 3–5). Defining infidelity and identifying cheaters: An inductive approach with a factoral design. Poster presented at the annual meeting of the National Council on Family Relations, Minneapolis, MN.

Tannen, D. (1990). *You just don't understand: Women and men in conversation.* London, England: Virago.

Tannen, D. (2006). *You're wearing that? Understanding mothers and daughters in conversation.* New York, NY: Random House.

Vail-Smith, K., D. Knox, & L. Whetstone. (2010). The illusion of safety in "monogamous" undergraduate relationships. *American Journal of Health Behavior, 34,* 12–20.

Weisskirch, R. S. and R. Delevi. (2011). "Sexting" and adult romantic attachment. *Computers in Human Behavior, 27,* 1697–1701.

Wickrama, K. A. S., C. M. Bryant, & T. K. Wickrama. (2010). Perceived community disorder, hostile marital interactions, and self-reported health of African American couples: An interdyadic process. *Personal Relationships, 17,* 515–531.

Chapter 4

Adams, M., et al. (2008, April). Rhetoric of alternative dating: Investment versus exploration. Southern Sociological Society, Richmond, VA.

Albright, J. M. (2007). How do I love thee and thee and thee? Self-presentation, deception, and multiple relationships online. In M. T. Whitty, A. J. Baker, & J. A. Inman (Eds.), *Online matchmaking* (pp. 81–93). New York, NY: Palgrave Macmillan.

Avellar, S., & P. J. Smock. (2005). The economic consequences of the dissolution of cohabiting unions. *Journal of Marriage and Family, 67,* 315–327.

Baker, A. J. (2007). Expressing emotion in text: Email communication of online couples. In M. T. Whitty, A. J. Baker, & J. A. Inman (Eds.), *Online matchmaking* (pp. 97–111). New York, NY: Palgrave Macmillan.

Blackstrom, L., E. A. Armstrong, & J. Puentes. (2012). Women's negotiation of cunnilingus in college hookups and relationships. *Journal of Sex Research, 49,* 1–12.

Bogle, K. A. (2002, Summer). From dating to hooking up: Sexual behavior on the college campus. Paper presented at the annual meeting of the Society for the Study of Social Problems.

Bogle, K. A. (2008). *Hooking up: Sex, dating, and relationships on campus.* New York, NY: New York University Press.

Booth, A., E. Rustenbach, & S. McHale. (2008). Early family transitions and depressive symptom changes from adolescence to early adulthood. *Journal of Marriage and Family, 70,* 3–14.

Bradshaw, C., A. S. Kahn, & B. K. Saville. (2010). To hook up or date: Which gender benefits? *Sex Roles, 62,* 661–669.

Brewster, C. D. D. (2006). African American male perspective on sex, dating, and marriage. *Journal of Sex Research, 43,* 32–33.

Brown, S. L., J. R. Bulanda, & G. R. Lee. (2005). The significance of nonmarital cohabitation: Marital status and mental health benefits among middle-aged and older adults. *Journals of Gerontology Series B—Psychological Sciences and Social Sciences, 60*(1), S21–S29.

Buchler, S., J. Baxter, M. Haynes, & M. Western. (2009). The social and demographic characteristics of cohabiters in Australia: Towards a typology of cohabiting couples. *Family Matters, 82,* 22–29.

Chambers, A. L., & A. Kravitz. (2011). Understanding the disproportionately low marriage rate among African Americans: An amalgam of sociological and psychological constraints. *Family Relations, 60,* 648–660.

Cherlin, A. J. (2010). Demographic trends in the United States: A review of research in the 2000s. *Journal of Marriage and Family, 72,* 403–419.

Cross-Barnet, C., A. Cherlin, & L. Burton (2011). Bound by children: Intermittent cohabitation and living together apart. *Family Relations, 60,* 633–647.

Denney, J. T. (2010). Family and household formations and suicide in the United States. *Journal of Marriage and Family, 72,* 202–213.

Dew, J. (2011). Financial issues and relationship outcomes among cohabiting individuals. *Family Relations, 60,* 178–190.

Donn, J. (2005). Adult development and well-being of midlife never-married singles. (Unpublished dissertation). Miami University, Oxford, OH.

England, P., & R. J. Thomas. (2006). The decline of the date and the rise of the college hook up. In A. S. Skolnick & J. H. Skolnick (Eds.), *Family in transition* (14th ed., pp. 151–162). Boston, MA: Pearson Allyn & Bacon.

Fieldera, R. L., & M. P. Careya (2010). Prevalence and characteristics of sexual hookups among first-semester female college students. *Journal of Sex and Marital Therapy, 36,* 346–359.

Fokkema, T., J. D. Gierveld, & P. A. Dykstra. (2012). Cross-national differences in older adult loneliness. *Journal of Psychology, 146,* 201–228.

Ha, J. H. (2008). Changes in support from confidants, children, and friends following widowhood. *Journal of Marriage and Family, 70,* 306–329.

Hall, J. A., N. Park, M. J. Cody, & H. Song. (2010). Strategic misrepresentation in online dating: The effects of gender, self-monitoring, and personality traits. *Journal of Social and Personal Relationships, 27,* 117–135.

Hall, S. S., & R. Adams (2011). Newlyweds' unexpected adjustments to marriage. *Family and Consumer Sciences Research Journal, 39,* 375–387.

Hansen, T., T. Moum, & A. Shapiro. (2007). Relational and individual well-being among cohabitors and married individuals in midlife. *Journal of Family Issues, 28,* 910–933.

Heino, R. D., N. B. Ellison, & J. L. Gibbs. (2010). Relationshopping: Investigating the market metaphor in online dating. *Journal of Social and Personal Relationships, 27,* 427–447.

Hess, J. (2009). Personal communication. Appreciation is expressed to Judye Hess for the development of this section. For more information about Judye Hess, see http://www.psychotherapist.com/judyehess/

Hodge, A. (2003, February 28). Video chatting and the males who do it. Paper presented at the 73rd annual meeting of the Eastern Sociological Association, Philadelphia, PA.

Holtzman, M. (2011). Nonmarital unions, family definitions, and custody decision making. *Family Relations, 60,* 617–632.

Huang, P. M., P. J. Smock, W. D. Manning, & C. A. Bergstrom-Lynch. (2011). He says, she says: Gender and cohabitation. *Journal of Family Issues, 32,* 876–905.

Jayson, S. (2011, March 30). Is dating dead? *USA Today*, p. A1.

Jose, A., K. D. O'Leary, & A. Moyer. (2010). Does premarital cohabitation predict subsequent marital stability and marital quality? A meta-analysis. *Journal of Marriage and Family, 72,* 105–116.

Kamp Dush, C. M. (2011). Relationship-specific investments, family chaos, and cohabitation dissolution following a nonmarital birth. *Family Relations, 60,* 586–601.

Kasearu, K. (2010). Intending to marry . . . students' behavioral intention towards family forming. *TRAMES: A Journal of the Humanities and Social Sciences, 14,* 3–20.

Kiernan, K. (2000). European perspectives on union formation. In L. J. Waite (Ed.), *The ties that bind* (pp. 40–58). New York, NY: Aldine de Gruyter.

Knox, D., & U. Corte. (2007). "Work it out/See a counselor": Advice from spouses in the separation process. *Journal of Divorce and Remarriage, 48,* 79–90.

Knox, D., & S. Hall. (2010). Relationship and sexual behaviors of a sample of 2,922 university students. Unpublished data collected for this text. Department of Sociology, East Carolina University, and Department of Family and Consumer Sciences, Ball State University.

Lichter, D. T., R. N. Turner, & S. Sassler. (2010). National estimates of the rise in serial cohabitation. *Social Science Research, 39,* 754–765.

Manning, W. D., D. Trella, H. Lyons, & N. C. DuToit. (2010). Marriageable women: A focus on participants in a community Healthy Marriage Program. *Family Relations, 59,* 87–102.

McKenna, K. Y. A. (2007). A progressive affair: Online dating to real world mating. In M. T. Whitty, A. J. Baker, & J. A. Inman (Eds.), *Online matchmaking* (pp. 112–124). New York, NY: Palgrave Macmillan.

Miller, A. J., S. Sassler, S. & D. Kusi-Appouh. (2011). The specter of divorce: Views from working- and middle-class cohabitors. *Family Relations, 60,* 602–616.

Murray, C. 2012. *Coming apart: The state of white America, 1960-2010.* New York: Crown Forum.

Ohayon, M. (Director). (2005). *Cowboy del amor* [Motion picture]. United States.

Owen, J. J., G. K. Rhoades, S. M. Stanley, & F. D. Fincham. (2010). "Hooking up" among college students: Demographic and psychosocial correlates. *Archives of Sexual Behavior, 39,* 653–663.

Pew Research Center. (2010). The decline of marriage and rise of new families. Retrieved November 18, 2010, from http://pewresearch.org/pubs/1802/decline-marriage-rise-new-families

Pew Research Center. (2011). Barely half of U.S. adults are married—A record low. Retrieved December 19, 2011, from http://pewresearch.org/pubs/2147/marriage-newly-weds-record-low

Reinhold, S. (2010). Reassessing the link between premarital cohabitation and marital instability. *Demography, 47,* 719–733.

Rhoades, G. K., C. M. Kamp Dush, D. C. Atkins, S. M. Stanley, & H. J. Markman. (2011). Breaking up is hard to do: The impact of unmarried relationship dissolution on mental health and life satisfaction. *Journal of Family Psychology, 25,* 366–375.

Rosin, H. (2010, August). The end of men. *The Atlantic.*

Sassler, S., & A. Cunningham. (2008). How cohabitors view childrearing. *Sociological Perspectives, 51,* 3–29.

Sassler, S., & A. J. Miller (2011). Class differences in cohabitation processes. *Family Relations, 60,* 163–177.

Schmeer, K. K. (2011). The child health disadvantage of parental cohabitation. *Journal of Marriage and Family, 73,* 181–193.

Schoen, R., N. S. Landale, & K. Daniels. (2007). Family transitions in young adulthood. *Demography, 44,* 807–830.

Sharp, E. A., & L. Ganong. (2007). Living in the gray: Women's experiences of missing the marital transition. *Journal of Marriage and Family, 69,* 831–844.

Sobal, J. and K. L. Hanson. 2011. Marital status, marital history, body weight, and obesity. *Marriage & Family Review, 47,* 474–504.

Spitzberg, B. H., & W. R. Cupach. (2007). Cyberstalking as (mis)matchmaking. In M. T. Whitty, A. J. Baker, & J. A. Inman (Eds.), *Online matchmaking,* (pp. 127–146). New York, NY: Palgrave Macmillan.

Statistical Abstract of the United States (2012, 131st ed.). Washington, DC: U.S. Census Bureau.

Stein, J. (2011, March 28). Can buy me love. *Time,* 82.

Stepp, L. S. (2007). *Unhooked: How young women pursue sex, delay love, and lose at both.* Riverhead, NY: Riverhead.

Thomson, E., & E. Bernhardt. (2010). Education, values, and cohabitation in Sweden. *Marriage and Family Review, 46,* 1–21.

Toma, C. L., & J. T. Hancock. (2010). Looks and lies: The role of physical attractiveness in online dating self-presentation and deception. *Communication Research, 37,* 335–351.

Uecker, J., & M. Regnerus. (2011). *Premarital sex in America: How young Americans meet, mate, and think about marrying.* United Kingdom: Oxford University Press.

Weden, M., & R. T. Kimbro. (2007). Racial and ethnic differences in the timing of first marriage and smoking cessation. *Journal of Marriage and Family, 69,* 878–887.

Whitty, M. T. (2007). The art of selling one's "self" on an online dating site: The BAR approach. In M. T. Whitty, A. J. Baker, & J. A. Inman (Eds.), *Online matchmaking* (pp. 37–69). New York, NY: Palgrave Macmillan.

Wienke, C., & G. J. Hill. (2009). Does the "marriage benefit" extend to partners in gay and lesbian relationships? Evidence from a random sample of sexually active adults. *Journal of Family Issues, 30,* 259–273.

Wilson, G. D., J. M. Cousins, & B. Fink. (2006). The CQ as a predictor of speed-date outcomes. *Sexual and Relationship Therapy, 21,* 163–169.

Wu, P., & W. Chiou. (2009). More options lead to more searching and worse choices in finding partners for romantic relationships online: An experimental study. *CyberPsychology and Behavior, 12,* 315–318.

Chapter 5

Ackerman, D. (1994). *A natural history of love.* New York, NY: Random House.

Alford, J. J., P. K. Hatemi, J. R. Hibbing, N. G. Martin, & L. J. Eaves. (2011). The politics of mate choice. *Journal of Politics, 73,* 362–379.

Algoe, S. B., S. L. Gable, & N. C. Maisel. (2010). It's the little things: Everyday gratitude as a booster shot for romantic relationships. *Personal Relationships, 17,* 217–233.

Amato, P. R., A. Booth, D. R. Johnson, & S. F. Rogers. (2007). *Alone together: How marriage in America is changing.* Cambridge, MA: Harvard University Press.

Ambwani, S., & J. Strauss. (2007). Love thyself before loving others? A qualitative and quantitative analysis of gender differences in body image and romantic love. *Sex Roles, 56,* 13–22.

Assad, K. K., M. B. Donnellan, & R. D. Conger. (2007). Optimism: An enduring resource for romantic relationships. *Journal of Personality and Social Psychology, 93,* 285–296.

Barelds, D. P., & P. Barelds-Dijkstra. (2007). Love at first sight or friends first? Ties among partner personality trait similarity, relationship onset, relationship quality, and love. *Journal of Social and Personal Relationships, 24,* 479–496.

Barelds-Dijkstra, D. P. H., & P. Barelds. (2007). Relations between different types of jealousy and self and partner perceptions of relationship quality. *Clinical Psychology and Psychotherapy, 14,* 176–188.

Berscheid, E. (2010). Love in the fourth dimension. *Annual Review of Psychology, 61,* 1–25.

Blair, S. L. (2010). The influence of risk-taking behaviors on the transition into marriage: An examination of the long-term consequences of adolescent behavior. *Marriage and Family Review, 46,* 126–146.

Bogg, R. A., & J. M. Ray. (2006). The heterosexual appeal of socially marginal men. *Deviant Behavior, 27,* 457–477.

Braithwaithe, S. R., R. Delevi, & F. D. Fincham. (2010). Romantic relationships and the physical and mental health of college students. *Personal Relationships, 17,* 1–12.

Brantley, A., D. Knox, & M. E. Zusman. (2002). When and why gender differences in saying "I love you" among college students. *College Student Journal, 36,* 614–615.

Bratter, J. L., & R. B. King. (2008). "But Will It Last?" Marital instability among interracial and same-race couples. *Family Relations, 57,* 160–171.

Bredow, C. A., T. L. Huston, & G. Noval. (2011). Market value, quality of the pool of potential mates, and singles' confidence about marrying. *Personal Relationships, 18,* 39–57.

Burr, B. K, J. Viera, B. Dial, H. Fields, K. Davis, & D. Hubler. (2011, November 18). Influences of personality on relationship satisfaction through stress. Poster session presented at the annual meeting of the National Council on Family Relations, Orlando, FL.

Buunk, A. P., J. Goor, & A. C. Solano. (2010). Intrasexual competition at work: Sex differences in the jealousy-evoking effect of rival characteristics in work settings. *Journal of Social and Personal Relationships, 27,* 671–684.

Carey, A. R., & K. Gellers. (2010, August 31). Do you believe in the idea of soulmates? *USA Today,* p. 1A.

Cassidy, M. L., & G. Lee. (1989). The study of polyandry: A critique and synthesis. *Journal of Comparative Family Studies, 20,* 1–11.

Clarkwest, A. (2007). Spousal dissimilarity, race, and marital dissolution. *Journal of Marriage and Family, 69,* 639–653.

Crowell, J. A., D. Treboux, & E. Waters. (2002). Stability of attachment representations: The transition to marriage. *Developmental Psychology, 38,* 467–479.

DeCuzzi, A., D. Knox, & M. Zusman. (2006). Racial differences in perceptions of women and men. *College Student Journal, 40,* 343–349.

Devall, E., M. Montanez, & D. Vanleeuwen. (2009, November). Effectiveness of relationship education with single, cohabiting, and married parents. Poster session presented at the meeting of the National Council on Family Relations.

Diamond, L. M. (2003). What does sexual orientation orient? A biobehavioral model distinguishing romantic love and sexual desire. *Psychological Review, 110,* 173–192.

Dotson-Blake, K., D. Knox, & A. Holman. (2008). College student attitudes toward marriage, family, and sex therapy. Unpublished data from 288 undergraduate/graduate students. East Carolina University, Greenville, NC.

Dubbs, S. L., & A. P. Buunk. (2010). Sex differences in parental preferences over a child's mate choice: A daughter's perspective. *Journal of Social and Personal Relationships, 27,* 1051–1059.

Duncan, S. F., T. B. Holman, & C. Yang. (2007). Factors associated with involvement in marriage preparation programs. *Family Relations, 56,* 270–278.

Edwards, T. M. (2000, August 28). Flying solo. *Time,* 47–53.

Elliott, L., B. Easterling, & D. Knox. (2012, March). Taking chances in romantic relationships. Poster session presented at the annual meeting of the Southern Sociological Society, New Orleans, LA.

Fisher, H. (2009). *Why him? Why her?* New York, NY: Holt.

Fisher, M., & A. Cox. (2011). Four strategies used during intrasexual competition for mates. *Personal Relationships, 18,* 20–38.

Foster, J. (2010). How love and sex can influence recognition of faces and words: A processing model account. *European Journal of Social Psychology, 40,* 524–535.

Futris, T. G., A. W. Barton, T. M. Aholou, & D. M. Seponski. (2011). The impact of PREPARE on engaged couples: Variations by delivery format. *Journal of Couple and Relationship Therapy, 10,* 69–86.

Gallmeier, C. P., M. E. Zusman, D. Knox, & L. Gibson. (1997). Can we talk? Gender differences in disclosure patterns and expectations. *Free Inquiry in Creative Sociology, 25,* 219–225.

Gatzeva, M., & A. Paik. (2011). Emotional and physical satisfaction in noncohabiting, cohabiting, and marital relationships: The importance of jealous conflict. *Journal of Sex Research, 48,* 29–42.

Gonzaga, G. C., S. Carter, & J. Galen Buckwalter. (2010). Assortative mating, convergence, and satisfaction in married couples. *Personal Relationships, 17,* 634–644.

Graham, J. M. (2011). Measuring love in romantic relationships: A meta-analysis. *Journal of Social and Personal Relationships, 28,* 748–771.

Greitemeyer, T. (2010). Effects of reciprocity on attraction: The role of a partner's physical attractiveness. *Personal Relationships, 17,* 317–330.

Halpern, C. T., R. B. King, S. G. Oslak, & J. R. Udry. (2005). Body mass index, dieting, romance, and sexual activity in adolescent girls: Relationships over time. *Journal of Research on Adolescence, 15,* 535–559.

Haring, M., P. L. Hewitt, & G. L. Flett. (2003). Perfectionism, coping, and quality of relationships. *Journal of Marriage and Family, 65,* 143–159.

Hatfield, E., L. Bensman and R. L. Rapson. (2012). A brief history of social scientists' attempts to measure passionate love. *Journal of Social and Personal Relationships, 29,* 143–164.

Hesse, C., & K. Floyd. (2011). The impact of alexithymia on initial interactions. *Personal Relationships, 18,* 453–470.

Hohmann-Marriott, B. E., & P. Amato. (2008). Relationship quality in interethnic marriages and cohabitation. *Social Forces, 87,* 825–855.

Huston, T. L., J. P. Caughlin, R. M. Houts, S. E. Smith, & L. J. George. (2001). The connubial crucible: Newlywed years as predictors of marital delight, distress, and divorce. *Journal of Personality and Social Psychology, 80,* 237–252.

Ingoldsby, B., P. Schvaneveldt, & C. Uribe. (2003). Perceptions of acceptable mate attributes in Ecuador. *Journal of Comparative Family Studies, 34,* 171–186.

Jackson, J. (2011, November). Premarital Counseling: An evidence-informed treatment protocol. Presentation, Annual Meeting of the National Council on Family Relations, Orlando, FL.

Johnson, M. D. and J. R. Anderson (2011, November). The longitudinal association of marital confidence, time spent together and marital satisfaction. Presentation, National Council on Family Relationship, Orlando, FL.

Johnson, R. A., H. G. Barney, J. H. Parry, & J. S. Boden. (2011, November 18). "Just be yourself": A content analysis of dating advice among university students. Poster session presented at the annual meeting of the National Council on Family Relations, Orlando, FL.

Kennedy, D. P., J. S. Tucker, M. S. Pollard, M. Go, & H. D. Green. (2011). Adolescent romantic relationships and change in smoking status. *Addictive Behaviors, 36,* 320–326.

Kem, J. (2010). Fatal lovesickness in Marguerite de Navarre's *Heptaméron. Sixteenth Century Journal, 41,* 355–370.

Knox, D., R. Breed, & M. Zusman. (2007). College men and jealousy. *College Student Journal, 41,* 494–498.

Knox, D., & S. Hall. (2010). Relationship and sexual behaviors of a sample of 2,922 university students. Unpublished data collected for this text. Department of Sociology, East Carolina University, and Department of Family and Consumer Sciences, Ball State University.

Knox, D., & K. McGinty. (2009). Searching for homogamy: An in-class exercise. *College Student Journal, 43,* 243–247.

Knox, D., M. E. Zusman, & W. Nieves. (1997). College students' homogamous preferences for a date and mate. *College Student Journal, 31,* 445–448.

Knox, D., M. E. Zusman, & W. Nieves. (1998). What I did for love: Risky behavior of college students in love. *College Student Journal, 32,* 203–205.

Lee, J. A. (1973). *The colors of love: An exploration of the ways of loving.* Don Mills, ON: New Press.

Lee, J. A. (1988). Love-styles. In R. Sternberg & M. Barnes (Eds.), *The psychology of love* (pp. 38–67). New Haven, CN: Yale University Press.

Leno, J. (1996). *Leading with my chin.* New York, NY: HarperCollins.

Luo, S. H., & E. C. Klohnen. (2005). Assortative mating and marital quality in newlyweds: A couple-centered approach. *Journal of Personality and Social Psychology, 88*, 304–326.

MacArthur, S. (2010, November 3–5). Adolescent religiosity, religious affiliation, and premarital predictors of marital quality and stability. Poster session presented at the annual meeting of the National Council on Family Relations, Minneapolis, MN.

Medora, N. P., J. H. Larson, N. Hortacsu, & P. Dave. (2002). Perceived attitudes towards romanticism: A cross-cultural study of American, Asian-Indian, and Turkish young adults. *Journal of Comparative Family Studies, 33*, 155–178.

Meehan, D., & C. Negy. (2003). Undergraduate students' adaptation to college: Does being married make a difference? *Journal of College Student Development, 44*, 670–690.

Meier, A., K. E. Hull, & T. A. Orty. (2009). Young adult relationship values at the intersection of gender and sexuality. *Journal of Marriage and Family, 71*, 510–525.

Ozay, B., D. Knox, & B. Easterling. (2012, March). You're dating who? Parental attitudes toward interracial dating. Poster session presented at the annual meeting of the Southern Sociological Society, New Orleans, LA.

Perilloux, C., D. S. Fleischman, & D. M. Buss. (2011). Meet the parents: Parent-offspring convergence and divergence in mate preferences. *Personality and Individual Differences, 50*, 253–258.

Petrie, J., J. A. Giordano, & C. S. Roberts. (1992). Characteristics of women who love too much. *Affilia: Journal of Women and Social Work, 7*, 7–20.

Pew Research Center. (2010). Religion among the millennials. Retrieved from http://www.pewforum.org/Age/Religion-Among-the-Millennials.aspx

Pew Research Center. (2011). The burden of student debt. Retrieved from http://pewresearch.org/databank/dailynumber/?NumberID=1257

Picca, L. H., & J. R. Feagin. (2007). *Two-faced racism.* New York, NY: Routledge.

Radmacher, K., & M. Azmitia. (2006). Are there gendered pathways to intimacy in early adolescents' and emerging adults' friendships? *Journal of Adolescent Research, 21*, 415–448.

Regan, P. C., & A. Joshi. (2003). Ideal partner preferences among adolescents. *Social Behavior and Personality, 31*, 13–20.

Ridley, C. (2009). Personal communication with Dr. Ridley, who retired from the University of Arizona in 2009.

Riela, S., G. Rodriguez, A. Aron, X. Xu, & B. P. Acevedo. (2010). Experiences of falling in love: Investigating culture, ethnicity, gender, and speed. *Journal of Social and Personal Relationships, 27*, 473–493.

Ross, C. B. (2006, March 24). An exploration of eight dimensions of self-disclosure on relationship. Paper presented at the meeting of the Southern Sociological Society, New Orleans, LA.

Sack, K. (2008, August 12). Health benefits inspire rush to marry, or divorce. *The New York Times.*

Saint, D. J. (1994). Complementarity in marital relationships. *Journal of Social Psychology, 134*, 701–704.

Sobal, J., & K. L. Hanson (2011). Marital status, marital history, body weight, and obesity. *Marriage and Family Review, 47*, 474–504.

Sprecher, S. (2002). Sexual satisfaction in premarital relationships: Associations with satisfaction, love, commitment, and stability. *Journal of Sex Research, 39*, 190–196.

Sprecher, S., & P. C. Regan. (2002). Liking some things (in some people) more than others: Partner preferences in romantic relationships and friendships. *Journal of Social and Personal Relationships, 19*, 463–481.

Statistical Abstract of the United States. (2012, 131st ed.). Washington, DC: U.S. Census Bureau.

Sternberg, R. J. (1986). A triangular theory of love. *Psychological Review, 93*, 119–135.

Toro-Morn, M., & S. Sprecher. (2003). A cross-cultural comparison of mate preferences among university students: The United States versus the People's Republic of China (PRC). *Journal of Comparative Family Studies, 34*, 151–162.

Trachman, M., & C. Bluestone. (2005). What's love got to do with it? *College Teaching, 53*, 131–136.

Tsunokai, G. T., & A. R. McGrath. (2011). Baby boomers and beyond: Crossing racial boundaries in search for love. *Journal of Aging Studies, 25*, 285–294.

Twenge, J. (2006). *Generation me.* New York, NY: Free Press.

Tzeng, O. C. S., K. Wooldridge, & K. Campbell. (2003). Faith love: A psychological construct in intimate relations. *Journal of the Indiana Academy of the Social Sciences, 7*, 11–20.

U.S. Census Bureau. (2010, January 15). *Census Bureau reports families with children increasingly face unemployment* (Press release). Retrieved from http://www.census.gov/newsroom/releases/archives/families_households/cb10-08.html

Waller, W., & R. Hill. (1951). *The family: A dynamic interpretation.* New York, NY: Holt, Rinehart and Winston.

Walster, E., & G. W. Walster. (1978). *A new look at love.* Reading, MA: Addison-Wesley.

Wiersma, J. D., J. L. Fischer, B. C. Bray and J. P. Clifton. (2011) Young adult drinking partnerships: Where are couples 6 years later? Poster, Annual Meeting of National Council on Family Relations, Orlando, November 18.

Winch, R. F. (1955). The theory of complementary needs in mate selection: Final results on the test of the general hypothesis. *American Sociological Review, 20*, 552- 555.

Zusman, M. E., J. Gescheidler, D. Knox, & K. McGinty. (2003). Dating manners among college students. *Journal of Indiana Academy of Social Sciences, 7*, 28–32.

Chapter 6

Ali, L. (2011). Sexual self disclosure predicting sexual satisfaction. Poster, National Council on Family Relations, Orlando, FL, November.

Backstrom, L., E. A. Armstrong, & J. Puentes. (2012). Women's negotiation of cunnilingus in college hookups and relationships. *Journal of Sex Research, 49*, 1–12.

Ballard, S. M., C. Sugita & K. H. Gross. (2011, November 17). A qualitative investigation of the sexual socialization of emerging adults. Paper presented at the 73rd Conference of the National Council on Family Relations, Orlando, FL.

Beckman, N. M., M. Waern, I. Skoog, & The Sahlgrenska Academy at Goteborg University, Sweden. (2006). Determinants of sexuality in 70 year olds. *Journal of Sex Research, 43*, 2–3.

Bersamin, M., B. L. Zamboanga, S. J. Schwartz, M. B. Donnellan, M. Hudson, R. S. Weisskirch, et al. (2012). One night stands may not be good for your mental health: Casual sex and psychological well-being and distress. *Submitted for publication.*

Boislard, P. & F. Poulin. (2011). Individual, familial, friends-related and contextual predictors of early sexual intercourse. *Journal of Adolescence, 34*, 289-300.

Brucker, H., & P. Bearman. (2005). After the promise: The STD consequences of adolescent virginity pledges. *Journal of Adolescent Health, 36*, 271–276.

Carpenter, L. M. (2010). Like a virgin . . . again? Secondary virginity as an ongoing gendered social construction. *Sexuality and Culture, 14*, 253–270.

Colson, M., A. Lemaire, P. Pinton, K. Hamidi, & P. Klein. (2006). Sexual behaviors and mental perception, satisfaction, and expectations of sex life in men and women in France. *Journal of Sexual Medicine, 3*, 121–131.

Cutler, W. B., E. Friedmann, & N. L. McCoy. (1996). Pheromonal influences on sociosexual behavior in men. *Archives of Sexual Behavior, 27*, 1–13.

Davidson, J. K., Sr., N. B. Moore, J. R. Earle, & R. Davis. (2008). Sexual attitudes and behavior at four universities: Do region, race, and/or religion matter? *Adolescence, 43*, 189–223.

DeLamater, J., & M. Hasday. (2007). The sociology of sexuality. In C. D. Bryant & D. L. Peck (Eds.), *21st century sociology: A reference handbook* (pp. 254–264). Thousand Oaks, CA: Sage.

Dotson-Blake, K. P., D. Knox and M. E. Zusman. (2012). Exploring social sexual scripts related to oral sex: A profile of college student perceptions. *The Professional Counselor, 2,* 1–11.

Eisenman, R., & M. L. Dantzker. (2006). Gender and ethnic differences in sexual attitudes at a Hispanic-serving university. *Journal of General Psychology, 133,* 153–163.

England, P., & R. J. Thomas. (2006). The decline of the date and the rise of the college hook up. In A. S. Skolnick & J. H. Skolnick (Eds.), *Family in transition* (14th ed., pp. 151–62). Boston, MA: Pearson Allyn and Bacon.

Foster, D. G., J. A. Higgins, M. A. Briggs, C. McCain, S. Holtby, & C. D. Brindis. (2012). Willingness to have unprotected sex. *Journal of Sex Research, 49,* 61–68.

Grammer, K., F. Bernard, & N. Neave. (2005). Human pheromones and sexual attraction. *European Journal of Obstetrics and Gynecology and Reproductive Biology, 118,* 135–142.

Greene, K., & S. Faulkner. (2005). Gender, belief in the sexual double standard, and sexual talk in heterosexual dating relationships. *Sex Roles, 53,* 239–251.

Hattori, M. K., & F. N. Dodoo. (2007). Cohabitation, marriage and sexual monogamy in Nairobi's slums. *Social Science and Medicine, 64,* 1067–1072.

Heintz, A. J., & R. M. Melendez. (2006). Intimate partner violence and HIV/STD risk among lesbian, gay, bisexual and transgender individuals. *Journal of Interpersonal Violence, 21,* 193–206.

Herbenick, D., M. Reece, V. Schick, K. N. Jozkowski, S. E. Middelstadt, S. A. Sanders, et al. (2011). Beliefs about women's vibrator use: Results from a nationally representative probability survey in the United States. *Journal of Sex and Marital Therapy, 37,* 329–345.

Hollander, D. (2006). Many teenagers who say they have taken a virginity pledge retract that statement after having intercourse. *Perspectives on Sexual and Reproductive Health, 38,* 168–173.

Impett, E. A., L. A. Peplau, & S. L. Gable. (2005). Approach and avoidance sexual motives: Implications for personal and interpersonal well-being. *Personal Relationships, 12,* 465–482.

Kim, J. L., C. L. Sorsoli, K. Collins, B. A. Zylbergold, D. Schooler, & D. L. Tolman. (2007). From sex to sexuality: Exposing the heterosexual script on prime-time network television. *Journal of Sex Research, 44,* 145–157.

Knox, D., & S. Hall. (2010). Relationship and sexual behaviors of a sample of 2,922 university students. Unpublished data collected for this text. Department of Sociology, East Carolina University, and Department of Family and Consumer Sciences, Ball State University.

Knox, D., M. Zusman, & A. McNeely. (2008). University student beliefs about sex: Men vs. women. *College Student Journal, 42,* 181–185.

Kornreich, J. L., K. D. Hern, G. Rodriguez, and L. F. O'Sullivan. (2003). Sibling influence, gender roles, and the sexual socialization of urban early adolescent girls. *Journal of Sex Research, 40,* 101–110.

Landor, A., & L. G. Simons. (2010, November 3–6). The impact of virginity pledges on sexual attitudes and behaviors among college students. Paper presented at the annual meeting of the National Council on Family Relations, Minneapolis, MN.

Laumann, E. O., A. Paik, D. B. Glasser, J.-H. Kang, T. Wang, B. Levinson, et al. (2006, April). A cross-national study of subjective sexual well-being among older women and men: Findings from the global study of sexual attitudes and behaviors. *Archives of Sexual Behavior.*

Lee, J. T., C. L. Lin, G. H. Wan, & C. C. Liang. (2010). Sexual positions and sexual satisfaction of pregnant women. *Journal of Sex and Marital Therapy, 36,* 408–420.

Levin, R. (2004). Smells and tastes: Their putative influence on sexual activity in humans. *Sexual and Relationship Therapy, 19,* 451–462.

Lykins, A. D., E. Janssen, & C. A. Graham. (2006). The relationship between negative mood and sexuality in heterosexual college women and men. *Journal of Sex Research, 43,* 136–144.

Masters, W. H., & V. E. Johnson. (1970). *Human sexual inadequacy.* Boston, MA: Little, Brown.

Mathy, R. M. (2007). Sexual orientation moderates online sexual activity. In M. T. Whitty, A. J. Baker, & J. A. Inman (Eds.), *Online matchmaking* (pp. 159–177). New York, NY: Palgrave MacMillan.

McCabe, M. P., & D. L. Goldhammer. (2012). Demographic and psychological factors related to sexual desire among heterosexual women in a relationship. *Journal of Sex Research, 49,* 78–87.

McGinty, K., D. Knox, & M. Zusman. (2007). Friends with benefits: Women want "friends," men want "benefits." *College Student Journal, 41,* 1128–1131.

Merrill, J., & D. Knox. (2010). *Finding love from 9 to 5: Secrets of office romance.* Santa Barbara, CA: Praeger.

Michael, R. T., J. H. Gagnon, E. O. Laumann, & G. Kolata. (1994). *Sex in America.* Boston, MA: Little, Brown.

Miller, S., A. Taylor, & D. Rappleyea. (2011, April 4–8). The influence of religion on young adults' attitudes of dating events. Poster session presented at the Fifth Annual Research and Creative Achievement Week, East Carolina University.

Miller, S. A., & E. S. Byers. (2004). Actual and desired duration of foreplay and intercourse: Discordance and misperceptions within heterosexual couples. *Journal of Sex Research, 41,* 301–309.

Nobre, P. J., & J. Pinto-Gouveia. (2006). Dysfunctional sexual beliefs as vulnerability factors for sexual dysfunction. *Journal of Sex Research, 43,* 68–74.

Paik, A. (2010). The contexts of sexual involvement and concurrent sexual partnerships. *Perspectives on Sexual and Reproductive Health, 42,* 33–43.

Puentes, J., D. Knox, & M. Zusman. (2008). Participants in "friends with benefits" relationships. *College Student Journal, 42,* 176–180.

Raley, R. K. (2000). Recent trends and differentials in marriage and cohabitation: The United States. In L. J. Waite (Ed.), *The ties that bind* (pp. 19–39). New York, NY: Aldine de Gruyter.

Reid, R. C., D. S. Li, R. Gilliland, J. A. Stein, & T. Fong. (2011). Reliability, validity, and psychometric development of Pornography Consumption Inventory in a sample of hypersexual men. *Journal of Sex and Marital Therapy, 37,* 359–385.

Reissing, E. D., H. L. Andruff, & J. J. Wentland. (2012). Looking back: The experience of first sexual intercourse and current sexual adjustment in young heterosexual adults. *Journal of Sex Research, 49,* 27–35.

Simon, W., & J. Gagnon. (1998). Psychosexual development. *Society, 35,* 60–66.

Stulhofer, A., & D. Ajdukovic. (2011). Should we take anodyspareunia seriously? A descriptive analysis of pain during receptive anal intercourse in young heterosexual women. *Journal of Sex and Marital Therapy, 37,* 346–358.

Toews, M. L., & A. Yazedjian. (2011, November 18). College students' knowledge, attitudes, and behaviors regarding sexuality. Poster session presented at the annual meeting of the National Council on Family Relations, Orlando, FL.

True Love Waits. (2012). http://www.lifeway.com/True-Love-Waits/c/N-1z13wiu

Uecker, J. E. (2008). Religion, pledging, and the premarital sexual behavior of married young adults. *Journal of Marriage and Family, 70,* 728–744.

Vail-Smith, K., D. Knox, & L. Whetstone. (2010). The illusion of safety in "monogamous" undergraduate relationships. *American Journal of Health Behavior, 34,* 12–20.

Vazonyi, A. I., & D. D. Jenkins. (2010). Religiosity, self control, and virginity status in college students from the "Bible belt": A research note. *Journal for the Scientific Study of Sex, 49,* 561–568.

Walsh, J. L., & L. M. Ward. (2010). Magazine reading and involvement and young adults' sexual health knowledge, efficacy, and behaviors. *Journal of Sex Research, 47,* 285–300.

Chapter 7

Aducci, A. J., J. A. Baptist, J. George, P. M.Barros, & B. S. Nelson Goff. (2011, November). Military wives' experience during OIF/OEF deployment. Paper presented at the annual meeting of the National Council on Family Relations, Orlando, FL.

Allen, E. S., S. M. Stanley, G. K. Rhoades, H. J. Markman, & B. A. Loew. (2011). Marriage education in the Army: Results of randomized clinical trial. *Journal of Couple and Relationship Therapy, 10*, 309–326.

Allgood, S., & A. Bakker. (2009, November). Connection rituals and marital satisfaction. Poster session presented at the annual meeting of the National Council on Family Relations, San Francisco, CA.

Amato, P. R., A. Booth, D. R. Johnson, & S. F. Rogers. (2007). *Alone together: How marriage in America is changing*. Cambridge, MA: Harvard University Press.

Anderson, J. R., M. J. Van Ryzin, & W. J. Doherty. (2010). Developmental trajectories of marital happiness in continuously married individuals: A group-based modeling approach. *Journal of Family Psychology, 24*, 587–596.

Arnold, A. L., C. W. O'Neal, C. Bryant, K. A. S. Wickrama, & C. Cutrona. (2011, November). Influences of intra and interpersonal factors in marital success. Presentation at the annual meeting of the National Council on Family Relations, Orlando, FL.

Ballard, S. M., & A. C. Taylor. (2011). *Family life education with diverse populations*. Thousand Oaks, CA: Sage.

Barnes, K., & J. Patrick. (2004). Examining age-congruency and marital satisfaction. *Gerontologist, 44*, 185–187.

Bartle-Haring, S. (2010). Using Bowen theory to examine progress in couple therapy. *Family Journal, 18*, 106–115.

Billingsley, S., M. Lim, & G. Jennings. (1995). Themes of long-term, satisfied marriages consummated between 1952 and 1967. *Family Perspective, 29*, 283–295.

Cadden, M., & D. Merrill. (2007). What married people miss most (based on *Reader's Digest* study of 1,001 married adults). *USA Today*, p. D1.

Calligas, A., F. Adler-Baeder, M. Keiley, T. Smith, & S. Ketring. (2010, November 3–5). Examining change in parenting dimensions in relation to change in couple dimensions. Poster session presented at the annual meeting of the National Council on Family Relations, Minneapolis, MN.

Campbell, K., J. C. Kaufman, T. D. Ogden, T. T. Pumaccahua, & H. Hammond. (2011, November). Wedding rituals: How cost and elaborateness relate to marital outcomes. Poster session presented at the annual meeting of the National Council on Family Relations, Orlando, FL.

Campbell, K., L. C. Silva, & D. W. Wright. (2011). Rituals in unmarried couple relationships: An exploratory study. *Family and Consumer Sciences Research Journal, 40*, 45–57.

Carroll, J. S., L. R. Dean, L. L. Call, & D. M. Busby. (2011). Materialism and marriage: Couple profiles of congruent and incongruent spouses. *Journal of Couple and Relationship Therapy, 10*, 287–308.

Coontz, S. (2000). Marriage: Then and now. *Phi Kappa Phi Journal, 80*, 16–20.

Corra, M., S. Carter, J. S. Carter, & D. Knox. (2009). Trends in marital happiness by sex and race, 1973–2006. *Journal of Family Issues, 30*, 1379–1404.

DeOllos, I. Y. (2005). Predicting marital success or failure: Burgess and beyond. In V. L. Bengtson, A. C. Acock, K. R. Allen, P. Dilworth-Anderson, & D. M. Klein (Eds.), *Sourcebook of family theory and research* (pp. 134–136). Thousand Oaks, CA: Sage.

Dew, J. (2008). Debt change and marital satisfaction change in recently married couples. *Family Relations, 57*, 60–71.

Dowd, D. A., M. J. Means, J. F. Pope, & J. H. Humphries. (2005). Attributions and marital satisfaction: The mediated effects of self-disclosure. *Journal of Family and Consumer Sciences, 97*, 22–27.

Drefahl, S. (2010). How does the age gap between partners affect their survival? *Demography, 47*, 313–326.

Easterling, B. A. (2005). *The invisible side of military careers: An examination of employment and well-being among military spouses.* (Doctoral dissertation). University of North Florida.

Easterling, B. A., & D. Knox. (2010). Left behind: How military wives experience the deployment of their husbands. *Journal of Family Life.* Retrieved from http://www.journaloffamilylife.org/militarywives

Fisher, T. D., & J. K. McNulty. (2008). Neuroticism and marital satisfaction: The mediating role played by the sexual relationship. *Journal of Family Psychology, 22*, 112–123.

Fu, X. (2006). Impact of socioeconomic status on inter-racial mate selection and divorce. *Social Science Journal, 43*, 239–258.

Gaines, S. O., Jr., & J. Leaver. (2002). Interracial relationships. In R. Goodwin & D. Cramer (Eds.), *Inappropriate relationships: The unconventional, the disapproved, and the forbidden* (pp. 65–78). Mahwah, NJ: Lawrence Erlbaum.

Gibson, V. (2002). *Cougar: A guide for older women dating younger men.* Boston, MA: Firefly Books.

Gottman, J., & S. Carrere. (2000, September/October). Welcome to the love lab. *Psychology Today*, 42.

Hall, S. S., & R. Adams. (2011). Newlyweds' unexpected adjustments to marriage. *Family and Consumer Sciences Research Journal, 39*, 375–387.

Hill, M. R., & V. Thomas. (2000). Strategies for racial identity development: Narratives of black and white women in interracial partner relationships. *Family Relations, 49*, 193–200.

Huyck, M. H., & D. L. Gutmann. (1992). Thirty something years of marriage: Understanding experiences of women and men in enduring family relationships. *Family Perspective, 26*, 249–265.

Jorgensen, B. L., J. Yorgason, R. Day, & J. A. Mancini. (2011, November 18). The influence of religious beliefs and practices on marital commitment. Poster session presented at the annual meeting of the National Council on Family Relations, Orlando, FL.

Kennedy, R. (2003). *Interracial intimacies.* New York, NY: Pantheon.

Knox, D. (2010). The wedding night. Unpublished data. Department of Sociology, East Carolina University.

Knox, D., & S. Hall. (2010). Relationship and sexual behaviors of a sample of 2,922 university students. Unpublished data collected for this text. Department of Sociology, East Carolina University, and Department of Family and Consumer Sciences, Ball State University.

Kreaer, D. A. (2008). Guarded borders: Adolescent interracial romance and peer trouble at school. *Social Forces, 87*, 887–910.

Lavner, J. A., & T. N. Bradbury. (2010). Patterns of change in marital satisfaction over the newlywed years. *Journal of Marriage and Family, 72*, 1171–1187.

Lundquist, J. H. (2007). A comparison of civilian and enlisted divorce rates during the early all-volunteer force era. *Journal of Political and Military Sociology, 35*(2), 199-217.

Marklein, M. B. (2008, November 17). High mark for foreign students here. *USA Today*, p. 4D.

McNulty, J. K. (2008). Forgiveness in marriage: Putting the benefits into context. *Journal of Family Psychology, 22*, 171–183.

Michael, R. T., J. H. Gagnon, E. O. Laumann, & G. Kolata. (1994). *Sex in America: A definitive survey.* Boston, MA: Little, Brown.

Morr Serewicz, M. C., & D. J. Canary. (2008). Assessments of disclosure from the in-laws: Links among disclosure topics, family privacy orientations, and relational quality. *Journal of Social and Personal Relationships, 25*, 333–357.

Murdock, G. P. (1949). *Social structure.* New York, NY: Free Press.

Musick, K. and L. Bumpass. 2012. Reexamining the case for marriage: Union formation and changes in well-being. *Journal of Marriage and Family, 74*, 1–18.

NCFR Policy Brief. (2004). *Building strong communities for military families.* Minneapolis, MN: National Council on Family Relations.

North, R. J., C. J. Holahan, R. H. Moos, & R. C. Cronkite. (2008). Family support, family income, and happiness: A 10-year perspective. *Journal of Family Psychology, 22*, 475–483.

Nuner, J. E. (2004). A qualitative study of mother-in-law/daughter-in-law relationships. *Dissertation Abstracts International: Section A. The Humanities and Social Sciences, 65*(August), 712A–713A.

Orthner, D. K., & R. Roderick. (2009). Work separation demands and spouse psychological well-being. *Family Relations, 58,* 392–403.

Passel, J. S., W. Wang, & P. Taylor. (2010, June 4). Marrying out: One-in-seven new U.S. marriages is interracial or interethnic. Pew Research Center. Retrieved from http://pewresearch.org/pubs/1616/american-marriage-interracial-interethnic

Patrick, S., J. N. Sells, F. G. Giordano, & T. Tollerud. (2012). Intimacy, differentiation, and personality variables as predictors of marital satisfaction. *Family Journal: Counseling and Therapy for Couples and Families, 20*(2), in press.

Pew Research Center. (2008). The U.S. religious landscape survey. Pew Forum on Religion and Public Life. Retrieved from *http://pewresearch.org/pubs/743/united*-states-religion

Plouffe, D. (2009). *The audacity to win.* New York, NY: Viking.

Proulx, C. M., & L. A. Snyder. (2009). Families and health: An empirical resource guide for researchers and practitioners. *Family Relations, 58,* 489–504.

Qian, Z., & D. T. Lichter. (2011). Changing patterns of interracial marriage in a multiracial society. *Journal of Marriage and Family, 73,* 1065–1084.

Sayer, L. C., S. M. Bianchi, & J. P. Robinson. (2004). Are parents investing less in children? Trends in mothers' and fathers' time with children. *American Journal of Sociology, 110,* 1–43.

Scott, L. S. (2009). *Two is enough.* Berkeley, CA: Seal Press.

Sobal, J., C. F. Bove, & B. S. Rauschenbach. (2002). Commensal careers at entry into marriage: Establishing commensal units and managing commensal circles. *Sociological Review, 50,* 378–397.

Solomon, Z., S. Debby-Aharon, G. Zerach, & D. Horesh. (2011). Marital adjustment, parental functioning, and emotional sharing in war veterans. *Journal of Family Issues, 32,* 127–147.

Statistical Abstract of the United States. (2012, 131st ed.). Washington, DC: U.S. Census Bureau.

Teten, A. L., J. A. Schumacher, C. T. Taft, M. A. Stanley, T. A. Kent, S. D. Bailey, et al. (2010). Intimate partner aggression perpetrated and sustained by male Afghanistan, Iraq, and Vietnam veterans with and without posttraumatic stress disorder. *Journal of Interpersonal Violence, 25,* 1612–1630.

Tomkins, S. A., J. Rosa, & J. Benavente. (2010, November 3–5). A holistic approach to relationship enrichment. Poster session presented at the annual meeting of the National Council on Family Relations, Minneapolis, MN.

Totenhagen, C., M. Katz, E. Butler, & M. Curran. (2011, November). Daily variability in relational experiences. Presentation at the annual meeting of the National Council on Family Relations, Orlando, FL.

Vaillant, C. O., & G. E. Vaillant. (1993). Is the U-curve of marital satisfaction an illusion? A 40-year study of marriage. *Journal of Marriage and Family, 55,* 230–239.

Van Laningham, J., & D. R. Johnson. (2009, November). Measuring marital quality: Comparing single and multiple item measures. Poster session presented at the annual meeting of the National Council on Family Relations, San Francisco, CA.

Wallerstein, J., & S. Blakeslee. (1995). *The good marriage.* Boston, MA: Houghton-Mifflin.

Weigel, D. J. (2010). Mutuality of commitment in romantic relationships: Exploring a dyadic model. *Personal Relationships, 17,* 495–513.

Weisfeld, G. E., N. T. Nowak, T. Lucas, C. C. Weisfeld, E. O. Imamoglu, M. Butovskaya, et al. (2011). Do women seek humorousness in men because it signals intelligence? A cross-cultural test. *International Journal of Humor Research, 24,* 435–462.

Wilmoth, J., & G. Koso. (2002). Does marital history matter? Marital status and wealth outcomes among preretirement adults. *Journal of Marriage and Family, 64,* 254–268.

Wilson, G., & J. Cousins. (2005). Measurement of partner compatibility; further validation and refinement of the CQ test. *Sexual and Relationship Therapy, 20,* 421–429.

Wilson, S. M., L. W. Ngige, & L. J. Trollinger. (2003). Kamba and Maasai paths to marriage in Kenya. In R. R. Hamon & B. B. Ingoldsby (Eds.), *Mate selection across cultures* (pp. 95–117). Thousand Oaks, CA: Sage.

Wright, R., C. H. Mindel, T. Van Tran, & R. W. Habenstein. (2012). *Ethnic families in America: Patterns and variations* (5th ed.). Boston, MA: Pearson.

Chapter 8

Amato, P. R. (2004). Tension between institutional and individual views of marriage. *Journal of Marriage and Family, 66,* 959–965.

American Psychological Association. (2004). *Sexual Orientation, Parents, and Children.* APA Policy Statement on Sexual Orientation, Parents, and Children. Retrieved from APA Online, http://www.apa.org

Bartholomew, K., K. V. Regan, M. A. White, & D. Oram. (2008). Patterns of abuse in male same-sex relationships. *Violence and Victims, 23,* 617–637.

Besen, W. (2000). Introduction. In *Feeling free: Personal stories—How love and self-acceptance saved us from "ex-gay" ministries.* Washington, DC: Human Rights Campaign Foundation, 7.

Biblarz, T. J. and J. Stacey. 2010. How does the gender of parents matter? *Journal of Marriage and Family, 72,* 3-22.

Blackwell, C. W., & S. F. Dziegielewski. (2012). Using the Internet to meet sexual partners: Research and practice implications. *Journal of Social Service Research, 38,* 46–55.

Bos, H. & N. Gartrell. 2010. Adolescents of the USA National Longitudinal Lesbian Family Study: Can family characteristics counteract the negative effects of stigmatization? *Family Process, 49,* 559–572

Buxton, A. P. (2004). Paths and pitfalls: How heterosexual spouses cope when their husbands or wives come out. *Journal of Couple and Relationship Therapy, 3.*

Cahill, S., & S. Slater. (2004). *Marriage: Legal protections for families and children. Policy brief.* Washington, DC: National Gay and Lesbian Task Force Policy Institute.

Carpenter, C., & G. J. Gates. (2008). Gay and lesbian partnership: Evidence from California. *Demography, 45,* 573–691.

Centers for Disease Control and Prevention. (2012). HIV incidence. Retrieved March 21, 2012, from http://www.cdc.gov/hiv/topics/surveillance/incidence.htm

Cianciotto, J., and S. Cahill. 2006. Youth in the crosshairs: The third wave of ex-gay activism. National Gay and Lesbian Task Force Policy Institute. http:www.thetaskforce.org

Clunis, D. M., & G. Dorsey Green. (2003). *The lesbian parenting book* (2nd ed.). Emeryville, CA: Seal Press.

Crowl, A., S. Ahn, & J. Baker. (2008). A meta-analysis of developmental outcomes for children of same-sex and heterosexual parents. *Journal of GLBT Family Studies, 4,* 385–407.

Custody and Visitation. (2000). *Human Rights Campaign FamilyNet.* Retrieved from http://familynet.hrc.org

Diamond, L. M. (2003). What does sexual orientation orient? A biobehavioral model distinguishing romantic love and sexual desire. *Psychological Review, 110,* 173–192.

Gartrell, N., H. Bos, H. Peyser, A. Deck, & C. Rodas. (2011). Family characteristics, custody arrangements, and adolescent psychological well-being after lesbian mothers break up. *Family Relations, 60,* 572–585.

Goldberg, A. E., Downing, J. B. and Moyer, A. M. 2012, Why Parenthood, and Why Now? Gay Men's Motivations for Pursuing Parenthood. *Family Relations, 61,* 157–174.

Gotta, G., R. J. Green, E. Rothblum, S. Solomon, K. Balsam, & P. Schwartz. (2011). Heterosexual, lesbian, and gay male relationships: A comparison of couples in 1975 and 2000. *Family Process, 50,* 353–376.

Gottman, J. M., R. W. Levenson, C. Swanson, K. Swanson, R. Tyson, & D. Yoshimoto. (2003). Observing gay, lesbian, and heterosexual couples' relationships: Mathematical modeling of conflict interaction. *Journal of Homosexuality, 45,* 65–91.

Green, J. C. (2004). The American religious landscape and political attitudes: A baseline for 2004. Pew Forum on Religion and Public Life. Retrieved from http://pewforum.org

Green, R. J., J. Bettinger, & E. Sacks. (1996). Are lesbian couples fused and gay male couples disengaged? In J. Laird & R. J. Green (Eds.), *Lesbians and gays in couples and families* (pp. 185–230). San Francisco, CA: Jossey-Bass.

Held, M. (2005, March 16). Mix it up: T-shirts and activism. Retrieved from http://www.tolerance.org/teens

Holthouse, D. (2005). Curious cures. *Intelligence Report, 117*(Spring), 14.

Howard, J. (2006). *Expanding resources for children: Is adoption by gays and lesbians part of the answer for boys and girls who need homes?* New York, NY: Evan B. Donaldson Adoption Institute.

Human Rights Campaign. (2012). Workplace. Retrieved March 23, 2012, from http://www.hrc.org/resources/category/workplace

Israel, T., and J. J. Mohr. 2004. Attitudes toward bisexual women and men: Current research, future directions. In *Current research on bisexuality*, ed. R. C. Fox, 117–134. New York: Harrington Park Press.

Jones, S. L., & M. A. Yarhouse. (2011). A longitudinal study of attempted religiously mediated sexual orientation change. *Journal of Sex and Marital Therapy, 37*(5), 404–427.

Kamen, C., M. Burns, & S. R. H. Beach. (2011). Minority stress in same-sex male relationships: When does it impact relationship satisfaction? *Journal of Homosexuality, 58*, 1372–1390.

Karten, E. Y., & J. C. Wade. (2010). Sexual orientation change efforts in men: A client perspective. *Journal of Men's Studies, 18*, 84–102.

Kinsey, A. C., W. B. Pomeroy, & C. E. Martin. (1948). *Sexual behavior in the human male*. Philadelphia, PA: Saunders.

Kinsey, A. C., W. B. Pomeroy, C. E. Martin, & P. H. Gebhard. (1953). *Sexual behavior in the human female*. Philadelphia, PA: Saunders.

Kirkpatrick, R. C. (2000). The evolution of human sexual behavior. *Current Anthropology, 41*, 385–414.

Knox, D., T. Beaver, & V. Kriskute. (2011). "I KISSED A GIRL": Heterosexual Women who Report Same-Sex Kissing (and more). *Journal of GLBT Family Studies, 7*, 217–225.

Knox, D., & S. Hall. (2010). Relationship and sexual behaviors of a sample of 2,922 university students. Unpublished data collected for this text. Department of Sociology, East Carolina University, and Department of Family and Consumer Sciences, Ball State University.

Kurdek, L. A. (1994). Conflict resolution styles in gay, lesbian, heterosexual nonparent, and heterosexual parent couples. *Journal of Marriage and Family, 56*, 705–722.

———. (2004). Gay men and lesbians: The family context. In M. Coleman & L. H. Ganong (Eds.), *Handbook of contemporary families: Considering the past, contemplating the future* (pp. 96–115). Thousand Oaks, CA: Sage.

———. (2005). What do we know about gay and lesbian couples? *Current Directions in Psychological Science, 14*, 251–254.

———. (2008). Change in relationship quality for partners from lesbian, gay male, and heterosexual couples. *Journal of Family Psychology, 22*, 701–711.

Lalicha, J., & K. McLarena. (2010). Inside and outcast: Multifaceted stigma and redemption in the lives of gay and lesbian Jehovah's Witnesses. *Journal of Homosexuality, 57*, 1303–1333.

Landis, D. (1999, February 17). Mississippi Supreme Court made a tragic mistake in denying custody to gay father, experts say. American Civil Liberties Union News. Retrieved from http://www.aclu.org

Lee, T., & G. R. Hicks. (2011). An analysis of factors affecting attitudes toward same-sex marriage: Do the media matter? *Journal of Homosexuality, 58*, 1391–1408.

Lever, J. (1994, August 23). The 1994 *Advocate* survey of sexuality and relationships: The men. *The Advocate*, pp. 16–24.

McKinney, J. (2004, February 8). *The Christian case for gay marriage.* Pullen Memorial Baptist Church. Retrieved from http://www.pullen.org

McLean, K. (2004). Negotiating (non)monogamy: Bisexuality and intimate relationships. In R. C. Fox (Ed.), *Current research on bisexuality* (pp. 82–97). New York, NY: Harrington Park Press.

Meyer, I. H., J. Dietrich, & S. Schwartz. (2008). Lifetime prevalence of mental disorders and suicide attempts in diverse lesbian, gay, and bisexual populations. *American Journal of Public Health, 98*, 1004–1006.

Mock, S. E., & R. P. Eibach. (2011, May). Stability and change in sexual orientation identity over a 10-year period in adulthood. *Archives of Sexual Behavior*. Online.

Moltz, K. (2005, April 13). Testimony of Kathleen Moltz, given before the United States Senate Judiciary Committee, Subcommittee on the Constitution, Civil Rights and Property Rights. Human Rights Campaign. Retrieved from http://www.hrc.org

Moskowitz, D. A., G. Rieger, & M. E. Roloff. (2010). Heterosexual attitudes toward same-sex marriage. *Journal of Homosexuality, 57*, 325–336.

National Coalition of Anti-Violence Programs. (2011). *National hate crimes report: Anti-lesbian, gay, bisexual and transgender violence in 2010*. New York, NY: Author.

National Conference of State Legislatures. (2011). States issuing same-sex marriage licenses. Retrieved from http://www.ncsl.org/default.aspx?tabid=16430

Ochs, R. (1996). Biphobia: It goes more than two ways. In B. A. Firestein (Ed.), *Bisexuality: The psychology and politics of an invisible minority* (pp. 217–239). Thousand Oaks, CA: Sage.

Oswalt, S. B., & T. J. Wyatt. (2011). Sexual orientation and differences in mental health, stress, and academic performance in a national sample of U.S. college students. *Journal of Homosexuality, 58*, 1255–1280.

Otis, M. D., S. S. Rostosky, E. D. B. Riggle, & R. Hamrin. (2006). Stress and relationship quality in same-sex couples. *Journal of Social and Personal Relationships, 23*, 81–99.

Page, S. (2011, July 20). Gay candidates gain acceptance. *USA Today*, p. 1.

Palmer, R., & R. Bor. (2001). The challenges to intimacy and sexual relationships for gay men in HIV serodiscordant relationships: A pilot study. *Journal of Marital and Family Therapy, 27*, 419–431.

Parelli, S. (2007). Why ex-gay therapy doesn't work. *The Gay and Lesbian Review Worldwide, 14*, 29–32.

Paul, J. P. (1996). Bisexuality: Exploring/exploding the boundaries. In R. Savin-Williams & K. M. Cohen (Eds.), *The lives of lesbians, gays, and bisexuals: Children to adults* (pp. 436–461). Fort Worth, TX: Harcourt Brace.

Pearcey, M. (2004). Gay and bisexual married men's attitudes and experiences: Homophobia, reasons for marriage, and self-identity. *Journal of GLBT Family Studies, 1*, 21–42.

Peplau, L. A., R. C. Veniegas, & S. N. Campbell. (1996). Gay and lesbian relationships. In R. C. Savin-Williams & K. M. Cohen (Eds.), *The lives of lesbians, gays, and bisexuals: Children to adults* (pp. 250–273). Fort Worth, TX: Harcourt Brace.

Pew Research Center. (2011). 39%—support for gay marriage. Retrieved August 26, 2011, from http://pewresearch.org/databank/dailynumber/?NumberID=881

Pinello, D. R. (2008). Gay marriage: For better or for worse? What we've learned from the evidence. *Law and Society Review, 42*, 227–230.

Porter, J., & L. M. Williams. (2011). Intimate violence among underrepresented groups on a college campus. *Journal of Interpersonal Violence, 26*, 3210–3224.

Potok, M. (2005). Vilification and violence. *Intelligence Report, 117*(Spring), 1.

———. (2010). Anti-gay hate crimes: Doing the math. *Intelligence Report, 140*(Winter), 29.

Pryor, J. H., S. Hurtado, L. DeAngelo, L. P. Blake, & S. Tran. (2011). *The American freshman: National norms Fall 2010*. Los Angeles, CA: Higher Education Research Institute, UCLA.

Ricks, J. L. (2012). Lesbians and alcohol abuse: Identifying factors for future research. *Journal of Social Service Research, 38*, 37–45.

Rosario, M., E. W. Schrimshaw, J. Hunter, & L. Braun. (2006). Sexual identity development among lesbian, gay, and bisexual youth: Consistency and change over time. *Journal of Sex Research, 43*, 46–58.

Rosin, H., & R. Morin. (1999, January 11). In one area, Americans still draw a line on acceptability. *The Washington Post National Weekly Edition, 16*, p. 8.

Rubinstein, G. (2010). Narcissism and self-esteem among homosexual and heterosexual male students. *Journal of Sex and Marital Therapy, 36*, 24–34.

Rust, P. (1993). Neutralizing the political threat of the marginal woman: Lesbians' beliefs about bisexual women. *Journal of Sex Research, 30*(3), 214–218.

Rutledge, S. E., D. C. Siebert, & J. Chonody. (2012). Attitudes toward gays and lesbians: A latent class analysis of university students. *Journal of Social Service Research, 38*, 18–28.

Sanday, P. R. (1995). Pulling train. In P. S. Rothenberg (Ed.), *Race, class, and gender in the United States* (3rd ed., pp. 396–402). New York, NY: St. Martin's Press.

Savin-Williams, R. C. (2006). Who's gay? Does it matter? *Current Directions in Psychological Science, 15*, 40–44.

Serovich, J. M., S. M. Craft, P. Toviessi, R. Gangamma, et al. (2008). A systematic review of the research base on sexual reorientation therapies. *Journal of Marital and Family Therapy, 34*, 227–239.

SIECUS (Sexuality Information and Education Council of the United States). (2012). Position statement on sexual orientation. Retrieved March 23, 2012, from http://www.siecus.org/index.cfm?fuseaction=page.viewPage&pageId=494&parentID=472

Statistical Abstract of the United States. (2012, 131st ed.). Washington, DC: U.S. Census Bureau.

Sullivan, A. (1997). The conservative case. In A. Sullivan (Ed.), *Same sex marriage: Pro and con* (pp. 146–154). New York, NY: Vintage Books.

Tobias, S., & S. Cahill. (2003). School lunches, the Wright brothers, and gay families. National Gay and Lesbian Task Force. Retrieved from http://www.thetaskforce.org

Tyagart, C. E. (2002). Legal rights to homosexuals in areas of domestic partnerships and marriages: Public support and genetic causation attribution. *Educational Research Quarterly, 25*, 20–29.

Van Eeden-Moorefield, B., C. R. Martell, M. Williams, & M. Preston. (2011). Same-sex relationships and dissolution: The connection between heteronormativity and homonormativity. *Family Relations, 60*, 562–571.

Wagner, C. G. (2006). Homosexual relationships. *Futurist, 40*, 6.

Whitley, B., C. Childs, and J. Collins. 2011. Differences in black and white American college students' attitudes toward lesbians and gay men. *Sex Roles, 64*, 299-310

Yost, M. R., & G. D. Thomas. (2011). Gender and binegativity: Men's and women's attitudes toward male and female bisexuals. *Archives of Sexual Behavior* (May).

Zaleski, R. M. (2007, March 30). What's in a name? "Civil union" fails the test. *New Jersey Law Journal*.

Chapter 9

Ahnert, L., & M. E. Lamb. (2003). Shared care: Establishing a balance between home and child care settings. *Child Development, 74*, 1044–1049.

Bennetts, L. (2007). *The feminine mistake: Are we giving up too much?* New York, NY: Voice/Hyperion.

Carriero, R. (2011). Perceived fairness and satisfaction with the division of housework among dual-earner couples in Italy. *Marriage and Family Review, 47*, 436–458.

Carroll, J. S., L. Call, D. M. Busby, & L. R. Dean. (2011). Materialism and marriage: Couple profiles of congruent and incongruent spouses. *Journal of Couple and Relationship Therapy, 10*, 287–308.

Cinamon, R. G. (2006). Anticipated work-family conflict: Effects of gender, self-efficacy, and family background. *Career Development Quarterly, 54*, 202–216.

Davies, J. B., S. Sandstrom, A. Shorrocks, & E. N. Wolff. (2006, December 5). *The World Distribution of Household Wealth*. United Nations University—World Institute for Development Economics Research.

De Schipper, J. C., L. W. C. Tavecchio, & M. H. Van IJzendoorn. (2008). Children's attachment relationships with day care caregivers: Associations with positive caregiving and the child's temperament. *Social Development, 17*, 454–465.

Deutsch, F. M., A. P. Kokot, & K. S. Binder. (2007). College women's plans for different types of egalitarian marriages. *Journal of Marriage and Family, 69*, 916–929.

Dew, J. (2008). Debt change and marital satisfaction change in recently married couples. *Family Relations, 57*, 60–72.

Diem, C., & J. M. Pizarro. (2010). Social structure and family homicides. *Journal of Family Violence, 25*, 521–532.

Etzioni, A. (2011). The new normal. *Sociological Forum, 26*, 779–789.

Gibb, S. J., D. M. Fergusson, & L. J. Horwood. (2012). Working hours and alcohol problems in early adulthood. *Addiction, 107*, 81–88.

Glorieux, I., J. Minnen, & T. Tienoven. (2011). Spouse "together time": Quality time within the household. *Social Indicators Research, 101*, 281–287.

Gordon, J. R., & K. S. Whelan-Berry. (2005). Contributions to family and household activities by the husbands of midlife professional women. *Journal of Family Studies, 26*, 899–923.

Gordon, R. A., M. L. Usdansky, X. Wang, & A. Gluzman. (2011). Child care and mothers' mental health: Is high-quality care associated with fewer depressive symptoms? *Family Relations, 60*, 446–460.

Gunnar, M. R., E. Kryzer, M. J. Van Ryzin, & D. A. Phillips. (2010). The rise in cortisol in family day care: Associations with aspects of care quality, child behavior, and child sex. *Child Development, 81*, 851–869.

Hochschild, A. R. (1989). *The second shift*. New York, NY: Viking.

———. (1997). *The time bind*. New York, NY: Metropolitan Books.

Kahneman, D., & A. Deaton. (2010, September 6). High income improves evaluation of life but not emotional well-being. *Proceedings of the National Academy of Sciences, Early Edition*.

Kiecolt, K. J. (2003). Satisfaction with work and family life: No evidence of a cultural reversal. *Journal of Marriage and Family, 65*, 23–35.

Kulik, L. (2011). Developments in spousal power relations: Are we moving toward equality? *Marriage and Family Review, 47*, 419–435.

Lavee, Y., & A. Ben-Ari. (2007). Relationship of dyadic closeness with work-related stress: A daily diary study. *Journal of Marriage and Family, 69*, 1021–1035.

Morin, R., & D. Cohn. (2008, September 25). Women call the shots at home: Public mixed on gender roles in jobs, gender and power. Pew Research Center.

Nazarinia Roy, R. R., & S. Britt. (2011, November). The gendered nature of household tasks. Presentation at the annual meeting of the National Council on Family Relations, Orlando, FL.

National Marriage Project. http://www.virginia.edu/marriageproject/

Noonan, M. C., & M. E. Corcoran. (2004). The mommy track and partnership: Temporary delay or dead end? *Annals of the American Academy of Political and Social Science, 596*, 130–150.

Opree, S. J., & M. Kalmijn. (2012). Exploring causal effects of combining work and intergenerational support on depressive symptoms among middle-aged women. *Ageing and Society, 32*, 130–146.

Pazol, K., L. Warner, L. Gavin, W. M. Callaghan, A. M. Spitz, J. E. Anderson, et al. (2011). Vital signs—United States, 1991–2009. *Morbidity and Mortality Weekly Report, 60*, 414–420.

Perry-Jenkins, M., J. Smith, A. E. Goldberg, & J. Logan. (2011). Working-class jobs and new parents' mental health. *Journal of Marriage and Family, 73*, 1117–1132.

Presser, H. B. (2000). Nonstandard work schedules and marital instability. *Journal of Marriage and Family, 62,* 93–110.

Rose, K., & K. J. Elicker. (2008). Parental decision making about child care. *Journal of Family Issues, 29,* 1161–1179.

Rosen, E. (2006, February 12). Derailed on the mommy track? There's help to get going again. *The New York Times,* pp. 1–3.

Schlehofer, M. (2012). Practicing what we teach? An autobiographical reflection on navigating academia as a single mother. *Journal of Community Psychology, 40,* 112–128.

Schoen, R., N. M. Astone, K. Rothert, N. J. Standish, & Y. J. Kim. (2002). Women's employment, marital happiness, and divorce. *Social Forces, 81,* 643–662.

Sefton, B. W. (1998). The market value of the stay-at-home mother. *Mothering, 86,* 26–29.

Shapiro, M. (2007). Money: A therapeutic tool for couples therapy. *Family Process, 46,* 279–291.

Snyder, K. A. (2007). A vocabulary of motives: Understanding how parents define quality time. *Journal of Marriage and Family, 69,* 320–340.

Sonnenberg, S. J., C. B. Burgoyne, & D. A. Routh. (2011). Income disparity and norms relating to intra-household financial organisation in the UK: A dimensional analysis. *Journal of Socio-Economics, 40,* 573–582.

Spark, C. (2011). Gender trouble in town: Educated women eluding male domination, gender violence and marriage in PNG. *Asia Pacific Journal of Anthropology, 12,* 164–179.

Stanfield, J. B. (1998). Couples coping with dual careers: A description of flexible and rigid coping styles. *Social Science Journal, 35,* 53–62.

Stanley, S. M., & L. A. Einhorn. (2007). Hitting pay dirt: Comment on "Money: A therapeutic tool for couples therapy." *Family Process, 46,* 293–299.

Statistical Abstract of the United States. (2012, 131st ed.). Washington, DC: U.S. Census Bureau.

Stone, P. (2007). *Opting out?* Berkeley: University of California Press.

Sun, Y., & J. Sundell. (2011). Early daycare attendance increases the risk for respiratory infections and asthma for children. *Journal of Asthma, 48,* 790–796.

Treas, J., T. Van der Lippe, & T. C. Tai. (2011). The happy homemaker? Married women's well-being in cross-national perspective. *Social Forces, 90,* 111–132.

Vandell, D. L., J. Belsky, M. Burchinal, L. Steinberg, N. Vandergrift, & NICHD Early Child Care Research Network. (2010). Do effects of early child care extend to age 15 years? Results from the NICHD study of early child care and youth development. *Child Development, 81,* 737–756.

Vincent, C. D., & B. L. Neis. (2011). Work and family life: Parental work schedules and child academic achievement. *Community, Work and Family, 14,* 449–468.

Walker, S. K. (2000, November). Making home work: Family factors related to stress in family child care providers. Poster session presented at the annual conference of the National Council on Family Relations, Minneapolis, MN.

Warash, B. G., C. A. Markstrom, & B. Lucci. (2005). The Early Childhood Environment Rating Scale–Revised as a tool to improve child care centers. *Education, 126,* 240–250.

Whitehead, B. D., & D. Popenoe. (2004). *The state of our union: The social health of marriage in America.* The National Marriage Project—Rutgers University. http://www.marriage.rutgers.edu/

Zimmerman, K. J. (2012). The influence of a money management course on couples' relationship quality. *Journal of Financial Counseling and Planning* (in press).

Chapter 10

Ames, C. M., & W. V. Norman. (2012). Preventing repeat abortion in Canada: Is the immediate insertion of intrauterine devices postabortion a cost-effective option associated with fewer repeat abortions? *Contraception, 85,* 51–55.

Baden, A. L., & M. O. Wiley. (2007). Counseling adopted persons in adulthood: Integrating practice and research. *Counseling Psychologist, 35,* 868–879.

Begue, L. 2001. Social judgment of abortion: A black-sheep effect in a Catholic sheepfold. *Journal of Social Psychology, 141,* 640–50.

Berger, R., & M. Paul. (2008). Family secrets and family functioning: The case of donor assistance. *Family Process, 47,* 553–566.

Block, S. (2008, August 19). Adopting domestically can lower hurdles to claiming tax credit. *USA Today,* p. 3B.

Brandes, M., C. Hamilton, J. van der Steen, J. de Bruin, R. Bots, W. Nelen, et al. (2011). Unexplained infertility: Overall ongoing pregnancy rate and mode of conception. *Human Reproduction, 26,* 360–368.

Brown, J. D. (2008). Foster parents' perceptions of factors needed for successful foster placements. *Journal of Child and Family Studies, 17,* 538–555.

Department of Commerce et al. (2011). *Women in America: Indicators of social and economic well-being.* Retrieved September 8, 2011, from http://www.whitehouse.gov/sites/default/files/rss_viewer/Women_in_America.pdf

Finer, L. B., L. F. Frohwirth, L. A. Dauphinne, S. Singh, & A. M. Moore. (2005). Reasons U.S. women have abortions: Quantitative and qualitative reasons. *Perspectives on Sexual and Reproductive Health, 37,* 110–118.

Flower Kim, K. M. (2003, February 27). We are family. Paper presented at the 73rd annual meeting of the Eastern Sociological Society, Philadelphia, PA.

Garrett, T. M., H. W. Baillie, & R. M. Garrett. (2001). *Health care ethics* (4th ed.). Upper Saddle River, NJ: Prentice Hall.

Ge, X., M. N. Natsuaki, D. Martin, J. M. Neiderhiser, D. S. Shaw, L. Scaramella, et al. (2008). Bridging the divide: Openness in adoption and postadoption psychosocial adjustment among birth and adoptive parents issue; Public health perspectives on family interventions. American Psychological Association. Sage Periodicals Press.

Geller, P., C. Psaros, & S. L. Kornfield. (2010). Satisfaction with pregnancy loss aftercare: Are women getting what they want? *Archives of Women's Mental Health, 13,* 111–124.

Goldberg, A. E., L. A. Kinkler, H. B. Richardson, & J. B. Downing. (2011). Lesbian, gay, and heterosexual couples in open adoption arrangements: A qualitative study. *Journal of Marriage and Family, 73,* 502–518.

Gross, J., & W. Connors. (2007, June 4). Ethopia: Open doors for foreign adoptions. *The New York Times,* p. A1.

Guttmacher Institute. (2012). Abortion in the United States. Retrieved January 13, 2010, from http://www.guttmacher.org/media/press kits/abortion-US/statsandfacts.html

Haag, P. 2011. *Marriage confidential.* New York: Harper Collins.

Hollingsworth, L. D. (1997). Same race adoption among African Americans: A ten-year empirical review. *African American Research Perspectives, 13,* 44–49.

Huggins, R., & K. Gelles. (2011, December 19). Waiting for adoption. *USA Today,* p. A1.

Huh, N. S., & W. J. Reid. (2000). Intercountry, transracial adoption and ethnic identity: A Korean example. *International Social Work, 43,* 75–87.

Jones, R. K., & K. Kooistra. (2011). Abortion incidence and access to services in the United States, 2008. *Perspectives on Sexual and Reproductive Health, 43,* 41–50.

Katz, P., J. Showstack, J. F. Smith, R. D. Nachtigall, S. G. Millstein, H. Wing, et al. (2011). Costs of infertility treatment: Results from an 18-month prospective cohort study. *Fertility and Sterility, 95,* 915–921.

Kennedy, R. (2003). *Interracial intimacies.* New York, NY: Pantheon.

Kero, A., & A. Lalos. (2004). Reactions and reflections in men, 4 and 12 months post-abortion. *Journal of Psychosomatic Obstetrics and Gynecology, 25,* 135–143.

Knox, D., & S. Hall. (2010). Relationship and sexual behaviors of a sample of 2,922 university students. Unpublished data collected for this text. Department of Sociology, East Carolina University, and Department of Family and Consumer Sciences, Ball State University.

Lee, D. (2006). Device brings hope for fertility clinics. Retrieved February 22, 2006, from http://www.indystar.com/apps/pbcs.dll/article?AID=/20060221/BUSINESS/602210365/1003

Leung, P., S. Erich, & H. Kanenberg. (2005). A comparison of family functioning in gay/lesbian, heterosexual and special needs adoptions. *Children and Youth Services Review, 27*, 1031–1044.

Lipman, E. L., K. Georgiades, & M. Boyle. (2011). Young adult outcomes of children born to teen mothers: Effects of being born during their teen or later years. *Journal of the American Academy of Child and Adolescent Psychiatry, 50*, 232–241.

Livermore, M. M., & R. S. Powers. (2006). Unfulfilled plans and financial stress: Unwed mothers and unemployment. *Journal of Human Behavior in the Social Environment, 13*, 1–17.

Major, B., M. Appelbaum, & C. West. (2008, August 13). Report of the APA task force on mental health and abortion. Washington, DC: American Psychological Association.

McDermott, E., & H. Graham. (2005). Resilient young mothering: Social inequalities, late modernity and the problem of teenage motherhood. *Journal of Youth Studies, 8*, 59–79.

Meyer, A. S., L. M. McWey, W. McKendrick, & T. L. Henderson. (2010). Substance using parents, foster care, and termination of parental rights: The importance of risk factors for legal outcomes. *Children and Youth Services Review, 32*, 639–649.

Mollborn, S. (2007). Making the best of a bad situation: Material resources and teenage parenthood. *Journal of Marriage and Family, 69*, 92–104.

Nickman, S. L., A. A. Rosenfeld, P. Fine, J. C. MacIntyre, D. J. Pilowsky, R. Howe, et al. (2005). Children in adoptive families: Overview and update. *Journal of the American Academy of Child and Adolescent Psychiatry, 44*, 987–995.

Pew Research Center. (2008). The U.S. religious landscape survey reveals a fluid and diverse pattern of faith. Pew Forum on Religion and Public Life. Retrieved from http://pewresearch.org/pubs/743/united-states-religion

Rochman, B. (2009, February 16). The ethics of octuplets. *Time*, 43–44.

Rolfe, A. (2008). "You've got to grow up when you've got a kid": Marginalized young women's accounts of motherhood. *Journal of Community and Applied Social Psychology, 18*, 299–312.

Ross, R., D. Knox, M. Whatley, & J. N. Jahangardi. (2003). Transracial adoption: Some college student data. Paper presented at the 73rd annual meeting of the Eastern Sociological Society, Philadelphia, PA.

Sanders, J. 2012. Family of origin and interest in childbearing. *Marriage & Family Review, 48*, 20–39

Sandler, L. (2010, July 19). One and done. *Time*, 34–41.

Scheib, J. E., M. Riordan, & S. Rubin. (2005). Adolescents with openidentity sperm donors: Reports from 12–17-year-olds. *Human Reproduction, 20*, 239–252.

Schmidt, L., T. Sobotka, J. G. Bentzen, & A. Nyboe Andersen. (2012). Demographic and medical consequences of the postponement of parenthood. *Human Reproduction Update, 18*, 29–43.

Schwarz, E. B., R. Smith, J. Steinauer, M. F. Reeves, & A. B. Caughey. (2008). Measuring the effects of unintended pregnancy on women's quality of life. *Contraception, 78*, 204–210.

Scott, L. S. (2009). *Two is enough*. Berkeley, CA: Seal Press.

Simon, R. J., & R. M. Roorda. (2000). *In their own voices: Transracial adoptees tell their stories*. New York, NY: Columbia University Press.

Smith, T. (2007, August 5). Four kids is the new standard. National Public Radio.

Smock, P. J., & F. R. Greenland. (2010). Diversity in pathways to parenthood: Patterns, implications, and emerging research directions. *Journal of Marriage and Family, 72*, 576–593.

Statistical Abstract of the United States. (2012, 131st ed.). Washington, DC: U.S. Census Bureau.

Steinberg, J. R., & N. F. Russo. (2008). Abortion and anxiety: What's the relationship? *Social Science and Medicine, 67*, 238–242.

Stone, A. (2006, February 21). Drives to ban gay adoption heat up. *USA Today*, p. A1.

Thomas, K. A., & R. C. Tessler. (2007). Bicultural socialization among adoptive families: Where there is a will, there is a way. *Journal of Family Issues, 28*, 1189–1219.

Wang, W., & P. Taylor. (2011, March 9). For millennials, parenthood trumps marriage. Pew Research Center Publications. Retrieved from http://www.pewsocialtrends.org/2011/03/09/for-millennials-parenthood-trumps-marriage/

Wang, Y. A., D. Healy, D. Black, & E. A. Sullivan. (2008). Age-specific success rate for women undertaking their first assisted reproduction technology treatment using their own oocytes in Australia, 2002–2005. *Human Reproduction, 23*, 1533–1639.

Waterman, J., E. Nadeem, E. Paczkowski, J. Foster, J. Lange, T. Belin et al. (2011, November). Behavior problems and parenting stress for children adopted from foster care over the first five years of placement. National Council on Family Relations, Orlando, FL.

Wildsmith, E., K. B. Guzzo, & S. R. Hayford. (2010). Repeat unintended, unwanted and seriously mistimed childbearing in the United States. *Perspectives on Sexual and Reproductive Health, 42*, 14–23.

Wilson, H., & A. Huntington. (2006). Deviant mothers: The construction of teenage motherhood in contemporary discourse. *Journal of Social Policy, 35*, 59–76.

Wirtberg I., A. Möller, L. Hogström, S. E. Tronstad, & A. Lalos. (2007). Life 20 years after unsuccessful infertility treatment. *Human Reproduction, 22*, 598–604.

Zachry, E. M. (2005). Getting my education: Teen mothers' experiences in school before and after motherhood. *Teachers College Record, 107*, 2566–2598.

Chapter 11

Ali, M. M., & D. S. Dwyer. (2010). Social network effects in alcohol consumption among adolescents. *Addictive Behaviors, 35*, 337–342.

Baltazar, A. M., B. Johnson, D. McBride, G. Hopkins, & C. Vanderwaal. (2011, November 18). Parental influence on substance and inhalant use. Poster session presented at the meeting of the National Council on Family Relations, Orlando, FL.

Baumrind, D. (1966). Effects of authoritative parental control on child behavior. *Child Development, 37*, 887–907.

Becker, K. (2010). Risk behavior in adolescents: What we need to know. *European Psychiatry, 25*, 85.

Biehle, S. N. and K. D. Michelson. 2012. First-time parent's expectations about the division of childcare and play. *Journal of Family Psychology, 26*, 36–45.

Bock, J. D. (2000). Doing the right thing? Single mothers by choice and the struggle for legitimacy. *Gender and Society, 14*, 62–86.

Booth, C. L., K. A. Clarke-Stewart, D. L. Vandell, K. McCartney, & M. T. Owen. (2002). Child-care usage and mother-infant "quality time." *Journal of Marriage and Family, 64*, 16–26.

Bost, K. K., M. J. Cox, M. R. Burchinal, & C. Payne. (2002). Structural and supporting changes in couples' family and friendships networks across the transition to parenthood. *Journal of Marriage and Family, 64*, 517–531.

British Columbia Reproductive Mental Health Program. (2005). Reproductive mental health: Psychosis. Retrieved June 15, 2005, from http://www.bcrmh.com/disorders/psychosis.htm

Bronte-Tinkew, J., J. Carrano, A. Horowitz, & A. Kinukawa. (2008). Involvement among resident fathers and links to infant cognitive outcomes. *Journal of Family Issues, 29*, 1211–1231.

Byrd-Craven, J., B. J. Auer, D. A. Granger and A. R. Massey. (2012). The father-daughter dance: The relationship between father-daughter relationship quality and daughters' stress response. *Journal of Family Psychology, 26*, 87-94.

Carr, K. and T. R. Wang. (2012) "Forgiveness isn't a simple process: It's a vast undertaking": Negotiating and communicating forgivenss in nonvoluntary family relationships. *Journal of Family Communication 12,* 40-56.

Castrucci, B. C., J. F. Culhane, E. K. Chung, I. Bennett, & K. F. McCollum. (2006). Smoking in pregnancy: Patient and provider risk reduction behavior. *Journal of Public Health Management and Practice, 12,* 68–76.

Chen, J. J., T. Chen, & X. X. Zheng. (2012). Parenting styles and practices among Chinese immigrant mothers with young children. *Early Child Development and Care, 182,* 1–21.

Clarke, J. I. (2004, November). The overindulgence research literature: Implications for family life educators. Poster session presented at the annual meeting of the National Council on Family Relations, Orlando, FL.

Cornelius-Cozzi, T. (2002). Effects of parenthood on the relationships of lesbian couples. *PROGRESS: Family Systems Research and Therapy, 11,* 85–94.

Crosnoe, R., & S. E. Cavanagh. (2010). Families with children and adolescents: A review, critique, and future agenda. *Journal of Marriage and Family, 72,* 594–611.

Crutzen, R., E. Nijhuis, & S. Mujakovic. (2012). Negative associations between primary school children's perception of being allowed to drink at home and alcohol use. *Mental Health and Substance Use, 5,* 64–69.

Cui, M., F. D. Fincham, & B. Kay Pasley. (2008). Young adult romantic relationships: The role of parents' marital problems and relationship efficacy. *Personality and Social Psychology Bulletin, 34,* 1226–1235.

Dawson, C., D. Bredehoft, & J. I. Clarke. (2003). *How much is enough?* Boston, MA: De Capo Press.

Diamond, A., J. Bowes, & G. Robertson. (2006). Mothers' safety intervention strategies with toddlers and their relationship to child characteristics. *Early Child Development and Care, 176,* 271–284.

Doherty, W. J., & S. M. Craft. (2011). Single mothers raising children with "male-positive" attitudes. *Family Process, 50,* 63–76.

Facer, J., & R. Day. (2004, November). Explaining diminished marital satisfaction when parenting adolescents. Poster session presented at the annual meeting of the National Council on Family Relations, Orlando, FL.

Flouri, E., & A. Buchanan. (2003). The role of father involvement and mother involvement in adolescents' psychological well-being. *British Journal of Social Work, 33,* 399–406.

Forbes, E. E., & R. E. Dahl. (2012). Research review: Altered reward function in adolescent depression: what, when and how? *Journal of Child Psychology and Psychiatry, 53,* 3–15.

Galambos, N. L., E. T. Barker, & D. M. Almeida. (2003). Parents do matter: Trajectories of change in externalizing and internalizing problems in early adolescents. *Child Development, 74,* 578–595.

Gavin, L. E., M. M. Black, S. Minor, Y. Abel, & M. E. Bentley. (2002). Young, disadvantaged fathers' involvement with their infants: An ecological perspective. *Journal of Adolescent Health, 31,* 266–276.

Gelabert, E., S. Subirà, L.García-Esteve, P. Navarro, A. Plaza, E. Cuyàs et al. (2012). Perfectionism dimensions in major postpartum depression. *Journal of Affective Disorders, 136,* 17–25.

Goldberg, A. E., J. Z. Smith, & D. A. Kashy. (2010). Preadoptive factors predicting lesbian, gay, and heterosexual couples' relationship quality across the transition to adoptive parenthood. *Journal of Family Psychology, 24,* 221–232.

Hammarberg, K., J. R. Fisher, & K. H. Wynter. (2008). Psychological and social aspects of pregnancy, childbirth and early parenting after assisted conception: A systematic review. *Human Reproduction, 14,* 395–415.

Harris-McKoy, D., & M. Cui. (2011, November). Parental control and delinquent behavior: A longitudinal study from adolescence to emerging adulthood. Poster session presented at the annual meeting of the National Council on Family Relations, Orlando, FL.

Hersh, J., J. Sawada, & B. J. Willoughby. (2011, November). Helicopter parenting and family formation values in emerging adulthood. Paper presented at the 73rd annual meeting of the National Council on Family Relations, Orlando, FL.

Huggins, R., & S. Ward. (2011, October 12). Time working moms and dads say they spend parenting. *USA Today,* p. A1.

Jordan, E. F., & M. E. Curtner-Smith. (2011, November). Multiple indicators of corporal punishment and current attachment to mother. Paper presented at the 73rd annual meeting of the National Council on Family Relations, Orlando, FL.

Keim, R. E., & A. L. Jacobson. (2011). *Wisdom for parents: Key ideas from parent educators.* Ontario, Canada: de Sitter Publications.

Knog, G., D. Camenga, & S. Krishnan-Sarin. (2012). Parental influence on adolescent smoking cessation: Is there a gender difference? *Addictive Behaviors, 37,* 211–216.

Knox, M., K. Burkhart, & T. Howe. (2011). Effects of the ACT Raising Safe Kids Parenting Program on children's externalizing problems. *Family Relations, 60,* 491–503.

Kolko, D. J., L. D. Dorn, O. Bukstein, & J. D. Burke. (2008). Clinically referred ODD children with or without CD and healthy controls: Comparisons across contextual domains. *Journal of Child and Family Studies, 17,* 714–734.

Kouros, C. D., C. E. Merrilees, & E. M. Cummings. (2008). Marital conflict and children's emotional security in the context of parental depression. *Journal of Marriage and Family, 70,* 684–697.

Lee, J. (2008). "A Kotex and a smile": Mothers and daughters at menarche. *Journal of Family Issues, 29,* 1325–1347.

Levy, S. (2009). *Paul Newman: A life.* New York, NY: Harmony Books.

Louv, R. (2006). *Last child in the woods.* Chapel Hill, NC: Algonquin Books.

Mashoa, S. W., D. Chapmana, & M. Ashbya. (2010). The impact of paternity and marital status on low birth weight and preterm births. *Marriage and Family Review, 46,* 243–256.

Mayall, B. (2002). *Toward a sociology of childhood.* Philadelphia, PA: Open University Press.

McClain, L. R. (2011). Better parents, more stable partners: Union transitions among cohabiting parents. *Journal of Marriage and Family, 73,* 889–901.

McKinney, C., & K. Renk. (2008). Differential parenting between mothers and fathers: Implications for late adolescents. *Journal of Family Issues, 29,* 806–827.

McLanahan, S. S. (1991). The long-term effects of family dissolution. In Brice J. Christensen (Ed.), *When families fail: The social costs* (pp. 5–26). New York, NY: University Press of America for the Rockford Institute.

McLanahan, S. S., & K. Booth. (1989). Mother-only families: Problems, prospects, and politics. *Journal of Marriage and Family, 51,* 557–580.

Nelson, M. C. (2010). *Parents out of control.* New York, NY: New York Press.

Parker, K. The boomerang generation. Pew Research Center. Retrieved on March 31, 2012. *http://www.pewsocialtrends.org/2012/03/15/the-boomerang*-generation/2/#who-are-the-boomerang-kids

Pew Research Center. (2007, May 2). Motherhood today: Tougher challenges, less success. Retrieved from http://pewresearch.org/pubs/468/motherhood

Pinheiro, R. T., R. A. da Silva, P. V. S. Magalhaes, B. L. Hortam, & K. A. T. Pinheiro. (2008). Two studies on suicidality in the postpartum. *Acta Psychiatrica Scandinavica, 118,* 160–162.

Pong, S. L., & B. Dong. (2000). The effects of change in family structure and income on dropping out of middle and high school. *Journal of Family Issues, 21,* 147–169.

Qing, M., Z. Li-xia, & S. Xiao-yin. (2011). A comparison of postnatal depression and related factors between Chinese new mothers and fathers. *Journal of Clinical Nursing, 20,* 645–652.

Rapoport, B., & C. Le Bourdais. (2008). Parental time and working schedules. *Journal of Population Economics, 21,* 903–933.

Ready, B., L. Asare, & E. Long. (2011, November). Factors that impact a father's philosophy on parenting. Presentation at the annual meeting of the National Council on Family Relations, Orlando, FL.

Schindler, H. S. (2010). The importance of parenting and financial contributions in promoting fathers' psychological health. *Journal of Marriage and Family, 72*, 318–332.

Schoppe-Sullivan, S. J., G. L. Brown, E. A. Cannon, S. C. Mangelsdorf, & M. S. Sokolowski. (2008). Maternal gatekeeping, coparenting quality, and fathering behavior in families with infants. *Journal of Family Psychology, 22*, 389–397.

Shook, S. E., D. J. Jones, R. Forehand, S. Dorsey, & G. Brody. (2010). The mother–coparent relationship and youth adjustment: A study of African American single-mother families. *Journal of Family Psychology, 24*, 243–251.

Slinger, M. R., & D. J. Bredehoft. (2010, November 3–5). Relationships between childhood overindulgence and adult attitudes and behavior. Poster session presented at the annual meeting of the National Council on Family Relations, Minneapolis, MN.

Solem, M., K. Christophersen, & M. Martinussen. (2011). Predicting parenting stress: Children's behavioral problems and parents' coping. *Infant and Child Development, 20*, 162–180.

Stanley, S. M., & H. J. Markman. (1992). Assessing commitment in personal relationships. *Journal of Marriage and Family, 54*, 595–608.

Statistical Abstract of the United States. (2012, 131st ed.). Washington, DC: U.S. Census Bureau.

Suitor, J. J., & K. Pillemer. (2007). Mothers' favoritism in later life: The role of children's birth order. *Research on Aging, 29*, 32–42.

Sulloway, F. J. (1996). *Born to rebel: Birth order, family dynamics, and creative lives.* New York, NY: Vintage Books.

———. (2007). Birth order and intelligence. *Age and Intelligence, 316*, 1711–1721.

Talwar, V., & K. Lee. (2008). Social and cognitive correlates of children's lying behavior. *Child Development, 79*, 866–881.

Tan, T. X., L. A. Camras, H. Deng, M. Zhang, & Z. Lu. (2012). Family stress, parenting styles, and behavioral adjustment in preschool-age adopted Chinese girls. *Early Childhood Research Quarterly, 27*, 128–136.

Thomas, P. A., E. M. Krampe, & R. R. Newton. (2008). Father presence, family structure, and feelings of closeness to the father among adult African American children. *Journal of Black Studies, 38*, 529–541.

Tucker, C. J., S. M. McHale and C. A. Foster. 2003. Dimensions of mothers' and fathers' differential treatment of siblings: Links with adolescents' sex-typed personal qualities. *Family Relations, 52*, 82-89.

Twenge, J. M., W. K. Campbell, & C. A. Foster. (2003). Parenthood and marital satisfaction: A meta-analytic review. *Journal of Marriage and Family, 65*, 574–583.

Van den Heuvel, A., R. J. J. M. Van den Eijnden, A. Van Rooij, & D. Van de Mheen. (2012). Meeting online contacts in real life among adolescents: The predictive role of psychosocial well-being and Internet-specific parenting. *Computers in Human Behavior, 28*, 465–472.

Walcheski, M. J., & D. J. Bredehoft. (2010, November 3–5). Exploring the relationship between overindulgence and parenting styles. Poster session presented at the annual meeting of the National Council on Family Relations, Minneapolis, MN.

Ward, R. A., & G. D. Spitze. (2007). Nestleaving and coresidence by young adult children: The role of family relations. *Research on Aging, 29*, 257–271.

Wilson, E. K., B. T. Dalberth, H. P. Koo, & J. C. Gard. (2010). Parents' perspectives on talking to preteenage children about sex. *Perspectives on Sexual and Reproductive Health, 42*, 56–64.

Wilson, K. R., S. S. Havighurst and A. E. Harley. 2012. Tuning in to Kids: An effectiveness trial of a parenting program targeting emotion socialization of preschoolers. *Journal of Family Psychology, 26*, 56-65.

Chapter 12

Alam, R., M. Barrera, N. D'Agostino, D. B. Nicholas, & G. Schneiderman. (2012). Bereavement experiences of mothers and fathers over time after the death of a child due to cancer. *Death Studies, 36*, 1–22.

Allen, E. S., G. K. Rhoades, S. M. Stanley, H. J. Markman, et al. (2008). Premarital precursors of marital infidelity. *Family Process, 47*, 243–260.

Bagarozzi, D. A. (2008). Understanding and treating marital infidelity: A multidimensional model. *American Journal of Family Therapy, 36*, 1–17.

Barnes, H., D. Knox, & J. Brinkley. (2012, March). CHEATING: Gender differences in reactions to discovery of a partner's cheating. Paper presented at the annual meeting of the Southern Sociological Society, New Orleans, LA.

Barton, A. L., & M. S. Kirtley. (2012). Gender differences in the relationships among parenting styles and college student mental health. *Journal of American College Health, 60*, 21–26.

Bermant, G. (1976). Sexual behavior: Hard times with the Coolidge Effect. In M. H. Siegel & H. P. Zeigler (Eds.), *Psychological research: The inside story.* New York, NY: Harper and Row.

Black, K., & M. Lobo. (2008). A conceptual review of family resilience factors. *Journal of Family Nursing, 14*, 1–33.

Bodenmann, G., D. C. Atkins, M. Schär, & V. Poffet (2010). The association between daily stress and sexual activity. *Journal of Family Psychology, 24*, 271–279.

Burke, M. L., G. G. Eakes, & M. A. Hainsworth. (1999). Milestones of chronic sorrow: Perspectives of chronically ill and bereaved persons and family caregivers. *Journal of Family Nursing, 5*, 387–384.

Burr, W. R., & S. R. Klein. (1994). *Reexamining family stress: New theory and research.* Thousand Oaks, CA: Sage.

Butterworth, P., & B. Rodgers. (2008). Mental health problems and marital disruption: Is it the combination of husbands' and wives' mental health problems that predicts later divorce? *Social Psychiatry and Psychiatric Epidemiology, 43*, 758–764.

Ciro, C. A., K. J. Ottenbacher, J. E. Graham, S. Fisher, I. Berges, & G. V. Ostir. (2012). Patterns and correlates of depression in hospitalized older adults. *Archives of Gerontology and Geriatrics, 54*, 202–205.

Collins, J. (1998). *Singing lessons: A memoir of love, loss, hope, and healing.* New York, NY: Pocket Books.

Confer, J. C., & M. D. Cloud. (2011). Sex differences in response to imagining a partner's heterosexual or homosexual affair. *Personality and Individual Differences, 50*, 129–134.

Cramer, R. E., R. E. Lipinski, J. D. Meteer, and J. A. Houska. 2008. Sex differences in subjective distress to unfaithfulness: Testing competing evolutionary and violation of infidelity expectations hypotheses. *The Journal of Social Psychology, 148*, 389–406.

Dean, C. J. (2011). Psychoeducation: A first step to understanding infidelity-related systemic trauma and grieving. *Family Journal, 19*, 15–21.

De Castro, S., & J. T. Guterman. (2008). Solution-focused therapy for families with suicide. *Journal of Marital and Family Therapy, 34*, 93–107.

Derby, K., B. Easterling, & D. Knox. (2012). Snooping in romantic relationships. *College Student Journal* (in press).

Dethier, M., C. Counerotte, & S. Blairy. (2011). Marital satisfaction in couples with an alcoholic husband. *Journal of Family Violence, 26*, 151–162.

Doosje, S., J. Landsheer, & M. Goede. (2012). Humorous coping scales and their fit to a stress and coping framework. *Quality and Quantity, 46*, 267–279.

Dotson-Blake, K., D. Knox, & A. R. Holman. (2010, Fall). Reaching out: College student perceptions of counseling. *Professional Issues in Counseling.* Retrieved September 13, 2011, from http://www.shsu.edu/~piic/CollegeStudentPerceptions.htm

Druckerman, P. (2007). *Lust in translation.* New York, NY: Penguin Group.

Ellison, C. G., A. K. Henderson, N. D. Glenn, & K. E. Harkrider. (2011). Sanctification, stress, and marital quality. *Family Relations, 60*, 404–420.

Elmslie, B., & E. Tebaldi. (2008). So, what did you do last night? The economics of infidelity. *Kyklos, 61*, 391–406.

Enright, E. (2004). A house divided. *AARP The Magazine,* July/August, 60.

Euser, E. M., M. H. van Ijzendoorn, P. Prinzie, & M. J. Bakermans-Kranenburg. (2010). Prevalence of child maltreatment in the Netherlands. *Child Maltreatment, 15*, 5–17.

Falba, T. M. T., J. L. Sindelar, & W. T. Gallo. (2005). The effect of involuntary job loss on smoking intensity and relapse. *Addiction, 100*, 1330–1339.

Field, T., M. Diego, M. Pelaez, O. Deeds and J. Delgado. 2012. Depression and related problems of university students. *College Student Journal, 46*, 193-202.

Fisher, A. D., E. Bandini, G. Corona, M. Monami, M. Cameron Smith, C. Melani et al. (2012). Stable extramarital affairs are breaking the heart. *International Journal of Andrology, 35*, 11–17.

Friedrich, R. M., S. Lively, & L. M. Rubenstein. (2008). Siblings' coping strategies and mental health services: A national study of siblings of persons with schizophrenia. *Psychiatric Services, 59*, 261–273.

Goode, E. (1999, July 17). New study finds middle age is prime of life. *The New York Times,* p. D6.

Gould, T. E., & A. Williams. (2010). Family homelessness: An investigation of structural effects. *Journal of Human Behavior in the Social Environment, 20*, 170–192.

Green, M., & M. Elliott. (2010). Religion, health, and psychological well-being. *Journal of Religion and Health, 49*, 149–163.

Hall, J. H., W. Fals-Stewart, & F. D. Fincham. (2008). Risky sexual behavior among married alcoholic men. *Journal of Family Psychology, 22*, 287–299.

Hughes, M. (2011). Hey, we love animals! *Industrial Engineer, 43*, 6.

Insel, T. R. (2008). Assessing the economic costs of serious mental illness. *American Journal of Psychiatry, 165*, 663–666.

Knox, D., & S. Hall. (2010). Relationship and sexual behaviors of a sample of 2,922 university students. Unpublished data collected for this text. Department of Sociology, East Carolina University, and Department of Family and Consumer Sciences, Ball State University.

Kottke, J. (2008). The Eliot Spitzer affair and the business of sex. Retrieved December 6, 2008, from http://kottke.org/08/03/the-eliot-spitzer-affair-and-the-business-of-sex

Linquist, L., & C. Negy. (2005). Maximizing the experiences of an extrarelational affair: An unconventional approach to a common social convention. *Journal of Clinical Psychology/In Session, 61*, 1421–1428.

Mahoney, D. (2005). Mental illness prevalence high, despite advances. *Clinical Psychiatry News, 33*, 1–2.

Mattingly, M. J., & K. E. Smith. (2010). Changes in wives' employment when husbands stop working: A recession–prosperity comparison. *Family Relations, 59*, 343–357.

Melhem, N. M., D. A. Brent, M. Ziegler, S. Iyengar et al. (2007). Familial pathways to early-onset suicidal behavior: Familial and individual antecedents of suicidal behavior. *American Journal of Psychiatry, 164*, 1364–1371.

Merline, A. C., J. E. Schulenberg, P. M. O'Malley, J. G. Bachman, & L. D. Johnston. (2008). Substance use in marital dyads: Premarital assortment and change over time. *Journal of Studies on Alcohol and Drugs, 69*, 352–365.

Merrill, J., & D. Knox. (2010). *Finding love from 9 to 5: Trade secrets of an office love.* Santa Barbara, CA: Praeger.

Miers, D., D. Abbott and P. R. Springer. 2012. A phenomenological study of family needs following the suicide of a teenager. *Death Studies, 36*, 118-133.

Moens, E., & C. Braet. (2012). Training parents of overweight children in parenting skills: A 12-month evaluation. *Behavioral and Cognitive Psychology, 40*, 1–18.

Mordoch, E., & W. A. Hall. (2008). Children's perceptions of living with a parent with a mental illness: Finding the rhythm and maintaining the frame. *Qualitative Health Research, 18*, 1127–1135.

Morell, V. (1998). A new look at monogamy. *Science, 281*, 1982.

Neuman, M. G. (2008). *The truth about cheating: Why men stray and what you can do to prevent it.* New York, NY: Wiley.

Olson, M. M., C. S. Russell, M. Higgins-Kessler, & R. B. Miller. (2002). Emotional processes following disclosure of an extramarital affair. *Journal of Marital and Family Therapy, 28*, 423–434.

Peilian, C., S. K. M. Tsang, S. C. Kin, X. Xiaoping, P. S. F. Yip, T. C. Yee et al. (2011). Marital satisfaction of Chinese under stress: Moderating effects of personal control and social support. *Asian Journal of Social Psychology, 14*, 15–25.

Phillips, L. J., J. Edwards, N. McMurray, & S. Francey. (2012). Comparison of experiences of stress and coping between young people at risk of psychosis and a nonclinical cohort. *Behavioural and Cognitive Psychotherapy, 40*, 69–88.

Pietrzak, R. H., C. A. Morgan, & S. M. Southwick. (2010). Sleep quality in treatment-seeking veterans of Operations Enduring Freedom and Iraqi Freedom: The role of cognitive coping strategies and unit cohesion. *Journal of Psychosomatic Research, 69*, 441–448.

Schneider, J. P. (2000). Effects of cybersex addiction on the family: Results of a survey. *Sexual Addiction and Compulsivity, 7*, 31–58.

———. (2003). The impact of compulsive cybersex behaviors on the family. *Sexual and Relationship Therapy, 18*, 329–355.

Schum, T. R. (2007). Dave's dead! Personal tragedy leading a call to action in preventing suicide. *Ambulatory Pediatrics, 7*, 410–412.

Sharpe, L., & L. Curran. (2006). Understanding the process of adjustment to illness. *Social Science and Medicine, 62*, 1153–1166.

Shin, S. H., H. G. Hong, & S. M. Jeon. (2012). Personality and alcohol use: The role of impulsivity. *Addictive Behaviors, 37*, 102–107.

Smedema, S. M., D. Catalano, & D. J. Ebener. (2010). The relationship of coping, self-worth, and subjective well-being: A structural equation model. *Rehabilitation Counseling Bulletin, 53*, 131–142.

Statistical Abstract of the United States. (2012, 131st ed.). Washington, DC: U.S. Census Bureau.

Stefansson, J., P. Nordström, & J. Jokinen. (2012). Suicide intent scale in the prediction of suicide. *Journal of Affective Disorders, 136*, 167–171.

Sweet, S., & P. Moen. (2012). Dual earners preparing for job loss: Agency, linked lives, and resilience. *Work and Occupations, 39*, 35–70.

Szabo, A., S. E. Ainsworth, & P. K. Danks. (2005). Experimental comparison of the psychological benefits of aerobic exercise, humor, and music. *International Journal of Humor Research, 18*, 235–246.

Taliaferro, L. A., B. A. Rienzo, M. D. Miller, R. M. Pigg, & V. J. Dodd. (2008). High school youth and suicide risk: Exploring protection afforded through physical activity and sport participation. *Journal of School Health, 78*, 545–556.

Teitler, J. O., & N. E Reichman. (2008). Mental illness as a barrier to marriage among unmarried mothers. *Journal of Marriage and Family, 70*, 772–783.

Tetlie, T., N. Eik-Nes, T. Palmstierna, P. Callaghan, & J. A Nøttestad. (2008). The effect of exercise on psychological and physical health outcomes: Preliminary results from a Norwegian forensic hospital. *Journal of Psychosocial Nursing and Mental Health Services, 46*, 38–44.

Termini, K. A. (2006, April 21). Reducing the negative psychological and physiological effects of chronic stress. Paper presented at the 4th annual ECU Research and Creative Activities Symposium, East Carolina University, Greenville, NC.

Tomlinson, K. L., & S. A. Brown. (2012). Self-medication or social learning? A comparison of models to predict early adolescent drinking. *Addictive Behaviors, 37*, 179–186.

Waller, M. R. (2008). How do disadvantaged parents view tensions in their relationships? Insights for relationship longevity among at-risk couples. *Family Relations, 57*, 128–143.

Chapter 13

Anderson, K. M., & E. Bang. (2012). Assessing PTSD and resilience for females who during childhood were exposed to domestic violence. *Child and Family Social Work, 17*, 55–65.

Becker, K. D., J. Stuewig, & L. A. McCloskey. (2010). Traumatic stress symptoms of women exposed to different forms of childhood victimization and intimate partner violence. *Journal of Interpersonal Violence, 25*, 1699–1715.

Brownridge, D. A. (2010). Does the situational couple violence–intimate terrorism typology explain cohabitors' high risk of intimate partner violence? *Journal of Interpersonal Violence, 25*, 1264–1283.

Burke, S., M. Wallen, K. Vail-Smith, & D. Knox. (2011). Using technology to control intimate partners: An exploratory study of college undergraduates. *Computers in Human Behavior, 27*, 1162–1167.

Busby, D. M., T. B. Holman, & E. Walker. (2008). Pathways to relationship aggression between adult partners. *Family Relations, 57*, 72–83.

Cattaneo, L. B., S. Cho, & S. Botuck. (2011). Describing intimate partner stalking over time: An effort to inform victim-centered service provision. *Journal of Interpersonal Violence, 26*, 3428–3454.

CDC (Centers for Disease Control and Prevention). (2011). Sexual violence, stalking, and intimate partner violence widespread in the U.S. [Press release]. Retrieved December 14, 2011, from http://www.cdc.gov/media/releases/2011/p1214_sexual_violence.html

Chapleau, K. M., D. L. Oswald, & B. L. Russell. (2008). Male rape myths: The role of gender, violence, and sexism. *Journal of Interpersonal Violence, 23*, 600–615.

Daigle, L. E., B. S. Fisher, & F. T. Cullen. (2008). The violent and sexual victimization of college women: Is repeat victimization a problem? *Journal of Interpersonal Violence, 23*, 1296–1313.

DiLillo, D., S. A. Hayes-Skelton, M. A. Fortier, A. R. Perry, S. E. Evans, T. L. Messman Moore et al. (2010). Development and initial psychometric properties of the Computer Assisted Maltreatment Inventory (CAMI): A comprehensive self-report measure of child maltreatment history. *Child Abuse and Neglect, 34*, 305–317.

Eaton, L., S. Kalichman, D. Skinner, M. Watt, D. Pieterse, & E. Pitpitan. (2012). Pregnancy, alcohol intake, and intimate partner violence among men and women attending drinking establishments in a Cape Town, South Africa, Township. *Journal of Community Health, 37*, 208–216.

Eke, A., N. Hilton, G. Harris, M. Rice, & R. Houghton. (2011). Intimate partner homicide: Risk assessment and prospects for prediction. *Journal of Family Violence, 26*, 211–216.

Fernandez-Montalvo, J., J. J. Lopez-Goni, & A. Arteaga. (2012). Violent behaviors in drug addiction: Differential profiles of drug-addicted patients with and without violence problems. *Journal of Interpersonal Violence, 27*, 142–157.

Few, A. L., & K. H. Rosen. (2005). Victims of chronic dating violence: How women's vulnerabilities link to their decisions to stay. *Family Relations, 54*, 265–279.

Flack, W. F., Jr., M. L. Caron, S. J. Leinen, K. G. Breitenbach, A. M. Barber, E. N. Brown et al. (2008). "The red zone": Temporal risk for unwanted sex among college students. *Journal of Interpersonal Violence, 23*, 1177–1196.

Follingstad, D. R., & M. Edmundson. (2010). Is psychological abuse reciprocal in intimate relationships? Data from a national sample of American adults. *Journal of Family Violence, 25*, 495–508.

Gidycz, C. A., A. V. Wynsberghe, & K. M. Edwards. (2008). Prediction of women's utilization of resistance strategies in a sexual assault situation: A prospective study. *Journal of Interpersonal Violence, 23*, 571–588.

Gormley, B., & F. G. Lopez. (2010). Psychological abuse perpetration in college dating relationships: Contributions of gender, stress, and adult attachment orientations. *Journal of Interpersonal Violence, 25*, 204–218.

Gottman, J. (2007). The mathematics of love. Retrieved August 23, 2007, from http://www.edge.org/3rd_culture/gottman05/gottman05_index.html

Hall, S. and D. Knox. (2012). Double victims: Sexual coercion by a dating partner and a stranger. *Journal of Aggression, Maltreatment & Trauma.* In press.

Katz, J., & L. Myhr. (2008). Perceived conflict patterns and relationship quality associated with verbal sexual coercion by male dating partners. *Journal of Interpersonal Violence, 23*, 798–804.

Knox, D., & S. Hall. (2010). Relationship and sexual behaviors of a sample of 2,922 university students. Unpublished data collected for this text. Department of Sociology, East Carolina University, and Department of Family and Consumer Sciences, Ball State University.

Kress, V. E., J. J. Protivnak, & L. Sadlak. (2008). Counseling clients involved with violent intimate partners: The mental health counselor's role in promoting client safety. *Journal of Mental Health Counseling, 30*, 200–211.

Larsen, C. D., J. G. Sandberg, J. M. Harper, & R. Bean. (2011). The effects of childhood abuse on relationship quality: Gender differences and clinical implications. *Family Relations, 60*, 435–445.

Lawyer, S., H. Resnick, V. Bakanic, T. Burkett, & D. Kilpatrick. (2010). Forcible, drug-facilitated, and incapacitated rape and sexual assault among undergraduate women. *Journal of American College Health, 58*, 453–460.

LeCouteur, A., & M. Oxlad. (2011). Managing accountability for domestic violence: Identities, membership categories and morality in perpetrators' talk. *Feminism and Psychology, 21*, 5–28.

McGee, H., M. O'Higgins, R. Garavan, & R. Conroy. (2011). Rape and child sexual abuse: What beliefs persist about motives, perpetrators, and survivors? *Journal of Interpersonal Violence, 26*, 3580–3593.

Oswald, D. L., & B. L. Russell. (2006). Perceptions of sexual coercion in heterosexual dating relationships: The role of aggressor gender and tactics. *Journal of Sex Research, 43*, 87–98.

Palmer, R. S., T. J. McMahon, B. J. Rounsaville, & S. A. Ball. (2010). Coercive sexual experiences, protective behavioral strategies, alcohol expectancies and consumption among male and female college students. *Journal of Interpersonal Violence, 25*, 1563–1578.

Pendry, P., F. Henderson, J. Antles, & E. Conlin. (2011, November 18). Parents' use of everyday conflict tactics in the presence of children: Predictors and implications for child behavior. Poster session presented at the annual meeting of the National Council on Family Relations, Orlando, FL.

Rhatigan, D. L., & A. M. Nathanson. (2010). The role of female behavior and attributions in predicting behavioral responses to hypothetical male aggression. *Violence Against Women, 16*, 621–637.

Robboy, J., & K. G. Anderson. (2011). Intergenerational child abuse and coping. *Journal of Interpersonal Violence, 26*, 3526–3541.

Rothman, E., & J. Silverman. (2007). The effect of a college sexual assault prevention program on first year students' victimization rates. *Journal of American College Health, 55*, 283–290.

Russell, D., K. W. Springer, & E. A. Greenfield. (2010). Witnessing domestic abuse in childhood as an independent risk factor for depressive symptoms in young adulthood. *Child Abuse and Neglect, 34*, 448–453.

Shamai, M., & E. Buchbinder. (2010). Control of the self: Partner-violent men's experience of therapy. *Journal of Interpersonal Violence, 25*, 1338–1362.

Spitzberg, B. H., & W. R. Cupach. (2007). Cyberstalking as (mis)matchmaking. In M. T. Whitty, A. J. Baker, & J. A. Inman (Eds.), *Online Matchmaking* (pp. 127–146). New York, NY: Palgrave Macmillan.

Stockl, H., L. Hertlein, I. Himsl, M. Delius, U. Hasbargen, K. Friese et al. (2012). Intimate partner violence and its association with pregnancy loss and pregnancy planning. *Acta Obstetricia et Gynecologica Scandinavica, 91*, 128–133.

Swan, S. C., L. J. Gambone, J. E. Caldwell, T. P. Sullivan, & D. L Snow. (2008). A review of research on women's use of violence with male intimate partners. *Violence and Victims, 23*, 301–315.

Zinzow, H. M., H. S. Resnick, A. B. Amstadter, J. L. McCauley, K. J. Ruggiero, & D. G. Kilpatrick. (2010). Drug- or alcohol-facilitated, incapacitated, and forcible rape in relationship to mental health among a national sample of women. *Journal of Interpersonal Violence, 25,* 2217–2236.

Chapter 14

Adler-Baeder, F., A. Robertson, & D. G. Schramm. (2010). Conceptual framework for marriage education programs for stepfamily couples with considerations for socioeconomic context. *Marriage and Family Review, 46,* 300–322.

Amato, P. R. (2010). Research on divorce: Continuing trends and new developments. *Journal of Marriage and Family, 72,* 650–666.

Amato, P. R., A. Booth, D. R. Johnson, & S. F. Rogers. (2007). *Alone together: How marriage in America is changing.* Cambridge, MA: Harvard University Press.

Amato, P. R., J. B. Kane & S. James. (2011). Reconsidering the "Good Divorce." *Family Relations, 60,* 511–524.

Baker, A. J. L. (2006). Patterns of parental alienation syndrome: A qualitative study of adults who were alienated from a parent as a child. *American Journal of Family Therapy, 34,* 63–78.

Baker, A. J. L., & D. Darnall. (2007). A construct study of the eight symptoms of severe parental alienation syndrome: A survey of parental experiences. *Journal of Divorce and Remarriage, 47,* 55–62.

Benton, S. D. (2008, November 10). Divorce mediation [Lecture]. East Carolina University, Greenville, NC.

Bray, J. H., & J. Kelly. (1998). *Stepfamilies: Love, marriage and parenting in the first decade.* New York, NY: Broadway Books.

Breed, R., D. Knox, & M. Zusman. (2007). "Hell hath no fury": Legal consequences of having Internet child pornography on one's computer. Southern Sociological Society, Atlanta, GA.

Brimhall, A. S. and M. L. Engblom-Deglmann. (2011). Starting over: A tentative theory exploring the effects of past relationships on post-bereavement remarried couples. *Family Process, 50,* 47–62.

Brown, S. L. and I. Fen Lin. (2012, March). The Gray Divorce Revolution: Rising divorce among middle aged and older adults, 1990-2009. National Center for Marriage and Family Research. Bowling Green State University. Working Paper Series.

Cashmore, J., P. Parkinson, & A. Taylor. (2008). Overnight stays and children's relationships with resident and nonresident parents after divorce. *Journal of Family Issues, 29,* 707–714.

Cherlin, A. J. (2009). *The marriage-go-round: The state of marriage and the family in America today.* New York, NY: Knopf.

Chiu, H., & D. Busby. (2010, November 3–5). Parental influence in adult children's marital relationships. Poster session presented at the annual meeting of the National Council on Family Relations, Minneapolis, MN.

Clarke, S. C., & B. F. Wilson. (1994). The relative stability of remarriages: A cohort approach using vital statistics. *Family Relations, 43,* 305–310.

Clarke-Stewart, A., & C. Brentano. (2006). *Divorce: Causes and consequences.* New Haven, CT: Yale University Press.

Coltrane, S., & M. Adams. (2003). The social construction of the divorce "problem": Morality, child victims, and the politics of gender. *Family Relations, 52,* 363–372.

Covizzi, I. (2008). Does union dissolution lead to unemployment? A longitudinal study of health and risk of unemployment for women and men undergoing separation. *European Sociological Review, 24,* 347–362.

Cui, M., F. D. Fincham, & J. A. Durtschi. (2011). The effect of parental divorce on young adults' romantic relationship dissolution: What makes a difference? *Personal Relationships, 18,* 410–426.

Dixon, L. J., K. C. Gordon, N. N. Frousakis, & J. A. Schumm. (2012). Expectations and the marital quality of participants of a marital enrichment seminar. *Family Relations, 61,* 75–89.

Doyle, M., C. O'Dywer, & V. Timonen. (2010). "How can you just cut off a whole side of the family and say move on?" The reshaping of paternal grandparent–grandchild relationships following divorce or separation in the middle generation. *Family Relations, 59,* 587–598.

Economist. (2005, February 10). "Yes, I really do." Retrieved from http://www.economist.com/node/3646161

Enright, E. (2004, July/August). A house divided. *AARP The Magazine,* 60.

Eubanks Fleming, C. J., & J. V. Córdova. (2012). Predicting relationship help seeking prior to a marriage checkup. *Family Relations, 61,* 90–100.

Fackrell, T. A., F. O. Paulsen, D. M. Busby, & D. C. Dollahite. (2011). Coming to terms with parental divorce: Associations with marital outcomes and the role of gender and religiosity. *Journal of Divorce and Remarriage, 52,* 444–463.

Faircloth, M., D. Knox and J. Brinkley. (2012, March). The good, the bad and technology mediated communication in romantic relationships. Paper, Southern Sociology Society Annual Meeting, New Orleans.

Finley, G. E. (2004). Divorce inequities. *NCFR Family Focus Report,* 49(3), F7.

Fish, J., A. Brackett and D. Knox. (2012). *Breaking up in emerging adulthood: Reasons and reactions.* Unpublished paper.

Ganong, L. H., & M. Coleman. (1994). *Remarried family relationships.* Thousand Oaks, CA: Sage.

Ganong, L. H., M. Coleman, & T. Jamison. (2011). Patterns of stepchild–stepparent relationship development. *Journal of Marriage and Family, 73,* 396–413.

Gardner, J., & A. J. Oswald. (2006). Do divorcing couples become happier by breaking up? *Journal of the Royal Statistical Society: Series A (Statistics and Society), 169,* 319–336.

Gardner, R. A. (1998). *The parental alienation syndrome* (2nd ed.). Cresskill, NJ: Creative Therapeutics.

Gelatt, V. A., F. Adler-Baeder, & J. R. Seeley. (2010). An interactive web-based program for stepfamilies: Development and evaluation of efficacy. *Family Relations, 59,* 572–586.

Gershon, I. (2010). *The breakup 2.0.* New York, NY: Cornell University Press.

Godbout, E., & C. Parent. (2012). The life paths and lived experiences of adults who have experienced parental alienation: A retrospective study. *Journal of Divorce and Remarriage, 53,* 34–54.

Goetting, A. (1982). The six stations of remarriage: The developmental tasks of remarriage after divorce. *Family Coordinator, 31,* 213–222.

Gordon, R. M. (2005). The doom and gloom of divorce research: Comment on Wallerstein & Lewis (2004). *Psychoanalytic Psychology, 22,* 450–451.

Greeff, A. P., & C. Du Toit. (2009). Resilience in remarried families. *American Journal of Family Therapy, 37,* 114–126.

Hawkins, A. J., S. L. Nock, J. C. Wilson, L. Sanchez, & J. D. Wright. (2002). Attitudes about covenant marriage and divorce: Policy implications from a three state comparison. *Family Relations, 51,* 166–175.

Hawkins, D. N., & A. Booth. (2005). Unhappily ever after: Effects of long-term, low-quality marriages on well-being. *Social Forces, 84,* 445–465.

Healy, M., & V. Salazar. (2010, April 27). Dealbreakers. *USA Today,* p. D1.

Hetherington, E. M. (2003). Intimate pathways: Changing patterns in close personal relationships across time. *Family Relations, 52,* 318–331.

Higginbotham, B. J., P. Davis, L. Smith, L. Dansie, L. Skogrand, & K. Reck. (2012). Stepfathers and stepfamily education. *Journal of Divorce and Remarriage, 53,* 76–90.

Higginbotham, B. J., & L. Skogrand. (2010). Relationship education with both married and unmarried stepcouples: An exploratory study. *Journal of Couple and Relationship Therapy, 9,* 133–148.

Higginbotham, B. J., L. Skogrand, & E. Torres. (2010). Stepfamily education: Perceived benefits for children. *Journal of Divorce and Remarriage, 51,* 36–49.

Imhoff, R., & R. Banse. (2011). Implicit and explicit attitudes toward ex-partners differentially predict breakup adjustment. *Personal Relationships, 18,* 427–438.

Kelly, J. B., & R. E. Emery. (2003). Children's adjustment following divorce: Risk and resilience perspectives. *Family Relations, 52,* 352–362.

Kesselring, R. G., & D. Bremmer. (2006). Female income and the divorce decision: Evidence from micro data. *Applied Economics, 38,* 1605–1617.

Kim, H. (2011). Exploratory study on the factors affecting marital satisfaction among remarried Korean couples. *Families in Society, 91,* 193–200.

Knox, D., & S. Hall. (2010). Relationship and sexual behaviors of a sample of 2,922 university students. Unpublished data collected for this text. Department of Sociology, East Carolina University, and Department of Family and Consumer Sciences, Ball State University.

Knox, D., M. E. Zusman, M. Kaluzny, & C. Cooper. (2000). College student recovery from a broken heart. *College Student Journal, 34,* 322–324.

Knox, D., M. E. Zusman, K. McGinty, & B. Davis. (2002). College student attitudes and behaviors toward ending an unsatisfactory relationship. *College Student Journal, 36,* 630–634.

Lambert, A. N. (2007). Perceptions of divorce: Advantages and disadvantages; A comparison of adult children experiencing one parental divorce versus multiple parental divorces. *Journal of Divorce and Remarriage, 48,* 55–77.

Lavner, J. A., & T. N. Bradbury. (2010). Patterns of change in marital satisfaction over the newlywed years. *Journal of Marriage and Family, 72,* 1171–1187.

Le, B., N. L. Dove, C. R. Agnew, M. S. Korn, & A. A. Mutso. (2010). Predicting nonmarital romantic relationship dissolution: A meta-analytic synthesis. *Personal Relationships, 17,* 377–390.

Leidy, M. S., T. J. Schofield, M. A. Miller, R. D. Parke, S. Coltrane, S. Braver et al. (2011). Fathering and adolescent adjustment: Variations by family structure and ethnic background. *Fathering: A Journal of Theory, Research, and Practice About Men as Fathers, 9,* 44–68.

Lewis, K. (2008). Personal communication. Dr. Lewis is also the author of *Five Stages of Child Custody* (Glenside, PA: CCES Press).

Licata, N. (2002). Should premarital counseling be mandatory as a requisite to obtaining a marriage license? *Family Court Review, 40,* 518–532.

Lucier-Greer, M., F. Adler-Baeder, S. A. Ketring, K. T. Harcourt, & T. Smith. (2012). Comparing the experiences of couples in first marriages and remarriages in couple and relationship education. *Journal of Divorce and Remarriage, 53,* 55–75.

Luscombe, B. (2010, May 5). Revoking the marriage license. *Time,* 64.

Luscombe, B. (2011, September 28). Latchkey parents. *Time,* 50–51.

Manning, W. D., & P. J. Smock. (2000). "Swapping" families: Serial parenting and economic support for children. *Journal of Marriage and Family, 62,* 111–122.

Marano, H. E. (2000, March/April). Divorced? Don't even think of remarrying until you read this. *Psychology Today,* 56–64.

Markham, M. S., & M. Coleman. (2011, November 18). "Part-time parent": Divorced mothers' experiences with sharing physical custody. Poster session presented at the annual meeting of the National Council on Family Relations, Orlando, FL.

Masheter, C. (1999). Examples of commitment in postdivorce relationships between spouses. In J. M. Adams & W. H. Jones (Eds.), *Handbook of interpersonal commitment and relationship stability* (pp. 293–306). New York, NY: Academic/Plenum Publishers.

McCarthy, B. W., & R. L. Ginsberg. (2007). Second marriages: Challenges and risks. *Family Journal, 15,* 119–123.

Meier, J. S. (2009). A historical perspective on parental alienation syndrome and parental alienation. *Journal of Child Custody, 6,* 232–257.

Menning, C. L. (2008). "I've kept it that way on purpose": Adolescents' management of negative parental relationship traits after divorce and separation. *Journal of Contemporary Ethnography, 37,* 586–597.

Mitchell, B. A. (2010). Midlife marital happiness and ethnic culture: A life course perspective. *Journal of Comparative Family Studies, 41,* 167–183.

Morgan, E. S. (1944). *The Puritan family.* Boston, MA: Public Library.

Murray, C. (2012). *Coming apart: The state of white America, 1960-2010.* New York: Crown Forum.

Musick, K., & L. Bumpass. (2012). Reexamining the case for marriage: Union formation and changes in well-being. *Journal of Marriage and Family, 74,* 1–18.

Nielsen, L. (2011). Divorced fathers and their daughters: A review of recent research. *Journal of Divorce and Remarriage, 52,* 77–93.

Nielsen, L. (2004). *Embracing your father: How to build the relationship you always wanted with your dad.* New York, NY: McGraw-Hill.

Nunley, J. M., & A. Seals. (2010). The effects of household income volatility on divorce. *American Journal of Economics and Sociology, 69,* 983–1011.

Papernow, P. L. (1988). Stepparent role development: From outsider to intimate. In W. R. Beer (Ed.), *Relative strangers* (pp. 54–82). Lanham, MD: Rowman and Littlefield.

Parker, K. (2011). A portrait of stepfamilies. Pew Research Center. Retrieved January 13, 2011, from http://pewsocialtrends.org/2011/01/13/a-portrait-of-stepfamilies/

Riffe, J., D. Brandon, M. Mulroy, & A. Faulkner. (2010, November 3–5). Parent education for separating or divorcing families: Results of a national extension survey. Poster session presented at the annual meeting of the National Council on Family Relations, Minneapolis, MN.

Riggio, H. R., & A. M. Valenzuela. (2011). Parental marital conflict and divorce, parent-child relationships, and social support among Latino-American young adults. *Personal Relationships, 18,* 392–409.

Saint-Jacques, M.-C., C. Robitaille, E. Godbout, C. Parent, S. Drapeau, & M. H. Gagne. (2011). The processes distinguishing stable from unstable stepfamily couples: A qualitative analysis. *Family Relations, 60,* 545–561.

Sakraida, T. (2005). Divorce transition differences of midlife women. *Issues in Mental Health Nursing, 26,* 225–249.

Schacht, T. E. (2000). Protection strategies to protect professionals and families involved in high-conflict divorce. *UALR Law Review, 22*(3), 565–592.

Schrodt, P., & D. O. Braithwaite. (2011). Coparental communication, relational satisfaction, and mental health in stepfamilies. *Personal Relationships, 18,* 352–369.

Schrodt, P., J. Soliz, & D. O. Braithwaite. (2008). A social relations model of everyday talk and relational satisfaction in stepfamilies. *Communication Monographs, 75,* 190–202.

Segal-Engelchin, D., & Y. Wozner. (2005). Quality of life to single mothers in Israel: A comparison to single mothers and divorced mothers. *Marriage and Family Review, 37,* 7–28.

Sever, I., J. Guttmann, & A. Lazar. (2007). Positive consequences of parental divorce among Israeli young adults: A long-term effect model. *Marriage and Family Review, 42,* 7–21.

Shafer, K. (2010). [Review of the book *Yours, mine, and hours: Relationship skills for blended families,* by John Penton & Shona Welsh]. *Journal of Comparative Family Studies, 41,* 185–186.

Shapiro, D. N., & A. J. Stewart. (2011). Parenting stress, perceived child regard, and depressive symptoms among stepmothers and biological mothers. *Family Relations, 60,* 533–544.

Siegler, I., & P. Costa. (2000). Divorce in midlife. Paper presented at the annual meeting of the American Psychological Association, Boston, MA.

Sobolewski, J. M., & P. R. Amato. (2007). Parents' discord and divorce, parent-child relationships and subjective well-being in early adulthood: Is feeling close to two parents always better than feeling close to one? *Social Forces*, 1105–1125.

Stork-Hestad, N., J. Hans, & R. J. Werner-Wilson. (2011, November 18). A non-marital, romantic dissolution study. Poster session presented at the annual meeting of the National Council on Family Relations, Orlando, FL.

Sweeney, M. M. (2010). Remarriage and stepfamilies: Strategic sites for family scholarship in the 21st century. *Journal of Marriage and Family*, 72, 667–684.

Teachman, J. (2008). Complex life course patterns and the risk of divorce in second marriages. *Journal of Marriage and Family*, 70, 294–306.

Teich, M. (2007). A divided house. *Psychology Today*, 40, 96–102.

Tillman, K. H. (2008). "Non-traditional" siblings and the academic outcomes of adolescents. *Social Science Research*, 37, 88–101.

Trinder, L. (2008). Maternal gate closing and gate opening in postdivorce families. *Journal of Family Issues*, 29, 1298–1324.

Tulane, S., L. Skogrand, & J. DeFrain. (2011). Couples in great marriages who considered divorcing. *Marriage and Family Review*, 47, 289–310.

Turkat, I. D. (2002). Shared parenting dysfunction. *American Journal of Family Therapy*, 30, 385–393.

Vinick, B. (1978). Remarriage in old age. *Family Coordinator*, 27, 359–363.

Vukalovich, D., & N. Caltabiano. (2008). The effectiveness of a community group intervention program on adjustment to separation and divorce. *Journal of Divorce and Remarriage*, 48, 145–168.

Waite, L. J., Y. Luo, & A. C. Lewin. (2009). Marital happiness and marital stability: Consequences for psychological well-being. *Social Science Research*, 28, 201–217.

Walters, B. (2008). *Barbara Walters: A memoir*. New York, NY: Knopf.

Whitehurst, D. H., S. O'Keefe, & R. A. Wilson. (2008). Divorced and separated parents in conflict: Results from a true experiment effect of a court mandated parenting education program. *Journal of Divorce and Remarriage*, 48, 127–144.

Wilson, A. C., & T. L. Huston. (2011, November). Shared reality in courtship: Does it matter for marital success? Paper presented at the annual meeting of the National Council on Family Relations, Orlando, FL.

Wiseman, R. S. (1975). Crisis theory and the process of divorce. *Social Casework*, 56, 205–212

Chapter 15

Agahi, N. (2008). Leisure activities and mortality: Does gender matter? *Journal of Aging and Health*, 20, 855–871.

Alford-Cooper, F. (2006, March 23–26). Where has all the sex gone? Sexual activity in lifetime marriage. Paper presented at the meeting of the Southern Sociological Society, New Orleans, LA.

Aoyagi, Y., & R. Shephard. (2011). Habitual physical activity and health in the elderly: The Nakanojo study. *Geriatrics and Gerontology International*, 10, 236–243.

Arab, L., M. L. Biggs, E. S. O'Meara, W. T. Longstreth, P. K. Crane, & A. L. Fitzpatrick. (2012). Gender differences in tea, coffee, and cognitive decline in the elderly: The cardiovascular health study. *Journal of Alzheimer's Disease*, 27, 553–566.

Aumann, K., E. Galinsky, K. Sakai, M. Brown, & J. T. Bond. (2010). The Elder Care Study: Everyday realities and wishes for change. Families and Work Institute. Available at http://www.familiesandwork.org

Ben-Zur, H. (2012). Loneliness, optimism, and well-being among married, divorced, and widowed individuals. *Journal of Psychology*, 146, 23–36.

Brown-Borg, H., R. Anderson, R. McCarter, J. Morley, N. Musi et al. (2012). Nutrition in aging and disease: Update on biological sciences. *Aging Health*, 8, 13–16.

Bryant, C. D. (2007). The sociology of death and dying. In C. D. Bryant & D. L. Peck (Eds.), *21st century sociology: A reference handbook* (pp. 156–166). Thousand Oaks, CA: Sage.

Caro, F. G., C. Yee, S. Levien, A. S. Gottlieb, J. Winter, D. L. McFadden et al. (2012). Choosing among residential options: Results of a vignette experiment. *Research on Aging*, 34, 3–33.

Carroll, D. D., H. M. Blanck, M. K. Serdula, & D. R. Brown. (2010). Obesity, physical activity, and depressive symptoms in a cohort of adults 51 to 61. *Journal of Aging and Health*, 22, 384–398.

Chao, J. K., Y. Lin, M. Ma, C. Lai, Y. Ku, W. Kuo et al. (2011). Relationship among sexual desire, sexual satisfaction, and quality of life in middle-aged and older adults. *Journal of Sex and Marital Therapy*, 37, 386–403.

Cook, K., D. Dranove and A. Sfekas. 2010. Does major illness cause financial catastrophe? *Health Services Research*, 45, 418-436

Cozijnsen, R., N. L. Stevens, & T. G. Van Tilburg. (2010). Maintaining work-related personal ties following retirement. *Personal Relationships*, 17, 345–356.

Cutler, N. E. 2002. *Advising mature clients*. New York: Wiley.

Danigelis, N. L., M. Hardy, and S. J. Cutler. (2007). Population aging, intracohort aging, and sociopolitical attitudes. *American Sociological Review*, 72, 812–30.

Dillion C. F., Q. Gu, H. Hoffman and C.W. Ko. 2010. Vision, hearing, balance, and sensory impairment in Americans aged 70 years and over: United States, 1999-2006. NCHS data brief, no 31. Hyattsville, MD: National Center for Health Statistics.

Fabian, N. (2007). Rethinking retirement—And a footnote on diversity. *Journal of Environmental Health*, 69, 86.

Felding, R. A., W. J. Rejeski, S. Blair, T. Church, M. A. Espeland et al. (2011). The Lifestyle Interventions and Independence for Elders Study: Design and methods. *Journals of Gerontology, Series A: Biological Sciences and Medical Sciences*, 66(A): 1226–1237.

Field, D., & S. Weishaus. (1992). Marriage over half a century: A longitudinal study. In M. Bloom (Ed.), *Changing lives* (pp. 269–273). Columbia: University of South Carolina Press.

Franks, M. M., T. Lucas, M. A. Stephens, K. S. Rook, & R. Gonzalez. (2010). Diabetes distress and depressive symptoms: A dyadic investigation of older patients and their spouses. *Family Relations*, 59, 599–610.

Gadalla, T. M. (2010). Relative body weight and disability in older adults: Results from a national survey. *Journal of Aging and Health*, 22, 403–418.

Haboush, A., C. S. Warren, & L. Benuto. (2012). Beauty, ethnicity, and age: Does internalization of mainstream media ideals influence attitudes towards older adults? *Sex Roles*, 66, 3–20.

Haley, W. E. (2011, November 18). Family caregiving for older adults in a changing economic world. Presentation at the annual meeting of the National Council on Family Relations, Orlando, FL.

Hays, J., J. K. Ockene, R. L. Brunner, J. M. Kotchen, J. E. Manson, R. E. Patterson et al. (2003). Effects of estrogen plus progestin on health-related quality of life. *New England Journal of Medicine*, 348, 1839–1854.

Henry, B. W., A. D. Ozier, & A. Johnson. (2011). Empathetic responses and attitudes about older adults: How experience with the aging game measures up. *Educational Gerontology*, 37, 924–941.

Hensley, B., P. Martin, J. A. Margrett, M. MacDonald, I. C. Siegler, & L. W. Poon. (2012). Life events and personality predicting loneliness among centenarians: Findings from the Georgia Centenarian Study. *Journal of Psychology*, 146, 173–188.

Hostetler, A. J. (2011). Senior centers in the era of the "Third Age:" Country clubs, community centers, or something else? *Journal of Aging Studies*, 25, 166-176.

Ikegami, N. 1998. Growing old in Japan. *Age and Ageing*, 27, 277–78.

Jackson, M. (2011, November). Nursing home placement and caregiver burden. Poster session presented at the annual meeting of the National Council on Family Relations, Orlando, FL.

Johnson, C. L., & B. M. Barer. (1997). *Life beyond 85 years: The aura of survivorship*. New York, NY: Springer.

Jorm, A. F., H. Christensen, A. S. Henderson, P. A. Jacomb, A. E. Korten, and A. Mackinnon. (1998). Factors associated with successful ageing. *Australian Journal of Ageing, 17,* 33–37.

Kalish, N. (1997). *Lost and found lovers: Facts and fantasies of rekindled romances.* New York, NY: William Morrow.

Kelley-Moore, J. A., J. G. Schumacher, E. Kahana, and B. Kahana. (2006). When do older adults become 'disabled'? Social and health antecedents of perceived disability in a panel study of the oldest old. *Journal of Health and Social Behavior, 47,* 126–42.

Knox, D., S. Kimuna, and M. Zusman. (2005) College student views of the elderly: Some gender differences. *College Student Journal, 39,* 14-16.

Largier, S. (2010, July 22). Seven secrets to a happy retirement. *U.S. News and World Report.*

Lee, J. A., P. Foos, & C. Clow. (2010). Caring for one's elders and family-to-work conflict. *Psychologist-Manager Journal, 13,* 15–39.

Liang, J., A. R. Quinones, J. M. Bennett, Y. Wen, X. Xiao, B. Shaw et al. (2010). Evolving self-rated health in middle and old age: How does it differ across Black, Hispanic, and White Americans? *Journal of Aging and Health, 22,* 3–26.

Lindau, S. T., L. P. Schumm, E. O. Laumann, W. Levinson, C. A. O'Muircheartaigh, & L. J. Waite. (2007). A study of sexuality and health among older adults in the United States. *New England Journal of Medicine, 357,* 762–774.

Madden, M. (2010). Older adults and social media. Pew Internet and American Life Project. Retrieved August 27, 2010, from http://pewresearch.org/pubs/1711/older-adults-social-networking-facebook-twitter

Maher, D., & C. Mercer (Eds.). (2009). Introduction. In *Religion and the implications of radical life extension.* New York, NY: Palgrave Macmillan.

McGowan, C. M. (2011). Legal aspects of end of life care. *Critical Care Nurse, 31,* 64–69.

Moser, C., J. Spagnoli, & B. Santos-Eggimann. (2011). Self-perception of aging and vulnerability to adverse outcomes at the age of 65–70 years. *Journals of Gerontology, Series B: Psychological Sciences and Social Sciences, 66*(B), 675–680.

Mutran, E. J., D. Reitzes, & M. E. Fernandez. (1997). Factors that influence attitudes toward retirement. *Research on Aging, 19,* 251–273.

Nordenmark, M., & M. Stattin. (2009). Psychosocial well-being and reasons for retirement in Sweden. *Ageing and Society, 29,* 413–441.

O'Reilly, E. M. (1997). *Decoding the cultural stereotypes about aging: New perspectives on aging talk and aging issues.* New York: Garland.

Paraschakis, A., A. Douzenis, I. Michopoulos, C. Christodoulou, K. Vassilopoulou, F. Koutsaftis et al. (2012). Late onset suicide: Distinction between "young-old" vs. "old-old" suicide victims; How different populations are they? *Archives of Gerontology and Geriatrics, 54,* 136–139.

Peterson, C., N. Park, & E. S. Kim. (2012). Can optimism decrease the risk of illness and disease among the elderly? *Aging Health, 8,* 5–8.

Pnina, R. (2007). Elderly people's attitudes and perceptions of aging and old age: The role of cognitive dissonance. *International Journal of Geriatric Psychiatry, 22,* 656–672.

Popham, L. E., S. M. Kennison, & K. I. Bradley. (2011). Ageism and risk-taking in young adults: Evidence for a link between death and anxiety and ageism. *Death Studies, 35,* 751–763.

Potter, J. F. (2010). Aging in America: Essential considerations in shaping senior care policy. *Aging Health, 63,* 289–300.

Powell, E. (2011, November 18). Sexuality in aging adults from a biosocial perspective: A literature review. Poster session presented at the annual meeting of the National Council on Family Relations, Orlando, FL.

Ryan, J., I. Carrière, J. Scali, K. Ritchie, & M. Ancelin. (2008). Lifetime hormonal factors may predict late-life depression in women. *International Psychogeriatrics, 20,* 1203–1229.

Sandberg, J. G., R. B. Miller, & J. M. Harper. (2002). A qualitative study of marital process and depression in older couples. *Family Relations, 51,* 256–264.

Schairer, C., J. Lubin, R. Troisi, S. Sturgeon, L. Brinton, & R. Hoover. (2000). Menopausal estrogen and estrogen-progestin replacement therapy and breast cancer risk. *Journal of the American Medical Association, 283,* 485–491.

Sheehy, G. (2010). *Passages in caregiving.* New York: William Morrow.

Silverstein, M., Z. Cong, & S. Li. (2006). Intergenerational transfers and living arrangements of older people in rural China: Consequences for psychological well-being. *Journals of Gerontology, Series B: Psychological Sciences and Social Sciences, 61*(B), S256–S267.

Smith, J., M. Borchelt, H. Maier, & D. Jopp. (2002). Health and well-being in the young old and oldest old. *Journal of Social Issues, 58,* 715–733.

Statistical Abstract of the United States. (2012, 131st ed.). Washington, DC: U.S. Census Bureau.

Stone, E. (2008). The last will and testament in literature: Rupture, rivalry, and sometimes rapprochement from Middlemarch to Lemony Snicket. *Family Process, 47,* 425–439.

Stroebe, M., & H. Schut. (2005). To continue or relinquish bonds: A review of consequences for the bereaved. *Death Studies, 29,* 477–495.

Vaillant, G. E. (2002). *Aging well: Surprising guideposts to a happier life from the Landmark Harvard Study of Adult Development.* New York, NY: Little, Brown.

Walker, R. B., & M. A. Luszcz. (2009). The health and relationship dynamics of late-life couples: A systematic review of the literature. *Ageing and Society, 29,* 455–481.

Weintraub, P. (2006). Guess who's back? *Psychology Today, 39*(4), 79–84.

Willson, A. E. (2007). The sociology of aging. In C. D. Bryant & D. L. Peck (Eds.), *21st century sociology: A reference handbook* (pp. 148–155). Thousand Oaks, CA: Sage.

Winterich, J. A. (2003). Sex, menopause, and culture: Sexual orientation and the meaning of menopause for women's sex lives. *Gender and Society, 17,* 627–642.

Wolkove, N., O. Elkholy, M. Baltzan, and M. Palayew. (2007). Sleep and aging. *Canadian Medical Association Journal, 176,* 1299–1304.

Wrosch, C., R. Schulz, G. E. Miller, S. Lupien, & E. Dunne. (2007). Physical health problems, depressive mood, and cortisol secretion in old age: Buffer effects of health engagement control strategies *Health Psychology, 26,* 341–349.

Wu, Z., & R. Hart. (2002). The mental health of the childless elderly. *Sociological Inquiry, 72,* 21–42.

Wu, Z., C. M. Schimmele, & N. L. Chappell. (2012). Aging and late-life depression. *Journal of Aging and Health, 24,* 3–28.

NAME INDEX

A

Abowitz, D., 36
Abowitz, D. A., 35
Ackerman, D., 92
Adams, 79
Adams, M., 265
Adams, R., 136, 137, 138
Adler-Baeder, F., 276
Aducci, A. J., 143
Agahi, N., 295
Ahnert, L., 173
Ajdukovic, D., 127
Alam, R., 234
Albright, J. M., 77, 78
Alford, J. J., 99
Alford-Cooper, F., 294
Algoe, S. B., 91
Ali, L., 127
Ali, M. M., 210
Allen, E. S., 144, 231
Allen, K. R., 18
Allen, W., 121
Amato, P., 97
Amato, P. R., 10, 96, 109, 110, 138,
    145, 146, 162, 255, 257, 264, 266,
    267, 268, 269
Ambwani, S., 90
American Psychological
    Association, 164
Ames, C. M., 195
Anderson, J. R., 105, 147
Anderson, K. G., 246
Anderson, K. M., 244
Anderson, S., 6
Aoyagi, Y., 292
Arab, L., 292
Arnold, A. L., 145
Asch-Goodkin, J., 164
Assad, K. K., 102
Auerbach, C. F., 8
Aulette, J. R., 4
Aumann, K., 289
Axelson, L., 9
Azmitia, M., 91

B

Backstrom, L., 118
Baden, A. L., 193
Bagarozzi, D. A., 227, 231
Baker, A. J., 77
Baker, A. J. L., 263, 264
Ballard, S. M., 119, 139
Baltazar, A. M., 204
Bang, E., 244
Banse, R., 260
Barelds, D. P., 85
Barelds, P., 92, 94
Barelds-Dijkstra, D. P. H., 92, 94
Barelds-Dijkstra, P., 85

Barer, B. M., 292, 296, 298, 299,
    300–301
Barnes, H., 43, 44, 230
Barnes, K., 139
Barr, R., 166
Bartholomew, K., 157
Bartle-Haring, S., 134
Barton, A. L., 218
Bauerlein, M., 52
Baumrind, D., 211
Bearman, P., 114
Becker, K., 215
Becker, K. D., 248
Beckman, N. M., 129
Beecher, H. W., 65
Begue, L., 197
Behringer, A. M., 61
Bem, S. L., 47
Ben-Ari, A., 174
Benjamin, 255
Bennetts, L., 171
Benton, S. D., 269
Ben-Zur, H., 298
Berger, R., 190
Berle, W., 67
Bermant, G., 228
Bernhardt, E., 79
Bersamin, M., 117
Berscheid, E., 85
Besen, W., 153
Bian, 300
Biblarz, T. J., 163
Biehle, S. N., 207
Billingsley, S., 145
Black, K., 220
Blackstrom, L., 75
Blackwell, C. W., 157
Blair, S. L., 103
Blakeslee, S., 145
Bloch, K., 35
Block, S., 194
Bluestone, C., 88
Blumer, H. G., 15
Bock, J. D., 215
Bodenmann, G., 219
Boehnke, M., 39
Bogg, R. A., 102
Bogle, K. A., 76
Boislard, P., 115
Bono, C., 34
Booth, A., 259
Booth, C. L., 202
Booth, K., 217
Bor, R., 158
Bos, H., 164
Boskin, W., 294, 295
Bost, K. K., 208
Bowles, L., 297
Bradbury, T. N., 62, 145, 255
Bradshaw, C., 76
Braet, C., 225
Braithwaite, D. O., 282

Braithwaite, S. R., 83
Brandes, M., 189
Brantley, A., 87
Bratter, J. L., 97
Braun, M., 50
Bray, J. H., 275, 276, 279, 282, 283
Bredehoft, D. J., 211, 212
Bredow, C. A., 101
Breed, R., 262
Bremmer, D., 253
Brentano, C., 254, 257, 260
Brewster, C. D. D., 72
Brimhall, A. S., 273
British Columbia Reproductive Men-
    tal Health Program, 207
Britt, S., 170
Bronte-Tinkew, J., 207
Brown, G. L., 15
Brown, J. D., 194
Brown, S. A., 232
Brown, S. L., 254
Brown-Borg, H., 295
Brownridge, D. A., 239
Brucker, H., 114
Bryant, C. D., 298
Buchanan, A., 207
Buchbinder, E., 251
Buchler, S., 81
Bulanda, J. R., 42
Bumpass, L., 137, 260
Burke, M. L., 234
Burke, S., 240
Burr, B. K., 99
Burr, W. R., 220, 223
Busby, D., 7, 257
Busby, D. M., 244
Butterworth, P., 225
Buunk, A. P., 94, 110
Buxton, A. P., 158
Byers, E. S., 120
Byrd-Craven, J., 207
Byron, L., 132

C

Cadden, M., 137–138
Cahill, S., 154, 161, 162
Calligas, A., 147
Caltabiano, N., 271
Campbell, K., 135, 136
Canary, D. J., 139
Caputo, 176
Carey, A., 52
Carey, A. R., 59, 85
Careya, M. P., 75
Carlson-Catalano, J., 223
Caro, F. G., 291
Carpenter, C., 152, 159
Carpenter, L. M., 115
Carr, K., 214
Carrere, S., 145

Carriero, R., 174
Carroll, D. D., 291
Carroll, J. S., 146, 178
Carter, J., 168, 299
Cartland, B., 112
Cashmore, J., 266
Cassidy, M. L., 92
Castrucci, B. C., 204
Cattaneo, L. B., 239
Caughlin, J. P., 64
Cavanagh, S. E., 213
Centers for Disease Control and Prevention, 157, 191, 221, 239–240
Chambers, A. L., 70
Chang, I., 189
Chao, J. K., 294
Chapleau, K. M., 244
Chemtob, N., 254
Chen, J. J., 211
Cheng, C., 47
Cherlin, A. J., 12, 23, 81, 252
Chiou, W., 78
Chiu, H., 7, 257
Cianciotto, J., 154
Cinamon, R. G., 174
Ciro, C. A., 225
Clarke, J. I., 212
Clarke, S. C., 274
Clarke-Stewart, A., 254, 257, 260
Clarkwest, A., 96
Cloud, M. D., 227
Clunis, D. M., 165
Cohen-Kettenis, P. T., 33
Cohn, D., 169
Colapinto, J., 33
Coleman, M., 266, 272
Collins, J., 235
Colson, M., 121, 127
Coltrane, S., 265
Confer, J. C., 227
Congreve, W., 106
Connors, W., 193
Cook, K., 290
Cooley, C.H., 15
Coontz, S., 132, 133, 152
Corcoran, M. E., 171
Cordova, J. V., 44
Córdova, J. V., 43, 259
Corelli, M., 68
Cornelius-Cozzi, T., 207
Corra, M., 42, 145
Corte, U., 73
Costa, P., 260
Couric, K., 22
Cousins, J., 146
Covizzi, I., 262
Cox, A., 92
Cox, M., 212
Coyne, S. M., 52
Cozijnsen, R., 293
Craft, S. M., 215
Craig, L., 42
Cramer, R. E., 227
Crawley, S. L., 32, 33
Crosnoe, R., 213
Cross-Barnet, C. A., 81

Crowell, J. A., 103
Crowl, A., 163
Crutzen, R., 213
Cui, M., 202, 204, 264
Cupach, W. R., 76, 240
Curran, L., 221
Curran, M. A., 15
Curtner-Smith, M. E., 213
Custody and Visitation, 164
Cutler, N. E., 286
Cutler, W. B., 122

D

Dabbous, Y., 47
Dahl, R. E., 215
Dahlberg, E., 127
Daigle, L. E., 244
Danigelis, N. L., 287
Dantzker, M. L., 121
Darnall, D., 264
Darton, K., 27
Davidson, J. K., Sr., 121
Davies, J. B., 178
Davis, 171
Dawson, C., 212
Day, R., 208
Dean, C. J., 230
Deaton, A., 178
De Bruyne, T., 54
De Castro, S., 235
DeCuzzi, A., 97
de Grey, A., 286
De La Baume, M., 7
DeLamater, J., 120
Delevi, R., 50
Denney, J. T., 74
DeOllos, I. Y., 145
Department of Commerce, 186
Derby, K., 230
De Schipper, J. C., 173
Dethier, M., 232
Detzner, 300
Deutsch, F. M., 172
Devall, E., 106
Dew, J., 81, 139, 176
Dewel, D., 144
Diamond, A., 204
Diamond, L. M., 88, 152
Diem, C., 177
DiLillo, D., 248
Dillion, C. F., 286
Dixon, L. J., 255
Dodoo, F. N., 125
Doherty, W. J., 215
Dong, B., 217
Donn, J., 73
Doosje, S., 222
Dorsey Green, G., 165
Dotson-Blake, K., 45, 85, 221
Dotson-Blake, K. P., 115
Dowd, D. A., 146
Dowd, M., 82
Doyle, M., 262
Drefahl, S., 139
Drucker, P. F., 134

Druckerman, P., 229
Dubbs, S. L., 110
Duncan, S. F., 105
Du Toit, C., 281
Dwyer, D. S., 210
Dysart-Gale, D., 32
Dziegielewski, S. F., 157

E

East, L., 42
Easterling, B., 59
Easterling, B. A., 142, 143, 144
Eaton, L., 243
Economist, 271
Edmundson, M., 238
Edwards, T. M., 102
Ehrmann, M., 224
Eibach, R. P., 152
Einhorn, L. A., 178
Eisenman, R., 121
Eke, A., 243
Elicker, K. J., 173
Elliott, L., 85
Elliott, M., 222
Ellison, C. G., 222
Elmslie, B., 228
Emerson, R. W., 284
Emery, R. E., 265
Engblom-Deglmann, M. L., 273
England, P., 39, 75, 76, 118
Enright, E., 226, 259
Ephron, N., 209
Etzioni, A., 177
Eubanks Fleming, C. J., 43, 259
Euser, E. M., 231

F

Fabian, N., 292
Facer, J., 208
Fackrell, T. A., 265
Faircloth, M., 52, 259
Falba, T. M. T., 231
Faulkner, S., 118
Feagin, J. R., 97
Feinstein, D., 239
Felding, R. A., 291
Fernandez-Montalvo, J., 243
Few, A. I., 249
Field, D., 296
Field, T., 225
Fieldera, R. L., 75
Finer, L. B., 195
Finley, G. E., 262
Fish, J., 24, 255
Fisher, A. D., 230
Fisher, H., 87, 99, 100
Fisher, M., 92
Fisher, T. D., 138
Flack, W. F., Jr., 244
Flouri, E., 207
Flower Kim, K. M., 192
Floyd, K., 91
Fokkema, T., 73

Follingstad, D. R., 238
Fonda, J., 228
Forbes, E. E., 215
Foster, D. G., 123
Foster, J., 84
Foster, J. D., 257
Fox, S., 19
Franks, M. M., 291
Friedrich, R. M., 223
Fu, X., 141
Futris, T. G., 106

G

Gabor, Z., 258
Gadalla, T. M., 291
Gagnon, J., 120
Gaines, S. O., Jr., 141
Galambos, N. L., 204
Gallmeier, C. P., 60, 91
Gallup Poll, 152
Ganong, L., 73
Ganong, L. H., 52, 272, 279
Gardner, J., 259
Gardner, R. A., 263
Garfield, R., 45
Garrett, T. M., 197
Gartrell, N., 157, 164
Gates, G. J., 152, 159
Gatzeva, M., 92
Gavin, L. E., 207
Ge, X., 194
Gelabert, E., 206
Gelatt, V. A., 283
Geller, P., 195
Gellers, K., 85
Gelles, K., 194
Gershon, I., 52, 259
Gibb, S. J., 174
Gibson, V., 140
Gidycz, C. A., 246
Ginsberg, R. L., 274
Girgis, S. R. P., 13
Glorieux, I., 172
Godbout, E., 263, 264
Goetting, A., 272
Goldberg, A. E., 164, 194, 208
Goldberg, N., 219
Goldhammer, D. L., 128
Gonzaga, G. C., 99
Goode, E., 226
Gordon, J. R., 172
Gordon, R. A., 173
Gordon, R. M., 264
Gormley, B., 242
Gotta, G., 157
Gottman, J., 54, 56, 145, 237
Gottmann, J. M., 157
Gould, T. E., 231
Graham, H., 188
Graham, J. M., 83
Grammer, K., 121
Greeff, A. P., 54, 281
Green, J. C., 163
Green, M., 222
Green, R. J., 157

Greene, K., 118
Greenland, F. R., 186, 192
Gregory, J. D., 7
Greitemeyer, T., 90
Grief, G. L., 44
Grogan, S., 42
Gross, J., 193
Grundy, E., 42
Gunnar, M. R., 173
Guterman, J. T., 235
Gutmann, D. L., 138
Guttmacher Institute, 195

H

Ha, J. H., 74
Haag, P., 186, 206
Haboush, A., 287
Haley, W. E., 289
Hall, J. A., 77
Hall, J. H., 228
Hall, S., 4, 23, 57, 59, 70, 74, 75, 76,
    79, 85, 92, 93, 97, 99, 108, 115,
    116, 117, 118, 125, 132, 133, 137,
    141, 154, 161, 163, 197, 227, 238,
    244, 252, 255, 256
Hall, S. S., 136, 137, 138
Hall, W. A., 226
Halpern, C. T., 89
Hammarberg, K., 207
Hancock, J. T., 78
Hans, J. D., 60
Hanson, K. L., 72, 90
Haring, M., 103
Harris-McKoy, D., 204
Hart, R., 292
Hasday, M., 120
Hatfield, 83
Hattori, M. K., 125
Hawkins, A. J., 271
Hawkins, D. N., 259
Hays, J., 294
Healy, M., 256
Heino, R. D., 76, 78
Heintz, A. J., 125
Held, M., 165
Heller, N., 34
Henry, B. W., 287
Hensley, B., 298
Hepp, U., 47
Herbenick, D., 121
Hersh, J., 209
Hertenstein, M. J., 56
Hesse, C., 91
Hetherington, E. M., 269
Hicks, G. R., 155
Higginbotham, B. J., 272, 281
Hill, E. W., 65
Hill, G. J., 72
Hill, M. R., 141
Hill, R., 102
Hochschild, A. R., 174
Hodge, A., 78
Hogue, M., 39
Hohmann-Marriott, B. E., 97, 257
Hollander, D., 114

Hollingsworth, L. D., 193
Holthouse, D., 154
Holtz, L., 213
Horban, D., 171
Hostetler, A. J., 286
Howard, J., 2, 164
Huang, P. M., 80
Huggins, R., 194, 205
Hughes, M., 223
Huh, N. S., 193
Human Rights Campaign, 160
Huntington, A., 188
Huston, T. L., 85, 109, 255
Huyck, M. H., 138

I

Ikegami, N., 300
Imhoff, R., 260
Impett, E. A., 127
Ingoldsby, B., 101
Insel, T. R., 225
Irvolino, A. C., 32
Israel, T., 156

J

Jackson, J., 105
Jackson, M., 290
Jacobson, A. L., 211
James, S., 267
James, S. D., 13
Jang, 300
Jayson, S., 75
Jenkins, D. D., 115
Johnson, 289
Johnson, C. L., 292, 296, 298, 299,
    300–301
Johnson, D. R., 145
Johnson, M. D., 105
Johnson, R. A., 92
Johnson, V. E., 129
Jones, R. K., 195
Jones, S. L., 154
Jordan, E. F., 213
Jorgensen, B. L., 146
Jorm, A. F., 286
Joshi, A., 102
Jung, C., 227

K

Kahneman, D., 178
Kaiser Family Foundation, 23
Kalish, N., 296
Kalmijn, M., 174
Kalush, W., 50
Kamen, C., 156
Kamp Dush, C. M., 79
Kane, J. B., 267
Karten, E. Y., 154
Kasearu, K., 79
Katz, J., 244
Katz, P., 192
Kaufman, J. C., 135, 136

Kefalas, M., 15
Keim, R. E., 211
Keller, H., 274
Kelley-Moore, J. A., 286
Kelly, J., 275, 276, 279, 282, 283
Kelly, J. B., 265
Kem, J., 93
Kennedy, D. P., 93
Kennedy, R., 140, 193
Kero, A., 198
Kerr, J., 184
Kesselring, R. G., 253
Kiecolt, K. J., 174
Kiernan, K., 79
Kim, H., 272, 281
Kim, J. L., 39, 118
Kimmel, M. S., 44
King, A. R., 150
King, M. L., Jr., 238
King, R. B., 97
King, T., 47
Kinsey, A. C., 151
Kirkpatrick, R. C., 149
Kirtley, M. S., 218
Klein, D. M., 12, 16
Klein, S. R., 220, 223
Klinenberg, E., 2
Klohnen, E. C., 99
Knog, G., 204
Knox, D., 4, 23, 32, 36, 57, 59, 64,
    70, 73, 74, 75, 76, 79, 85, 92, 93,
    95, 97, 98, 99, 108, 112, 115, 116,
    117, 118, 119, 125, 132, 133, 136,
    137, 141, 142, 143, 154, 155, 161,
    163, 197, 227, 228, 238, 244, 252,
    255, 256, 258, 260, 287
Knox, M., 214
Kohlberg, L., 37
Kolko, D. J., 204
Kooistra, K., 195
Kornreich, J. L., 118
Koso, G., 132
Kottke, J., 228
Kouros, C. D., 201
Kravitz, A., 70
Kreaer, D. A., 141
Kress, V. E., 250
Kuehl, S., 158
Kulik, L., 169
Kurdek, L. A., 65, 156, 157, 159
Kurzweil, R., 286

L

Ladley, A., 47
LaLanne, J., 222, 295
Lalicha, J., 155
Lalos, A., 198
Lamb, M. E., 173
Lambert, A. N., 264
Landis, D., 164
Landor, A., 115
Lareau, A., 35
Largier, S., 293
Larsen, C. D., 241
Laumann, E. O., 127

Lavee, Y., 174
Lavner, J. A., 62, 145, 255
Lawyer, S., 246
Lease, S. H., 44
Leaver, J., 141
Lebob, M. J., 40
Le Bourdais, C., 202
LeCouteur, A., 251
Lee, D., 191
Lee, G., 92
Lee, J., 214
Lee, J. A., 84, 289
Lee, J. T., 122
Lee, K., 212
Lee, T., 155
Leidy, M. S., 263
Leno, J., 98, 100
Leung, P., 192
Lever, J., 153
Levin, R., 121
Levy, S., 200
Lewis, K., 266
Lewis, P., 267
Lewis, R., 39
Liang, J., 291
Licata, N., 271
Lichter, D. T., 79, 140
Lin, I. F., 254
Lindau, S. T., 294
Linquist, L., 231
Lipman, E. L., 188
Lively, K., 42
Livermore, M. M., 188
Lobo, M., 220
Lodge, D., 206
Logan, 300
Longfellow, H. W., 235
Looi, C., 52
Loomans, D., 210
Lopez, F. G., 242
Lopoo, L. M., 257
Lorber, J., 16
Louv, R., 214
Lucier-Greer, M., 270
Lundquist, J. H., 144
Luo, S. H., 99
Luscombe, B., 255, 268
Luszcz, M. A., 35, 296
Lykins, A. D., 121

M

Ma, M. K., 47
MacArthur, S., 99
Madden, M., 298
Maher, D., 286
Mahoney, D., 225
Major, B., 198
Malinen, K., 16
Maltby, L. E., 39, 42
Mandara, J., 38
Manning, W. D., 69, 70, 72, 268
Marano, H. E., 272
Markham, M. S., 266
Marklein, M. B., 141
Markman, H. J., 56, 62, 208

Marx, G., 169
Masheter, C., 272
Mashoa, S. W., 217
Masters, W. H., 129
Mathy, R. M., 120
Mattingly, M. J., 231
Maume, D. J., 44
Mayall, B., 209
McCabe, M. P., 128
McCarthy, B. W., 274
McClain, L. R., 207
McDermott, E., 188
McGee, H., 244
McGillis, K., 244
McGinty, K., 117
McGowan, C. M., 301
McGrath, A. R., 88
McKenna, K. Y. A., 77
McKinney, C., 211
McKinney, J., 162
McLanahan, S. S., 217
McLarena, K., 155
McLaughlin, M., 185
McLean, K., 158
McNulty, J. K., 138, 146
McPherson, M., 44
Mead, G. H., 15
Mead, M., 33, 202
Medora, N. P., 102
Meehan, D., 108
Meier, A., 83
Meier, J. S., 263
Meinhold, J. L., 8, 38
Melendez, R. M., 125
Melhem, N. M., 235
Mellor, D., 42
Menning, C. L., 268
Mercer, C., 286
Merline, A. C., 232
Merolla, A. J., 65
Merrill, D., 137–138
Merrill, J., 112, 228
Meyer, A. S., 194
Meyer, I. H., 166
Michael, R. T., 120, 122, 138
Michelson, K. D., 207
Miers, D., 235
Mill, J. S., 221
Miller, A. J., 80, 81
Miller, S., 118
Miller, S. A., 120
Minnottea, K. L., 42
Mirchandani, R., 48
Mitchell, B. A., 256
Mock, S. E., 152
Moen, P., 231
Moens, E., 225
Mohr, J. J., 156
Mollborn, S., 188
Moltz, K., 161
Monro, S., 48
Moore, M. M., 52
Moore, O., 47
Moore, T., 86
Mordoch, E., 226
Morell, V., 228

Morgan, E. S., 252
Morin, R., 153, 169
Morr Serewicz, M. C., 139
Moser, C., 287
Moskowitz, D. A., 161
Mullan, K., 42
Murdock, G. P., 8, 133
Murray, C., 72, 255
Musick, K., 137, 260
Mutran, E. J., 293
Myhr, L., 244

N

Nathanson, A. M., 248
National Coalition of Anti-Violence
    Programs, 148, 166
National Conference of State Legisla-
    tures, 160
Nazarinia Roy, R. R., 170
NCFR Policy Brief, 142
Negy, C., 108, 231
Neis, B. L., 173
Nelms, B. J., 43
Nelson, M. C., 209
Neuman, M. G., 229, 231
Nickman, S. L., 193
Nielsen, L., 263, 268
Nobre, P. J., 129
Nock, S. L., 35
Noonan, M. C., 17 1
Nordenmark, M., 293
Norman, W. V., 195
Nuner, J. E., 139
Nunley, J. M., 257

O

Obama, B., 161
O'Brien, M., 60
Ochs, R., 158
Ohayon, M., 79
Olson, M. M., 230
Opree, S. J., 174
O'Reilly, E. M., 286
Orthner, D. K., 144
Oswald, A. J., 259
Oswald, D. L., 244
Oswalt, S. B., 155
Otis, M. D., 156
Ovid, 96
Owen, J. J., 75
Oxlad, M., 251
Oyamot, C. M., 57
Ozay, B., 97

P

Page, S., 152
Paik, A., 92, 117
Palmer, R., 158
Palmer, R. S., 244
Papernow, P. L., 279
Paraschakis, A., 292
Parelli, S., 153

Parent, C., 263, 264
Parker, K., 201, 274
Pascal, B., 25
Passel, J.S., 140
Patrick, J., 139
Paul, J. P., 158
Paul, M., 190
Pazol, K., 178
Pearcey, M., 159
Peilian, C., 220
Pendry, P., 244
Peoples, J. G., 33
Peplau, L. A., 156
Perilloux, C., 110
Perry, M. S., 52
Perry-Jenkins, M., 176
Peter, L., 200
Peterson, C., 292
Petrie, J., 91
Pew Research Center, 4, 39, 69, 70,
    80, 99, 141, 160, 163, 182, 186,
    202
Phillips, L. J., 221
Picca, L. H., 97
Pietrzak, R. H., 222
Pillemer, K., 209
Pinello, D. R., 162
Pinheiro, R. T., 206
Pinina, R., 287
Pinsof, W., 80
Pinto-Gouveia, J., 129
Pizarro, J. M., 177
Planned Parenthood, 182
Plouffe, D., 130
Pogarsky, G., 39
Pollack, W. S., 38
Pong, S. L., 217
Popenoe, D., 178
Popham, L. E., 287
Porter, J., 157
Potok, M., 154, 166
Potter, J. F., 284, 291
Poulin, F., 115
Poverty Guidelines, 177
Powell, E., 294
Powers, R. S., 188
Presser, H. B., 176
Proulx, C. M., 146
Pryor, J. H., 2, 21, 26, 165
Puentes, J., 117
Punyanunt-Carter, N. N., 60

Q

Qian, Z., 140
Qing, M., 207

R

Radmacher, K., 91
Raley, R. K., 116
Ramey, M. B., 64
Randall, B., 21
Rapoport, B., 202
Rawson, G., 168

Ray, J. M., 102
Read, S., 42
Ready, B., 207
Rees, C., 39
Regan, P. C., 102
Regnerus, M., 75
Reichman, N. E., 226
Reid, R. C., 120
Reid, W. J., 193
Reissing, E. D., 117
Renk, K., 211
Rhatigan, D. L., 248
Ricks, J. L., 155
Ridley, C., 109
Riela, S., 90, 91, 92
Riffe, J., 266
Riggio, H. R., 265
Rivadeneyraa, R., 40
Robbins, C. A., 58
Robboy, J., 246
Robnett, R. D., 32
Rochman, B., 191
Roderick, R., 144
Rodgers, B., 225
Rohrer, L., 74
Rolfe, A., 188
Romney, G., 50
Roorda, R. M., 193
Rosario, M., 151
Rose, K., 173
Rosen, E., 171
Rosen, K. H., 249
Rosin, H., 70, 153
Ross, C., 35, 45
Ross, C. B., 91
Ross, R., 193
Rothman, E., 251
Rowland, H., 268
Rubinstein, G., 155
Russell, B. L., 244
Russell, D., 248
Russo, N. F., 198
Rust, P., 155, 156
Rutledge, S. E., 155
Ryan, J., 292

S

Sack, K., 107
Saint, D. J., 100
Saint-Jacques, M.-C., 276, 279
Sakraida, T., 260
Salazar, V., 52, 256
Sammons, L., 216
Samuels, C. A., 47
Sanday, P. R., 167
Sandberg, J. G., 292
Sanders, J., 186
Sandler, L., 186
Sanford, K., 56
Sassler, S., 80, 81
Savin-Williams, R. C., 151
Sawyer, F., 297
Sayare, S., 7
Sayer, L. C., 132
Schacht, T. E., 263

Schairer, C., 294
Scheib, J. E., 190
Schindler, H. S., 207
Schlehofer, M., 174
Schmidt, L., 189
Schneider, J. P., 227, 230
Schoen, R., 79, 170
Schoppe-Sullivan, S. J., 207
Schrodt, P., 281, 282
Schum, T. R., 235
Schut, H., 299
Schwarz, E. B., 183
Scott, L. S., 132, 186
Seals, A., 257
Sefton, B. W., 170
Segal-Engelchin, D., 262
Serovich, J. M., 153
Sever, I., 264
Shafer, K., 272
Shami, M., 251
Shapiro, D. N., 279
Shapiro, M., 180
Sharp, E. A., 73
Sharpe, A., 60
Sharpe, L., 221
Sheehy, G., 290
Shephard, R., 292
Shin, K., 47
Shin, S. H., 232
Shinn, L. K., 60
Shook, S. E., 217
SIECUS, 155
Siegler, I., 260
Silverman, J., 251
Silverstein, L. B., 8
Silverstein, M., 292
Simon, R. J., 193
Simon, R. W., 42
Simon, W., 120
Simons, L. G., 115
Skogrand, L., 272
Slater, S., 161
Slinger, M. R., 212
Sloman, L., 50
Smedema, S. M., 222
Smith, J., 292
Smith, K. E., 231
Smith, T., 187
Smock, P. J., 186, 192, 268
Snyder, K. A., 172
Snyder, L. A., 146
Sobal, J., 72, 90, 132
Sobolewski, J. M., 264
Solem, M., 216
Solomon, Z., 144
Sonnenberg, S. J., 168
Spark, C., 172
Spencer, H., 130
Spitzberg, B. H., 76, 240
Spitze, G. D., 214
Sprecher, S., 102, 110
Stacey, J., 163
Stanfield, J. B., 174
Stanley, S. M., 178, 208
Starnes, T., 26
Starr, R., 177

Statistical Abstract of the United States, 2, 8, 40, 41, 42, 45, 70, 72, 73, 74, 79, 88, 98, 140, 142, 152, 169, 170, 177, 181, 185, 217, 226, 232, 284, 286, 291
Stattin, M., 293
Stefansson, J., 235
Stein, J., 79
Steinberg, J. R., 198
Stepp, L. S., 76
Sternberg, R. J., 86
Stevens, W., 293
Stewart, A. J., 279
Stockl, H., 248
Stone, A., 192
Stone, E., 301
Stone, P., 170, 171, 172, 174
Stork-Hestad, N., 255
Strauss, J., 90
Strickler, B. L., 60
Stroebe, M., 299
Stulhofer, A., 127
Suitor, J. J., 209
Sullivan, A., 162
Sulloway, F. J., 209–210
Sun, Y., 173
Sundell, J., 173
Susskind, J. E., 32
Swan, S. C., 239
Sweeney, M., 71
Sweeney, M. M., 271, 274, 275, 277
Sweet, S., 231
Szabo, A., 222

T

Taliaferro, L. A., 221, 235
Talwar, V., 212
Tan, T. X., 211
Tannen, D., 60
Taylor, A. C., 139
Taylor, P., 182
Taylor, R. L., 38
Taylor, T., 35
Teachman, J., 274
Tebaldi, E., 228
Teich, M., 263
Teitler, J. O., 226
Ter Gogt, T. F. M., 39
Termini, K. A., 223
Tessler, R. C., 193
Teten, A. L., 143
Tetlie, T., 221
Thomas, G. D., 155
Thomas, K. A., 193
Thomas, P. A., 208
Thomas, R. J., 76, 118
Thomas, V., 141
Thompson, M., 44
Thompson, P., 257
Thomson, E., 79
Tillman, K. H., 280
Tobias, S., 162
Toews, M. L., 114, 118, 123
Toma, C. L., 78
Tomkins, S. A., 147

Tomlinson, K. L., 232
Toro-Morn, M., 102
Totenhagen, C., 137
Trachman, M., 88
Trap, P., 59
Treas, J., 171
Trinder, L., 262
Trucco, E. M., 38
True Love Waits, 114
Trust, C., 160
Tsunokai, G. T., 88
Tucker, C. J., 209
Tulane, S., 256
Turkat, I. D., 263
Twenge, J., 102
Twenge, J. M., 208
Tyagart, C. E., 153
Tyler, R., 151
Tzeng, O. C. S., 84

U

Uecker, J., 75
Uecker, J. E., 121
U. S. Census Bureau, 40, 70

V

Vail-Smith, K., 44, 60, 125
Valenzuela, A. M., 265
Valliant, C. O., 147
Valliant, G. E., 147, 295
Van Buren, A., 148, 187
Vandell, D. L., 173
Van den Heuvel, A., 204
Van Eeden-Moorefield, B., 157
Van Laningham, J., 145
Vazonyi, A. I., 115
Veenhoven, R., 20
Vincent, C. D., 173
Vinick, B., 273
Von Geothe, J. W., 236
Vukalovich, D., 271

W

Wade, J. C., 154
Wagner, C. G., 162
Waite, L. J., 259
Waitley, D., 172
Walcheski, M. J., 211
Walker, R. B., 35, 296
Walker, S. K., 173
Waller, M. R., 221
Waller, W., 102
Wallerstein, J., 145
Wallis, C., 40
Walsh, J. L., 119
Walster, E., 92
Walster, G. W., 92
Walters, B., 256
Walzer, S., 35
Wang, T. R., 214
Wang, W., 182
Wang, Y. A., 191

Warash, B. G., 173
Ward, H., 243
Ward, L. M., 119
Ward, R. A., 214
Ward, S., 205
Waterman, J., 192
Weigel, D. J., 134
Weiner, 289
Weininger, E. B., 35
Weintraub, P., 296
Weisfeld, G. E., 146
Weisgram, E. S., 39
Weishaus, S., 296
Weisskirch, R. S., 50
Wells, R. S., 40
Werner-Wilson, R. J., 52
West, M., 114
Western, B., 257
Whatley, M., 124
Whelan-Berry, K. S., 172
White, J. M., 12, 16
Whitehead, B. D., 178
Whitehurst, D. H., 266
Whitley, B., 155
Whitty, M. T., 77
Wickrama, K. A. S., 53
Wienke, C., 72
Wiersma, J. D., 103
Wilcox, W. B., 35
Wildsmith, E., 183
Wiley, M. O., 193
Williams, A., 231

Williams, L. M., 157
Willson, A. E., 289, 292
Wilmoth, J., 132
Wilson, A. C., 255
Wilson, B. F., 274
Wilson, E. K., 214
Wilson, G., 146, 252
Wilson, G. D., 79
Wilson, H., 188
Wilson, K. R., 214
Wilson, S. M., 134
Winch, R. F., 100
Winterich, J. A., 294
Wirtberg, I., 189
Wiseman, R. S., 270
Wolkove, N., 286
Woodhill, B. M., 47
Wozner, Y., 262
Wright, R., 139
Wrosch, C., 295
Wu, P., 78
Wu, Z., 292
Wyatt, T. J., 155

Y

Yarhouse, M. A., 154
Yazedjian, A., 114, 118, 123
Yelsma, P., 293
Yost, M. R., 155
Yu, L., 42

Z

Zachry, E. M., 188
Zaleski, R. M., 160
Zeitzen, M. K., 5
Zhang, S., 65
Zie, D., 42
Zimmerman, K. J., 178
Zinzow, H. M., 246
Zusman, M., 36
Zusman, M. E., 102

SUBJECT INDEX

A

abortion, 194–198
absolutism, 114–115
abstinence, 114–115
abuse and violence in relationships
    abuse seen as intense love, 82, 93
    choices related to, 17
    community factors influencing, 242
    conflict theory on, 14
    cultural influences on, 240–242, 251
    cycle of, 248–251
    effects of, 246–248
    elder, 290
    entrapment in abusive
        relationship, 249
    explanations for, 240–244
    family factors influencing, 243–244
    female abuse of partner, 239
    jealousy leading to, 95, 242
    leaving abusive relationship, 249–250
    marital, 246, 247
    nature of, 236–240
    parental teaching of nonviolence,
        214
    personality factors influencing,
        242–243
    poverty impacting, 178, 242, 251
    power displays as, 14, 138, 238,
        241, 242
    prevention of, 237, 250–251
    PTSD related to, 143, 246, 248
    in same-sex relationships, 157, 167
    self-esteem lack and, 90–91
    sexual, 167, 244–246
    stalking as, 93, 95, 239–240
    substance abuse influencing, 243,
        245, 246
    treatment of abusers, 251
    types of, 237–240
    uxoricide as result of, 95, 236,
        238, 250
    wedding cancellation due to, 109
accommodating style of conflict, 54
acquaintance and date rape, 244–246
activity theory, 288
ACT Raising Safe Kids Parenting
    Program, 214
ADHD (attention deficit/hyperactivity
    disorder), 58, 209, 226
adolescents. See children and youth;
    teenagers
adoptions, 164–165, 192–194
affairs. See extradyadic involvement
    or infidelity
Afghanistan, arranged marriage in, 87
African Americans. See also race and
    ethnicity
    divorce among, 255
    elderly, 291
    extradyadic involvement among, 229
    gender roles among, 38
    interracial relationships of, 88, 95,
        97, 140–141
    parenting issues among, 208, 213
    secrets kept by, 60
    sexual behavior among, 121, 229
    singlehood among, 70
    transracial adoption of/by, 192, 193
agape love style, 85
age. See also children and youth;
    elderly adults
    age-discrepant marriages,
        139–140, 273
    ageism, 287–288
    concept of, 285–287
    end of life issues, 298–301
    life expectancy, 42, 45, 286
    mating gradient related to, 22, 45,
        97, 98, 272

research age and cohort effects, 27
    successful aging, 295–296
    theories of aging, 288, 289
    at time of marriage, 69, 70, 108,
        139–140, 273–274
AIDS (acquired immunodeficiency
    syndrome), 123, 157. See also
    HIV infections
Al-Anon, 232
Alcoholics Anonymous, 232
alcohol use. See substance use/abuse
alexithymia, 91
alimony, 262, 273
alone time, lack of, 137
Alternatives to Marriage Project, 72
Alzheimer's disease, 289, 292
ambivalence, 18
androgyny, 46–47
Angelil, René, 139
animals, 7, 223, 301
antigay bias and discrimination,
    148–149, 154–156, 164–167
antinatalism, 186
anxiety and anxiety disorders
    anxious jealousy, 92, 95
    stress and crisis related to, 225
    substance abuse related to, 232
Arapesh people, 33
arousal, 92
arranged marriages, 87, 88–89
artificial insemination, 190
asceticism, 115
Asian Americans. See also race and
    ethnicity
    divorce among, 255
    extended families of, 9
    interracial relationships of, 88, 97
asset distribution. See property
    division
assisted reproductive technology,
    189–192
attachment, mate selection influenced
    by, 99
Australia, Aboriginal child removal
    in, 21
authoritarian parenting style, 211
authoritative parenting style, 211
avoiding style of conflict, 54

B

baby blues, 206
bachelor/bachelorette parties, 134–135
bankruptcy, 177, 221. See also
    foreclosures and evictions
battered-woman syndrome, 238
Beatie, Thomas, 34
behavior
    abusive (see abuse and violence in
        relationships)
    conflict arising from, 53
    criminal (see criminal behavior)
    divorce due to decreasing
        positive, 256
    new desired, identifying, 63
    parental monitoring of children's,
        202, 204, 210, 212–213, 214, 215
    risk taking, 85–86, 93
    secretive, 59–60
    sexual (see sexual behavior and
        sexuality)
Belgium, same-sex marriage in,
    148–149
beliefs, choices influenced by, 23. See
    also religion; values
bias. See also discrimination
    antigay, 148–149, 154–156,
        164–167
    gender, 36
    researcher, 27

binuclear families, 8, 215, 275
biofeedback, 222–223
biosocial theory, 36
birth control, 118. See also condoms;
    family planning
birth order, 209–210
bisexuality, 150, 151–152, 155–156,
    158. See also homosexuality and
    GLBT individuals
blackmail, psychological, 107
blended families, 8, 274–275. See also
    stepfamilies
blood transfusions, 123
"blue" wedding artifacts, 135
body image, 42, 90, 126
Boehner, John, 44
Bono, Chaz, 34
boomerang generation, 201
"borrowed" wedding artifacts, 135
boundaries, 16, 272–273
Boyle, Susan, 68
brainstorming, 64
breaking up, 258–259. See also
    divorce
bride wealth, 134
broken heart, recovering from, 260, 262
Brown, Chris, 25, 236
Buddhism, 6, 221
Bullock, Sandra, 215
Bush administration, 161

C

Canada
    domestic partnerships in, 7
    polygamy in, 5
cancer, 224–225
careers. See work, occupations and
    careers
caregiving
    day-care options, 173, 175, 202
    elderly adult, 287, 289–290, 292,
        298, 300, 301
    gender differences in, 35, 37, 40–41,
        42, 43, 170, 174, 176, 289, 300
    parental caregiving role, 200–201
        (see also parents and parenthood)
    physical care by families, 13, 200–201
Carlin, George, 258
Catholicism. See Roman Catholic
    Church
Channing, Carol, 296
Chaplin, Oona and Charles, 140
cheating. See extradyadic involvement
    or infidelity
children and youth. See also families
    abuse of (see abuse and violence in
        relationships)
    adoption of, 164–165, 192–194
    biracial, 140
    birth order of, 209–210
    child support for, 262, 268, 273, 276
    conflicts with (see conflicts)
    custody of (see custody and
        visitation)
    day-care options for, 173, 175, 202
    death of, 15, 234, 235
    desire for, 184–185
    discipline of, 209, 213, 241
    divorce impacting, 261, 262–264,
        264–268
    education of (see schools and
        education)
    elderly adult relationships with, 287,
        289–290, 292, 298, 300, 301
    emotional competence of, 214
    extradyadic relationships
        impacting, 230
    fathers' separation from, 217,
        262–263

friends of (see friends, peers, and
    social groups)
gender differences in caregiving
    for, 35, 37, 40–41, 42, 43, 170,
    174, 176
gender of (see gender)
Generation Y, 19, 71
historical view of, 209
HIV transmission to, 123–124
juvenile delinquency of, 10, 204, 213
legal responsibility for, 4–5, 133,
    165, 217
mother-child bonding, 43, 206
number of, 186–187
overindulgence of, 212
parenting of (see parents and
    parenthood)
planning for, 182–198
quality time with, 172, 202
remarriage impacting, 268, 273,
    274–283 (see also stepfamilies)
research consent for, 25–26
responsibility of, 213–214
siblings of (see siblings)
substance abuse among, 232
suicide, suicide attempts, and
    threats by, 15, 235
teen pregnancy and parenthood
    among, 178, 185, 187–188
uniqueness of, 209
work-related impacts on, 172–173,
    175, 202
China
    adopted children from, 192, 193
    arranged marriage in, 87
    elderly adult caregiving in, 300
choices
    abortion-related positions on, 197
    facts of, 17–20
    influences on, 20–25
    parenting, 205
    in relationships, 17–25
chronic fatigue syndrome, 225
Church of Jesus Christ of Latter-day
    Saints, 5–6, 39, 172, 184, 186
civil unions and domestic
    partnerships, 7, 81, 159–160
Clinton, Chelsea, 80
Clinton, William/Clinton
    administration, 161
cognition and perception
    age-related issues of, 286, 289, 292
    changes to, in conflict resolution, 63
    cognitive developmental theory, 37
    cognitive restructuring of work
        stress, 174
    conflict arising from, 53, 63
    divorce perceptions, 258, 269
    love influenced by, 92
    marital success influenced by, 146
    stress management changes of,
        220–221
cohabitation, 7, 79–81, 300
cohousing, 291
collaborating style of conflict, 54
Collins, Judy, 235
colonial America, 89
commensality, 132
commitment
    love as element of, 86–87
    marital success influenced by, 145
    marriage as, 133–134
    negative, 272–273
    parenthood influencing marital, 208
common-law marriage, 4
communes. See intentional
    communities
communication
    conflict and, 50, 52–54, 57, 62–67
    deceptive, 58–60

effective, 55–58
electronic (*see* Internet; technology; texting)
fighting fairly using safe communication guidelines, 62–67
gender differences in, 59, 60–61
interpersonal, 51–52
marital success influenced by, 145–146
money-related, 180–181
nonverbal, 51–52, 55, 56, 57
sexual satisfaction influenced by open, 127
technology impacting, 50, 52 (*see also* Internet; texting)
theories applied to, 61
community
  community factors influencing abuse and violence, 242
  divorce influenced by, 255
  elderly adults in, 291
  intentional communities, 6, 16
  remarriage impacting, 273
companionship, marrying for, 132, 133
competing style of conflict, 54
competitive birthing, 187
complementary-needs theory, 100–101
compromising style of conflict, 54
computer-mediated communication. *See* Internet; technology
concurrent sexual partnerships, 5–6, 117. *See also* extradyadic involvement or infidelity
condoms, 17, 20, 24, 42, 114, 118, 123, 125, 157, 216. *See also* family planning
conflicts
  benefits of, 52–53
  conflict framework, 14–15, 278, 288
  conflict resolution, 52, 57, 62–67, 157, 256, 269
  divorce due to, 256
  inevitability of, 52
  money as source of, 178
  parent-teenager, 215, 216
  in relationships, 50, 52–54, 57, 62–67, 157
  role, 23
  sources of, 53–54
  styles of, 54
conjugal love, 86
consummate love, 86
continuity theory, 288
conversion therapy, 153–154
Coolidge effect, 228
Cooper, Sue Ellen, 226
Cooperative Parenting and Divorce Program, 266
corporal punishment, 241
costs. *See* money and finances
"cougars," 140
covenant marriage, 271
credit cards and credit rating, 179–180
criminal behavior
  abusers', 95, 236, 238, 243, 249, 250 (*see also* abuse and violence in relationships)
  decriminalization of sodomy, 159
  familial deterrents of, 14
  hate crimes as, 148, 165–166
  honor crime/killing, 242
  juvenile, 10, 204, 213
  marital deterrents of, 5
  rape and sexual assault as, 167, 244–246
  stalking as, 93, 95, 239–240
crisis. *See* stress and crisis
cross-dressers, 34
cross-national issues. *See* international issues
cultural influences. *See also* race and ethnicity
  on abuse and violence in relationship, 240–242, 251

on age, 286–287
on choices, 23
on cohabitation, 79–80
on divorce, 252–253, 255, 270–271
on extradyadic involvement, 229
on familial relationships, 9
on family planning, 182, 184
on gender identity, 30, 33
globalization and, 20
on love, 4, 87–89, 89–90, 95–96
on marriage, 4, 5–6, 87, 95–96, 133, 134, 135, 142
on mate selection, 87–89, 89–90, 95–96
on same-sex relationships, 88
on sexuality, 24
on singlehood, 73
on superperson strategy, 174
on types of marriage, 5–6, 87
on weddings, 135
custody and visitation
  divorce leading to, 258, 259, 262–263, 265–266
  gender-based awards of, 45, 258, 259, 262
  of pets, 7
  remarriage impacting, 271
  same-sex parents', 164, 165
  stepfamily issues with, 277
cybersex, 227
cybervictimization, 240

D

Dangerfield, Rodney, 217
Darwin, Charles, 104
date rape, 244–246
day-care, 173, 175, 202
death
  abuse leading to, 95, 236, 238, 250
  of children, 15, 234, 235
  of elderly adults, 286, 292, 298–299, 300–301
  end of life issues, 225, 298–301
  inheritance after, 4, 136, 160, 277
  life expectancy and, 42, 45, 286
  of parents, 234
  preparing for, 300–301
  of spouse, 4, 74, 136, 273–274, 286, 298–299
  stress and crisis caused by, 234–235, 298–299
  suicide as cause of, 15, 73–74, 87, 107, 155, 166, 234–235, 292
  widowed singlehood due to, 74, 273–274, 298–299
debt, marital, 139, 176–181
deception. *See also* extradyadic involvement or infidelity; lies/lying
  communication-related, 58–60
  in Internet partner searches, 77–78
  research distortion and deception, 27–28
decision making
  choices made through, 17, 18, 19–20
  leadership ambiguity related to, 54
  skills, 19–20
  styles of, 18
deep muscle relaxation, 223, 269
defense mechanisms, 66–67
Defense of Marriage Act (DOMA), 161
DeGeneres, Ellen, 7, 71, 139
dementia, 292. *See also* Alzheimer's disease
De Niro, Robert, 97
depression
  abuse and violence leading to, 246, 248
  death causing, 234
  divorce leading to, 260, 270
  elderly adults', 291, 292
  postpartum, 206–207

stress and crisis created by, 225
substance abuse related to, 232
de Rossi, Portia, 7, 139
Dion, Celine, 139
disabilities
  adoption of children with, 192, 194
  age-induced, 286, 289, 299
  insurance coverage for, 231
  stress and crisis related to, 224–225, 231
discipline, 209, 213, 241
discrimination
  ageism, 287–288
  antinatalism, 186
  divorcism, 270–271
  heterosexism, 154
  homophobic and antigay, 148–149, 154–156, 164–167
  racism, 97
  sexism, 42
  stepism, 276–277
disenchantment, after marriage, 137
disengagement theory, 288
displacement, 66–67
divorce
  broken hearts from, 260, 262
  children impacted by, 261, 262–264, 264–268
  cultural influences on, 252–253, 255, 270–271
  custody and visitation after (*see* custody and visitation)
  discrimination based on, 270–271
  ending unsatisfactory relationship, 258–259
  fathers' separation from children in, 262–263
  gender differences in matters related to, 258, 259, 260
  infidelity as grounds for, 4, 256
  legal dissolution in, 134, 254, 261, 269
  macro factors in, 252–255
  micro factors in, 255–258
  military couples' rate of, 144
  monetary considerations in, 4, 132, 136, 170, 253–254, 259, 261, 262, 268, 273, 276
  no-fault, 254
  parental alienation in, 263–264
  parenting influencing rate of, 208
  prevention of, 270–271
  property division in, 4, 136, 262
  remarriage after (*see* remarriage)
  sexual behavior after, 122–123
  shared parenting dysfunction in, 263
  singlehood due to, 73–74
  spousal consequences of, 260–264
  stages of, 261, 270
  stress in, 219, 254, 269–270
  substance abuse and, 233, 269
  "successful," 268–270
  telling children about, 261, 267
domestic partnerships and civil unions, 7, 81, 159–160
domestic violence. *See* abuse and violence in relationships
double standards, sexual, 117–118, 120
Douglas, Cameron, 232
down low, 229
dowry, 89
Drugfree.org, 213
Drug-Induced Rape Prevention Punishment Act, 246
drug use. *See* substance use/abuse

E

Eastwood, Clint, 286
economic factors. *See also* money and finances
  cohabitation for, 80–81
  divorce impacted by, 253–254

families based on, 9–10, 13
family planning influenced by, 184
gender roles influenced by, 39, 40–41
global inequality of, 178
globalization impacting, 20
marriage based on, 6, 84, 132–133
mate selection based on, 36, 88–89, 99, 104
parental economic resource role, 202
poverty as, 41, 177, 178, 188, 217, 242, 251
singlehood influenced by, 71
social exchange framework on, 12, 61, 101–102
unemployment and, 168, 178, 221, 231, 251
education. *See* schools and education
Edwards, John, 59, 228, 229
elderly adults
  abuse of, 290
  age related issues for, 285–290, 295–296, 298–301
  caregiving for, 287, 289–290, 292, 298, 300, 301
  discrimination against, 287–288
  end of life issues for, 298–301
  health of, 286, 289, 291–292, 295–296, 299, 300–301
  housing issues for, 291
  life expectancy for, 42, 45, 286
  monetary considerations for, 290–291, 292–293, 300, 301
  relationships of, 273–274, 296–298, 299–300, 301
  remarriage and new partner involvement among, 273–274, 299–300
  retirement of, 292–293
  sexual behavior and sexuality of, 129, 294–295
  statistics on, 284
  successful aging of, 295–296
  suicide among, 292
  widowed, 74, 273–274, 298–299
ElderSpirit Community, 291
electromyographic biofeedback, 222
electronic communication. *See* Internet; technology; texting
e-mail, 298
emotions
  communication using "soft," 56
  defense mechanisms related to, 66–67
  emotional abuse, 238–239 (*see also* abuse and violence in relationships)
  emotional competence, 214
  emotional remarriage, 273
  familial emotional support, 13
  gender differences in expression of, 44
  grief as, 259, 266, 272
  jealousy as, 85, 92–95, 242
  lack of, 91, 92
  love as (*see* love)
  marriage as emotional relationship, 4, 133
  parental emotional resource role, 201–202
  single-parents' emotional needs, 216
empty love, 86
endogamy, 96
engagement, 80, 104–106, 108–110
England, extradyadic involvement in, 229
equal relationships and egalitarianism
  dual career marriages and, 171–172
  gender inequality leading to abuse and violence, 241–242
  marital success influenced by, 146
  in modern vs. traditional marriage, 133

eros love style, 84–85
escapism, 66, 106
Ethiopia, adopted children from, 193
ethnicity. See race and ethnicity
exchange theory, 101–102. See also
    social exchange framework
exercise, 221, 270, 293, 295
exogamy, 96
experience, choices influenced by,
    24–25
extended families, 8–9, 10. See also
    grandparents
extradyadic involvement or infidelity
    concurrent sexual partnerships
        and, 117
    demographics related to, 60
    divorce as result of, 4, 256
    effects of, 230
    familial patterns of, 24
    jealousy based on, 94, 95
    military separations leading
        to, 143
    prevention of, 231
    reasons for, 228–230
    sexually transmitted infections due
        to, 124
    simultaneous loves and, 93
    stress and crisis caused by, 226–231
    successful recovery from, 230–231
    types of, 226–227
eye contact, communication via, 55

F

Facebook, 59, 298
families. See also children and youth;
    parents and parenthood
    abuse and violence influenced by,
        243–244
    binuclear, 8, 215, 275
    blended, 8, 274–275 (see also
        stepfamilies)
    changes in, 9–11, 254
    choices influenced by, 21, 24
    conflict in (see conflicts)
    definition of, 6–7, 10
    elderly adult caregiving by, 287,
        289–290, 292, 298, 300, 301
    extended, 8–9, 10 (see also
        grandparents)
    family life course/cycle, 12–13, 19,
        27, 39
    family systems framework, 16
    gender role influences of, 10,
        36–37, 38
    homosexual relationship response
        of, 149, 166
    incest in, 96
    marriage as family-to-family
        commitment, 134
    marriage differences from, 9
    nuclear, 8, 275–277
    of origin, 7–8, 24, 243
    pets in, 7, 223, 301
    planning for children in, 182–198
    poverty impacting, 178 (see also
        poverty)
    of procreation, 8
    research on, 25–29
    same-sex, 7 (see also same-sex
        couples)
    siblings in (see siblings)
    stress and crisis in (see stress and
        crisis)
    theoretical frameworks of,
        12–17, 278
    types of, 7–9
    work issues impacting (see work,
        occupations and careers)
family planning
    abortion, 194–198
    adoption planning, 192–194
    assisted reproductive technology,
        189–192
    cultural and social influences on,
        182, 184

desire for children, 184–185
foster parenting planning, 194
infertility, 188–192
number of children, 186–187
overview of, 182–183
Family Success program, 147
fathers and fatherhood. See also
    parents and parenthood
    absence of/separation from children,
        217, 262–263
    familial changes to role of, 10, 172
    Father-Daughter Purity Ball, 115
    full-time parental role of/Mr.
        Mom, 172
    gender role influences of, 38
    parenting styles of, 211
    transition to fatherhood, 207–208
fatuous love, 86
feminism
    feminist framework, 16–17
    pro-life vs. pro-choice stance in, 197
    sexual values influenced by, 119
    singlehood acceptance influenced
        by, 71
Fertell (fertility kit), 189
filial piety, 300
finances. See money and finances
First Wives World, 270
Fisher, Herbert and Zelmyra, 297
flirting, 52
foreclosures and evictions, 178. See
    also bankruptcy
foreign issues. See international issues
forgiveness, 65–66, 146, 214,
    230–231, 248–249
Foster, Jodie, 215
foster families, 194
France
    civil unions in, 7
    cohabitation in, 81
    extradyadic involvement in, 229
    marriage delays in, 69
    sexual satisfaction in, 121, 127
Frank, Barney, 71
freedom, loss of, 137
freedom of movement, 45
Freud, Sigmund, 37, 102
friends, peers, and social groups
    changes to, after marriage, 136–137
    child development influences of, 210
    choices influenced by, 22
    divorce impacting, 260, 261, 270
    family planning influenced by, 184
    friends with benefits, 117
    gender roles influenced by, 38–39
    jealous individuals turning to, 95
    parental monitoring of, 204
    remarriage impacting, 273
    stress management assistance
        from, 221
Fundamentalist Church of Jesus
    Christ of Latter-day Saints
    (FLDS), 5–6

G

galvanic skin response biofeedback, 222
garriage, 160
gay/lesbian couples. See
    homosexuality and GLBT
    individuals; same-sex couples
gender
    androgynous, 46–47
    biological sex and, 31–32, 35
    cultural influences on, 30, 33
    definition of, 32
    differences based on (see gender
        differences)
    double standards based on,
        117–118, 120
    family balancing by, 191
    gender dysphoria, 34 (see also
        transgenderism)
    gender identity, 30, 33, 34
    gender roles (see gender roles)
    intersex, 30, 32, 46

nature/nurture issues with, 32–33
    terminology related to, 31–36
    transgenderism, 34–35, 60, 150 (see
        also homosexuality and GLBT
        individuals)
gender differences
    in body image, 42
    in child care, 35, 37, 40–41, 42, 43,
        170, 174, 176
    in cohabitation meaning, 80
    in communication, 59, 60–61
    in custody awards, 45, 258, 259, 262
    in death responses, 234, 299
    in divorce matters, 258, 259, 260
    in elderly adult caregiving, 289, 300
    in elderly adults' aging issues, 287,
        290–291, 299
    in emotional expression, 44
    in freedom of movement, 45
    in grief, 234
    in hooking up, 75–76
    in income, 39, 40–41, 45,
        168–170, 181
    in intimacy, 44–45
    in jealousy, 95
    in life expectancy, 42, 45
    in love, 84, 85
    in marital satisfaction, 42
    in power, 39, 41, 42, 45, 168–169
    in psychological and mental health
        issues, 42
    in relationship focus and
        maintenance, 42–43, 45
    in romantic relationship views, 35
    in sexual behavior, 115, 117, 118,
        119–121, 128, 157
    in sexually transmitted infections,
        42, 152
    in stress management, 223
    in substance use/abuse, 45
    in work, occupations and careers,
        39, 40, 41, 43–44, 231–232
gender roles
    abuse and violence based on
        inequality of, 241–242
    changes in, 46–48
    consequences of traditional, 40–46
    definition of, 35
    familial and parental influences on,
        10, 36–37, 38
    identifying with parental, 37
    ideology on, 35–36
    influences on, 10, 33, 36–40
    military-influenced, 143, 241
    postmodern, 48
    in same-sex relationships, 156–157
    socialization of, 33, 37–46
    terminology related to, 31–36
    theories of development, 36–37
    transcendence of, 47
Generation Y, 19, 71
Genie, 13
Germany
    cohabitation in, 81
    corporal punishment in, 241
    marriage delays in, 69
globalization, 20
government, choices influenced by, 21
grandparents, 8, 85, 144, 196, 261,
    262, 283, 290, 292
gratitude, 91
grief, 234–235, 259, 266, 272. See
    also death
guardianship, 217

H

habits, choices influenced by, 24
hanging out, 74–75
hate crimes, 148, 165–166
Hawn, Goldie, 81
health
    abortion health impact, 195, 197–198
    age-induced issues with, 286,
        289, 291–292, 295–296, 299,
        300–301

androgyny impacting, 47
conflict impacting, 53
disabilities impacting, 192, 194,
    224–225, 231, 286, 289, 299
gender differences in, 42, 45
marital success influencing,
    146–147
mental (see psychological and
    mental health issues)
parental health promotion role, 204
physical care of, 13, 200–201 (see
    also caregiving)
poverty impacting, 177, 178
sexual (see sexually transmitted
    infections)
sexual satisfaction impacted by, 126
stress and crisis related to, 224–226
teenage parenthood impacting, 188
therapeutic abortion to protect
    mother's, 195, 197
health insurance. See insurance benefits
Healthy Marriage Initiative, 147
hedonism, 117, 120
helicopter parents, 209
heterosexuality, 150, 151–152, 154,
    158–159
Hightower, Grace, 97
Hispanic Americans. See also race and
    ethnicity
    divorce among, 255
    elderly, 291, 292
    family planning among, 187
    gender roles among, 38
    interracial relationships of, 88,
        97, 140
    sexual behavior among, 121
HIV infections, 118, 123–124, 125, 157
homogamy theory of mate selection,
    96–99
homosexuality and GLBT individuals.
    See also same-sex couples
    adoption by, 164–165, 192
    conflict resolution among, 157
    definition of, 150
    discrimination and antigay bias
        against, 148–149, 154–156,
        164–167
    extradyadic involvement of, 229
    family planning for, 190, 191, 192
    gay liberation movement, 71
    hate crimes against, 148, 165–166
    Internet partner searches based
        on, 77
    legal issues for, 95, 159–163
    mixed-orientation relationships of,
        158–159
    monogamy and sexuality of, 157–158
    parenting issues for, 161, 162,
        163–165, 192, 207
    prevalence of, 152
    relationship satisfaction of,
        156–157
    secretive behavior related to, 60
    sexual orientation as, 149–150,
        151–152, 153–154, 155, 229
    stepfamilies of, 275
    terminology related to, 20, 150
honeymoons, 136
honor crime/killing, 242
hooking up, 75–76
hormone therapy, 189–190, 294–295
hostage situations, 238
Houdini, Harry and Bess, 50
housing, 291
Huguely, George, 95, 236, 238
humor, stress management via, 222

I

identity
    gender, 30, 33, 34 (see also gender
        roles)
    identity theft, 180
    personality and (see personality)
    self-worth/self-concept and, 15, 61
        (see also self-esteem)

illness. *See* health
immigration status, 142, 162
impulse control disorders, 225
incest, 96
income. *See* money and finances
India
  arranged marriage in, 87
  surrogacy in, 190
individualism, 10, 19, 23, 88, 255
Indonesia, arranged marriage in, 87
Industrial Revolution, 9–10, 254
infatuation, defined, 84, 86
infertility, 188–192
infidelity. *See* extradyadic involvement
  or infidelity
inheritance, 4, 136, 160, 277
in-laws, 138–139
insecurity. *See* safety and security
institutions
  choices influenced by, 20–21
  educational (*see* schools and
    education)
  Institutional Review Boards of, 25
  marriage as (*see* marriage)
insurance benefits
  elderly adults', 289, 290–291
  inability to afford, 177
  marriage entitling, 4, 6, 162
  marriage to acquire, 107, 142
  same-sex couples' access to, 160, 162
  unemployment impacting, 231
intentional communities, 6, 16
international issues. *See also specific*
  countries
  international adoptions, 192,
    193, 194
  international/cross-national
    relationships, 79, 141–142
Internet
  child development influenced by, 210
  cybersex via, 227
  cybervictimization via, 240
  divorce rate influenced by, 254
  elderly adult use of, 298
  extradyadic involvement via, 227
  finding a partner via, 76–78,
    157, 300
  Generation Y as Internet
    Generation, 19, 71
  parental monitoring of children's
    use of, 204, 210
  sexual values influenced by, 119
  social networking via, 59, 298
  stepfamily education program
    via, 283
  video chatting via, 78
  weddings broadcasted via, 136
interracial relationships, 88, 95, 97,
  140–141
intersex gender issues, 30, 32, 46
intimacy, 44–45, 86–87
intimate partner violence, 237. *See*
  also abuse and violence in
  relationships
in vitro fertilization, 190–191, 192
isolation, 242, 243
iSpQ software, 78
"I" statements, 56
Italy
  cohabitation in, 79–80
  corporal punishment in, 241
  marriage delays in, 69

J

Japan, sexual satisfaction in, 127
jealousy, 85, 92–95, 242
Jessop, Joe, 5–6
jobs. *See* work, occupations and
  careers
Jolie, Angelina, 98, 192, 193, 228
Jordan, abuse and violence in
  relationships in, 241–242
Judaism, 88, 135, 141, 184
Judd, Naomi and Wynonna, 197
juvenile delinquency, 10, 204, 213

K

Kenya
  antigay discrimination in, 149
  bride wealth in, 134
King, Larry, 255
Korea, adopted children from, 192
Kulijian, Harry, 296
Kutcher, Ashton, 140

L

Lady Gaga, 48
LaLanne, Jack, 222, 284, 286, 295
Lambert, Miranda, 98
latchkey parents, 268
Latinos. *See* Hispanic Americans
Lawrence v. Texas, 159
leadership. *See* power and leadership
legal issues
  abortion legality, 195
  custody as (*see* custody and
    visitation)
  divorce as legal dissolution, 134,
    254, 261, 269
  elderly adult, 301
  legal blurring of married/
    unmarried, 72
  legal changes after marriage, 136
  legal responsibility for children,
    4–5, 133, 165, 217
  liberal divorce laws, 254
  marriage as legal contract, 4, 131,
    134, 163
  remarriage creating, 273
  same-sex relationship-based, 95,
    159–163
  stepfamily legal
    considerations, 277
  transracial adoption legality, 193
Leno, Jay and Mavis, 100, 138
lesbian/gay couples. *See*
  homosexuality and GLBT
  individuals; same-sex couples
Levine, Robert, 140
licenses, marriage, 4, 135
lies/lying, 59, 77–78, 212. *See also*
  deception
life course/cycles, family, 12–13, 19,
  27, 39
life expectancy, 42, 45, 286
lifestyle, parenthood impacting, 185
LinkedIn, 298
listening, reflective, 55–56
living together. *See* cohabitation
loneliness, 73
looking-glass self, 15, 61
Lopez, Jennifer, 84
love
  abuse seen as intense, 82, 93
  conceptualization of, 84–87
  cultural influences on, 4, 87–89,
    89–90, 95–96
  falling out of, 255
  jealousy based on, 85, 92–95
  marriage based on, 4, 131, 133
  mate selection based on, 81–92
  monetary spending as sign of, 170
  new relationship development
    through, 89–92
  parental, 212
  polyamorous, 6
  problems caused by, 93
  remarriage based on, 273
  renewing old love relationships, 296
  simultaneous, 93 (*see also*
    concurrent sexual partnerships)
  social and historical context of,
    87–89
  stress management via, 221
  styles of, 84–86
  triangular view of, 86–87
  unrequited or unfulfilling, 93
Love, Yeardley, 95, 236, 238
ludic love style, 84
lust, defined, 84

M

mania love style, 85
manners, good, 102
marital status, 98
marriage
  abuse and violence in, 246, 247
    (*see also* abuse and violence in
    relationships)
  age at time of, 69, 70, 108,
    139–140, 273–274
  age-discrepant, 139–140, 273
  Alternatives to Marriage Project, 72
  arranged, 87, 88–89
  benefits of, 5
  ceremony, 5 (*see also* weddings)
  changes after, 136–139
  changes in history of, 9–11
  child-free, 186
  cohabitation in lieu of, 7, 79–81, 300
  as commitment, 133–134
  common-law, 4
  conflict in (*see* conflicts)
  covenant, 271
  cross-national, 141–142
  cultural influences on, 4, 5–6, 87,
    95–96, 133, 134, 135, 142
  definition of, 3
  delaying timing of, 69–70, 108
  diversity in, 139–145
  divorce ending (*see* divorce)
  economic considerations in, 6, 84,
    132–133
  elderly adults', 273–274, 296–297
  elements of, 4–5
  engagement prior to, 80, 104–106,
    108–110
  family differences from, 9
  honeymoons after, 136
  infidelity in (*see* extradyadic
    involvement or infidelity)
  interracial, 88, 95, 97, 140–141
  interreligious, 141
  as legal contract, 4, 131, 134, 163
  licenses, 4, 135
  love in (*see* love)
  marital satisfaction in, 42, 65,
    122, 139, 144, 145–147, 208,
    296–297
  marriage-resilience perspective
    on, 10
  mate selection for (*see* mate
    selection)
  May-December, 139–140, 273
  military, 142–145
  monetary issues in (*see* money and
    finances)
  motivations for, 131–133
  polygamous, 5–6
  poverty impacting, 178 (*see also*
    poverty)
  rebound, 106
  remaining in unhappy, 259
  research on, 25–29
  as rite of passage, 134–136
  same-sex, 3, 4, 7, 71, 95, 148–149,
    160–163 (*see also* same-sex
    couples)
  sexual behavior in, 122, 138, 146
  societal functions of, 133
  statistics on, 2–3
  stress and crisis in (*see* stress and
    crisis)
  subsequent (*see* remarriage)
  success of, 145–147
  theoretical frameworks of, 12–17
  traditional *vs.* egalitarian, 133
  types of, 5–6, 87, 139–145
  weddings for, 5, 108–110, 134,
    135–136
  for wrong reasons, 106–108
mass media. *See* media
matchmaking, 79, 87
mate selection
  arranged, 87, 88–89
  biosocial theory of, 36

mania love style, 85
cultural influences on, 87–89,
  89–90, 95–96
economic influences on, 36, 88–89,
  99, 104
engagement as declaration of,
  104–106
jealousy in, 85, 92–95
love in, 81–92
marriage for wrong reasons,
  106–108
mating gradient in, 22, 45, 97,
  98, 272
new relationship development in,
  89–92, 299–300
parental influences on, 87–88, 89,
  97, 102, 110
psychological factors in, 100–104
race and ethnicity influencing, 88,
  95, 96–97
sociobiological factors in, 104
sociological factors in, 96–99
wedding cancellation, 108–110
May-December marriages,
  139–140, 273
McBride, Martina, 241
McCain, John, 262
Mead, Margaret, 201, 256
meals, 132, 290
media. *See also* Internet
  abuse and violence in, 236,
    241, 251
  child development influenced
    by, 210
  choices influenced by, 23–24
  gender roles influenced by, 39–40
  sexual values influenced by, 118–119
  social, 59, 298
mediation, 269
medieval Europe, 88–89
men. *See also specific topics such as*
  fathers and fatherhood, gender,
  marriage, *or* sexuality
  abuse of, by female partners, 239
  identity of, tied to work, 43–44,
    231–232
  postabortion male attitudes, 198
  rape of, 244
mental health. *See* psychological and
  mental health issues
Mexico
  adopted children from, 192
  cost of living in, 177
midlife crisis, 226
military
  extradyadic involvement of military
    spouses, 230
  gender roles in, 143, 241
  marriage in, 142–145
Milk, Harvey, 152, 167
minorities. *See* cultural influences;
  race and ethnicity; *specific groups*
  by name
miscarriages, 195
mistrust, 94
modeling
  gender role behavior, 36–37
  modeling theory of mate
    selection, 102
modern families, 8
modernization theory, 288
money and finances. *See also*
  economic factors
  adoption costs impacting, 193
  assisted reproductive technology
    costs impacting, 192
  bankruptcy impacting, 177, 221
  bride wealth as evidence of, 134
  communication about, 180–181
  cost of living, 177
  credit cards and credit rating,
    179–180
  day-care costs impacting, 173, 175
  debt issues, 139, 176–181
  divorce impacting, 4, 132, 136, 170,
    253, 254, 259, 261, 262, 268,
    273, 276

money and finances *(continued)*
  dual-earner families, 8, 10, 170–171, 172–173
  educational costs impacting, 175, 185
  elderly adult issues with, 290–291, 292–293, 300, 301
  family planning impacting, 183, 185, 192
  gender influences on, 39, 40–41, 45, 168–170, 181
  global inequality of, 178
  income distribution, 177
  marital considerations related to, 6, 84, 132–133, 136, 137–138, 139, 146, 168–170
  mate selection influenced by, 36, 88–89, 99, 104
  money savers cohabitating, 80–81
  nonmaterialistic approach to, 146
  parental economic resource role, 202
  poverty from lack of, 41, 177, 178, 188, 217, 242, 251
  power from access to, 168–170
  property division of, 4, 136, 262
  remarriage impacting, 132, 273
  social exchange framework on, 12, 61, 101–102
  stepfamily issues with, 276, 277, 278, 282
  stress and crisis related to, 178, 221, 231
  tax issues impacting, 161, 162, 184, 194
  wedding expenses, 135–136
monogamy, 4, 124–125, 157, 158
mood disorders, 225, 292. *See also* depression
Moore, Demi, 140
Moore, Mary Tyler, 140
Mormon Church, 5–6, 39, 172, 184, 186
mothers and motherhood. *See also* parents and parenthood
  mommy track for, 171
  mother-child bonding, 43, 206
  mother-child HIV transmission, 123–124
  mothers-in-law, 139
  parenting styles of, 211
  single, 215–217 (*see also* single-parents and single-parenthood)
  supermoms, 174
  surrogate, 190, 191
  teenage, 178, 185, 187–188
  transition to motherhood, 205–207
motivations
  child bearing, 185, 187
  marriage, 131–133
  unconscious, 24
Mundugumor people, 33
Muslim/Islamic religion, 135, 186
Myers, Deborah, 139
myths
  sexual, 129
  stepfamily, 275

N

Nabozny, Jamie, 166
Namath, Joe, 139
Narcotics Anonymous, 232
National Abortion Federation Hotline, 195
National Association for Research and Therapy of Homosexuality (NARTH), 153–154
National Association of Black Social Workers, 193
National Domestic Violence Hotline, 250
National Family Caregiver Support Program, 290
National Organization for Women, 197
nature, children's contact with, 214–215

negative commitment, 272–273
negative communication, 56
Nepal, arranged marriage in, 87
Netherlands, The
  extradyadic involvement in, 229
  same-sex marriage in, 148–149
neurofeedback, 222–223
never-married singles, 72–73, 122
Newman, Paul, 200
"new" wedding artifacts, 135
no-fault divorce, 254
nonverbal communication, 51–52, 55, 56, 57
nonviolence, parental teaching of, 214
Norway, same-sex marriage in, 148–149
nuclear families, 8, 275–277

O

Obama, Barack/Obama administration, 38, 98, 130, 140, 161, 206, 217
Obama, Michelle, 38, 98, 130, 206
obesity and overweight, 225, 291
obsessive relational intrusion, 240
occupations. *See* work, occupations and careers
older adults. *See* elderly adults
"old" wedding artifacts, 135
Oneida Community, 6
online activities. *See* Internet
open-mindedness, 98
"opposites attract" theory, 100–101
oppositional defiant disorder, 204
organ transplants, 124
overindulgence, avoiding, 212
ovum transfer, 191

P

Pakistan
  antigay discrimination in, 149
  marriage in, 142
palliative care, 225
pantagamy, 6
parallel style of conflict, 54
Parenting Toolkit: Skills for Stepfamilies, 283
parents and parenthood. *See also* children and youth; families; fathers and fatherhood; grandparents; mothers and motherhood
  abortion parental consent/ notification, 195, 196
  abuse and violence between, 244 (*see also* abuse and violence in relationships)
  birth order influencing, 209–210
  caregiving for, 287, 289–290, 292, 298, 300, 301 (*see also* elderly adults)
  choices perspective of, 205
  conflicts of (*see* conflicts)
  custody and visitation of (*see* custody and visitation)
  death of, 234
  discipline by, 209, 213, 241
  divorce impacting, 261, 262–264, 264–268
  external influences on, 210
  families based on, 7–8
  family planning for, 182–198
  foster, 194
  gender differences in, 35, 37, 40–41, 42, 43, 170, 174, 176
  gender role development influenced by, 10, 36–37, 38
  helicopter, 209
  historical views of, 209
  homosexual relationship response of, 149, 166
  latchkey, 268
  lifestyle changes of, 185
  as marital role models, 146
  marriage changing relationship with, 138–139

as marriage motivation, 132
monitoring children's behavior, 202, 204, 210, 212–213, 214, 215
parental alienation, 263–264
parental influence on mate selection, 87–88, 89, 97, 102, 110
parental support of marriage, 133
parenting facts, 208–211
parenting styles, 211
  principles of effective, 211–215
  quality time with, 172, 202
  roles involved in, 200–204
  same-sex, 161, 162, 163–165, 192, 207
  shared parenting dysfunction, 263
  single-parenting, 8, 41, 188, 215–217
  step- (*see* stepfamilies)
  teenage, 178, 185, 187–188
  transition to parenthood, 205–208
  uniqueness of children impacting, 209
  visiting mate's, 105
  work issues impacting (*see* work, occupations and careers)
partners
  domestic, 7, 81, 159–160
  finding, 74–79, 157, 300
  same-sex (*see* same-sex couples)
  selecting (*see* mate selection)
  terminology identifying, 20
passion, love as element of, 86–87
peers. *See* friends, peers, and social groups
Pelosi, Nancy, 47
pensions. *See* retirement benefits
perception. *See* cognition and perception
permissive parenting style, 211
personal fulfillment, 10, 132
personality
  abuse and violence influenced by, 242–243
  addiction-prone, 232
  birth order influencing, 209–210
  choices influenced by, 24
  love influenced by, 90
  mate selection influenced by, 99, 102–104
pets, 7, 223, 301
pheromones, 121–122
physical appearance, 77–78, 89–90, 98, 225, 291
physical health. *See* health
Pitt, Brad, 98, 192, 228
pity, 107
political thinking, 99, 197, 284
polyamory, 6
polyandry, 6
polygamy, 5–6
polygyny, 5–6
pornography, 120
Porter, Cole, 15, 229
Porter, Linda Lee, 229
positive communication, 56
Positive Parenting Through Divorce program, 266
possessive jealousy, 92, 95
POSSLQs (people of the opposite sex sharing living quarters), 80
postmodernism
  gender postmodernism, 48
  postmodern families, 8 (*see also* binuclear families; same-sex couples; single-parents and single-parenthood)
postnuptial agreements, 262
postpartum depression/psychosis, 206–207
posttraumatic stress disorder (PTSD), 143, 246, 248
Potti, Anil, 27
poverty
  abuse and violence due to, 178, 242, 251
  definition of, 177

effects of, 178
feminization of, 41
global inequality of, 178
income distribution and, 177
single-parenthood and, 41, 188, 217
teenage parenthood and, 188
power and leadership. *See also* status
  abuse as expression of, 14, 138, 238, 241, 242
  gender differences in, 39, 41, 42, 45, 168–169
  leadership ambiguity as conflict source, 54
  marital distribution of, 138, 168–170
  money as source of, 168–170
  sharing, in relationships, 57–58
pragma love style, 84
praise, parental, 212
pregnancy
  aborted, 194–198
  abuse during, 248
  birth control to prevent, 118 (*see also* condoms)
  family planning for, 182–198
  HIV transmission during, 123–124
  marriage due to unplanned, 106–107
  substance use/abuse during, 188, 202, 204
  teen, 178, 185, 187–188
  wastage, infertility and, 189
prejudice, 149, 154. *See also* discrimination
premarital education programs, 105–106
prenuptial agreements, 254, 262
projection, 66
pro-life *vs.* pro-choice stance, 197
pronatalism, 184
property division, in divorce, 4, 136, 262
prostate cancer, 224–225
protector role, parental, 202, 204. *See also* safety and security
Protestant religions, 88, 135, 141, 163
psychological and mental health issues
  abortion-related, 197–198
  ADHD as, 58, 209, 226
  adoptive children's, 193
  age-induced issues, 286, 291, 292, 299
  androgyny impacting, 47
  anxiety and anxiety disorders as, 92, 95, 225, 232
  choices influenced by, 24
  conflict impacting, 53
  depression as, 206–207, 225, 232, 234, 246, 248, 260, 270, 291, 292
  divorce creating, 260, 263–264, 270
  gender differences in, 42
  impulse control disorders as, 225
  mood disorders as, 225, 292
  oppositional defiant disorder as, 204
  parental alienation as, 263–264
  parenthood impacting, 206–207
  posttraumatic stress disorder as, 143, 246, 248
  poverty impacting, 178
  psychological abuse, 238–239 (*see also* abuse and violence in relationships)
  psychological blackmail, 107
  psychological conditions for love, 90–91
  psychological factors in mate selection, 100–104
  sexual orientation impacting, 155, 166
  shared parenting dysfunction as, 263
  stress and crisis related to, 225–226
  substance abuse as (*see* substance use/abuse)
  suicide as, 15, 73–74, 87, 107, 155, 166, 234–235, 292
Pugach, Burt, 82

Q

quality time, 172, 202
queer theory, 150
questions
  engaged couples asking, 105
  open- vs. closed-ended, 55

R

race and ethnicity. See also cultural
  influences; specific groups by name
  divorce influenced by, 255
  elder issues influenced by, 291, 292
  family planning influenced by, 187
  family structure influenced by, 9
  gender roles influenced by, 38
  Internet partner searches based
    on, 77
  interracial relationships, 88, 95, 97,
    140–141
  mate selection influenced by, 88,
    95, 96–97
  parenting issues influenced by,
    208, 213
  racism based on, 97
  secretive behavior related to, 60
  sexual behavior influenced by,
    121, 229
  sexual orientation and, 155
  transracial adoptions, 192, 193
Radical Parenting website, 216
rape and sexual assault, 167, 244–246
rationalization, 66
reactive jealousy, 92, 94
realistic love, 86
rebound marriages, 106
reciprocal liking, 90
Red Hat Society, 226, 227
reflective listening, 55–56
registered partnerships. See civil
  unions and domestic partnerships
Reimer, David, 33
relativism, 116–117
relaxation, 223, 269
religion
  abortion influenced by, 197
  choices influenced by, 21
  cohabitation influenced by, 80
  divorce views in, 135, 254
  extended families based on, 9
  extradyadic involvement influenced
    by, 228
  family planning influenced by, 184,
    186, 197
  gender roles influenced by, 39
  Internet partner searches based
    on, 77
  interracial relationship views in,
    88, 140
  interreligious marriages, 141
  marital success influenced by, 146
  marriage views in, 5–6, 135,
    141, 146
  mate selection influenced by, 96,
    98–99
  sexual behavior influenced by, 114,
    115, 118
  sexual orientation and same-sex
    relationship issues influenced by,
    153, 154, 155, 161, 162, 163
  stress management influenced by,
    221, 222
  types of marriage practiced under,
    5–6
remarriage
  blended families after, 8, 274–275
    (see also stepfamilies)
  children born after, 268
  for divorced individuals, 271–272
  issues of, 272–273
  mental health issues impacting, 226
  monetary considerations in, 132, 273
  preparation for, 272
  stability of, 274
  stages of involvement in, 274
  for widowed individuals, 273–274

reparative therapy, 153–154
research, 25–29
resiliency, 10, 220
responsibility
  delegation of, to balance work and
    family, 174, 176
  divorce-related, 269
  increased, after marriage, 137
  parental encouraging of, 213–214
retirement, 292–293
retirement benefits. See also Social
  Security benefits
  cohabitation vs. marriage to
    retain, 81
  elderly adults', 290–291, 292
  marriage entitling, 6, 81, 162
  same-sex couples' access to, 160, 162
revenge, 229, 240, 262
Rihanna, 25, 236, 241
risk taking behavior, 85–86, 93
Riss, Linda, 82
rite of passage, marriage as, 134–136
rituals, 204, 282
Rivera, Geraldo, 255
roles
  choices influenced by, 22–23
  compartmentalization of, 176
  gender (see gender roles)
  parenting, 200–204
  role conflicts, 23
Roman Catholic Church
  abortion influenced by, 197
  cohabitation influenced by, 80
  family planning influenced by,
    184, 197
  gender roles influenced by, 39
  interracial relationship views in, 88
  marriage views in, 135
  same-sex relationship views of, 163
  sexual behavior influenced by, 115
romantic love style, 84–85, 85–86, 88
Rophypnol, 246
rules, familial, 16, 54
Russell, Kurt, 81
Russia, adopted children from, 192

S

safety and security
  abuse and violence creating lack
    of (see abuse and violence in
    relationships)
  ACT Raising Safe Kids Parenting
    Program, 214
  fighting fairly using safe
    communication guidelines,
    62–67
  jealousy due to insecurity, 94
  mate selection influenced by
    insecurity, 103
  parental protection and monitoring
    providing, 202, 204, 210,
    212–213, 214, 215
  safe sex, 123–125 (see also
    condoms)
Safe Zone programs, 148
same-sex couples. See also
  homosexuality and GLBT
  individuals
  adoption by, 164–165, 192
  civil unions and domestic
    partnerships of, 7, 81, 159–160
  conflict resolution among, 157
  cultural influences on, 88
  discrimination and antigay bias
    against, 148–149, 154–156,
    164–167
  as family, 7
  family planning for, 190, 191, 192
  hate crimes against, 148, 165–166
  legal issues for, 95, 159–163
  marriage of, 3, 4, 7, 71, 95,
    148–149, 160–163
  mate selection in, 88
  monogamy and sexuality of,
    157–158, 229

parenting issues for, 161, 162,
  163–165, 192, 207
  partner terminology in, 20
  prevalence of, 152
  relationship satisfaction of,
    156–157
  sexual orientation of, 149–150,
    151–152, 153–154, 155
  stepfamilies of, 275
sandwich generation, 289–290
satiation, 122, 256, 258
satisfaction
  extradyadic involvement due to lack
    of, 228–229
  GLBT relationship satisfaction,
    156–157
  level of life, 20
  marital, 42, 65, 122, 139, 144,
    145–147, 208, 296–297
  marital success creating, 145–147
  sexual, 110, 121, 122, 126–129,
    138, 229, 294–295
schools and education
  abuse and violence addressed
    in, 251
  child development influenced
    by, 210
  cost of, 175, 185
  divorce education programs,
    266, 268
  educational institution influences on
    choice, 20–21
  gender roles influenced by, 39
  mate selection based on education,
    97–98, 102
  parental teacher role, 202, 203,
    213, 214
  premarital education programs,
    105–106
  sex education, 214
  sexual values influenced by, 118
  shootings and violence in, 167
  stepfamily education programs,
    272, 283
  stress management educational
    support, 223
  teenage parenthood impacting, 188
  workforce return after
    additional, 171
Schwarzennegger, Arnold, 24, 59, 228
second-parent adoptions, 165
secrets, 59–60
Selective Search, 79
self-disclosure, 58–59, 60, 91, 127
self-esteem
  jealousy due to low, 94
  love influenced by, 90–91
  marriage boosting, 136
  self-worth/self-concept impacting,
    15, 61
  sexual orientation impacting, 155, 156
  sexual satisfaction impacted by, 126
self-fulfilling prophecies, 15–16
self-fulfillment, 10, 132
Semenya, Caster, 30
sex, biological, 31–32, 35. See also
  gender
sexism, 42
sexual behavior and sexuality
  abstinence from, 114–115
  abusive, 167, 244–246 (see
    also abuse and violence in
    relationships)
  arousal and, 92
  body image influencing, 90, 126
  cultural influences on, 24
  double standards related to,
    117–118, 120
  dysfunctional, 167
  elderly adults', 129, 294–295
  extramarital (see extradyadic
    involvement or infidelity)
  familial regulation of, 14
  friends with benefits including, 117
  gender differences in, 115, 117, 118,
    119–121, 128, 157

health issues due to (see sexually
    transmitted infections)
  homosexual (see homosexuality and
    GLBT individuals)
  hooking up including, 75–76
  marital success influenced by sexual
    desire, 146
  marriage altering, 138
  monogamous, 4, 124–125, 157, 158
  orientation of (see sexual
    orientation)
  pheromones and, 121–122
  polygamous, 5–6
  pregnancy following (see
    pregnancy)
  in relationships, 122–125
  safe, 123–125 (see also condoms)
  same-sex, 157–158, 229 (see
    also homosexuality and GLBT
    individuals)
  satisfaction from (see sexual
    satisfaction)
  sex education on, 214
  sexual myths about, 129
  sexual revolution on, 71
  sexual values and, 112–119, 124
  single-parents', 217
  stress and crisis impacting, 219
  substance use influencing, 24, 75,
    76, 117, 125, 126
sexually transmitted infections (STIs)
  gender differences in vulnerability
    to, 42, 152
  HIV/AIDS as, 118, 123–124,
    125, 157
  hooking up as exposure to, 76
  infertility impacted by, 189
  safe sex to avoid, 123–125
  sexual orientation and, 152
  sexual values influenced by, 118
  virginity pledgers contracting, 114
sexual orientation. See also
  heterosexuality; homosexuality
  and GLBT individuals
  "causes" of, 153
  changing, 153–154
  classification by, 149–150,
    151–152
  definitions related to, 149–150
  discrimination based on, 148–149,
    154–156, 164–167
  extradyadic involvement based
    on, 229
  legal issues based on, 95, 159–163
  mixed-orientation relationships,
    158–159
  prevalence of homosexuality/
    bisexuality, 152
  terminology related to, 150
sexual satisfaction
  elderly adults', 294–295
  extradyadic involvement due to lack
    of, 229
  gender differences in, 121
  marital, 122, 138
  prerequisites for, 126–129
  wedding cancellation due to
    low, 110
shared parenting dysfunction, 263
Shelton, Blake, 98
Shepard, Matthew, 166
Shields, Brooke, 207
siblings
  birth order among, 209–210
  child development influenced
    by, 210
  elderly adult relationships with, 298
  families including, 7, 8
  gender roles influenced by, 38
  half-, 268
  sexual values influenced by, 118
  step-, 280 (see also stepfamilies)
silence, 67
Silver Sage Village, 291
simultaneous loves, 93. See also
  concurrent sexual partnerships

singlehood
acceptance of, 71–72
Alternatives to Marriage Project supporting, 72
categories of singles, 72–74
cohabitation and, 7, 79–81, 300
finding a partner in, 74–79, 157, 300
legal blurring of married/ unmarried, 72
liabilities of, 5
marriage delays for, 69–70, 108
reasons for, 69–70
statistics on, 2
single-parents and single-parenthood, 8, 41, 188, 215–217
sleep, 222, 286
smoking. *See* substance use/abuse
social class, 14, 96, 98, 255. *See also* status
social exchange framework, 12, 61, 101–102
social groups. *See* families; friends, peers, and social groups
social isolation, 242, 243
socialization, gender identity/roles influenced by, 33, 37–46
social learning theory, 36–37
social networking/media, 59, 298
Social Security benefits. *See also* retirement benefits
elderly adults', 290–291, 293
marriage entitling, 4, 6, 162
post spousal death, 74
same-sex couples' access to, 162
sociobiological perspective, 104
socioeconomic factors. *See* economic factors; money and finances; poverty; social class
sociological imagination, 23
sodomy, decriminalization of, 159
South Africa, same-sex marriage in, 148–149
Spain
elder care in, 287
same-sex marriage in, 148–149
spectatoring, avoiding, 128–129
speed-dating, 78–79
Spitzer, Eliot, 214, 227, 230
spousal abuse. *See* abuse and violence in relationships
spousal support, 262, 273
spouses. *See* marriage
stalking, 93, 95, 239–240
status, 14, 22. *See also* power and leadership; social class
stepfamilies
definition and types of, 275
developmental tasks for, 280–283
monetary issues for, 276, 277, 278, 282
myths about, 275
remarriage creating (*see* remarriage)
stages in becoming, 279
stepfamily education programs, 272, 283
stepism against, 276–277
strengths of, 279–280
theoretical perspectives on, 278
unique aspects of, 275–277
storge love style, 85
Straight Spouse Network, 159
Strengthening Families Program, 232, 234
stress and crisis
abuse and violence caused by, 242, 251
death/suicide of family member causing, 234–235, 298–299
definitions of, 218–220
divorce impacted by, 219, 254, 269–270
elderly adult caregiving creating, 289
extramarital affairs creating, 226–231
family crisis examples, 224–235

health issues related to, 45, 224–226
midlife, 226
monetary concerns creating, 178, 221, 231
resilience to, 10, 220
stress management strategies, 220–223, 269–270
substance use/abuse causing, 225, 232–234
unemployment creating, 231, 251
work-related, 174, 176, 231, 251
structure-function framework, 13–14, 278
substance abuse
divorce and, 233, 269
substance use/abuse
abuse and violence influenced by, 243, 245, 246
elderly adults', 295, 299
gender differences in, 45
HIV transmission via, 123
jealousy leading to, 95
mate selection influenced by, 103
parental influence on, 204, 212–213
peer influence on, 38–39
poverty impacting, 178
prenatal, 188, 202, 204
sexual behavior influenced by, 24, 75, 76, 117, 125, 126
sexual orientation linked to, 155, 166
stress and crisis created by, 225, 232–234
support groups for, 232, 234
suicide, suicide attempts, and threats
as arranged marriage alternative, 87
conflict theory on, 15
elderly adults', 292
psychological blackmail using, 107
sexual orientation linked to, 155, 166
singlehood increasing likelihood of, 73–74
stress and crisis caused by, 234–235
Suleman, Nadya, 191
superperson strategy, 174
support groups, 232, 234, 235
surrogacy, 190, 191
Survivors of Suicide, 235
Sweden
cohabitation in, 79
corporal punishment in, 241
elderly adults in, 293
family planning in, 187
symbolic interaction framework, 15–16, 61, 278, 288

T

Taiwan, extradyadic involvement in, 229
tax issues, 161, 162, 184, 194
Tchambuli people, 33
technology. *See also* Internet; texting
communication impacted by, 50, 52
conflict resolution via, 52
elderly adult use of, 298
parental use and monitoring of, 204, 210
teenagers. *See also* children and youth
parental response to, 215, 216
substance abuse among, 232
suicide, suicide attempts, and threats by, 15, 235
teen pregnancy and parenthood, 178, 185, 187–188
terminology
age-discrepant relationship, 139–140, 273
cohabitation-related, 80
gender role, 31–36
male-biased, 36
partner identification, 20
research use of, 27
same-sex marriage, 160
sexual orientation-related, 150

texting, 52, 204, 298
theoretical frameworks, 12–17, 36–37, 61, 96–99, 100–102, 104, 278, 288, 289
therapeutic abortions, 195, 197
thermal or temperature biofeedback, 222
time periods/timing
alone time, lack of, 137
conflict resolution discussion timing, 63
divorce adjustment, 260, 266, 270, 272
divorce after lack of time together, 256
length of time knowing partner, 108–109
marital happiness across time, 147
marriage delays, 69–70, 108
quality time with children, 172, 202
research publication time lags, 27
stepfamily time considerations, 278, 281
time management, 176
time-outs, disciplinary, 213
tobacco use. *See* substance use/abuse
touch, communication via, 56
trade-offs, 17–18
traditional families, 8. *See also* nuclear families
transgenderism, 34–35, 60, 150. *See also* homosexuality and GLBT individuals
transracial adoptions, 192, 193
True Love Waits campaign, 114
trust
extradyadic involvement impacting, 230
jealousy and mistrust, 94
marital success influenced by, 146
remarriage building, 273
Twin Oaks Intentional Community, 6, 16
Twitter, 59, 298

U

unconscious motivations, 24
unemployment, 168, 178, 221, 231, 251
uninvolved parenting style, 211
unrequited love, 93
uxoricide, 95, 236, 238, 250

V

values
choices influenced by, 23
differences in, as conflict source, 53–54
divorce due to changing, 256
double standards of, 117–118, 120
mate selection influenced by, 99
sexual, 112–119, 124
sources of, 118–119
stress management changes of, 220–221
Van Dyke, Dick, 286
video chatting, 78
violence. *See* abuse and violence in relationships; hate crimes
virginity, 114–115

W

Walt Disney Company, 7
weddings, 5, 108–110, 134, 135–136
weight, 77, 89–90, 225, 291
Weiner, Anthony, 23–24
Welles, Orson, 15
White, Betty, 286
widowed singles, 74, 273–274, 298–299
Williams, Charles Andrew, 167
Winehouse, Amy, 234
Winfrey, Oprah, 132, 243
win-win *vs.* win-lose relationships, 64–65

women. *See also specific topics such as* mothers and motherhood, gender, marriage, *or* sexuality
as abusers, 239
battered-woman syndrome, 238 (*see also* abuse and violence in relationships)
"cougar" term applied to, 140
feminism of, 16–17, 71, 119, 197
feminization of poverty among, 41
work-related issues facing (*see* work, occupations and careers)
Woods, Tiger, 24, 59, 103
Woodward, Joanne, 200
word choice. *See* terminology
work, occupations and careers
balancing work and family, 174, 176
children impacted by, 172–173, 175, 202
debt issues and, 176–181
divorce and, 253–254, 262
dual-career marriages, 171–172
dual-earner families, 8, 10, 170–171, 172–173
extradyadic involvement in workplace, 228
Industrial Revolution impacting, 9–10, 254
male identity tied to, 43–44, 231–232
mate selection influenced by, 98, 102
military spouse challenges with, 142, 144
mommy track impacting, 171
opting out of, 171
parental influence in, 209
retirement from, 292–293 (*see also* retirement benefits)
same-sex partners' right to benefits in, 159, 160
sex segregation in, 39, 40, 41
shift work, 176
spouses impacted by employment, 168–172
stress and crisis caused by, 174, 176, 231, 251
superperson strategy in, 174
unemployment, 168, 178, 221, 231, 251

Y

"you" statements, 56
youth. *See* children and youth; teenagers

REVIEWcard

LEARNING OUTCOMES

To help you succeed, we've designed a review card for each chapter.

2-1 What are the important terms

Sex refers to the biological distinction between female biological sex is identified on the basis of one's chromosomes, gonads, hormones, internal sex organs, and external genitals, and exists on a continuum rather than being a dichotomy. *Gender* refers to the social and psychological characteristics often associated with being fer ~~~~~~ gender include *gender identity, gender role, g* ~~~~~~ d *transgenderism*.

In this column, you'll find summary points supported by diagrams or tables when relevant to help you better understand important concepts.

Feminine

Masculine

© Hasan Kursad Ergan/iStockphoto.com

© Hans-Peter Merten/Mauritius Die Bildagentur Gmbh/Photolibrary

2-2 What theories explain gender role development?

Biosocial theory emphasizes that social behaviors (for example, gender roles) are biologically based and have an evolutionary survival func~~~~ women stayed close to shelter or gathered food nearby, wher~~~~ to find food. Such a conceptualization focuses on the divisio~~~~ women and men as functional for the survival of the species. S~~~~ theory emphasizes the roles of reward and punishment in explaining how children learn gender role behavior. Identification theory says that children acquire the characteristics and behaviors of their same-sex parent through a process of identification. Boys identify with their fathers; girls identify with their mothers. Cognitive-developmental theory emphasizes biological readiness, in terms of cognitive development, of the child's responses to gender cues in the environment. Once children learn the concept of gender permanence, they seek to become compete~~~~~~~~~~~~~~~~~~~~~~~~~~

Here you'll find key terms and definitions in alphabetical order

2-3 ~~~~~~~~~~ socialization?

Vario~~~~~~~~~~~~~~~~~~~ siblings (representing d~~~~ races ~~~~~~~~~~~~~~~~~~~ ducation, and mass me~~~~ Thes~~~~~~~~~~~~~~~~~~~ er roles and influence w~~~~ peop~~~~~~~~~~~~~~~~~~~ men. For example, the fa~~~~ a ger~~~~~~~~~~~~~~~~~~ hly structured by gender. The name~~~~~~~~~~~~~~~~~~ ey dress them in, and the toys they ~~~~~~~~~~~~~~~~~~ stricter with female children, deter~~~~~~~~~~~~~~~~~~ ve the house at night, setting a time ~~~~~~~~~~~~~~~~~~ hen you get to the party."

How to Use This Card

1. Look over the card to preview the new concepts you'll be introduced to in the chapter.
2. Read your chapter to fully understand the material.
3. Go to class (and pay attention).
4. Review the card one more time to make sure you've registered the key concepts.
5. Don't forget, this card is only one of many M&F learning tools available to help you succeed in your marriage and family course.

When it's time to prepare for exams, use the review card and the technique to the left to ensure successful study sessions.

2-4 ~~~~~~~~~~~~~~~ traditional gender

Traditional female role socialization may result in negative outcomes, such as less income, a negative body image, and less marital satisfaction, but also in positive outcomes, such as a longer life, a stronger relationship focus, keeping

KEY TERMS

androgyny a blend of traits that are stereotypically associated with masculinity and femininity.

biosocial theory (sociobiology) emphasizes the interaction of one's biological or genetic inheritance with one's social environment to explain and predict human behavior.

cross-dresser a generic term for individuals who may dress or present themselves in the gender of the opposite sex.

feminization of poverty the idea that women disproportionately experience poverty.

gender the social and psychological behaviors associated with being female or male.

gender dysphoria the condition in which one's gender identity does not match one's biological sex.

gender identity the psychological state of viewing oneself as a girl or a boy and, later, as a woman or a man.

gender role ideology the proper role relationships between women and men in a society.

~~~~~~~~~~**ndence** ~~~~eworks and ~~~~~~~~~~dependent ~~~~~~gories.

gender roles behaviors assigned to women and men in a society.

intersexed individuals people with mixed or ambiguous genitals

intersex development refers to congenital variations in the reproductive system, sometimes resulting in ambiguous genitals.

occupational sex segregation the concentration of women in certain occupations and men in other occupations.

parental investment any investment by a parent that increases the chance that the offspring will survive and thrive.

positive androgyny a view of androgyny that is devoid of the negative traits associated with masculinity and femininity.

sex the biological distinction between being female and being male.

sex roles behaviors defined by biological constraints.

sexism an attitude, action, or institutional structure that subordinates or discriminates against an individual or group because of their sex.

socialization the process through which we learn attitudes, values, beliefs, and behaviors appropriate to the social positions we occupy.

transgender a generic term for a person of one biological sex who displays characteristics of the opposite sex.

transsexual an individual who has the anatomical and genetic characteristics of one sex but the self-concept of the other.

relationships on track, and a closer emotional bond with children. Traditional male role socialization may result in the fusion of self and occupation, a more limited expression of emotion, disadvantages in child custody disputes, and a shorter lifespan but also in higher income, greater freedom of movement, a greater available pool of potential partners, and greater acceptance in initiating relationships. A recent research study of 335 undergraduate men revealed that about 31% of college males reported their preference for marrying a traditional wife (one who would stay at home to take care of children). These men believe that a w~~ife~~ ~~~~arriage.

For additional study tools such as flashcards, interactive quizzing, games, videos, study worksheets, note taking outlines, and Internet activities, visit www.cengagebrain.com.

...Socialization

...tive Consequences
~~Less education/income (more dependent)~~ Longer life
Feminization of poverty — Stronger relationship focus
Higher STD/HIV infection risk — Keep relationships on track
Negative body image — Bonding with children
Less marital satisfaction — Identity not tied to job

© Cengage Learning 2014

Consequences of Traditional Male Role Socialization

Negative Consequences	Positive Consequences
Identity tied to work role	Higher income and occupational status
Limited emotionality	More positive self-concept
Fear of intimacy; more lonely	Less job discrimination
Disadvantaged in getting custody	Freedom of movement; more partners to select from; more normative to initiate relationships
Shorter life	Happier marriage

© Cengage Learning 2014

2-5 How are gender roles changing?

Androgyny refers to a blend of traits that are stereotypically associated with both masculinity and femininity. The term may also imply flexibility of traits; for example, an androgynous individual may be emotional in one situation, logical in another, assertive in another, and so forth. The concept of gender role transcendence involves abandoning gender schema (that is, becoming "gender aschematic"), so that personality traits, social and occupational roles, and other aspects of our lives become divorced from gender categories. However, such transcendence is not equal for women and men. Although females are becoming more masculine, partly because our society values whatever is masculine, men are not becoming more feminine.

LEARNING OUTCOMES

1-1 What is marriage?

Marriage is a system of binding a man and a woman together for the reproduction, care (physical and emotional), and socialization of offspring. Marriage in the United States is a legal contract between a couple and their state that regulates their economic and sexual relationship. The federal government supports marriage education in the public school system with the intention of reducing divorce (which is costly to both individuals and society). The various types of marriage are polygyny, polyandry, polyamory, pantagamy, and domestic partnerships.

1-2 What is family?

In recognition of the diversity of families, the definition of family is increasing beyond the U.S. Census Bureau's definition to include two adult partners whose interdependent relationship is long-term and characterized by an emotional and financial commitment. Types of family include nuclear, extended, and blended. There are also traditional, modern, and postmodern families. See the table to the right for differences between marriage and family.

Differences Between Marriage and the Family in the United States

Marriage	Family
Usually initiated by a formal ceremony.	Formal ceremony not essential.
Involves two people.	Usually involves more than two people.
Ages of the individuals tend to be similar.	Individuals represent more than one generation.
Individuals usually choose each other.	Members are born or adopted into the family.
Ends when one spouse dies or the couple is divorced.	Continues beyond the life of the individual.
Sex between spouses is expected and approved.	Sex between near kin is neither expected nor approved.
Requires a license.	No license is needed to become a parent.
Procreation expected.	Consequence of procreation.
Spouses are focused on each other.	Focus changes with addition of children.
Spouses can voluntarily withdraw from marriage.	Parents cannot remove themselves from their obliation to children via divorce.
Money in unit is spent on the couple.	Money is used for the needs of the children.
Recreation revolves around adults.	Recreation revolves around children.

© Cengage Learning 2014

1-3 How have marriage and the family changed?

The advent of industrialization, urbanization, and mobility involved the demise of familism and the rise of individualism. When family members functioned together as an economic unit, they were dependent on one another for survival and were concerned about what was good for the family. The shift from familism to individualism is only one change; others include divorce replacing death as the endpoint for the majority of marriages, marriage and relationships emerging as legitimate objects of scientific study, the rise of feminism, changes in gender roles, increasing marriage age, and the acceptance of singlehood, cohabitation, and child-free marriages.

1-4 What are the theoretical frameworks for viewing marriage and the family?

Theoretical frameworks provide a set of interrelated principles designed to explain a particular phenomenon and provide a point of view. The following table gives an overview of the frameworks used in this text.

1-5 What is the view/theme of this text?

A central theme of this text is to encourage you to be proactive—to make conscious, deliberate relationship choices to enhance your own well-being and the well-being of those in your intimate groups. Though global, structural, cultural, and media influences are operative, a choices framework emphasizes that individuals have some control over their relationships. Important issues to keep in mind about a choices framework for viewing marriage and the family are that (1) not to decide is to decide, (2) some choices require correcting, (3) all choices involve trade-offs, (4) choices include selecting a positive or negative view, (5) making choices produces ambivalence, and (6) some choices are not revocable. Generation Yers (born in the early 1980s) are relaxed about relationship choices. Rather than pair bond, they "hang out," "hook up," and "live together." They are in no hurry to find "the one," to marry, and to begin a family.

Theoretical Frameworks for Marriage and the Family

Theory	Description	Concepts	Level of Analysis	Strengths	Weaknesses
Social Exchange	In their relationships, individuals seek to maximize their benefits and minimize their costs.	Benefits, Costs, Profit, Loss	Individual, Couple, Family	Provides explanations of human behavior based on outcome.	Assumes that people always act rationally and all behavior is calculated.
Family Life Course Development	All families have a life course that is composed of all the stages and events that have occurred within the family.	Stages, Transitions, Timing	Institution, Individual, Couple, Family	Families are seen as dynamic rather than static. Useful in working with families who are facing transitions in their life courses.	Difficult to adequately test the theory through research.
Structural-Function	The family has several important functions within society; within the family, individual members have certain functions.	Structure, Function	Institution	Emphasizes the relation of family to society, noting how families affect and are affected by the larger society.	Families with nontraditional structures (single-parent, same-sex couples) are seen as dysfunctional.
Conflict	Conflict in relationships is inevitable, due to competition over resources and power.	Conflict, Resources, Power	Institution	Views conflict as a normal part of relationships and as necessary for change and growth.	Sees all relationships as conflictual, and does not acknowledge cooperation.
Symbolic Interaction	People communicate through symbols and interpret the words and actions of others.	Definition of the situation, Looking-glass self, Self-fulfilling prophecy	Couple	Emphasizes the perceptions of individuals, not just objective reality or the viewpoint of outsiders.	Ignores the larger social interaction context and minimizes the influence of external forces.
Family Systems	The family is a system of interrelated parts that function together to maintain the unit.	Subsystem, Roles, Rules, Boundaries, Open system, Closed system	Couple, Family	Very useful in working with families who are having serious problems (violence, alcoholism). Describes the effect family members have on each other.	Based on work with systems, troubled families, and may not apply to nonproblem families.
Feminism	Women's experience is central and different from men's experience of social reality.	Inequality, Power, Oppression	Institution, Individual, Couple, Family	Exposes inequality and oppression as explanations for frustrations women experience.	Multiple branches of feminism may inhibit central accomplishment of increased equality.

© Cengage Learning 2014

1-6 What are some factors to keep in mind when evaluating research?

Caveats that are factors to be used in evaluating research include a random sample (the respondents providing the data reflect those who were not in the sample), a control group (the group not subjected to the experimental design for a basis of comparison), terminology (the phenomenon being studied should be objectively defined), researcher bias (present in all studies), time lag (takes two years from study to print), and distortion or deception (although rare, some researchers distort their data). Few studies avoid all research problems.

Potential Research Problems in Marriage and Family

Weakness	Consequences	Example
Sample not random	Cannot generalize findings	Opinions of college students do not reflect opinions of other adults.
No control group	Inaccurate conclusions	Study on the effect of divorce on children needs control group of children whose parents are still together.
Age differences between groups of respondents	Inaccurate conclusions	Effect may be due to passage of time or to cohort differences.
Unclear terminology	Inability to measure what is not clearly defined	What is living together, marital happiness, sexual fulfillment, good communication, quality time?
Researcher bias	Slanted conclusions	Male researcher may assume that, because men usually ejaculate each time they have intercourse, women should have an orgasm each time they have intercourse.
Time lag	Outdated conclusions	Often-quoted Kinsey sex research is more than 50 years old.
Distortion	Invalid conclusions	Research subjects exaggerate, omit information, and/or recall facts or events inaccurately. Respondents may remember what they wish had happened.

© Cengage Learning 2014

KEY TERMS

beliefs definitions and explanations about what is thought to be true.

binuclear family a family in which the members live in two separate households.

blended family (stepfamily) a family created when two individuals marry and at least one of them has a child or children from a previous relationship or marriage.

civil union a pair-bond, or monogamous relationship, given legal significance in terms of rights and privileges.

common-law marriage a marriage by mutual agreement between cohabitants without a marriage license or ceremony (recognized in some, but not all, states).

conflict framework view that individuals in relationships compete for valuable resources.

control group group used to compare with the experimental group that is not exposed to the independent variable being studied.

domestic partnership a relationship in which individuals who live together are emotionally and financially interdependent and are given some kind of official recognition by a city or corporation so as to receive partner benefits.

experimental group the group exposed to the independent variable.

extended family the nuclear family or parts of it plus other relatives.

familism philosophy in which decisions are made in reference to what is best for the family as a collective unit.

family a group of two or more people related by blood, marriage, or adoption.

family life course development the stages and process of how families change over time.

family life cycle stages which identify the various challenges faced by members of a family across time.

family of orientation the family of origin into which a person is born.

family of origin the family into which an individual is born or reared, usually including a mother, father, and children.

family of procreation the family a person begins by getting married and having children.

family systems framework theoretical perspective that views each member of the family as part of a system and the family as a unit that develops norms of interaction.

feminist framework theoretical perspective that views marriage and the family as contexts for inequality and oppression.

functionalists theorists who view the family as an institution with values, norms, and activities meant to provide stability for the larger society.

Generation Y children of the baby boomers, typically born between 1979 and 1984. Also known as the Millennial or Internet Generation.

hypothesis a suggested explanation for a phenomenon.

individualism philosophy in which decisions are made on the basis of what is best for the individual.

institutions established and enduring patterns of social relationships.

IRB (Institutional Review Board) approval the endorsement by one's college, university, or institution that the proposed research is consistent with research ethics standards and poses no undo harm to participants.

marriage a legal contract between a couple and the state in which they reside that regulates their economic and sexual relationship.

marriage-resilience perspective the view that changes in the institution of marriage are not indicative of a decline and do not have negative effects.

mating gradient the tendency for husbands to marry wives who are younger and have less education and less occupational success.

modern family the dual-earner family, in which both spouses work outside the home.

nuclear family family consisting of an individual, his or her spouse, and his or her children, or of an individual and his or her parents and siblings.

pantagamy a group marriage in which each member of the group is "married" to the others.

polyamory literally, multiple loves (*poly* means "many"; *amorous* means "love"); a lifestyle in which lovers embrace the idea of having multiple emotional and sexual partners.

polyandry a form of polygamy in which one wife has two or more husbands.

polygamy a generic term referring to a marriage involving more than two spouses.

polygyny a form of polygamy in which one husband has two or more wives.

postmodern family nontraditional family emphasizing that a healthy family need not be heterosexual or have two parents.

primary group small, intimate, informal group.

random sample research sample in which each person in the population being studied has an equal chance of being included in the study.

roles the behaviors in which individuals in certain status positions are expected to engage.

secondary group large or small group characterized by impersonal and formal interaction.

sequential ambivalence the individual experiences one wish and then another.

simultaneous ambivalence the person experiences two conflicting wishes at the same time.

social exchange framework marital perspective in which spouses exchange resources and decisions are made on the basis of perceived profit and loss.

sociological imagination the perspective of how powerful social structure and culture are in influencing personal decision making.

status a social position a person occupies within a social group.

structure-function framework theoretical perspective that emphasizes how marriage and family contribute to the larger society.

symbolic interaction framework theoretical perspective that views marriage and families as symbolic worlds in which the various members give meaning to each other's behavior.

theoretical framework a set of interrelated principles designed to explain a particular phenomenon and to provide a point of view.

traditional family the two-parent nuclear family with the husband as breadwinner and wife as homemaker.

utilitarianism the doctrine holding that individuals rationally weigh the rewards and costs associated with behavioral choices.

values standards regarding what is good and bad, right and wrong, desirable and undesirable.

GENDER

LEARNING OUTCOMES

2-1 What are the important terms related to gender?

Sex refers to the biological distinction between females and males. One's biological sex is identified on the basis of one's chromosomes, gonads, hormones, internal sex organs, and external genitals, and exists on a continuum rather than being a dichotomy. *Gender* refers to the social and psychological characteristics often associated with being female or male. Other terms related to gender include *gender identity, gender role, gender role ideology, transgender,* and *transgenderism.*

Feminine ⟨⟨⟨⟨⟩⟩ Masculine

2-2 What theories explain gender role development?

Biosocial theory emphasizes that social behaviors (for example, gender roles) are biologically based and have an evolutionary survival function. Traditionally, women stayed close to shelter or gathered food nearby, whereas men traveled far to find food. Such a conceptualization focuses on the division of labor between women and men as functional for the survival of the species. Social learning theory emphasizes the roles of reward and punishment in explaining how children learn gender role behavior. Identification theory says that children acquire the characteristics and behaviors of their same-sex parent through a process of identification. Boys identify with their fathers; girls identify with their mothers. Cognitive-developmental theory emphasizes biological readiness, in terms of cognitive development, of the child's responses to gender cues in the environment. Once children learn the concept of gender permanence, they seek to become competent, proper members of their gender group.

2-3 What are the various agents of socialization?

Various socialization influences include parents and siblings (representing different races and ethnicities), peers, religion, the economy, education, and mass media. These factors shape individuals toward various gender roles and influence what people think, feel, and do in their roles as women or men. For example, the family is a gendered institution with female and male roles highly structured by gender. The names parents assign to their children, the clothes they dress them in, and the toys they buy them all reflect gender. Parents may also be stricter with female children, determining the age at which they are allowed to leave the house at night, setting a time for curfew, and giving directives such as "call when you get to the party."

2-4 What are the consequences of traditional gender role socialization?

Traditional female role socialization may result in negative outcomes, such as less income, a negative body image, and less marital satisfaction, but also in positive outcomes, such as a longer life, a stronger relationship focus, keeping

KEY TERMS

androgyny a blend of traits that are stereotypically associated with masculinity and femininity.

biosocial theory (sociobiology) emphasizes the interaction of one's biological or genetic inheritance with one's social environment to explain and predict human behavior.

cross-dresser a generic term for individuals who may dress or present themselves in the gender of the opposite sex.

feminization of poverty the idea that women disproportionately experience poverty.

gender the social and psychological behaviors associated with being female or male.

gender dysphoria the condition in which one's gender identity does not match one's biological sex.

gender identity the psychological state of viewing oneself as a girl or a boy and, later, as a woman or a man.

gender role ideology the proper role relationships between women and men in a society.

gender role transcendence abandoning gender frameworks and looking at phenomena independent of traditional gender categories.

gender roles behaviors assigned to women and men in a society.

intersexed individuals people with mixed or ambiguous genitals.

intersex development refers to congenital variations in the reproductive system, sometimes resulting in ambiguous genitals.

occupational sex segregation the concentration of women in certain occupations and men in other occupations.

parental investment any investment by a parent that increases the chance that the offspring will survive and thrive.

positive androgyny a view of androgyny that is devoid of the negative traits associated with masculinity and femininity.

sex the biological distinction between being female and being male.

sex roles behaviors defined by biological constraints.

sexism an attitude, action, or institutional structure that subordinates or discriminates against an individual or group because of their sex.

socialization the process through which we learn attitudes, values, beliefs, and behaviors appropriate to the social positions we occupy.

transgender a generic term for a person of one biological sex who displays characteristics of the opposite sex.

transsexual an individual who has the anatomical and genetic characteristics of one sex but the self-concept of the other.

relationships on track, and a closer emotional bond with children. Traditional male role socialization may result in the fusion of self and occupation, a more limited expression of emotion, disadvantages in child custody disputes, and a shorter lifespan but also in higher income, greater freedom of movement, a greater available pool of potential partners, and greater acceptance in initiating relationships. A recent research study of 335 undergraduate men revealed that about 31% of college males reported their preference for marrying a traditional wife (one who would stay at home to take care of children). These men believe that a wife's working outside the home weakens the marriage.

Consequences of Traditional Female Role Socialization

Negative Consequences	Positive Consequences
Less education/income (more dependent)	Longer life
Feminization of poverty	Stronger relationship focus
Higher STD/HIV infection risk	Keep relationships on track
Negative body image	Bonding with children
Less marital satisfaction	Identity not tied to job

© Cengage Learning 2014

Consequences of Traditional Male Role Socialization

Negative Consequences	Positive Consequences
Identity tied to work role	Higher income and occupational status
Limited emotionality	More positive self-concept
Fear of intimacy; more lonely	Less job discrimination
Disadvantaged in getting custody	Freedom of movement; more partners to select from; more normative to initiate relationships
Shorter life	Happier marriage

© Cengage Learning 2014

2-5 How are gender roles changing?

Androgyny refers to a blend of traits that are stereotypically associated with both masculinity and femininity. The term may also imply flexibility of traits; for example, an androgynous individual may be emotional in one situation, logical in another, assertive in another, and so forth. The concept of gender role transcendence involves abandoning gender schema (that is, becoming "gender aschematic"), so that personality traits, social and occupational roles, and other aspects of our lives become divorced from gender categories. However, such transcendence is not equal for women and men. Although females are becoming more masculine, partly because our society values whatever is masculine, men are not becoming more feminine.

LEARNING OUTCOMES

3-1 What is the nature of interpersonal communication?

Communication is the exchange of information and feelings by two individuals. It involves both verbal and nonverbal messages. The nonverbal part of a message often carries more weight than the verbal part.

3-2 What are various issues related to conflict in relationships?

Conflict is both inevitable and desirable. Unless individuals confront and resolve issues over which they disagree, one or both may become resentful and withdraw from the relationship. Conflict may result from one partner's doing something the other does not like, having different perceptions, or having different values. Sometimes it is easier for one partner to view a situation differently or alter a value than for the other partner to change the behavior causing the distress.

3-3 What are some principles and techniques of effective communication?

Some basic principles and techniques of effective communication include making communication a priority, maintaining eye contact, asking open-ended questions, using reflective listening, using "I" statements, complimenting each other, and sharing power. Partners must also be alert to keeping the dialogue (process) going even when they don't like what is being said (content).

Judgmental and Nonjudgmental Responses to a Partner's Saying, "I'd Like to Spend One Evening a Week With My Friends"

Nonjudgmental, Reflective Statements	Judgmental Statements
You value your friends and want to maintain good relationships with them.	You only think about what you want.
You think it is healthy for us to be with our friends some of the time.	Your friends are more important to you than I am.
You really enjoy your friends and want to spend some time with them.	You just want a night out so that you can meet someone new.
You think it is important that we not abandon our friends just because we are involved.	You just want to get away so you can drink.
You think that our being apart one night each week will make us even closer.	You are selfish.

© Cengage Learning 2014

3-4 How are relationships affected by self-disclosure, lying, secrets, and cheating?

The levels of self-disclosure and honesty influence intimacy in relationships. High levels of self-disclosure are associated with increased intimacy. Despite the importance of honesty in relationships, deception occurs frequently in interpersonal relationships. Partners sometimes lie to each other about previous sexual relationships, how they feel about each other, and how they experience each other sexually. Telling lies is not the only form of dishonesty. People exaggerate, minimize, tell partial truths, pretend, and engage in self-deception. Partners may withhold information or keep secrets to protect themselves or to preserve the relationship, or both. The top reason reported in a 2012 study of

KEY TERMS

brainstorming suggesting as many alternatives as possible without evaluating them.

branching in communication, going out on different limbs of an issue rather than staying focused on the issue.

closed-ended question question that allows for a one-word answer and does not elicit much information.

competing style of conflict conflict style in which partners are both assertive and uncooperative. Each tries to force a way on the other so that there is a winner and a loser.

conflict the interaction that occurs when the behavior or desires of one person interfere with the behavior or desires of another.

congruent message one in which verbal and nonverbal behaviors match.

defense mechanisms unconscious techniques that function to protect individuals from anxiety and minimize emotional hurt.

displacement shifting one's feelings, thoughts, or behaviors from the person who evokes them onto someone else.

escapism the simultaneous denial of and withdrawal from a problem.

"I" statements statements that focus on the feelings and thoughts of the communicator without making a judgment on others.

lose-lose solution a solution to a conflict in which neither partner benefits.

431 undergraduates was "To avoid hurting the partner." Even in "monogamous" relationships, there is considerable cheating. However, the more intimate the relationship, the greater our desire to share our most personal and private selves with our partner and the greater the emotional consequences of not sharing. In intimate relationships, keeping secrets can block opportunities for healing, resolution, self-acceptance, and a deeper intimacy with your partner.

3-5 What are the gender differences in communication?

Women and men differ in their approach to and patterns of communication. Women are more communicative about relationship issues, view a situation emotionally, and initiate discussions about relationship problems. A woman's goal is to preserve intimacy and avoid isolation. To men, conversations are about winning and achieving the upper hand. Mothers and fathers speak differently to children. Fathers use more assertive speech than mothers.

3-6 How are interactionist and exchange theories applied to relationship communication?

Symbolic interactionists examine the process of communication between two actors in terms of the meanings each attaches to the actions of the other. Definition of the situation, the looking-glass self, and taking the role of the other are all relevant to understanding how partners communicate.

Exchange theorists suggest that the partners' communication can be described as a ratio of rewards to costs. Rewards are positive exchanges, such as compliments, compromises, and agreements. Costs refer to negative exchanges, such as critical remarks, complaints, and attacks. When the rewards are high and the costs are low, the outcome is likely to be positive for both partners (profit). When the costs are high and the rewards low, neither may be satisfied with the outcome (loss).

3-7 What are examples of fighting fair to resolve conflict?

The sequence of resolving conflict includes deciding to address recurring issues rather than suppressing them, asking the partner for help in resolving issues, finding out the partner's point of view, summarizing in a nonjudgmental way the partner's perspective, and finding alternative win-win solutions. Defense mechanisms that interfere with conflict resolution include escapism, rationalization, projection, and displacement.

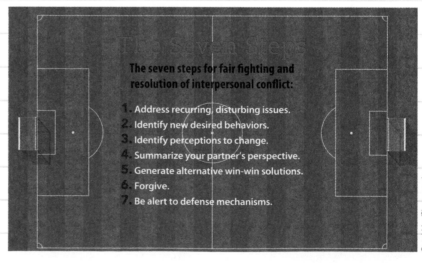

The Seven Steps

The seven steps for fair fighting and resolution of interpersonal conflict:

1. Address recurring, disturbing issues.
2. Identify new desired behaviors.
3. Identify perceptions to change.
4. Summarize your partner's perspective.
5. Generate alternative win-win solutions.
6. Forgive.
7. Be alert to defense mechanisms.

© archidea/Shutterstock

LEARNING OUTCOMES

4-1 What are the attractions of singlehood?

An increasing percentage of people are delaying marriage. Young Americans are not alone in their delay of marriage—individuals in France, Germany, and Italy are engaging in a similar pattern. The Alternatives to Marriage Project is designed to give visibility and credibility to the status of being unmarried. The primary attraction of singlehood is the freedom to do as one chooses. As a result of the sexual revolution, the women's movement, and the gay liberation movement, social approval of being unmarried has increased. Single people are those who have never married as well as those who are divorced or widowed.

Reasons to Remain Single

Benefits of Singlehood	Limitations of Marriage
Freedom to do as one wishes	Restricted by spouse or children
Variety of lovers	One sexual partner
Spontaneous lifestyle	Routine, predictable lifestyle
Close friends of both sexes	Pressure to avoid close other-sex friendships
Responsible for one person only	Responsible for spouse and children
Spend money as one wishes	Expenditures influenced by needs of spouse and children
Freedom to move as career dictates	Restrictions on career mobility
Avoid being controlled by spouse	Potential to be controlled by spouse
Avoid emotional and financial stress of divorce	Possibility of divorce

© Cengage Learning 2014

4-2 What are some categories of singles?

The term *singlehood* is most often associated with young, unmarried individuals. However, there are three categories of single people: the never-married, the divorced, and the widowed.

U.S. Adult Population by Relationship Status
N = 229,100,000

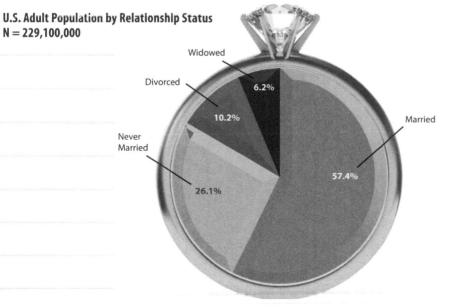

Widowed 6.2%
Divorced 10.2%
Never Married 26.1%
Married 57.4%

Source: *Statistical Abstract of the United States* (2012, 131th ed.), Table 56 (Washington, DC).

KEY TERMS

cohabitation (living together) two unrelated adults (by blood or by law) involved in an emotional and sexual relationship who sleep in the same residence at least four nights a week.

domestic partnership a relationship in which individuals who live together are emotionally and financially interdependent and are given some kind of official recognition by a city or corporation so as to receive partner benefits.

hanging out going out in groups where the agenda is to meet others and have fun.

hooking up having a one-time sexual encounter in which there are generally no expectations of seeing one another again.

loneliness the subjective evaluation that the number of relationships a person has is smaller than the person desires or that intimacy the person wants has not been realized.

POSSLQ an acronym used by the U.S. Census Bureau that stands for "people of the opposite sex sharing living quarters."

singlehood state of being unmarried.

iStockphoto.com

4-3 What are some ways to find a partner?

Besides the traditional way of meeting people at work or school or through friends and going on a date, couples today may also "hang out," which may lead to "hooking up." Internet dating, video dating, speed-dating, and international dating are newer forms for finding each other. Wealthy, busy clients looking for marriage partners pay Selective Search $20,000 to find them a mate.

Pros:
- Highly efficient
- Develop a relationship without visual distraction
- Avoid crowded, loud, uncomfortable locations, like bars
- The opportunity to try on new identities

Online Meeting

Cons:
- Ease of deception
- Relationship involvement escalates too quickly
- Inability to assess "chemistry" through the computer
- A lot of competition

© Kalim/Shutterstock

4-4 What is cohabitation like among today's youth?

Cohabitation, also known as living together, is becoming a "normative life experience," with almost 60% of American women reporting that they had cohabited before marriage. Reasons for an increase in living together include delaying marriage, fear of marriage, fear of divorce, increased social tolerance for living together, and a desire to avoid the legal entanglements of marriage. Some types of relationships in which couples live together include the here-and-now, testers, engaged couples, and cohabitants forever.

Although living together before marriage does not ensure a happy, stable marriage, it has some potential advantages. These include a sense of well-being, delayed marriage, learning about yourself and your partner, and being able to disengage with minimal legal hassle. Disadvantages include feeling exploited, feeling guilty about lying to parents, and not having the same economic benefits as those who are married. Social Security and retirement benefits are paid to spouses, not live-in partners.

LEARNING OUTCOMES

5-1 What are some ways that love has been described?

Love remains an elusive and variable phenomenon. Researchers have conceptualized love as a continuum from romanticism to realism, as a triangle consisting of three basic elements (intimacy, passion, and commitment), and as a style (from playful ludic love to obsessive and dangerous manic love).

5-2 How has love expressed itself in various social and historical contexts?

The society in which we live exercises considerable control over our love object or choice and conceptualizes it in various ways. Parents inadvertently influence the mate choice of their children by moving to certain neighborhoods, joining certain churches, and enrolling their children in certain schools. Doing so increases the chance that their offspring will "hang out" with, fall in love with, and marry people who are similar in race, education, and social class.

In the 1100s in Europe, marriage was an economic and political arrangement that linked two families. As aristocratic families declined after the French Revolution, love bound a woman and man together. Previously, Buddhists, Greeks, and Hebrews had their own views of love. Love in colonial America was also tightly controlled.

5-3 How does love develop in a new relationship?

Love occurs under certain conditions. Social conditions include a society that promotes the pursuit of love, peers who enjoy it, and a set of norms that link love and marriage. Psychological conditions involve high self-esteem, a willingness to disclose one's self to others, and gratitude. Researchers have also found that one of the most important psychological factors associated with falling in love is the perception of reciprocal liking. This factor is especially powerful if the person feels strong physical attraction. Individuals seeking emotional relationships are not attracted to those with little affect, a condition known as Alexithymia. Physiological and cognitive conditions imply that the individual experiences a stirred-up state and labels it "love." Other factors associated with love include appearance, common interests, and similar friends. Love sometimes provides a context for problems in that a young person in love will lie to parents and become distant from them so as to be with the lover. Also, lovers experience problems such as being in love with two people at the same time, being in love with someone who is abusive, and making risky, dangerous, or questionable choices while in love (for example, not using a condom) or reacting to a former lover who has become a stalker.

5-4 How do jealousy and love interface?

Jealousy is an emotional response to a perceived or real threat to a valued relationship. Types of jealousy are reactive, anxious, and possessive. Jealous feelings may have both internal and external causes and may have both positive and negative consequences for a couple's relationship.

5-5 What are the cultural restrictions that influence partnering?

Two types of cultural influences in mate selection are endogamy (to marry someone inside one's own social group—race, religion, social class) and exogamy (to marry someone outside one's own family).

5-6 What are the sociological factors operative in partnering?

Sociological aspects of mate selection involve homogamy. Variables include race, age, religion, education, social class, personal appearance, career, attachment, personality, and open-mindedness. Couples who have a lot in common are more likely to have a happy and durable relationship.

5-7 What are the psychological factors operative in partnering?

Psychological aspects of mate selection include complementary needs, exchange theory, and parental characteristics. Personality characteristics of a potential mate that are desired by both men and women include being warm, kind, and open and having a sense of humor. Negative personality characteristics to avoid in a potential mate include controlling behavior, narcissism, poor impulse control, hypersensitivity to criticism, inflated ego, perfectionism or insecurity, control by someone else (for example, parents), and substance abuse. Paranoid, schizoid, and borderline personalities are also to be avoided.

Personality Types Problematic in a Potential Partner

Type	Characteristics	Impact on Partner
Paranoid	Suspicious, distrustful, thin-skinned, defensive	Partners may be accused of everything.
Schizoid	Cold, aloof, solitary, reclusive	Partners may feel that they can never "connect" and that the person is not capable of returning love.
Borderline	Moody, unstable, volatile, unreliable, suicidal, impulsive	Partners will never know what their Jekyll-and-Hyde partner will be like, which could be dangerous.
Antisocial	Deceptive, untrustworthy, conscienceless, remorseless	Such a partner could cheat on, lie to, or steal from a partner and not feel guilty.
Narcissistic	Egotistical, demanding, greedy, selfish	Such a person views partners only in terms of their value. Don't expect such a person to see anything from a partner's point of view; expect such a person to bail in tough times.
Dependent	Helpless, weak, clingy, insecure	Such a person will demand a partner's full time and attention, and other interests will incite jealousy.
Obsessive-compulsive	Rigid, inflexible	Such a person has rigid ideas about how a partner should think and behave and may try to impose them on the partner.

© Cengage Learning 2014

5-8 What are the sociobiological factors operative in partnering?

The sociobiological view of mate selection suggests that men and women select each other on the basis of their biological capacity to produce and support healthy offspring. Men seek young women with healthy bodies, and women seek ambitious men who will provide economic support for their offspring. There is considerable controversy about the validity of this theory.

5-9 What factors should be considered when becoming engaged?

The engagement period is the time to ask specific questions about one's partner's values, goals, and marital agenda, to visit each other's parents to assess parental models, and to consider involvement in premarital educational programs or counseling or both.

5-10 What are some of the wrong reasons for getting married?

Negative reasons for getting married include being on the rebound, escaping from an unhappy home life, psychological blackmail, and pity.

5-11 What factors suggest you might consider calling off the wedding?

Getting married on the rebound is not a good reason to marry. One should wait until the negative memories of past relationships have been replaced by positive aspects of one's current relationship. One should not marry to escape an unhappy home. Getting married because a partner becomes pregnant is a bad idea. Factors suggesting that a couple may not be ready for marriage include being in their teens, having known each other less than two years, and having a relationship characterized by significant differences or dramatic parental disapproval or both. Some research suggests that partners with the greatest number of similarities in values, goals, and interests are most likely to have happy and durable marriages.

KEY TERMS

agape love style love style characterized by a focus on the well-being of the love object, with little regard for reciprocation; the love of parents for their children is agape love.

anxious jealousy obsessive rumination about the partner's alleged infidelity.

arranged marriage mate selection pattern whereby parents select the spouse of their offspring.

compersion the opposite of jealousy, whereby a person feels joy in having the person one loves and has sex with enjoying others both emotionally and sexually.

complementary-needs theory tendency to select mates whose needs are opposite and complementary to one's own needs.

conjugal (married) love the love between married people characterized by companionship, calmness, comfort, and security.

dowry (trousseau) the amount of money or valuables a woman's father pays a man's father for the man to marry his daughter. It functioned to entice the man to marry the woman, because an unmarried daughter stigmatized the family of the woman's father.

educational homogamy the tendency to select a cohabitant or marital partner with similar education.

endogamous pressures cultural attitudes reflecting approval for selecting a partner within one's social group and disapproval for selecting a partner outside one's social group.

endogamy the cultural expectation to select a marriage partner within one's social group.

engagement period of time during which committed, monogamous partners focus on wedding preparations and systematically examine their relationship.

eros love style love style characterized by passion and romance.

exchange theory theory that emphasizes that relations are formed and maintained between individuals offering the greatest rewards and least costs to each other.

exogamous pressures cultural attitudes reflecting approval for selecting a partner outside one's family group.

exogamy the cultural expectation that one will marry outside the family group.

extradyadic relationship emotional or sexual involvement between a member of a couple and someone other than the partner.

homogamy tendency to select someone with similar characteristics.

homogamy theory of mate selection theory that individuals tend to be attracted to and become involved with those who have similar characteristics.

infatuation emotional feelings based on little actual exposure to the love object.

jealousy an emotional response to a perceived or real threat to an important or valued relationship.

ludic love style love style that views love as a game in which the love interest is one of several partners, is never seen too often, and is kept at an emotional distance.

lust sexual desire.

mania love style an out-of-control love whereby the person "must have" the love object; obsessive jealousy and controlling behavior are symptoms of manic love.

marriage squeeze the imbalance of the ratio of marriageable-age men to marriageable-age women.

mating gradient the tendency for husbands to be more advanced than their wives with regard to age, education, and occupational success.

open-minded being open to understanding alternative points of view, values, and behaviors.

pool of eligibles the population from which a person selects an appropriate mate.

possessive jealousy attacking the partner who is perceived as being unfaithful.

pragma love style love style that is logical and rational; the love partner is evaluated in terms of assets and liabilities.

premarital education programs formal, systematized experiences designed to provide information to individuals and couples about how to have a good relationship.

principle of least interest principle stating that the person who has the least interest in a relationship controls the relationship.

racial homogamy tendency for individuals to marry someone of the same race.

reactive jealousy feelings that the partner may be straying.

religion specific fundamental set of beliefs and practices generally agreed upon by a number of people or sects.

religious homogamy tendency for people of similar religious or spiritual philosophies to seek out each other.

role (modeling) theory of mate selection emphasizes that children select partners similar to the one their same-sex parent selected.

romantic love an intense love whereby the lover believes in love at first sight, only one true love, and love conquers all.

sociobiology suggests a biological basis for all social behavior.

storge love style a love consisting of friendship that is calm and nonsexual.

© iStockphoto.com

LEARNING OUTCOMES

6-1 What are sexual values?

Sexual values are moral guidelines for sexual behavior in relationships. One's sexual values may be identical to one's sexual behavior, but sexual behavior does not always correspond with sexual values.

6-2 What are alternative sexual values?

Three sexual values are absolutism (rightness is defined by an official code of morality), relativism (rightness depends on the situation—who does what, with whom, in what context), and hedonism ("if it feels good, do it"). Relativism is the sexual value most college students hold, with women being more relativistic than men and men being more hedonistic than women. Virginity Pledge programs may result in youth experiencing a later "sexual debut," but most eventually engage in premarital sex and are less likely to use a condom when they first have intercourse. They are also more likley to substitute oral or anal sex, or both, in the place of vaginal sex. About 39% of undergraduates report involvement in a "friends with benefits" relationship. Women are more likely to focus on the "friendship" aspect, men on the "benefits" (sex) aspect. More males than females identified hedonism as their primary sexual value.

6-3 What is the sexual double standard?

The sexual double standard is the view that encourages and accepts the sexual expression of men more than that of women. For example, men may have more sexual partners than women without being stigmatized. The double standard is also reflected in movies.

6-4 What are sources of sexual values?

The sources of sexual values include one's school, family, and religion as well as technology, television, social movements, and the Internet.

6-5 What are gender differences in sexuality?

Men are more likely than women to believe that oral sex is not sex, that cybersex is not cheating, that men can't tell if a woman is faking orgasm, and that sex frequency drops in marriage. In regard to sexual behavior, men are more likely than women to report frequenting strip clubs, paying for sex, having anonymous sex with strangers, having casual sexual relations, and having more sexual partners.

6-6 How do pheromones affect sexual behavior?

Pheromones are chemical messengers emitted from the body that activate physiological and behavioral responses. The functions of pheromones include opposite-sex attractants, same-sex repellents, and mother-infant bonding. Researchers disagree about whether pheromones do in fact influence human sociosexual behaviors. In one study, men who applied a male hormone to their aftershave lotion reported significant increases in sexual intercourse and sleeping next to a partner in comparison with men who had a placebo in their aftershave lotion.

KEY TERMS

absolutism belief system based on unconditional allegiance to the authority of religion, law, or tradition.

AIDS acquired immunodeficiency syndrome; the last stage of HIV infection, in which the immune system of a person's body is so weakened that it becomes vulnerable to disease and infection.

asceticism the belief that giving in to carnal lusts is wrong and that one must rise above the pursuit of sensual pleasure to a life of self-discipline and self-denial.

concurrent sexual partnerships relationships in which the partners have sex with several individuals concurrently.

friends with benefits (FWB) a relationship between nonromantic friends who also have a sexual relationship.

hedonism belief that the ultimate value and motivation for human actions lie in the pursuit of pleasure and the avoidance of pain.

HIV human immunodeficiency virus, which attacks the immune system and can lead to AIDS.

pheromones body scents which activate physiological or behavioral responses in other individuals of the same species.

relativism belief system in which decisions are made in reference to the emotional, security, and commitment aspects of the relationship.

satiation the state in which a stimulus loses its value with repeated exposure.

secondary virginity the conscious decision of a sexually active person to refrain from intimate encounters for a specified period of time.

sexual double standard the view that encourages and accepts sexual expression of men more than women.

sexual values moral guidelines for sexual behavior in relationships.

social script the identification of the roles in a social situation, the nature of the relationship between the roles, and the expected behaviors of those roles.

spectatoring mentally observing one's own and one's partner's sexual performance.

STI sexually transmitted infection.

6-7 What are the sexual relationships of never-married, married, and divorced people?

Never-married and noncohabiting individuals report more sexual partners than do those who are married or living with a partner. Marital sex is distinctive for its social legitimacy, declining frequency, and satisfaction (both physical and emotional). Divorced individuals have a lot of sexual partners but are the least sexually fulfilled.

6-8 How does one avoid contracting or transmitting STIs?

The best way to avoid getting an STI is to avoid sexual contact or to have contact only with partners who are not infected. This means restricting your sexual contacts to those who limit their relationships to one person. The person most likely to get an STI has sexual relations with a number of partners or with a partner who has a variety of partners. Even if you are in a mutually monogamous relationship, you may be at risk for acquiring an STI, as 30% of male undergraduate students and 20% of female undergraduate students in "monogamous" relationships report having oral, vaginal, or anal sex with another partner outside the monogamous relationship.

6-9 What are the prerequisites of sexual fulfillment?

Fulfilling sexual relationships involve self-knowledge, self-esteem, health, a good nonsexual relationship, open sexual communication, safer sex practices, and making love with, not to, one's partner. Other variables include realistic expectations ("my partner will not always want what I want") and not buying into sexual myths ("masturbation is sick").

Did you take the self-assessment on page 124?

If so, what was your score? Read on to find out scores of other students who completed the scale.

Scores of Other Students Who Completed the Scale

This scale was completed by 252 student volunteers at Valdosta State University. The mean score of the students was 40.81 (standard deviation [SD] = 13.20), reflecting that the students were virtually at the midpoint between a very negative and a very positive attitude toward premarital sex. For the 124 males and 128 females in the total sample, the mean scores were 42.06 (SD = 12.93) and 39.60 (SD = 13.39), respectively (not statistically significant). In regard to race, 59.5% of the sample participants were White and 40.5% were non-White (35.3% Black, 2.4% Hispanic, 1.6% Asian, 0.4% American Indian, and 0.8% other). The mean scores of Whites, Blacks, and non-Whites were 41.64 (SD = 13.38), 38.46 (SD = 13.19), and 39.59 (SD = 12.90), respectively (not statistically significant). Finally, regarding year in college, 8.3% were freshmen, 17.1% sophomores, 28.6% juniors, 43.3% seniors, and 2.8% graduate students. Freshmen and sophomores reported more positive attitudes toward premarital sex (mean = 44.81; SD = 13.39) than did juniors (mean = 40.32; SD = 12.58) or seniors and graduate students (mean = 38.91; SD = 13.10; p = .05).

LEARNING OUTCOMES

7-1 What are individual motivations for marriage?

Individuals' motives for marriage include
- personal fulfillment,
- companionship,
- legitimacy of parenthood, and
- emotional and financial security.

Traditional Versus Egalitarian Marriages

Traditional Marriage	Egalitarian
There is limited expectation of husband to meet emotional needs of wife and children.	Husband is expected to meet emotional needs of wife and to be involved with children.
Wife is not expected to earn income.	Wife is expected to earn income.
Emphasis is on ritual and roles.	Emphasis is on companionship.
Couples do not live together before marriage.	Couples may live together before marriage.
Wife takes husband's last name.	Wife may keep her maiden name.
Husband is dominant; wife is submissive.	Neither spouse is dominant.
Roles for husband and wife are rigid.	Roles for spouses are flexible.
Husband initiates sex; wife complies.	Either spouse initiates sex.
Wife takes care of children.	Parents share child rearing.
Education is important for husband, not for wife.	Education is important for both spouses.
Husband's career decides family residence.	Career of either spouse determines family residence.

© Cengage Learning 2014

7-2 What are social functions of marriage?

Social functions include
- continuing to provide society with socialized members,
- regulating sexual behavior, and
- stabilizing adult personalities.

7-3 What are three levels of commitment in marriage?

Marriage involves three levels of commitment
1. person-to-person,
2. family-to-family, and
3. couple-to-state.

7-4 What are two rites of passage associated with marriage?

The wedding is a rite of passage signifying the change from the role of fiancé to the role of spouse. In general women, more than men, are invested in preparation for the wedding; the wedding is perceived to be more for the bride's family, and many women prefer a traditional wedding. The honeymoon is a time in which the couple recovers from the wedding and solidifies their new status as spouses.

KEY TERMS

artifact concrete symbol that reflects the existence of a phenomenon.

"blue" wedding artifact blue artifact worn by a bride that is symbolic of fidelity.

"borrowed" wedding artifact artifact worn by a bride that may be a garment or accessory owned by a currently happy bride.

bride wealth the amount of money or goods (e.g., cows) given by the groom or his family to the wife's family for giving her up; also known as bride price or bride payment.

commensality eating with others; most spouses eat together and negotiate who joins them.

commitment an intent to maintain a relationship.

cougar a woman, usually in her 30s or 40s, who is financially stable and mentally independent and who seeks a younger man with whom to have fun.

disenchantment the change in a relationship from a state of newness and high expectation to a state of mundaneness and boredom in the face of reality.

honeymoon time for the new spouses to recover from the wedding and to solidify their relationship.

marital success term for spouses in long-term marriages who are happy.

May-December marriage an age-discrepant marriage, usually in which the woman is in the spring of her life (May) and the man is in his later years (December).

military contract marriage
marriage in which a military person and a civilian participate to get more money and benefits from the government.

"new" wedding artifact artifact worn by a bride that symbolizes the new life she is about to begin (e.g., a new undergarment).

"old" wedding artifact artifact worn by a bride that symbolizes durability of the impending marriage (e.g., old gold locket).

rite of passage event that marks the transition from one status to another.

7-5 What changes might a person anticipate after marriage?

Changes after the wedding are

- legal,
- personal,
- social,
- economic,
- sexual, and
- parental.

7-6 What are examples of diversity in marriage relationships?

Mixed marriages include interracial, interreligious, and age-discrepant. When age-discrepant and age-similar marriages are compared, there are no differences in regard to marital happiness. There are three main types of military marriages: (1) those in which the soldier falls in love with a high school sweetheart, marries the person, and subsequently joins the military; (2) those in which the partners meet and marry after one of them has signed up for the military; and (3) the contract marriage in which a soldier will marry a civilian to get more money and benefits from the government. Military contract marriages are not common, but they do exist. Military families cope with deployment, the double standard, and limited income. Military marriages are particularly difficult for women.

7-7 What are the characteristics associated with successful marriages?

Marital success is defined in terms of both marital stability and marital happiness. Characteristics associated with marital success include commitment, common interests, communication, religiosity, trust, and nonmaterialism, and having positive role models, absence of negative attributions, health, and equitable relationships.

Trajectories of Marital Happiness

Source: Anderson et al. (2010).

LEARNING OUTCOMES

8-1 How prevalent are homosexuality/bisexuality and same-sex households and families?

In a national survey, fewer than 1% of women and 2% of men identified themselves as homosexual. The 2010 census revealed that about 1 in 9 (or 581,300) unmarried-partner households in the United States involved partners of the same sex.

8-2 Can people who are homosexual change their sexual orientation?

Some individuals believe that homosexual people choose their sexual orientation and think that, through various forms of reparative therapy, homosexual people should change their sexual orientation. However, many professional organizations (including the American Psychiatric Association, the American Psychological Association, and the American Medical Association) agree that homosexuality is not a mental disorder, that it needs no cure, and that efforts to change sexual orientation do not work and may, in fact, be harmful. These organizations also support the notion that changing societal reaction to those who are homosexual is a more appropriate focus.

8-3 Who tends to have more positive views toward homosexuality?

In general, individuals who are more likely to have positive attitudes toward homosexuality are younger, have advanced education, have no religious affiliation, have liberal political party affiliation, and have personal contact with homosexual individuals. Bisexual people experience "double discrimination" because in addition to rejection by heterosexual individuals, they experience rejection from many homosexual individuals. This response may be based on the belief that individuals who claim bisexuality are really in denial about their lesbian or gay sexual orientation.

8-4 How are gay and lesbian couples different from heterosexual couples?

Research suggests that gay and lesbian couples tend to be more similar to than different from heterosexual couples. Both gay and straight couples tend to value long-term monogamous relationships, experience high relationship satisfaction early in the relationship that decreases over time, disagree about the same topics, and have the same factors linked to relationship satisfaction.

However, same-sex couples' relationships are different from heterosexual relationships. Same-sex couples have more concern about when and how to disclose their relationships, are more likely to achieve a fair distribution of household labor, argue more effectively and resolve conflict in a positive way, and face prejudice and discrimination without much government support.

8-5 What forms of legal recognition of same-sex couples exist in the United States?

The website Infoplease provides a timeline about the gay rights movement in the United States from 1924 to the present (http://www.infoplease.com/ipa/

KEY TERMS

antigay bias any behavior or statement which reflects a negative attitude toward homosexuals.

biphobia (binegativity) refers to a parallel set of negative attitudes toward bisexuality and those identified as bisexual.

bisexuality emotional and sexual attraction to members of both sexes.

civil union a pair-bond given legal significance in terms of rights and privileges (more than a domestic partnership and less than a marriage).

Defense of Marriage Act (DOMA) legislation passed by Congress denying federal recognition of homosexual marriage and allowing states to ignore same-sex marriages licensed by other states.

discrimination behavior that denies individuals or groups equality of treatment.

domestic partnership a relationship in which individuals who live together and are emotionally and financially interdependent are given some kind of official recognition by a city or corporation so as to receive partner benefits (e.g., health insurance).

garriage term for relationship of gay individuals who are married or committed.

gay homosexual woman or man.

heterosexism the denigration and stigmatization of any behavior, person, or relationship that is not heterosexual.

heterosexuality the predominance of emotional and sexual attraction to individuals of the other sex.

homonegativity a construct that refers to antigay responses such as negative feelings (fear, disgust, anger), thoughts ("homosexuals are HIV carriers"), and behavior ("homosexuals deserve a beating").

homophobia negative (almost phobic) attitudes toward homosexuality.

homosexuality predominance of emotional and sexual attraction to individuals of the same sex.

internalized homophobia a sense of personal failure and self-hatred among lesbians and gay men resulting from the acceptance of negative social attitudes and feelings toward homosexuals.

lesbian homosexual woman.

lesbigay population collective term referring to lesbians, gays, and bisexuals.

LGBT (GLBT) refers collectively to lesbians, gays, bisexuals, and transgendered individuals.

prejudice negative attitudes toward others based on differences.

queer broad self-identifier term to indicate that the person has a sexual orientation other than heterosexual.

reparative therapy (conversion therapy) therapy designed to change a person's homosexual orientation to a heterosexual orientation.

second-parent adoption (also called co-parent adoption) a legal procedure that allows individuals to adopt their partner's biological or adoptive child without terminating the first parent's legal status as parent.

sexual orientation classification of individuals as heterosexual, bisexual, or homosexual, based on their emotional, cognitive, and sexual attractions and self-identity.

sodomy oral and anal sexual acts.

transgendered individuals who express their masculinity and femininity in nontraditional ways consistent with their biological sex.

A0761909.html). Included in this timeline are the states that recognize same-sex relationships. Also included is such information as the city of Berkeley, California, becoming the first city to offer its employees domestic-partnership benefits in 1984, and President Obama's endorsement of same-sex marriage on May 9, 2012.

8-6 What does research on gay and lesbian parenting conclude?

A growing body of credible, scientific research on gay and lesbian parenting concludes that children raised by gay and lesbian parents adjust positively and their families function well. Lesbian and gay parents are as likely as heterosexual parents to provide supportive and healthy environments for their children, and the children of lesbian and gay parents are as likely as those of heterosexual parents to flourish. Some research suggests that children raised by lesbigay parents develop in less gender-stereotypical ways, are more open to homoerotic relationships, contend with the social stigma of having gay parents, and have more empathy for social diversity than children of opposite-sex parents. There is no credible social science evidence that gay parenting negatively affects the well-being of children.

8-7 In what ways are heterosexual individuals victimized by heterosexism and antigay prejudice and discrimination?

Heterosexual individuals may be victims of antigay hate crimes. Many heterosexual family members and friends of homosexuals experience concern, fear, and grief over the mistreatment of and discrimination toward their gay or lesbian friends or family members or both. Because of the antigay social climate, heterosexuals, especially males, are hindered in their own self-expression and intimacy in same-sex relationships. Homophobia has also been linked to some cases of rape and sexual assault by males who are trying to prove they are not gay. The passage of state constitutional amendments that prohibit same-sex marriage can result in denial of rights and protections to opposite-sex unmarried couples.

REVIEWcard

LEARNING OUTCOMES

9-1 How does work or money affect a couple's relationship?

Generally, the more money a partner makes, the more power that person has in the relationship. Males make considerably more money than females and generally have more power in relationships. Seventy percent of all U.S. wives are in the labor force. The stereotypical family consisting of a husband who earns an income and a wife who stays at home with two or more children is no longer the norm. Only 13% are "traditional" in the sense of consisting of a breadwinning husband, a stay-at-home wife, and their children. In contrast, most marriages may be characterized as dual-earner. Employed wives in unhappy marriages are more likely to leave the marriage than unemployed wives.

> ### What do you think?
> Take the self-assessment for Chapter 9 in the self-assessment card deck to determine your attitudes towards maternal employment.

9-2 What is the effect of parents' work decisions on the children?

Children do not appear to suffer cognitively or emotionally from their parents' working as long as positive, consistent child-care alternatives are in place. However, less supervision of children by parents is an outcome of having two-earner parents. Leaving children to come home to an empty house is particularly problematic.

Parents view quality time as structured, planned activities, talking with their children, or just hanging out with them. Day care is typically mediocre but day-care workers who engage in high-frequency, positive behavior engender secure attachments with the children they work with. High-quality child care predicts higher cognitive-academic achievement at age 15, with increasing positive effects at higher levels of quality.

9-3 What are the various strategies for balancing the demands of work and family?

Strategies used for balancing the demands of work and family include the superperson strategy, cognitive restructuring, delegation of responsibility, planning and time management, and role compartmentalization. Government and corporations have begun to respond to the family concerns of employees by implementing work-family policies and programs. These policies are typically inadequate and cosmetic.

KEY TERMS

absolute poverty the lack of resources that leads to hunger and physical deprivation.

cognitive restructuring viewing a situation in positive terms.

dual-career marriage a marriage in which both spouses pursue careers and maintain a life together that may or may not include dependents.

dual-earner marriage both husband and wife work outside the home to provide economic support for the family.

HER/his career a wife's career is given precedence over a husband's career.

HIS/her career a husband's career is given precedence over a wife's career.

HIS/HER career a husband's and wife's careers are given equal precedence.

identity theft one person using the Social Security number, address, and bank account numbers of another, posing as that person to make purchases.

installment plan repayment plan whereby you sign a contract to pay for an item with regular installments over an agreed-upon period.

mommy track stopping paid employment to spend time with young children.

open charge repayment plan whereby you agree to pay the amount owed in full within the agreed amount of time.

opting out professional women leaving their careers and returning home to care for their children.

poverty the lack of resources necessary for material well-being.

relative poverty a deficiency in material and economic resources compared with some other population.

revolving charge the repayment plan whereby you may pay the total amount you owe, any amount over the stated minimum payment due, or the minimum payment.

role compartmentalization separating the roles of work and home so that an individual does not dwell on the problems of one role while physically being at the place of the other role.

shift work having one parent work during the day and the other parent work at night so that one parent can always be with the children.

supermom (superwoman) a cultural label that allows a mother who is experiencing role overload to regard herself as particularly efficient, energetic, and confident.

superperson strategy involves working as hard and as efficiently as possible to meet the demands of work and family.

THEIR career a career shared by a couple who travel and work together (e.g., journalists).

third shift the emotional energy expended by a spouse or parent in dealing with various family issues.

9-4 How do debt and poverty affect relationships?

The more in debt couples are, the less time they spend together, the more they argue, and the more they are unhappy. Poverty is devastating to couples and families. Those living in poverty have poorer physical and mental health, report lower personal and marital satisfaction, and die sooner. The table below reflects that a significant proportion of families in the United States continue to be affected by unemployment and low wages.

The stresses associated with low income also contribute to substance abuse, domestic violence, child abuse and neglect, and divorce. Couples with incomes of less than $25,000 are 30% more likely to divorce than are those with incomes of $50,000 or higher.

2012 DHHS Poverty Guidelines

People in Family or Household	48 Contiguous States and DC	Alaska	Hawaii
1	$11,170	$13,970	$12,860
2	15,130	18,920	17,410
3	19,090	23,870	21,960
4	23,050	28,820	26,510
5	27,010	33,770	31,060
6	30,970	38,720	35,610
7	34,930	43,670	40,160
8	38,890	48,620	44,170
For each additional person, add	3,960	4,950	4,550

Source: 2012 Poverty Guidelines, 77 Fed. Reg. 4034–4035 (Jan. 26, 2012), http://aspe.hhs.gov/poverty/12fedreg.shtml.

10-1 Why do people have children?

Having children continues to be a major goal of most young adults (women more than men). Among youth between the ages of 18 and 29, almost three fourths in a Pew Research Center report noted that they wanted to have children and most said that "being a good parent" was more important than "having a successful marriage." Social influences to have a child include family, friends, religion, government, favorable economic conditions, and cultural observances. The reasons people give for having children include love and companionship with one's own offspring, the desire to be personally fulfilled as an adult by having a child, and the desire to recapture one's youth. Having a child (particularly for women) reduces one's educational and career advancement. The cost for housing, food, transportation, clothing, health care, and child care for a child up to age 2 is more than $11,000 annually.

10-2 How many children do people have?

About 18% of women ages 40 to 44 do not have children. About 44% of these child-free women have chosen not to have children. Those who choose to be child-free are sometimes viewed with suspicion, avoidance, discomfort, rejection, and pity.

The most preferred type of family in the United States is the two-child family. Some of the factors in a couple's decision to have more than one child are the desire to repeat a good experience, the feeling that two children provide companionship for each other, and the desire to have a child of each sex.

10-3 Are teen mothers disadvantaged?

The reasons for teenagers having a child include not being socialized as to the importance of contraception, having limited parental supervision, and perceiving few alternatives to parenthood. Teens in the United States also view motherhood as one of the only viable roles open to them. Consequences associated with teenage motherhood include stigmatization and marginalization, poverty, alcohol or drug abuse, babies with lower birth weights, and lower academic achievement.

10-4 What are the causes of infertility?

Infertility is defined as the inability to achieve a pregnancy after at least one year of regular sexual relations without birth control, or the inability to carry a pregnancy to a live birth. Forty percent of infertility problems are attributed to the woman, 40% to the man, and 20% to both of them. The causes of infertility in women include blocked fallopian tubes, endocrine imbalance that prevents ovulation, dysfunctional ovaries, chemically hostile cervical mucus that may kill sperm, and effects of STIs. The psychological reaction to infertility is often depression over having to give up a lifetime goal. Some of the more common causes of infertility in men include low sperm production, poor semen motility, effects of STIs, and interference with passage of sperm through the genital ducts due to an enlarged prostate.

A number of technological innovations are available to assist women and couples in becoming pregnant. These include hormonal therapy, artificial insemination, ovum transfer, in vitro fertilization, gamete intrafallopian transfer, and zygote intrafallopian transfer. Being infertile (for the woman) may have a

abortion rate the number of abortions per 1,000 women ages 15 to 44.

abortion ratio the number of abortions per 1,000 live births.

antinatalism opposition to having children.

competitive birthing pattern in which a woman will want to have the same number of children as her peers.

cryopreservation procedure by which fertilized eggs are frozen and implanted at a later time.

Fertell an at-home fertility kit that allows women to measure the level of their follicle-stimulating hormone on the third day of their menstrual cycle and men to measure the concentration of motile sperm.

fertilization (conception) the fusion of the egg and sperm.

foster parent (family caregiver) a person who, either alone or with a spouse, takes care of and fosters a child taken into custody.

induced abortion the deliberate termination of a pregnancy through chemical or surgical means.

infertility the inability to achieve a pregnancy after at least one year of regular sexual relations without birth control, or the inability to carry a pregnancy to a live birth.

parental consent woman needs permission from parent to get an abortion if she is under a certain age, usually 18.

parental notification woman is required to tell parents she is getting an abortion if she is under a certain age, usually 18, but she does not need parental permission.

pregnancy a condition that begins five to seven days after conception, when the fertilized egg is implanted (typically in the uterine wall).

procreative liberty the freedom to decide whether or not to have children.

pronatalism view that encourages having children.

spontaneous abortion (miscarriage) an unintended termination of a pregnancy.

therapeutic abortion an abortion performed to protect the life or health of a woman.

transracial adoption the practice of parents adopting children of another race.

negative lifetime effect, both personal and interpersonal (half the women in one study were separated from their husbands or partners or reported a negative effect on their sex lives).

10-5 Why do people adopt?

Motives for adoption include a couple's inability to have a biological child (infertility), their desire to give an otherwise unwanted child a permanent loving home, or their desire to avoid contributing to overpopulation by having more biological children. Adoption is actually quite rare. Just over 1% of 18- to 44-year-old women reported having adopted a child.

Although those who typically adopt are currently White, educated, and of high income, adoptees are increasingly being placed in nontraditional families including with older, gay, and single individuals; it is recognized that these individuals may also be White, educated, and of high income. Most college students are open to transracial adoption.

10-6 Why do people choose foster parenting?

Some individuals seek the role of parent via foster parenting. A foster parent, also known as a family caregiver, is a person who, either alone or with a spouse, takes care of and fosters a child taken into custody in his or her home. A foster parent has made a contract with the state for the service, has judicial status, and is reimbursed by the state.

10-7 What are the types of and motives for an abortion?

An abortion may be either induced, which is the deliberate termination of a pregnancy through chemical or surgical means, or spontaneous (miscarriage), which is the unintended termination of a pregnancy. The most frequently cited reasons for induced abortion were that having a child would interfere with a woman's education, work, or ability to care for dependents (74%); that she could not afford a baby now (73%); and that she did not want to be a single mother or was having relationship problems (48%). Nearly 4 in 10 women said they had completed their childbearing, and almost one third of the women were not ready to have a child. Less than 1% said their parents' or partner's desire for them to have an abortion was the most important reason. In regard to the psychological effects of abortion, the American Psychological Association reviewed the literature and concluded that "among women who have a single, legal, first-trimester abortion of an unplanned pregnancy for nontherapeutic reasons, the relative risks of mental health problems are no greater than the risks among women who deliver an unplanned pregnancy."

> **Remember**
> There is a self-assessment card for this chapter in the self-assessment card deck. There are two tools for Chapter 10: the Childfree Lifestyles Scale and the Abortion Attitude Scale.

LEARNING OUTCOMES

11-1 What are the basic roles of parents?

Parenting includes providing physical care for children, loving them, being an economic resource, providing guidance as a teacher or model, and protecting them from harm. One of the biggest problems confronting parents today is the societal influence on their children. These influences include drugs and alcohol; peer pressure; TV, Internet, and movies; and crime or gangs.

11-2 What is a choices perspective of parenting?

Although both genetic and environmental factors are at work, the choices parents make have a dramatic impact on their children. Parents who don't make a choice about parenting have already made one. The five basic choices parents make include deciding (1) whether to have a child, (2) the number of children, (3) the interval between children, (4) methods of discipline and guidance, and (5) the degree to which they will be invested in the role of parent.

11-3 What is the transition to parenthood like for women, men, and couples?

Transition to parenthood refers to that period of time from the beginning of pregnancy through the first few months after the birth of a baby. The mother, father, and couple all undergo changes and adaptations during this period. Most mothers relish their new role; some may experience the transitory feelings of baby blues; a few report postpartum depression. Fathers may also experience depression following the birth of a baby. The preference for a male baby is associated with the fathers' depression.

The father's involvement with his children is sometimes predicted by the quality of the parents' romantic relationship. If the father is emotionally and physically involved with the mother, he is more likely to take an active role in the child's life. In recent years, there has been a renewed cultural awareness of fatherhood.

A summary of almost 150 studies involving almost 50,000 respondents on the question of how children affect marital satisfaction revealed that parents (both women and men) reported lower marital satisfaction than nonparents. In addition, the higher the number of children, the lower the marital satisfaction; the factors that depressed marital satisfaction were conflict and loss of freedom.

Percentage of Couples Getting Divorced by Number of Children

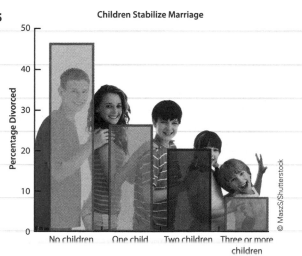

Children Stabilize Marriage

© MaszS/Shutterstock

KEY TERMS

baby blues transitory symptoms of depression in mothers 24 to 48 hours after the baby is born.

boomerang generation 18- to 34-year-old young adults who have moved back in with their parents after having lived on their own.

demandingness the manner in which parents place demands on children in regard to expectations and discipline.

emotional competence capacity to experience emotion, express emotion, and regulate emotion.

gatekeeper role term used to refer to the influence or control of the mother on the father's involvement and relationship with his children.

menarche first menstruation signaling a woman's fertility.

nature-deficit disorder children who are not encouraged and who have little opportunity to have direct contact with nature; instead they are overscheduled with activities such as soccer, swimming, etc.

oppositional defiant disorder disorder in which children fail to comply with requests of authority figures.

overindulgence defined as more than just giving children too much, includes overnurturing and providing too little structure.

oxytocin a hormone released from the pituitary gland during the expulsive stage of labor that has been associated with the onset of maternal behavior in lower animals.

parenting defined in terms of role including caregiver, emotional resource, teacher, and economic resource.

postpartum depression a severe reaction following the birth of a baby, which occurs in reference to a complicated delivery as well as numerous physiological and psychological changes; usually in the first month after birth but can be experienced after a couple of years have passed.

postpartum psychosis a reaction (rare) following the birth of a woman's baby where she wants to harm her baby.

responsiveness refers to the extent to which parents respond to and meet the needs of their children.

time-out a noncorporal form of punishment that involves removing the child from a context of reinforcement to a place of isolation.

transition to parenthood period from the beginning of pregnancy through the first few months after the birth of a baby, during which the mother and father undergo changes.

11-4 What are several facts about parenthood?

Parenthood will involve about 40% of the time a couple lives together, parents are only one influence on their children, each child is unique, and parenting styles differ. Research suggests that an authoritative parenting style, characterized by being both demanding and warm, is associated with positive outcomes. In addition, being emotionally connected to a child, respecting the child's individuality, and monitoring the child's behavior to encourage positive contexts have positive outcomes. Birth order effects include that firstborns are the first on the scene with parents and always have the "inside track." They want to stay that way so they are traditional and conforming to their parents' expectations. Children born later learn quickly that they entered an existing family constellation where everyone else is bigger and stronger. They cannot depend on having established territory, so they must excel in ways different from the firstborn. They are open to experience, adventurousness, and trying new things because their status is not already assured.

11-5 What are some of the principles of effective parenting?

Giving time, love, praise, and encouragement; being realistic; avoiding overindulgence; monitoring activities and drug use; setting limits and disciplining children for inappropriate behavior; providing security; encouraging responsibility; teaching emotional competence; providing sex education; teaching nonviolence; establishing norms of forgiveness; and keeping children connected with nature are aspects of effective parenting.

11-6 What are the issues of single parenting?

About 40% of births in the United States are to unmarried mothers. By adolescence, 20% of chldren have no contact with their father. The challenges of single parenthood for the parent include taking care of the emotional and physical needs of a child alone, meeting one's own adult emotional and sexual needs, earning money, and rearing a child without a father (the influence of whom can be positive and beneficial).

Remember
There is a self-assessment card for this chapter in the self-assessment card deck. For Chapter 11, there are two tools: the Traditional Motherhood Scale and the Spanking versus Time-Out Scale.

LEARNING OUTCOMES

12-1 What is stress and what is a crisis event?

Stress is a reaction of the body to substantial or unusual demands (physical, environmental, or interpersonal). Stress is often accompanied by irritability, high blood pressure, and depression. Stress also has an effect on a person's relationships and sex life. Stress is a process rather than a state. A crisis is a situation that requires changes in normal patterns of behavior. A family crisis is a situation that upsets the normal functioning of the family and requires a new set of responses to the stressor. Sources of stress and crises can be external (for example, a hurricane, a tornado, downsizing, military deployment) or internal (for example, alcoholism, an extramarital affair, Alzheimer's disease).

Family resilience is displayed when family members successfully cope under adversity, which enables them to flourish with warmth, support, and cohesion. Key factors include a positive outlook, spirituality, flexibility, communication, financial management, shared recreation, routines or rituals, and support networks.

12-2 What are positive stress-management strategies?

Changing one's basic values and perspective is the most helpful strategy in reacting to a crisis. Viewing ill health as a challenge, bankruptcy as an opportunity to spend time with one's family, and infidelity as an opportunity to improve communication are examples. Other positive coping strategies are exercise, adequate sleep, love, religion, friends or relatives, humor, education, and counseling. Still other strategies include intervening early in a crisis, not blaming each other, keeping destructive impulses in check, and seeking opportunities for fun. Pets are also useful in helping individuals cope with stress. They are associated with reducing blood pressure, preventing heart disease, and fighting depression.

12-3 What are harmful strategies for reacting to a crisis?

Some harmful strategies include keeping feelings inside, taking out frustrations on others, and denying or avoiding the problem.

© Juriah Mosin/Shutterstock.com

KEY TERMS

Al-Anon an organization that provides support for family members and friends of alcohol abusers.

biofeedback a process in which information that is relayed back to the brain enables people to change their biological state.

Coolidge effect term used to describe the waning of sexual excitement and the effect of novelty and variety on sexual arousal.

crisis a sharp change for which typical patterns of coping are not adequate and new patterns must be developed.

down low African American "heterosexual" man who has sex with men.

extradyadic (extrarelational) involvement emotional or sexual involvement between a member of a pair and someone other than the partner.

extramarital affair a spouse's sexual involvement with someone outside the marriage.

family resilience the successful coping of family members under adversity that enables them to flourish with warmth, support, and cohesion.

palliative care health care focused on the relief of pain and suffering of the individual who has a life-threatening illness and support for such individuals and their loved ones.

resiliency the ability of a family to respond to a crisis in a positive way.

sanctification viewing the marriage as having divine character or significance.

stress a nonspecific response of the body to demands made on it.

12-4 What are five of the major family crisis events?

Some of the more common crisis events that spouses and families face include physical illness and disability, mental illness, middle-age crazy (midlife crisis), an extramarital affair, unemployment, substance abuse, the death of a family member, and the suicide of a family member. When an affair is discovered, a sense of betrayal is pervasive for the partner. Surviving an affair involves forgiveness on the part of the offended spouse and the ability to grant a pardon to the offending spouse, to give up feeling angry, and to relinquish the right to retaliate against the offending spouse. In exchange, an offending spouse must take responsibility for the affair, agree not to repeat the behavior, and grant the partner the right to check up on the offending partner to regain trust. The best affair prevention is a happy and fulfilling marriage as well as avoidance of intimate conversations with members of the other sex and a context (for example, where alcohol is consumed and being alone), which are conducive to physical involvement. The occurrence of a "midlife crisis" is reported by less than a quarter of adults in the middle years. Those who did experience a crisis were going through a divorce.

Did you take the Attitudes toward Infidelity self-assessment in the self-assessment card deck?

Check out how other students scored on the scale. How do you compare?

Scores of Other Students Who Completed the Scale

The scale was completed by 150 male and 136 female student volunteers at Valdosta State University. The average score on the scale was 27.85 ($SD = 12.02$). Their ages ranged from 18 to 49, with a mean age of 23.36 ($SD = 5.13$). The ethnic backgrounds of the sample included 60.8% White, 28.3% African American, 2.4% Hispanic, 3.8% Asian, 0.3% American Indian, and 4.2% other. The college classification level of the sample included 11.5% freshmen, 18.2% sophomores, 20.6% juniors, 37.8% seniors, 7.7% graduate students, and 4.2% postbaccalaureate. Male participants reported more positive attitudes toward infidelity (mean = 31.53; $SD =11.86$) than did female participants (mean = 23.78; $SD = 10.86$; $p = .05$). White participants had more negative attitudes toward infidelity (mean = 25.36; $SD = 11.17$) than did non-White participants (mean = 31.71; SD 12.32; $p = .05$). There were no significant differences in regard to college classification.

Remember

Chapter 12 has another self-assessment in the self-assessment card deck. Take the Internality, Powerful Others, and Chance self-assessment and compare your results with other students'.

LEARNING OUTCOMES

13-1 What is the nature of relationship abuse?

Violence or physical abuse may be defined as the deliberate infliction of physical harm by either partner on the other. Violence may occur over an issue on which the partners disagree, or it may be a mechanism of control. Female violence is as prevalent as male violence. The difference is that female violence is often in response to male violence, whereas male violence is more often to control the partner. Therefore, women tend to be striking back rather than throwing initial blows.

Emotional abuse is designed to denigrate the partner, reduce the partner's status, and make the partner vulnerable, thereby giving the abuser more control. Stalking is unwanted following or harassment that induces fear in the target person. The stalker is most often a heterosexual male who has been rejected by someone who fails to return his advances. Stalking is typically designed either to seek revenge or to win a partner back. Cybervictimization includes being sent threatening e-mail, unsolicited obscene e-mail, computer viruses, or junk mail (spamming). It may also include flaming (online verbal abuse) and leaving improper messages on message boards designed to get back at the person. Obsessive relationship intrusion (ORI) is the relentless pursuit of intimacy with someone who does not want it. Unlike stalking, the goal of which is to harm, ORI involves hyperintimacy (telling people that they are beautiful or desirable to the point of making them uncomfortable), relentless mediated contacts (flooding people with e-mail messages, cell phone messages, or faxes), or interactional contacts (showing up at work or the gym, or joining the same volunteer groups as the pursued).

13-2 What are some explanations for violence in relationships?

Cultural explanations for violence include violence in the media, corporal punishment in childhood, gender inequality, and stress. Community explanations involve social isolation of individuals and spouses from extended family, poverty, inaccessible community services, and lack of violence prevention programs. Individual factors include psychopathology of the person (antisocial), personality (dependency or jealousy), and alcohol or substance abuse. Family factors include experiencing child abuse by one's parents and observing parents who abuse each other.

13-3 How does sexual abuse in undergraduate relationships manifest itself?

Some women experience sexual abuse in addition to a larger pattern of physical abuse, and the combination is associated with less general satisfaction, less sexual satisfaction, more conflict, and more psychological abuse from the partner. About 33% of undergraduates report being forced to have sex against their will. Acquaintance rape is defined as nonconsensual sex between adults who know each other.

One type of acquaintance rape is date rape, which refers to nonconsensual sex between people who are dating or on a date. Rohypnol, known as the date rape drug, causes profound sedation so that the person may not remember being raped. Most women do not report being raped by an acquaintance or date.

KEY TERMS

acquaintance rape nonconsensual sex between adults who know each other.

battered-woman syndrome general pattern of battering that a woman is subjected to, defined in terms of the frequency, severity, and injury.

corporal punishment the use of physical force with the intention of causing a child to experience pain, but not injury, for the purpose of correction or control of the child's behavior.

cybervictimization being sent unwanted e-mail, spam, or viruses, or being threatened online.

date rape nonconsensual sex between two people who are dating or on a date.

emotional abuse (psychological abuse; verbal abuse; symbolic aggression) the denigration of an individual with the purpose of reducing the victim's status and increasing the victim's vulnerability so that he or she can be more easily controlled by the abuser.

entrapped stuck in an abusive relationship and unable to extricate oneself from the abusive partner.

honor crime (honor killing) killing a daughter because she brought shame to the family by having sex while not married. The killing is typically overlooked by the society. Jordan is a country where honor crimes occur.

intimate-partner violence an all-inclusive term that refers to crimes committed against current or former spouses, boyfriends, or girlfriends.

marital rape forcible rape by one's spouse.

obsessive relational intrusion the relentless pursuit of intimacy with someone who does not want it.

physical abuse (violence) intentional infliction of physical harm by either partner on the other.

red zone the first month of the first year of college when women are particularly vulnerable to unwanted sexual advances.

Rohypnol date rape drug that causes profound, prolonged sedation and short-term memory loss.

stalking unwanted following or harassment that induces fear in a target person.

uxoricide the murder of a woman by her romantic partner.

13-4 How does abuse in marriage relationships manifest itself?

Abuse in marriage is born out of the need to control the partner and may include repeated rape. About half of the women raped by an intimate partner and two thirds of the women physically assaulted by an intimate partner have been victimized multiple times.

13-5 What are the effects of abuse?

The effects of IPV (intimate partner violence) include symptoms of PTSD—loss of interest in activities and life in general, a feeling of detachment from others, inability to sleep, and irritability. Women with low self-esteem are more likely to take some responsibility for their boyfriends' abusive behavior. In effect, they excuse the rape as their fault. Violence on pregnant women significantly increases the risk for miscarriage and birth defects. Negative effects may also accrue to children who witness abuse. These children are more likely to be depressed as adults.

13-6 What is the cycle of abuse and why do people stay in an abusive relationship?

The cycle of abuse begins when a person is abused and the perpetrator feels regret, asks for forgiveness, and starts acting nice (for example, gives flowers). The victim, who perceives few options and feels guilty terminating the relationship with the partner who asks for forgiveness, feels hope for the relationship at the contriteness of the abuser and does not call the police or file charges. The couple usually experiences a period of making up or honeymooning, during which the victim feels good again about the abusing partner. However, tensions mount again and are released in the form of violence. Such violence is followed by the familiar sense of regret and pleadings for forgiveness, accompanied by being nice (a new bouquet of flowers, and so on).

The reasons people stay in abusive relationships include love, emotional dependency, commitment to the relationship, hope, view of violence as legitimate, guilt, fear, economic dependency, and isolation. The catalyst for breaking free combines the sustained aversiveness of staying, the perception that they and their children will be harmed by doing so, and the awareness of an alternative path or of help in seeking one. One must be cautious in getting out of an abusive relationship because an abuser is most likely to kill his partner when she actually leaves the relationship.

The Cycle of Abuse

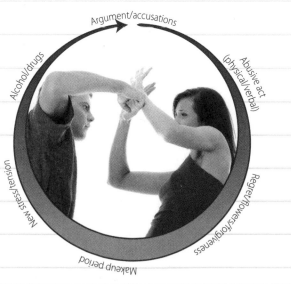

Argument/accusations

Abusive act (physical/verbal)

Regret/flowers/forgiveness

Makeup period

New stress/tension

Alcohol/drugs

© Cengage Learning 2011

LEARNING OUTCOMES

14-1 What are macro factors contributing to divorce?

Macro factors contributing to divorce include increased economic independence of women (women can afford to leave), changing family functions (companionship is the only remaining function), liberal divorce laws (it's easier to leave), prenuptial agreements, the Internet, fewer moral and religious sanctions (churches embrace single individuals), more divorce models (Hollywood models abound), mobility and anonymity, and social class, ethnicity, and culture. The United States has one of the highest divorce rates in the world, higher even than vanguard countries such as Sweden.

14-2 What are micro factors contributing to divorce?

Micro factors include having numerous differences, falling out of love, spending limited time together, decreasing positive behaviors, having an affair, having poor conflict resolution skills, changing values, experiencing the onset of satiation, and having the perception that one would be happier if divorced.

14-3 How might one go about ending a relationship?

About 30% of undergraduates in one study reported that they were unhappy in their present relationships; another 30% reported that they knew they were in relationships that should end. After considering that one might improve an unhappy relationship and deciding to end a relationship, telling a partner that one needs out "for one's space" without giving a specific reason helps a partner avoid feeling obligated to stay if the other partner changes. Although some couples can remain friends, others may profit from ending the relationship completely. And recovery can take time (12 to 18 months).

14-4 What are gender differences in filing for divorce?

Women are the first to seek help when there is trouble in a relationship. Two thirds of the women who filed for divorce in one study did so because they felt their lives would be better without their husbands. The top advantage women reported for divorce was feeling a sense of renewed "self-identity." Men are less likely to seek a divorce because they view the cost as separation from their children; women get primary custody of children in 80% of divorces.

14-5 What are the consequences of divorce for spouses?

The psychological consequences for divorcing spouses depend on how unhappy the marriage was. Spouses who were miserable while in a loveless conflictual marriage often regard the divorce as a relief. Spouses who were left (for example, a spouse leaves for another partner) may be devastated and suicidal. Women tend to fare better emotionally after separation and divorce than do men. Women are more likely than men not only to have a stronger network of supportive relationships but also to profit from divorce by developing a new sense of self-esteem and confidence, because they are thrust into a more independent role. Divorced men are more likely than divorced women to date more partners sooner and to remarry more quickly.

KEY TERMS

binuclear a family that spans two households, often because of divorce.

blended family a family in which new spouses both have children from previous marriages.

covenant marriage type of marriage that permits divorce only under specific conditions.

December marriage a new marriage in which both spouses are elderly.

developmental task a skill that allows a family to grow as a cohesive unit.

divorce the termination of a valid marriage contract.

divorce mediation process in which divorcing parties make agreements with a third party (mediator) about custody, visitation, child support, property settlement, and spousal support.

divorcism the belief that divorce is a disaster.

homogamy tendency to select someone with similar characteristics to marry.

legal custody decisional authority over major issues involving the child.

mating gradient the tendency for husbands to be more advanced than their wives with regard to age, education, and occupational success.

negative commitment also known as *negative attachment*, individuals who remain emotionally invested in their relationship with their former spouse, despite remarriage.

no-fault divorce a divorce in which neither party is identified as the guilty party or the cause of the divorce.

parental alienation an alliance between a parent and a child that isolates the other parent.

parental alienation syndrome a disturbance in which children are obsessively preoccupied with deprecation or criticism of a parent.

physical custody also called *visitation*, refers to distribution of parenting time following divorce.

postnuptial agreement specifies what is to be done with property and holdings at death or divorce; similar to a premarital agreement.

satiation the state in which a stimulus loses its value with repeated exposure.

shared parenting dysfunction the set of behaviors by both parents that are focused on hurting the other parent and are counterproductive for a child's well-being.

stepfamily (step relationships) a family in which partners bring children from previous marriages into the new home, where the partners may also have a child of their own.

stepism the assumption that stepfamilies are inferior to biological families.

A sample of 410 freshmen and sophomores at a large southeastern university revealed that though recovery was not traumatic for either men or women, men reported more difficulty than women did in adjusting to a breakup. Men more than women reported "a new partner" was more helpful in relationship recovery. Women more than men reported that "time" was more helpful in relationship recovery.

Getting divorced affects one financially, affects fathers' separation from children, and may result in shared parenting dysfunction and parental alienation.

14-6 What are the effects of divorce on children?

Although researchers agree that a civil, cooperative, co-parenting relationship between ex-spouses is the greatest predictor of a positive outcome for children, researchers disagree on the long-term negative effects of divorce on children. However, there is no disagreement that most children do not experience long-term negative effects. Divorce mediation encourages civility between divorcing spouses who negotiate the issues of division of property, custody, visitation, child support, and spousal support.

14-7 What are some conditions of a "successful" divorce?

Divorce is an emotionally traumatic event for everyone involved, but there are some steps that spouses can take to minimize the pain and help each other and their children with the transition. Some of these steps include mediating the divorce, co-parenting, sharing responsibility, creating positive thoughts, avoiding drugs and alcohol, being active, releasing anger, allowing time to heal, and progressing through the psychological stages of divorce.

14-8 What are strategies to prevent divorce?

Three states (Louisiana, Arizona, and Arkansas) offer covenant marriages, in which spouses agree to divorce only for serious reasons such as imprisonment on a felony or separation of more than two years. They also agree to see a marriage counselor if problems threaten the marriage. When given the option to choose a covenant marriage, few couples do so.

14-9 What is the nature of remarriage in the United States?

Two thirds of divorced females and three fourths of divorced males remarry. Ninety percent of remarriages are of people who are divorced rather than widowed. Most divorced individuals select someone who is divorced to remarry just as widowed individuals select someone who is also widowed to remarry.

Two years is the recommended time from the end of one marriage to the beginning of the next. Older divorced women (over age 40) are less likely than younger divorced women to remarry. National data reflect that remarriages are more likely than first marriages to end in divorce in the early years of remarriage. After 15 years, however, second marriages tend to be more stable and happier than first marriages. The reason for this is that remarried individuals tend not to be afraid of divorce and would divorce if unhappy in a second marriage. First-time married individuals may be fearful of divorce and stay married even though they are unhappy.

14-10 What is the nature of stepfamilies in the United States?

Stepfamilies represent the fastest-growing type of family in the United States. There is a movement away from the use of the term *blended* when referring to stepfamilies, because stepfamilies really do not blend. Although a stepfamily can be created when a never-married or a widowed parent with children marries a person with or without children, most stepfamilies today are composed of spouses who were once divorced.

Stepfamilies differ from nuclear families: the children in nuclear families are biologically related to both parents, whereas the children in stepfamilies are biologically related to only one parent. Also, in nuclear families, both biological parents live with their children, whereas only one biological parent in stepfamilies lives with the children. In some cases, the children alternate living with each parent. Stepism is the assumption that stepfamilies are inferior to biological families. Stepism, like racism, heterosexism, sexism, and ageism, involves prejudice and discrimination.

Stepfamilies go through a set of stages. Newly remarried couples often expect instant bonding between the new members of the stepfamily. It does not often happen. The stages are fantasy (everyone will love everyone), reality (possible bitter conflict), assertiveness (parents speak their mind), strengthening pair ties (spouses nurture their relationship), and recurring change (stepfamily members know there will continue to be change). Involvement in stepfamily discussion groups such as the Stepfamily Enrichment Program provides enormous benefits.

Differences Between Nuclear Families and Stepfamilies

Nuclear Families	Stepfamilies
1. Children are (usually) biologically related to both parents.	1. Children are biologically related to only one parent.
2. Both biological parents live together with children.	2. As a result of divorce or death, one biological parent does not live with the children. In the case of joint physical custody, children may live with both parents, alternating between them.
3. Beliefs and values of members tend to be similar.	3. Beliefs and values of members are more likely to be different because of different backgrounds.
4. The relationship between the adults has existed longer than the relationship between children and parents.	4. The relationship between children and parents has existed longer than the relationship between adults.
5. Children have one home they regard as theirs.	5. Children may have two homes they regard as theirs.
6. The family's economic resources come from within the family unit.	6. Some economic resources may come from an ex-spouse.
7. All money generated stays in the family.	7. Some money generated may leave the family in the form of alimony or child support.
8. Relationships are relatively stable.	8. Relationships are in flux: new adults adjusting to each other; children adjusting to a stepparent; a stepparent adjusting to stepchildren; stepchildren adjusting to each other.
9. No stigma is attached to the nuclear family.	9. Stepfamilies are stigmatized.
10. Spouses had a child-free period.	10. Spouses had no child-free period.
11. Inheritance rights are automatic.	11. Stepchildren do not automatically inherit from stepparents.
12. Rights to custody of children are assumed if divorce occurs.	12. Rights to custody of stepchildren are usually not available.
13. Extended family networks are smooth and comfortable.	13. Extended family networks become complex and strained.
14. Nuclear family may not have experienced loss.	14. Stepfamily has experienced loss.

© Cengage Learning 2014

14-11 What are the strengths of stepfamilies?

The strengths of stepfamilies include exposure to a variety of behavior patterns, happier parents, opportunity for a new relationship with stepsiblings, and greater objectivity on the part of the stepparent.

© PhotoAlto sas/Alamy

14-12 What are the developmental tasks of stepfamilies?

Developmental tasks for stepfamilies include nurturing the new marriage relationship, allowing time for partners and children to get to know each other, maintaining realistic expectations, deciding whose money will be spent on whose children, deciding who will discipline the children and how, accepting one's stepchildren, establishing one's own family rituals, and supporting the children's relationship with both parents and natural grandparents. Both sets of parents and stepparents should form a parenting coalition in which they cooperate and actively participate in child rearing.

> **Remember**
> Take the Chapter 14 self-assessment, called Children's Beliefs about Parental Divorce, located in your self-assessment card deck.

LEARNING OUTCOMES

15-1 What is meant by the terms *age* and *ageism?*

Age is defined chronologically (by time), physiologically (by capacity to see, hear, and so on), psychologically (by self-concept), sociologically (by social roles), and culturally (by the value placed on the elderly). Ageism is the denigration of the elderly, and gerontophobia is the fear or dread of the elderly. Theories of aging range from disengagement (individuals and societies mutually disengage from each other) to continuity (the habit patterns of youth are continued in old age), as you can see in the following table. Life course is the aging theory currently in vogue. This approach examines differences in aging across cohorts by emphasizing that "individual biography is situated within the context of social structure and historical circumstance."

Theories of Aging

Name of Theory	Level of Theory	Theorists	Basic Assumptions	Criticisms
Disengagement	Macro	Elaine Cumming, William Henry	The gradual and mutual withdrawal of the elderly and society from each other is a natural process. It is also necessary and functional for society that the elderly disengage so that new people can be phased in to replace them in an orderly transition.	Not all people want to disengage; some want to stay active and involved. Disengagement does not specify what happens when the elderly stay involved.
Activity	Macro	Robert Havighurst	People continue the level of activity they had in middle age into their later years. Though high levels of activity are unrelated to living longer, they are related to reporting high levels of life satisfaction.	Ill health may force people to curtail their level of activity. The older a person, the more likely the person is to curtail activity.
Conflict	Macro	Karl Marx, Max Weber	The elderly compete with youth for jobs and social resources such as government programs (Medicare).	The elderly are presented as disadvantaged. Their power to organize and mobilize political resources such as the American Association of Retired Persons is underestimated.
Age stratification	Macro	M. W. Riley	The elderly represent a powerful cohort of individuals passing through the social system that both affect and are affected by social change.	Too much emphasis is put on age, and little recognition is given to other variables within a cohort such as gender, race, and socioeconomic differences.
Modernization	Macro	Donald Cowgill	The status of the elderly is in reference to the evolution of the society toward modernization. The elderly in premodern societies have more status because what they have to offer in the form of cultural wisdom is more valued. The elderly in modern technologically advanced societies have low status because they have little to offer.	Cultural values for the elderly, not level of modernization, dictate the status of the elderly. Japan has high respect for the elderly and yet is highly technological and modernized.
Symbolic	Micro	Arlie Hochschild	The elderly socially construct meaning in their interactions with others and society. Developing social bonds with other elderly can ward off being isolated and abandoned. Meaning is in the interpretation, not in the event.	The power of the larger social system and larger social structures to affect the lives of the elderly is minimized.
Continuity	Micro	Bernice Neugarten	The earlier habit patterns, values, and attitudes of the individual are carried forward as a person ages. The only personality change that occurs with aging is the tendency to turn one's attention and interest on the self.	Other factors than one's personality affect aging outcomes. The social structure influences the life of the elderly rather than vice versa.

© Cengage Learning 2014

KEY TERMS

age term defined chronologically, physiologically, sociologically, and culturally.

age discrimination discriminating against a person because of age.

ageism systematic persecution and degradation of people because they are old.

blurred retirement the process of retiring gradually so that the individual works part-time before completely retiring or takes a bridge job that provides a transition between a lifelong career and full retirement.

cohousing an arrangement where elderly individuals live in group housing or shared living arrangements.

dementia the mental disorder most associated with aging, whereby the normal cognitive functions are slowly lost.

family caregiving care provided to the elderly by family members.

filial piety respect for one's parents.

frail elderly person who has difficulty with at least one personal care activity or other activities related to independent living.

gerontology the study of aging.

gerontophobia fear or dread of the elderly.

sandwich generation individuals who attempt to meet the needs of their children and elderly parents at the same time.

thanatology examination of the social dimensions of death, dying, and bereavement.

15-2 What is the "sandwich generation"?

Eldercare combined with child care is becoming common among the sandwich generation—adult children responsible for the needs of both their parents and their children. Two levels of eldercare include help with personal needs such as feeding, bathing, and toileting as well as instrumental care such as going to the grocery store, managing bank records, and so on. Members of the sandwich generation report feelings of exhaustion over the relentless demands, guilt over not doing enough, and resentment over feeling burdened.

Most adult children want to take care of their aging parents either in the parents' own home or in the adult child's home. Parents are usually resistant to full-time nursing home care, but become resigned to such care when there is no other option. But professional caregiving is expensive—$70,000 is the average annual cost. Home health care can reduce the strain of caring for an elderly parent.

15-3 What issues confront the elderly?

Issues of concern to the elderly include housing, health, retirement, and sexuality. Most elderly live in their own homes, which they have paid for. Most elderly housing is adequate, although repair becomes a problem when people age. Health concerns are paramount for the elderly. Good health is the single most important factor associated with an elderly person's perceived life satisfaction.

When elders' vision, hearing, physical mobility, and strength are markedly diminished, their sense of well-being is often significantly negatively impacted. Driving accidents also increase with aging. Mental problems may also occur with mood disorders; depression is the most common. Dementia, which includes Alzheimer's disease, is the mental disorder most associated with aging; however, only 3% of the aged population experience severe cognitive impairment.

Individuals who have a positive attitude toward retirement are those who have a pension waiting for them, are married, have planned for retirement, are in good health, and have high self-esteem.

For most elderly women and men, sexuality involves lower reported interest, activity, and capacity. Fear of the inability to have an erection is the sexual problem elderly men most frequently report (Viagra, Levitra, and Cialis have helped allay this fear). The absence of a sexual partner is the sexual problem elderly women most frequently report.

15-4 What factors are associated with successful aging?

Factors associated with successful aging include not smoking (or quitting early), developing a positive view of life and life's crises, avoiding alcohol and substance abuse, maintaining healthy weight, exercising daily, continuing to educate oneself, and having a happy marriage. Indeed, those who were identified as "happy and well" were six times more likely to be in good marriages than those who were identified as "sad and sick." Success in one's career is also associated with successful aging.

15-5 What are relationships like for the elderly?

Marriages that survive into old age (beyond age 85) tend to have limited conflict, considerable companionship, and mutual supportiveness. Marital satisfaction is related to equality of roles and marital communication. Some widowed or divorced elderly try to find and renew an earlier love relationship. Older adults and senior citizens are increasing their use of technology to stay connected by using networking sites such as Facebook and LinkedIn. Relationships with siblings are primarily emotional rather than functional. Relationships with children are emotional and expressive. Actual caregiving by one's children is rare.

15-6 How do the elderly face the end of life?

Thanatology is the examination of the social dimensions of death, dying, and bereavement. The end of life can involve adjusting to the death of one's spouse and to the gradual decline of one's health. Most elderly are satisfied with their lives, relationships, and health. Declines begin when people reach their 80s. Most fear the process of dying more than death itself.

NOTES

NOTES

The Relationship Involvement Scale

This scale is designed to assess the level of your involvement in a current relationship. Please read each statement carefully, and write the number next to the statement that reflects your level of disagreement to agreement, using the following scale.

1	2	3	4	5	6	7
Strongly Disagree						*Strongly Agree*

____ **1.** I have told my friends that I love my partner.

____ **2.** My partner and I have discussed our future together.

____ **3.** I have told my partner that I want to marry him/her.

____ **4.** I feel happier when I am with my partner.

____ **5.** Being together is very important to me.

____ **6.** I cannot imagine a future with anyone other than my partner.

____ **7.** I feel that no one else can meet my needs as well as my partner.

____ **8.** When talking about my partner and me, I tend to use the words "us," "we," and "our."

____ **9.** I depend on my partner to help me with many things in life.

____ **10.** I want to stay in this relationship no matter how hard times become in the future.

Scoring

Add the numbers you assigned to each item. A 1 reflects the least involvement and a 7 reflects the most involvement. The lower your total score (10 is the lowest possible score), the lower your level of involvement; the higher your total score (70), the greater your level of involvement. A score of 40 places you at the midpoint between a very uninvolved and very involved relationship.

Other Students Who Completed the Scale

Valdosta State University. The participants were 31 male and 86 female undergraduate psychology students haphazardly selected from Valdosta State University. They received course credit for their participation. These participants ranged in age from 18 to 59 with a mean age of 20.25 (*SD* = 4.52). The ethnic background of the sample included 70.9% white, 23.9% Black, 1.7% Hispanic, and 3.4% from other ethnic backgrounds. The college classification level of the sample included 46.2% freshmen, 36.8% sophomores, 14.5% juniors, and 2.6% seniors.

East Carolina University. Also included in the sample were 60 male and 129 female undergraduate students haphazardly selected from East Carolina University. These participants ranged in age from 18 to 43 with a mean age of 20.40 (*SD* = 3.58). The ethnic background of the sample included 76.2% white, 14.3% Black, 0.5% Hispanic, 1.6% Asian, 2.6% American Indian, and 4.8% from other ethnic backgrounds. The college classification level of the sample included 40.7% freshmen, 19.0% sophomores, 19.6% juniors, and 20.6% seniors. All participants were treated in accordance with the ethical guidelines of the American Psychological Association (1992).

Scores of Participants

When students from both universities were combined, the average score of the men was 50.06 ($SD = 14.07$) and the average score of the women was 52.93 ($SD = 15.53$), reflecting moderate involvement for both sexes. There was no significant difference between men and women in level of involvement. However, there was a significant difference ($p < .05$) between whites and non-whites, with whites reporting greater relationship involvement (M = 53.37; $SD = 14.97$) than non-whites (M = 48.33; $SD = 15.14$).

In addition, there was a significant difference between the level of relationship involvement of seniors compared with juniors ($p < .05$) and freshmen ($p < .01$). Seniors reported more relationship involvement (M = 57.74; $SD = 12.70$) than did juniors (M = 51.57; $SD = 15.37$) or freshmen (M = 50.31; $SD = 15.33$).

Source: "The Relationship Involvement Scale," 2004, by Mark Whatley, Ph.D., Department of Psychology, Valdosta State University, Valdosta, Georgia 31698-0100. Used by permission. Other uses of this scale by written permission of Dr. Whatley only (mwhatley@valdosta.edu). Information on the reliability and validity of this scale is available from Dr. Whatley.

The Beliefs about Women Scale (BAWS)

The following statements describe different attitudes toward men and women. There are no right or wrong answers, only opinions. Indicate how much you agree or disagree with each statement, using the following scale: (A) strongly disagree, (B) slightly disagree, (C) neither agree nor disagree, (D) slightly agree, or (E) strongly agree.

_____ **1.** Women are more passive than men.

_____ **2.** Women are less career-motivated than men.

_____ **3.** Women don't generally like to be active in their sexual relationships.

_____ **4.** Women are more concerned about their physical appearance than are men.

_____ **5.** Women comply more often than men.

_____ **6.** Women care as much as men do about developing a job or career.

_____ **7.** Most women don't like to express their sexuality.

_____ **8.** Men are as conceited about their appearance as are women.

_____ **9.** Men are as submissive as women.

_____ **10.** Women are as skillful in business-related activities as are men.

_____ **11.** Most women want their partner to take the initiative in their sexual relationships.

_____ **12.** Women spend more time attending to their physical appearance than men do.

_____ **13.** Women tend to give up more easily than men.

_____ **14.** Women dislike being in leadership positions more than men.

_____ **15.** Women are as interested in sex as are men.

_____ **16.** Women pay more attention to their looks than most men do.

_____ **17.** Women are more easily influenced than men.

_____ **18.** Women don't like responsibility as much as men.

_____ **19.** Women's sexual desires are less intense than men's.

_____ **20.** Women gain more status from their physical appearance than do men.

The Beliefs about Women Scale (BAWS) consists of fifteen separate subscales; only four are used here. The items for these four subscales and coding instructions are as follows:

1. Women are more passive than men (items 1, 5, 9, 13, 17).

2. Women are interested in careers less than men (items 2, 6, 10, 14, 18).

3. Women are less sexual than men (items 3, 7, 11, 15, 19).

4. Women are more appearance conscious than men (items 4, 8, 12, 16, 20).

Score the items as follows: strongly agree = +2; slightly agree = +1; neither agree nor disagree = 0; slightly disagree = −1; strongly disagree = −2.

Scores range from 0 to 40; subscale scores range from 0 to 10. The higher your score, the more traditional your gender beliefs about men and women.

Source: William E. Snell, Jr., Ph.D. 1997. College of Liberal Arts, Department of Psychology, Southeast Missouri State University. Reprinted with permission. Contact Dr. Snell for further use: wesnell@semo.edu

Effective Communication Scale

The purpose of this scale is to assess the degree to which you and your partner engage in behavior conducive to effective communication. After reading each statement, select the number that applies to you.

1	2	3	4	5
Never	*Rarely*	*Occasionally*	*Frequently*	*Very Frequently*

_____ **1.** My partner and I address issues that are bothering one or both of us.

_____ **2.** We look at each other when we are talking (e.g., we are not texting too).

_____ **3.** We ask open rather than closed-ended questions.

_____ **4.** We make reflective statements when we talk about an issue over which we disagree.

_____ **5.** We use "I" (rather than "you") statements.

_____ **6.** We compliment each other.

_____ **7.** We resolve disagreements in a way so that we both feel good.

_____ **8.** We share the power in our relationship.

_____ **9.** We make positive behavioral requests for change (e.g., rather than just complaining).

_____ **10.** Our discussion of a problem ends with a solution so that the issue does not recur.

Scoring

Add the numbers you selected. 1 (never) represents a behavior that is not conducive to effective communication. 5 (very frequently) represents a behavior that is conducive to effective communication. The lower your total score (10 is the lowest possible score), the less effective the communication between you and your partner. The higher your total score (50 is the highest possible score), the more effective your communication with your partner. A score of thirty places you at the midpoint between having an ineffective and effective communication pattern in your relationship.

Source: Developed exclusively for this text by the author. The scale is to be used for fun and to be thought provoking. It is not intended to be a clinical diagnostic instrument.

Relationships Dynamics Scale

Please answer each of the following questions in terms of your relationship with your "mate" if married, or your "partner" if dating or engaged. We recommend that you answer these questions by yourself (not with your partner), using the ranges following for your own reflection.

Use the following three-point scale to rate how often you and your mate or partner experience the following:

1 = Almost never
2 = Once in a while
3 = Frequently

1 2 3 Little arguments escalate into ugly fights with accusations, criticisms, name-calling, or bringing up past hurts.

1 2 3 My partner criticizes or belittles my opinions, feelings, or desires.

1 2 3 My partner seems to view my words or actions more negatively than I mean them to be.

1 2 3 When we have a problem to solve, it is like we are on opposite teams.

1 2 3 I hold back from telling my partner what I really think and feel.

1 2 3 I think seriously about what it would be like to date or marry someone else.

1 2 3 I feel lonely in this relationship.

1 2 3 When we argue, one of us withdraws (that is, doesn't want to talk about it anymore; or leaves the scene).

Who tends to withdraw more when there is an argument?

Male

Female

Both Equally

Neither Tend to Withdraw

Where Are You in Your Marriage?

We devised these questions based on seventeen years of research at the University of Denver on the kinds of communication and conflict management patterns that predict whether a relationship is headed for trouble. We have recently completed a nationwide, random phone survey using these questions. The average score was 11 on this scale. Although you should not take a higher score to mean that your relationship is somehow destined to fail, higher scores can mean that your relationship may be in greater danger unless changes are made. (These ranges are based only on your individual ratings—not a couple total.)

8 to 12 "Green Light"

If you scored in the 8 to 12 range, your relationship is probably in good or even great shape *at this time*, but we emphasize *"at this time"* because relationships don't stand still. In the next twelve months, you'll either have a stronger, happier relationship, or you could head in the other direction.

To think about it another way, it's like you are traveling along and have come to a green light. There is no need to stop, but it is probably a great time to work on making your relationship all it can be.

13 to 17 "Yellow Light"

If you scored in the 13 to 17 range, it's like you are coming to a "yellow light." You need to be cautious. Although you may be happy now in your relationship, your score reveals warning signs of patterns you don't want to let get worse. You'll want to take action to protect and improve what you have. Spending time to strengthen your relationship now could be the best thing you could do for your future together.

18 to 24 "Red Light"

Finally, if you scored in the 18 to 24 range, this is like approaching a red light. Stop and think about where the two of you are headed. Your score indicates the presence of patterns that could put your relationship at significant risk. You may be heading for trouble—you may already be there. But there is ***good news***. You can stop and learn ways to improve your relationship now!

1. *We wrote these items based on understanding of many key studies in the field.* The content or themes behind the questions are based on numerous in-depth studies on how people think and act in their marriages. These kinds of dynamics have been compared with patterns on many other key variables, such as satisfaction, commitment, problem intensity, and so on. Because the kinds of methods researchers can use in their laboratories are quite complex, this actual measure is far simpler than many of the methods we and others use to study marriages over time. However, the themes are based on many solid studies. Caution is warranted in interpreting scores.

2. *The discussion of the Relationships Dynamics Scale gives rough guidelines for interpreting the meaning of the scores.* The ranges we suggest for the measure are based on results from a nationwide, random phone survey of 947 people (85 percent married) in January 1996. These ranges are meant as a rough guideline for helping couples assess the degree to which they are experiencing key danger signs in their marriages. The measure as you have it here powerfully discriminated between those doing well in their marriages or relationships and those who were not doing well on a host of other dimensions (thoughts of divorce, low satisfaction, low sense of friendship in the relationship, lower dedication, and so on). Couples scoring higher on these items are truly more likely to be experiencing problems (or, based on other research, are more likely to experience problems in the future).

3. *This measure in and of itself should not be taken as a predictor of couples who are going to fail in their marriages.* No couple should be told they will "not make it" based on a higher score. That would not be in keeping with our intention in developing this scale or with the meaning one could take from it for any one couple. Although the items are based on studies that assess such things as the likelihood of a marriage working out, we would hate for any one person to take this and assume the worst about their future based on a high score. Rather, we believe that the measure can be used to motivate high- and moderately high-scoring people to take a serious look at where their relationships are heading—and take steps to turn such negative patterns around for the better.

Source: Dr. Howard Markman, Director Center for Marital and Family Studies at the University of Denver. Used by permission. Contact Dr. Markman at http://loveyourrelationship.com/ for information about Communication/Relationship Retreats. Also see, Markman, H. J., S. M. Stanley and S. L. Blumberg. (2010). Fighting for your marriage (3rd ed.) San Francisco, CA: Jossey-Bass.

Note: For more information on constructive tools for strong marriages, or for questions about the measure and the meaning of it, please write to us at PREP, Inc., P. O. Box 102530, Denver, Colorado 80250-2530.

The Love Attitudes Scale

This scale is designed to assess the degree to which you are romantic or realistic in your attitudes toward love. There are no right or wrong answers. After reading each sentence carefully, circle the number that best represents the degree to which you agree or disagree wlth the sentence.

1	2	3	4	5
Strongly Agree	Mildly Agree	Undecided	Mildly Disagree	Strongly Disagree

	SA	MA	U	MD	SD
1. Love doesn't make sense. It just is.	1	2	3	4	5
2. When you fall "head over heels" in love, it's sure to be the real thing.	1	2	3	4	5
3. To be in love with someone you would like to marry but can't is a tragedy.	1	2	3	4	5
4. When love hits, you know it.	1	2	3	4	5
5. Common interests are really unimportant; as long as each of you is truly in love, you will adjust.	1	2	3	4	5
6. It doesn't matter if you marry after you have known your partner for only a short time as long as you know you are in love.	1	2	3	4	5
7. If you are going to love a person, you will "know" after a short time.	1	2	3	4	5
8. As long as two people love each other, the educational differences they have really do not matter.	1	2	3	4	5
9. You can love someone even though you do not like any of that person's friends.	1	2	3	4	5
10. When you are in love, you are usually in a daze.	1	2	3	4	5
11. Love "at first sight" is often the deepest and most enduring type of love.	1	2	3	4	5
12. When you are in love, it really does not matter what your partner does because you will love him or her anyway.	1	2	3	4	5
13. As long as you really love a person, you will be able to solve the problems you have with the person.	1	2	3	4	5
14. Usually you can really love and be happy with only one or two people in the world.	1	2	3	4	5
15. Regardless of other factors, if you truly love another person, that is a good enough reason to marry that person.	1	2	3	4	5
16. It is necessary to be in love with the one you marry to be happy.	1	2	3	4	5
17. Love is more of a feeling than a relationship.	1	2	3	4	5
18. People should not get married unless they are in love.	1	2	3	4	5
19. Most people truly love only once during their lives.	1	2	3	4	5
20. Somewhere there is an ideal mate for most people.	1	2	3	4	5
21. In most cases, you will "know it" when you meet the right partner.	1	2	3	4	5
22. Jealousy usually varies directly with love; that is, the more you are in love, the greater your tendency to become jealous will be.	1	2	3	4	5
23. When you are in love, you are motivated by what you feel rather than by what you think.	1	2	3	4	5
24. Love is best described as an exciting rather than a calm thing.	1	2	3	4	5
25. Most divorces probably result from falling out of love rather than failing to adjust.	1	2	3	4	5
26. When you are in love, your judgment is usually not too clear.	1	2	3	4	5

	SA	MA	U	MD	SD
27. Love comes only once in a lifetime.	1	2	3	4	5
28. Love is often a violent and uncontrollable emotion.	1	2	3	4	5
29. When selecting a marriage partner, differences in social class and religion are of small importance compared with love.	1	2	3	4	5
30. No matter what anyone says, love cannot be understood.	1	2	3	4	5

Scoring

Add the numbers you circled. 1 (strongly agree) is the most romantic response and 5 (strongly disagree) is the most realistic response. The lower your total score (30 is the lowest possible score), the more romantic your attitudes toward love. The higher your total score (150 is the highest possible score), the more realistic your attitudes toward love. A score of 90 places you at the midpoint between being an extreme romantic and an extreme realist. Both men and women undergraduates typically score above 90, with men scoring closer to 90 than women.

A team of researchers (Medora et al. 2002) gave the scale to 641 young adults at three international universities in America, Turkey, and India. Female respondents in all three cultures had higher romanticism scores than male respondents (reflecting their higher value for, desire for, and thoughts about marriage). When the scores were compared by culture, American young adults were the most romantic, followed by Turkish students, with Indians having the lowest romanticism scores.

Reference: Medora, N. P., J. H. Larson, N. Hortacsu, and P. Dave. 2002. Perceived attitudes towards romanticism: A cross-cultural study of American, Asian-Indian, and Turkish young adults. *Journal of Comparative Family Studies* 33:155–78.

Source: David Knox. Department of Sociology, East Carolina University. Email Dr. Knox for permission to use. Knoxd@ecu.edu

Attitudes toward Premarital Sex Scale

Premarital sex is defined as engaging in sexual intercourse prior to marriage. The purpose of this survey is to assess your thoughts and feelings about intercourse before marriage. Read each item carefully and consider how you feel about each statement. There are no right or wrong answers to any of these statements, so please give your honest reactions and opinions. Please respond by using the following scale:

1	2	3	4	5	6	7
Strongly Disagree						Strongly Agree

_____ 1. I believe that premarital sex is healthy.

_____ 2. There is nothing wrong with premarital sex.

_____ 3. People who have premarital sex develop happier marriages.

_____ 4. Premarital sex is acceptable in a long-term relationship.

_____ 5. Having sexual partners before marriage is natural.

_____ 6. Premarital sex can serve as a stress reliever.

_____ 7. Premarital sex has nothing to do with morals.

_____ 8. Premarital sex is acceptable if you are engaged to the person.

_____ 9. Premarital sex is a problem among young adults.

_____ 10. Premarital sex puts unnecessary stress on relationships.

Scoring

Selecting a 1 reflects the most negative attitude toward premarital sex; selecting a 7 reflects the most positive attitude toward premarital sex. Before adding the numbers you assigned to each item, change the scores for items #9 and #10 as follows: replace a score of 1 with a 7; 2 with a 6; 3 with a 5; 4 with a 4; 5 with a 3; 6 with a 2; and 7 with a 1. After changing these numbers, add your ten scores. The lower your total score (10 is the lowest possible score), the less accepting you are of premarital sex; the higher your total score (70 is the highest possible score), the greater your acceptance of premarital sex. A score of 40 places you at the midpoint between being very disapproving of premarital sex and very accepting of premarital sex.

Scores of Other Students Who Completed the Scale

The scale was completed by 252 student volunteers at Valdosta State University. The mean score of the students was 40.81 (standard deviation [SD] = 13.20), reflecting that the students were virtually at the midpoint between a very negative and a very positive attitude toward premarital sex. For the 124 males and 128 females in the total sample, the mean scores were 42.06 (SD = 12.93) and 39.60 (SD = 13.39), respectively (not statistically significant). In regard to race, 59.5 percent of the sample was white and 40.5 percent was nonwhite (35.3 percent black, 2.4 percent Hispanic, 1.6 percent Asian, 0.4 percent American Indian, and 0.8 percent other). The mean scores of whites, blacks, and nonwhites were 41.64 (SD = 13.38), 38.46 (SD = 13.19), and 39.59 (SD = 12.90) (not statistically significant). Finally, regarding year in college, 8.3 percent were freshmen, 17.1 percent sophomores, 28.6 percent juniors, 43.3 percent seniors, and 2.8 percent graduate students. Freshmen and sophomores reported more positive attitudes toward premarital sex (mean = 44.81; SD = 13.39) than did juniors (mean = 40.32; SD = 12.58) or seniors and graduate students (mean = 38.91; SD = 13.10) ($p = .05$).

Source: "Attitudes toward Premarital Sex Scale," 2006, by Mark Whatley, Ph.D., Department of Psychology, Valdosta State University, Valdosta, Georgia 31698-0100. Used by permission. Other uses of this scale by written permission of Dr. Whatley only (mwhatley@valdosta.edu). Information on the reliability and validity of this scale is available from Dr. Whatley.

Attitudes toward Interracial Dating Scale

Interracial dating or marrying is the dating or marrying of two people from different races. The purpose of this survey is to gain a better understanding of what people think and feel about interracial relationships. Please read each item carefully, and in each space, score your response using the following scale. There are no right or wrong answers to any of these statements.

1	2	3	4	5	6	7
Strongly Disagree						Strongly Agree

_____ **1.** I believe that interracial couples date outside their race to get attention.
_____ **2.** I feel that interracial couples have little in common.
_____ **3.** When I see an interracial couple, I find myself evaluating them negatively.
_____ **4.** People date outside their own race because they feel inferior.
_____ **5.** Dating interracially shows a lack of respect for one's own race.
_____ **6.** I would be upset with a family member who dated outside our race.
_____ **7.** I would be upset with a close friend who dated outside our race.
_____ **8.** I feel uneasy around an interracial couple.
_____ **9.** People of different races should associate only in nondating settings.
_____ **10.** I am offended when I see an interracial couple.
_____ **11.** Interracial couples are more likely to have low self-esteem.
_____ **12.** Interracial dating interferes with my fundamental beliefs.
_____ **13.** People should date only within their race.
_____ **14.** I dislike seeing interracial couples together.
_____ **15.** I would not pursue a relationship with someone of a different race, regardless of my feelings for that person.
_____ **16.** Interracial dating interferes with my concept of cultural identity.
_____ **17.** I support dating between people with the same skin color, but not with a different skin color.
_____ **18.** I can imagine myself in a long-term relationship with someone of another race.
_____ **19.** As long as the people involved love each other, I do not have a problem with interracial dating.
_____ **20.** I think interracial dating is a good thing.

Scoring

First, reverse the scores for items 18, 19, and 20 by switching them to the opposite side of the spectrum. For example, if you selected 7 for item 18, replace it with a 1; if you selected 3, replace it with a 5, and so on. Next, add your scores and divide by 20. Possible final scores range from 1 to 7, with 1 representing the most positive attitudes toward interracial dating and 7 representing the most negative attitudes toward interracial dating.

Norms

The norming sample was based upon 113 male and 200 female students attending Valdosta State University. The participants completing the Attitudes toward Interracial Dating Scale (IRDS) received no compensation for their participation. All participants were U.S. citizens. The average age was 23.02 years (standard deviation [SD] = 5.09), and participants ranged in age from 18 to 50 years. The ethnic composition of the sample was 62.9 percent white, 32.6 percent black, 1 percent Asian, 0.6 percent Hispanic, and 2.2 percent other. The classification of the sample was 9.3 percent freshmen, 16.3 percent sophomores, 29.1 percent juniors, 37.1 percent seniors, and 2.9 percent graduate students. The average score on the IRDS was 2.88 (SD = 1.48), and scores ranged from 1.00 to 6.60, suggesting very positive views of interracial dating. Men scored an average of 2.97 (SD = 1.58), and women, 2.84 (SD = 1.42). There were no significant differences between the responses of women and men.

Source: The Attitudes toward Interracial Dating Scale. 2004. Mark Whatley, Ph.D., Department of Psychology, Valdosta State University, Valdosta, GA 31698. The scale is used by permission.

Satisfaction with Marriage Scale

Below are five statements with which you may agree or disagree. Using the 1-7 scale below, indicate your agreement with each item by circling the appropriate number on the line following that item. Please be open and honest in responding to each item.

1	2	3	4	5	6	7
Strongly Disagree	*Disagree*	*Slightly Disagree*	*Neither Agree nor Disagree*	*Slightly Agree*	*Agree*	*Strongly Agree*

	SD	D	SD	N	SA	A	SA
1. In most ways my married life is close to ideal	1	2	3	4	5	6	7
2. The conditions of my married life are excellent.	1	2	3	4	5	6	7
3. I am satisfied with my married life.	1	2	3	4	5	6	7
4. So far I have gotten the important things I want in my married life.	1	2	3	4	5	6	7
5. If I could live my married life over, I would change almost nothing.	1	2	3	4	5	6	7

Scoring

Add the numbers you circled. The marital satisfaction score will range from 5 to 35. For purposes of this study a couple's combined marital satisfaction score was calculated by summing both partners' scores, resulting in a possible score range of 10 to 70, with higher scores indicating greater marital satisfaction for the couple. The internal consistency of the SWML has been reported with a Cronbach's alpha of .92 along with some evidence of construct validity (Johnson et al., 2006).

Source: Johnson, H. A., Zabriskie, R. B. and Hill, B. (2006) The contribution of couple leisure involvement, leisure time, and leisure satisfaction to marital satisfaction. Marriage and Family Review 40: 69-91. Reprinted by permission of the publisher (Taylor & Francis Ltd, http://www.tandf.co.uk/journals)

The Self-Report of Behavior Scale

This questionnaire is designed to examine which of the following statements most closely describes your behavior during past encounters with people you thought were homosexuals. Rate each of the following self-statements as honestly as possible using the following scale. Write each value in the provided blank.

1	2	3	4	5
Never	*Rarely*	*Occasionally*	*Frequently*	*Always*

____ **1.** I have spread negative talk about someone because I suspected that the person was gay.

____ **2.** I have participated in playing jokes on someone because I suspected that the person was gay.

____ **3.** I have changed roommates and/or rooms because I suspected my roommate was gay.

____ **4.** I have warned people who I thought were gay and who were a little too friendly with me to keep a way from me.

____ **5.** I have attended antigay protests.

____ **6.** I have been rude to someone because I thought that the person was gay.

____ **7.** I have changed seat locations because I suspected the person sitting next to me was gay.

____ **8.** I have had to force myself to keep from hitting someone because the person was gay and very near me.

____ **9.** When someone I thought to be gay has walked toward me as if to start a conversation, I have deliberately changed directions and walked away to avoid the person.

____ **10.** I have stared at a gay person in such a manner as to convey my disapproval of the person being too close to me.

____ **11.** I have been with a group in which one (or more) person(s) yelled insulting comments to a gay person or group of gay people.

____ **12.** I have changed my normal behavior in a restroom because a person I believed to be gay was in there at the same time.

____ **13.** When a gay person has checked me out, I have verbally threatened the person.

____ **14.** I have participated in damaging someone's property because the person was gay.

____ **15.** I have physically hit or pushed someone I thought was gay because the person brushed against me when passing by.

____ **16.** Within the past few months, I have told a joke that made fun of gay people.

____ **17.** I have gotten into a physical fight with a gay person because I thought the person had been making moves on me.

____ **18.** I have refused to work on school and/or work projects with a partner I thought was gay.

____ **19.** I have written graffiti about gay people or homosexuality.

____ **20.** When a gay person has been near me, I have moved away to put more distance between us.

Scoring

Determine your score by adding your points together. The lowest score is 20 points, the highest 100 points. The higher the score, the more negative the attitudes toward homosexuals.

Comparison Data

Sunita Patel (1989) originally developed the Self-Report of Behavior Scale in her thesis research in her clinical psychology master's program at East Carolina University. College men (from a university campus and from a military base) were the original participants (Patel et al. 1995). The scale was revised by Shartra Sylivant (1992), who used it with a coed high school student population, and by Tristan Roderick (1994), who involved college students to assess its psychometric properties. The scale was found to have high internal consistency. Two factors were identifed: a passive avoidance of homosexuals and active or aggressive reactions.

In a study by Roderick et al. (1998), the mean score for 182 college women was 24.76. The mean score for 84 men was significantly higher, at 31.60. A similar-sex difference, although with higher (more negative) scores, was found in Sylivant's high school sample (with a mean of 33.74 for the young women, and 44.40 for the young men).

The following table provides detail for the scores of the college students in Roderick's sample (from a mid-sized state university in the southeast):

	N	Mean	Standard Deviation
Women	182	24.76	7.68
Men	84	31.60	10.36
Total	266	26.91	9.16

Sources: By Ms. Shartra Sylivant M.A., L.P.A. Clinical, H.S.P. The cognitive, affective, and behavioral components of adolescent homonegativity. Master's thesis, East Carolina University, 1992. The SBS-R is reprinted by the permission S. Sylivant.

Maternal Employment Scale

Directions

Using the following scale, please mark a number on the blank next to each statement to indicate how strongly you agree or disagree.

1	2	3	4	5	6
Disagree Very Strongly	Disagree Strongly	Disagree Slightly	Agree Slightly	Agree Strongly	Agree Very Strongly

_____ **1.** Children are less likely to form a warm and secure relationship with a mother who works full-time outside the home.

_____ **2.** Children whose mothers work are more independent and able to do things for themselves.

_____ **3.** Working mothers are more likely to have children with psychological problems than mothers who do not work outside the home.

_____ **4.** Teenagers get into less trouble with the law if their mothers do not work full-time outside the home.

_____ **5.** For young children, working mothers are good role models for leading busy and productive lives.

_____ **6.** Boys whose mothers work are more likely to develop respect for women.

_____ **7.** Young children learn more if their mothers stay at home with them.

_____ **8.** Children whose mothers work learn valuable lessons about other people they can rely on.

_____ **9.** Girls whose mothers work full-time outside the home develop stronger motivation to do well in school.

_____ **10.** Daughters of working mothers are better prepared to combine work and motherhood if they choose to do both.

_____ **11.** Children whose mothers work are more likely to be left alone and exposed to dangerous situations.

_____ **12.** Children whose mothers work are more likely to pitch in and do tasks around the house.

_____ **13.** Children do better in school if their mothers are not working full-time outside the home.

_____ **14.** Children whose mothers work full-time outside the home develop more regard for women's intelligence and competence.

_____ **15.** Children of working mothers are less well-nourished and don't eat the way they should.

_____ **16.** Children whose mothers work are more likely to understand and appreciate the value of a dollar.

_____ **17.** Children whose mothers work suffer because their mothers are not there when they need them.

_____ **18.** Children of working mothers grow up to be less competent parents than other children because they have not had adequate parental role models.

_____ **19.** Sons of working mothers are better prepared to cooperate with a wife who wants both to work and have children.

_____ **20.** Children of mothers who work develop lower self-esteem because they think they are not worth devoting attention to.

_____ **21.** Children whose mothers work are more likely to learn the importance of teamwork and cooperation among family members.

_____ **22.** Children of working mothers are more likely than other children to experiment with alcohol, other drugs, and sex at an early age.

_____ **23.** Children whose mothers work develop less stereotyped views about men's and women's roles.

_____ **24.** Children whose mothers work full-time outside the home are more adaptable; they cope better with the unexpected and with changes in plans.

Scoring

Items 1, 3, 4, 7, 11, 13, 15, 17, 18, 20, and 22 refer to "costs" of maternal employment for children and yield a Costs Subscale score. High scores on the Costs Subscale reflect strong beliefs that maternal employment is costly to children. Items 2, 5, 6, 8, 9, 10, 12, 14, 16, 19, 21, 23, and 24 refer to "benefits" of maternal employment for children and yield a Benefits Subscale score. To obtain a Total Score, reverse the score of all items in the Benefits Subscale so that 1 = 6, 2 = 5, 3 = 4, 4 = 3, 5 = 2, and 6 = 1. The higher one's Total Score, the more one believes that maternal employment has negative consequences for children.

Source: E. Greenberger, W. A. Goldberg, T. J. Crawford, and J. Granger, Beliefs about the consequences of maternal employment for children in Psychology of Women Quarterly, Maternal Employment Scale 12:35–59, COPYRIGHT © 1988 by SAGE Publications. Reprinted by Permission of SAGE Publications.

Childfree Lifestyle Scale

The purpose of this scale is to assess your attitudes toward having a childfree lifestyle. After reading each statement, select the number that best reflects your answer, using the following scale:

1	2	3	4	5	6	7
Strongly Disagree						Strongly Agree

_____ **1.** I do not like children.

_____ **2.** I would resent having to spend all my money on kids.

_____ **3.** I would rather enjoy my personal freedom than have it taken away by having children.

_____ **4.** I would rather focus on my career than have children.

_____ **5.** Children are a burden.

_____ **6.** I have no desire to be a parent.

_____ **7.** I am too "into me" to become a parent.

_____ **8.** I lack the nurturing skills to be a parent.

_____ **9.** I have no patience for children.

_____ **10.** Raising a child is too much work.

_____ **11.** A marriage without children is empty.

_____ **12.** Children are vital to a good marriage.

_____ **13.** You can't really be fulfilled as a couple unless you have children.

_____ **14.** Having children gives meaning to a couple's marriage.

_____ **15.** The happiest couples that I know have children.

_____ **16.** The biggest mistake couples make is deciding not to have children.

_____ **17.** Childfree couples are sad couples.

_____ **18.** Becoming a parent enhances the intimacy between spouses.

_____ **19.** A house without the "pitter patter" of little feet is not a home.

_____ **20.** Having a child means your marriage is successful.

Scoring

Reverse score items 11 through 20. For example if you wrote a 7 for item 20, change this to a 1. If you wrote a 7, change to a 1. If you wrote a 2, change to a 6, etc. Add the numbers. The higher the score (140 is the highest possible score), the greater the value for a childfree lifestyle. The lower the score (20 is the lowest possible score), the less the desire to have a childfree lifestyle. The midpoint between the extremes is 80: Scores below 80 suggest less preference for a childfree lifestyle and scores above 80 suggest a desire for a childfree lifestyle. The average score of 52 male and 138 female undergraduates at Valdosta State University was below the midpoint ($M = 68.78$, $SD = 17.06$), suggesting a tendency toward a lifestyle that included children. A significant difference was found between males, who scored 72.94 ($SD = 16.82$), and females, who scored 67.21 ($SD = 16.95$), suggesting that males are more approving of a childfree lifestyle. There were no significant differences between Whites and Blacks or between students in different ranks (freshman, sophomore, junior, senior, etc.).

Source: "The Childfree Lifestyle Scale," 2010, by Mark A. Whatley, Ph.D., Department of Psychology, Valdosta State University, Valdosta, Georgia 31698-0100. Used by permission. Other uses of this scale by written permission of Dr. Whatley only (mwhatley@valdosta.edu). Information on the reliability and validity of this scale is available from Dr. Whatley.

Abortion Attitude Scale

This is not a test. There are no wrong or right answers to any of the statements, so just answer as honestly as you can. The statements ask your feelings about legal abortion (the voluntary removal of a human fetus from the mother during the first three months of pregnancy by a qualified medical person). Tell how you feel about each statement by circling only one response. Use the following scale for your answers:

5	4	3	2	1
Strongly Agree	Slightly Agree	Agree	Slightly Disagree	Strongly Disagree

_____ 1.   The Supreme Court should strike down legal abortions in the United States.

_____ 2.   Abortion is a good way of solving an unwanted pregnancy.

_____ 3.   A mother should feel obligated to bear a child she has conceived.

_____ 4.   Abortion is wrong no matter what the circumstances are.

_____ 5.   A fetus is not a person until it can live outside its mother's body.

_____ 6.   The decision to have an abortion should be the pregnant mother's.

_____ 7.   Every conceived child has the right to be born.

_____ 8.   A pregnant female not wanting to have a child should be encouraged to have an abortion.

_____ 9.   Abortion should be considered killing a person.

_____ 10. People should not look down on those who choose to have abortions.

_____ 11. Abortion should be an available alternative for unmarried pregnant teenagers.

_____ 12. People should not have the power over the life or death of a fetus.

_____ 13. Unwanted children should not be brought into the world.

_____ 14. A fetus should be considered a person at the moment of conception.

Scoring and Interpretation

As its name indicates, this scale was developed to measure attitudes toward abortion. Sloan (1983) developed the scale for use with high school and college students. To compute your score, first reverse the point scale for items 1, 3, 4, 7, 9, 12, and 14. For example, if you selected a 5 for item one, this becomes a 0; if you selected a 1, this becomes a 4. After reversing the scores on the seven items specified, add the numbers you circled for all the items. Sloan provided the following categories for interpreting the results:

70–56   Strong proabortion

54–44   Moderate proabortion

43–27   Unsure

26–16   Moderate pro-life

15–0   Strong pro-life

Reliability and Validity

The Abortion Attitude Scale was administered to high school and college students, Right to Life group members, and abortion service personnel. Sloan (1983) reported a high total test estimate of reliability (0.92). Construct validity was supported in that the mean score for Right to Life members was 16.2; the mean score for abortion service personnel was 55.6; and other groups' scores fell between these values.

Source: "Abortion Attitude Scale" by L. A. Sloan. _Journal of Health Education_ Vol. 14, No. 3, May/June 1983. The _Journal of Health Education_ is a publication of the American Alliance for Health, Physical Education, Recreation and Dance, 1900 Association Drive, Reston, VA 20191. Reprinted by permission.

The Traditional Motherhood Scale

The purpose of this survey is to assess the degree to which students possess a traditional view of motherhood. Read each item carefully and consider what you believe. There are no right or wrong answers, so please give your honest reaction and opinion. After reading each statement, select the number that best reflects your level of agreement, using the following scale:

1	2	3	4	5	6	7
Strongly Disagree						Strongly Agree

____ 1. A mother has a better relationship with her children than a father does.

____ 2. A mother knows more about her child than a father, thereby being the better parent.

____ 3. Motherhood is what brings women to their fullest potential.

____ 4. A good mother should stay at home with her children for the first year.

____ 5. Mothers should stay at home with the children.

____ 6. Motherhood brings much joy and contentment to a woman.

____ 7. A mother is needed in a child's life for nurturance and growth.

____ 8. Motherhood is an essential part of a female's life.

____ 9. I feel that all women should experience motherhood in some way.

____ 10. Mothers are more nurturing than fathers.

____ 11. Mothers have a stronger emotional bond with their children than do fathers.

____ 12. Mothers are more sympathetic to children who have hurt themselves than are fathers.

____ 13. Mothers spend more time with their children than do fathers.

____ 14. Mothers are more lenient toward their children than are fathers.

____ 15. Mothers are more affectionate toward their children than are fathers.

____ 16. The presence of the mother is vital to the child during the formative years.

____ 17. Mothers play a larger role than fathers in raising children.

____ 18. Women instinctively know what a baby needs.

Scoring

After assigning a number from 1 (strongly disagree) to 7 (strongly agree), add the numbers and divide by 18. The higher your score (7 is the highest possible score), the stronger the traditional view of motherhood. The lower your score (1 is the lowest possible score), the less traditional the view of motherhood.

Norms

The norming sample of this self-assessment was based upon twenty male and eighty-six female students attending Valdosta State University. The average age of participants completing the scale was 21.72 years (SD = 2.98), and ages ranged from 18 to 34. The ethnic composition of the sample was 80.2 percent white, 15.1 percent black, 1.9 percent Asian, 0.9 percent American Indian, and 1.9 percent other. The classification of the sample was 16.0 percent freshmen, 15.1 percent sophomores, 27.4 percent juniors, 39.6 percent seniors, and 1.9 percent graduate students.

Participants responded to each of the eighteen items according to the 7-point scale. The most traditional score was 6.33; the score reflecting the least support for traditional motherhood was 1.78. The midpoint (average score) between the top and bottom score was 4.28 (SD = 1.04); thus, people scoring above this number tended to have a more traditional view of motherhood and people scoring below this number tended to have a less traditional view of motherhood.

There was a significant difference (p = .05) between female participants' scores (mean = 4.19; SD = 1.08) and male participants' score (mean = 4.60; SD = 0.72), suggesting that males had more traditional views of motherhood than females.

Source: Mark Whatley, Ph.D. 2004. The Traditional Motherhood Scale. Department of Psychology, Valdosta State University. Use of this scale is permitted only by prior written permission of Dr. Whatley (mwhatley@valdosta.edu).

Spanking versus Time-Out Scale

Parents discipline their children to help them develop self-control and correct misbehavior. Some parents spank their children; others use time-out. Spanking is a disciplinary technique whereby a mild slap (that is, a "spank") is applied to the buttocks of a disobedient child. Time-out is a disciplinary technique whereby, when a child misbehaves, the child is removed from the situation. The purpose of this survey is to assess the degree to which you prefer spanking versus time-out as a method of discipline. Please read each item carefully and select a number from 1 to 7, which represents your belief. There are no right or wrong answers; please give your honest opinion.

1	2	3	4	5	6	7
Strongly Disagree						Strongly Agree

_____ 1. Spanking is a better form of discipline than time-out.

_____ 2. Time-out does not have any effect on children.

_____ 3. When I have children, I will more likely spank them than use a time-out.

_____ 4. A threat of a time-out does not stop a child from misbehaving.

_____ 5. Lessons are learned better with spanking.

_____ 6. Time-out does not give a child an understanding of what the child has done wrong.

_____ 7. Spanking teaches a child to respect authority.

_____ 8. Giving children time-outs is a waste of time.

_____ 9. Spanking has more of an impact on changing the behavior of children than time-out.

_____ 10. I do not believe "time-out" is a form of punishment.

_____ 11. Getting spanked as a child helps you become a responsible citizen.

_____ 12. Time-out is only used because parents are afraid to spank their kids.

_____ 13. Spanking can be an effective tool in disciplining a child.

_____ 14. Time-out is watered-down discipline.

Scoring

If you want to know the degree to which you approve of spanking, reverse the number you selected for all odd-numbered items (1, 3, 5, 7, 9, 11, and 13) after you have selected a number from 1 to 7 for each of the fourteen items. For example, if you selected a 1 for item 1, change this number to a 7 (1 = 7; 2 = 6; 3 = 5; 4 = 4; 5 = 3; 6 = 2; 7 = 1). Now add these seven numbers. The lower your score (7 is the lowest possible score), the lower your approval of spanking; the higher your score (49 is the highest possible score), the greater your approval of spanking. A score of 21 places you at the midpoint between being very disapproving of or very accepting of spanking as a discipline strategy.

If you want to know the degree to which you approve of using time-out as a method of discipline, reverse the number you selected for all even-numbered items (2, 4, 6, 8, 10, 12, and 14. For example, if you selected a 1 for item 2, change this number to a 7 (that is, 1 = 7; 2 = 6; 3 = 5; 4 = 4; 5 = 3; 6 = 2; 7 = 1). Now add these seven numbers. The lower your score (7 is the lowest possible score), the lower your approval of time-out; the higher your score (49 is the highest possible score), the greater your approval of time-out. A score of 21 places you at the midpoint between being very disapproving of or very accepting of time-out as a discipline strategy.

Scores of Other Students Who Completed the Scale

The scale was completed by 48 male and 168 female student volunteers at East Carolina University. Their ages ranged from 18 to 34, with a mean age of 19.65 ($SD = 2.06$). The ethnic background of the sample included 73.1 percent white, 17.1 percent African American, 2.8 percent Hispanic, 0.9 percent Asian, 3.7 percent from other ethnic backgrounds; 2.3 percent did not indicate ethnicity. The college classification level of the sample included 52.8 percent freshman, 24.5 percent sophomore, 13.9 percent junior, and 8.8 percent senior. The average score on the spanking dimension was 29.73 ($SD = 10.97$), and the time-out dimension was 22.93 ($SD = 8.86$), suggesting greater acceptance of spanking than time-out.

Time-out differences. In regard to sex of the participants, female participants were more positive about using time-out as a discipline strategy ($M = 33.72$, $SD = 8.76$) than were male participants ($M = 30.81$, $SD = 8.97$; $p = .05$). In regard to ethnicity of the participants, white participants were more positive about using time-out as a discipline strategy ($M = 34.63$, $SD = 8.54$) than were nonwhite participants ($M = 28.45$, $SD = 8.55$; $p = .05$). In regard to year in school, freshmen were more positive about using spanking as a discipline strategy ($M = 34.34$, $SD = 9.23$) than were sophomores, juniors, and seniors ($M = 31.66$, $SD = 8.25$; $p = .05$).

Spanking differences. In regard to ethnicity of the participants, nonwhite participants were more positive about using spanking as a discipline strategy ($M = 35.09$, $SD = 10.02$) than were white participants ($M = 27.87$, $SD = 10.72$; $p = .05$). In regard to year in school, freshmen were less positive about using spanking as a discipline strategy ($M = 28.28$, $SD = 11.42$) than were sophomores, juniors, and seniors ($M = 31.34$, $SD = 10.26$; $p = .05$). There were no significant differences in regard to sex of the participants ($p = .05$) in the opinion of spanking.

Overall differences. There were no significant differences in overall attitudes to discipline in regards to sex of the participants, ethnicity, or year in school.

Source; "The Spanking vs. Time-Out Scale," 2004, by Mark Whatley, Ph.D., Department of Psychology, Valdosta State University, Valdosta, Georgia 31698-0100. Used by permission. Other uses of this scale by written permission of Dr. Whatley only (mwhatley@valdosta.edu). Information on the reliability and validity of this scale is available from Dr. Whatley.

Attitudes toward Infidelity Scale

The purpose of this survey is to gain a better understanding of what people think and feel about issues associated with infidelity. There are no right or wrong answers to any of these statements; we are interested in your honest reactions and opinions. Please read each statement carefully, and respond by using the following scale:

1	2	3	4	5	6	7
Strongly Disagree						Strongly Agree

_____ **1.** Being unfaithful never hurt anyone.

_____ **2.** Infidelity in a marital relationship is grounds for divorce.

_____ **3.** Infidelity is acceptable for retaliation of infidelity.

_____ **4.** It is natural for people to be unfaithful.

_____ **5.** Online/Internet behavior (for example, visiting sex chat rooms, porn sites) is an act of infidelity.

_____ **6.** Infidelity is morally wrong in all circumstances, regardless of the situation.

_____ **7.** Being unfaithful in a relationship is one of the most dishonorable things a person can do.

_____ **8.** Infidelity is unacceptable under any circumstances if the couple is married.

_____ **9.** I would not mind if my significant other had an affair as long as I did not know about it.

_____ **10.** It would be acceptable for me to have an affair, but not my significant other.

_____ **11.** I would have an affair if I knew my significant other would never find out.

_____ **12.** If I knew my significant other was guilty of infidelity, I would confront him/her.

Scoring

Selecting a 1 reflects the least acceptance of infidelity; selecting a 7 reflects the greatest acceptance of infidelity. Before adding the numbers you selected, reverse the scores for item numbers 2, 5, 6, 7, 8, and 12. For example, if you responded to item 2 with a "6," change this number to a "2"; if you responded with a "3," change this number to "5," and so on. After making these changes, add the numbers. The lower your total score (12 is the lowest possible), the less accepting you are of infidelity; the higher your total score (84 is the highest possible), the greater your acceptance of infidelity. A score of 48 places you at the midpoint between being very disapproving and very accepting of infidelity.

Scores of Other Students Who Completed the Scale

The scale was completed by 150 male and 136 female student volunteers at Valdosta State University. The average score on the scale was 27.85 ($SD = 12.02$). Their ages ranged from 18 to 49, with a mean age of 23.36 ($SD = 5.13$). The ethnic backgrounds of the sample included 60.8 percent white, 28.3 percent African American, 2.4 percent Hispanic, 3.8 percent Asian, 0.3 percent American Indian, and 4.2 percent other. The college classification level of the sample included 11.5 percent freshmen, 18.2 percent sophomores, 20.6 percent juniors, 37.8 percent seniors, 7.7 percent graduate students, and 4.2 percent postbaccalaureate. Male participants reported more positive attitudes toward infidelity (mean = 31.53; $SD = 11.86$) than did female participants (mean = 23.78; $SD = 10.86$; $p = .05$). White participants had more negative attitudes toward infidelity (mean = 25.36; $SD = 11.17$) than did nonwhite participants (mean = 31.71; $SD = 12.32$; $p = .05$). There were no significant differences in regard to college classification.

Source: Mark Whatley, Ph.D. 2006. Attitudes toward Infidelity Scale. Department of Psychology, Valdosta State University. Used by permission. Other uses of this scale by written permission of Dr. Whatley only. Information on the reliability and validity of this scale is available from Dr. Whatley (mwhatley@valdosta.edu).

Internality, Powerful Others, and Chance Scales

People have different feelings about their vulnerability to crisis events. The following scale addresses the degree to which you feel you have control, you feel others have control, or you feel that chance has control of what happens to you.

Directions

To assess the degree to which you believe that you have control over your own life (I = Internality), the degree to which you believe that other people control events in your life (P = Powerful Others), and the degree to which you believe that chance affects your experiences or outcomes (C = Chance), read each of the following statements and select a number from −3 to +3.

-3	-2	-1	+1	+2	+3
Strongly Disagree	Slightly Disagree	Neither Disagree	Slightly Agree	Agree	Strongly Agree

Subscale

I 1. Whether or not I get to be a leader depends mostly on my ability.

C 2. To a great extent, my life is controlled by accidental happenings.

P 3. I feel like what happens in my life is mostly determined by powerful people.

I 4. Whether or not I get into a car accident depends mostly on how good a driver I am.

I 5. When I make plans, I am almost certain to make them work.

C 6. Often there is no chance of protecting my personal interests from bad luck happenings.

C 7. When I get what I want, it's usually because I'm lucky.

P 8. Although I might have good ability, I will not be given leadership responsibility without appealing to those in positions of power.

I 9. How many friends I have depends on how nice a person I am.

C 10. I have often found that what is going to happen will happen.

P 11. My life is chiefly controlled by powerful others.

C 12. Whether or not I get into a car accident is mostly a matter of luck.

P 13. People like myself have very little chance of protecting our personal interests when they conflict with those of strong pressure groups.

C 14. It's not always wise for me to plan too far ahead because many things turn out to be a matter of good or bad fortune.

P 15. Getting what I want requires pleasing those people above me.

C 16. Whether or not I get to be a leader depends on whether I'm lucky enough to be in the right place at the right time.

P 17. If important people were to decide they didn't like me, I probably wouldn't make many friends.

I 18. I can pretty much determine what will happen in my life.

I 19. I am usually able to protect my personal interests.

P 20. Whether or not I get into a car accident depends mostly on the other driver.

I 21. When I get what I want, it's usually because I worked hard for it.

P 22. To have my plans work, I make sure that they fit in with the desires of other people who have power over me.

I 23. My life is determined by my own actions.

C 24. Whether or not I have a few friends or many friends is chiefly a matter of fate.

Scoring

Each of the subscales of Internality, Powerful Others, and Chance is scored on a six-point Likert format from −3 to +3. For example, the eight Internality items are 1, 4, 5, 9, 18, 19, 21, 23. A person who has strong agreement with all eight items would score +24; strong disagreement, −24. After adding and subtracting the item scores, add 24 to the total score to eliminate negative scores. Scores for Powerful Others and Chance are similarly derived.

Norms

For the Internality subscale, means range from the low 30s to the low 40s, with 35 being the modal mean (SD values approximating 7). The Powerful Others subscale has produced means ranging from 18 through 26, with 20 being characteristic of normal college student subjects (SD = 8.5). The Chance subscale produces means between 17 and 25, with 18 being a common mean among undergraduates (SD = 8).

Source: From *Research with the Locus of Control Construct*, by Hervert M. Lefcourt, 1981, pp. 57–59. Reprinted with permission from Elsevier.

Abusive Behavior Inventory

Circle the number that best represents your closest estimate of how often each of the behaviors have happened in the relationship with your current or former partner during the previous six months.

1.	Called you a name and/or criticized you	1	2	3	4	5
2.	Tried to keep you from doing something you wanted to do (for example, going out with friends or going to meetings)	1	2	3	4	5
3.	Gave you angry stares or looks	1	2	3	4	5
4.	Prevented you from having money for your own use	1	2	3	4	5
5.	Ended a discussion and made a decision without you	1	2	3	4	5
6.	Threatened to hit or throw something at you	1	2	3	4	5
7.	Pushed, grabbed, or shoved you	1	2	3	4	5
8.	Put down your family and friends	1	2	3	4	5
9.	Accused you of paying too much attention to someone or something else	1	2	3	4	5
10.	Put you on an allowance	1	2	3	4	5
11.	Used your children to threaten you (for example, told you that you would lose custody or threatened to leave town with the children)	1	2	3	4	5
12.	Became very upset with you because dinner, housework, or laundry was not done when or how it was wanted	1	2	3	4	5
13.	Said things to scare you (for example, told you something "bad" would happen or threatened to commit suicide)					
14.	Slapped, hit, or punched you	1	2	3	4	5
15.	Made you do something humiliating or degrading (for example, begging for forgiveness or having to ask permission to use the car or do something)	1	2	3	4	5
16.	Checked up on you (for example, listened to your phone calls, checked the mileage on your car, or called you repeatedly at work)	1	2	3	4	5
17.	Drove recklessly when you were in the car	1	2	3	4	5
18.	Pressured you to have sex in a way you didn't like or want	1	2	3	4	5
19.	Refused to do housework or child care	1	2	3	4	5
20.	Threatened you with a knife, gun, or other weapon	1	2	3	4	5
21.	Spanked you	1	2	3	4	5
22.	Told you that you were a bad parent	1	2	3	4	5
23.	Stopped you or tried to stop you from going to work or school	1	2	3	4	5
24.	Threw, hit, kicked, or smashed something	1	2	3	4	5
25.	Kicked you	1	2	3	4	5
26.	Physically forced you to have sex	1	2	3	4	5
27.	Threw you around	1	2	3	4	5
28.	Physically attacked the sexual parts of your body	1	2	3	4	5
29.	Choked or strangled you	1	2	3	4	5
30.	Used a knife, gun, or other weapon against you	1	2	3	4	5

Add the numbers you circled and divide the total by 30 to determine your score. The higher your score (five is the highest score), the more abusive your relationship.

The inventory was given to 100 men and 78 women equally divided into groups of abusers or abused and nonabusers or nonabused. The men were members of a chemical dependency treatment program in a veterans' hospital and the women were partners of these men. Abusing or abused men earned an average score of 1.8; abusing or abused women earned an average score of 2.3. Nonabusing, abused men and women earned scores of 1.3 and 1.6, respectively.

Source: M. F. Shepard and J. A. Campbell. The Abusive Behavior Inventory: A measure of psychological and physical abuse. *Journal of Interpersonal Violence* 7(3):291–305. © 1992 by Sage Publications Inc. Journals. Reproduced with permission of Sage Publications Inc. Journals in the format Other book via Copyright Clearance Center.

Children's Beliefs about Parental Divorce Scale

The following are some statements about children and their separated parents. Some of the statements are *true* about how you think and feel, so you will want to check *yes*. Some are *not true* about how you think or feel, so you will want to check *no*. There are no right or wrong answers. Your answers will just indicate some of the things you are thinking now about your parents' separation.

1.	It would upset me if other kids asked a lot of questions about my parents.	___ Yes	___ No
2.	It was usually my father's fault when my parents had a fight.	___ Yes	___ No
3.	I sometimes worry that both my parents will want to live without me.	___ Yes	___ No
4.	When my family was unhappy, it was usually because of my mother.	___ Yes	___ No
5.	My parents will always live apart.	___ Yes	___ No
6.	My parents often argue with each other after I misbehave.	___ Yes	___ No
7.	I like talking to my friends as much now as I used to.	___ Yes	___ No
8.	My father is usually a nice person.	___ Yes	___ No
9.	It's possible that both my parents will never want to see me again.	___ Yes	___ No
10.	My mother is usually a nice person.	___ Yes	___ No
11.	If I behave better, I might be able to bring my family back together.	___ Yes	___ No
12.	My parents would probably be happier if I were never born.	___ Yes	___ No
13.	I like playing with my friends as much now as I used to.	___ Yes	___ No
14.	When my family was unhappy, it was usually because of something my father said or did.	___ Yes	___ No
15.	I sometimes worry that I'll be left all alone.	___ Yes	___ No
16.	Often I have a bad time when I'm with my mother.	___ Yes	___ No
17.	My family will probably do things together just like before.	___ Yes	___ No
18.	My parents probably argue more when I'm with them than when I'm gone.	___ Yes	___ No
19.	I'd rather be alone than play with other kids.	___ Yes	___ No
20.	My father caused most of the trouble in my family.	___ Yes	___ No
21.	I feel that my parents still love me.	___ Yes	___ No
22.	My mother caused most of the trouble in my family.	___ Yes	___ No
23.	My parents will probably see that they have made a mistake and get back together again.	___ Yes	___ No
24.	My parents are happier when I'm with them than when I'm not.	___ Yes	___ No
25.	My friends and I do many things together.	___ Yes	___ No
26.	There are a lot of things I like about my father.	___ Yes	___ No
27.	I sometimes think that one day I may have to go live with a friend or relative.	___ Yes	___ No
28.	My mother is more good than bad.	___ Yes	___ No
29.	I sometimes think that my parents will one day live together again.	___ Yes	___ No
30.	I can make my parents unhappy with each other by what I say or do.	___ Yes	___ No
31.	My friends understand how I feel about my parents.	___ Yes	___ No
32.	My father is more good than bad.	___ Yes	___ No
33.	I feel my parents still like me.	___ Yes	___ No
34.	There are a lot of things about my mother I like.	___ Yes	___ No
35.	I sometimes think that my parents will live together again once they realize how much I want them to.	___ Yes	___ No
36.	My parents would probably still be living together if it weren't for me.	___ Yes	___ No

Scoring

The Children's Beliefs about Parental Divorce Scale (CBAPS) identifies problematic responding. A **yes** response on items 1, 2, 3, 4, 6, 9, 11, 12, 14–20, 22, 23, 27, 29, 30, 35, and 36, and a **no** response on items 5, 7, 8, 10, 13, 21, 24–26, 28, and 31–34 indicate a problematic reaction to one's parents divorcing. A total score is derived by adding the number of problematic beliefs across all items, with a total score of 36. The higher the score, the more problematic the beliefs about parental divorce.

Norms

A total of 170 schoolchildren whose parents were divorced completed the scale; of the children, 84 were boys and 86 were girls, with a mean age of 11. The mean for the total score was 8.20, with a standard deviation of 4.98.

Parental Status Inventory

The Parental Status Inventory (PSI) is a fourteen-item inventory that measures the degree to which respondents consider their stepfather to be a parent on an 11-point scale from 0 percent to 100 percent. Read each of the following statements and circle the percentage indicating the degree to which you regard the statement as true.

1. I think of my stepfather as my father. (0%, 10, 20, 30, 40, 50%, 60, 70, 80, 90, 100%)
2. I am comfortable when someone else refers to my stepfather as my father or dad. (0%, 10, 20, 30, 40, 50%, 60, 70, 80, 90, 100%)
3. I think of myself as his daughter/son. (0%, 10, 20, 30, 40, 50%, 60, 70, 80, 90, 100%)
4. I refer to him as my father or dad. (0%, 10, 20, 30, 40, 50%, 60, 70, 80, 90, 100%)
5. He introduces me as his son/daughter. (0%, 10, 20, 30, 40, 50%, 60, 70, 80, 90, 100%)
6. I introduce my mother and him as my parents. (0%, 10, 20, 30, 40, 50%, 60, 70, 80, 90, 100%)
7. He and I are just like father and son/daughter. (0%, 10, 20, 30, 40, 50%, 60, 70, 80, 90, 100%)
8. I introduce him as "my father" or "my dad." (0%, 10, 20, 30, 40, 50%, 60, 70, 80, 90, 100%)
9. I would feel comfortable if he and I were to attend a father-daughter/father-son function, such as a banquet, baseball game, or cookout, alone together. (0%, 10, 20, 30, 40, 50%, 60, 70, 80, 90, 100%)
10. I introduce him as "my mother's husband" or "my mother's partner." (0%, 10, 20, 30, 40, 50%, 60, 70, 80, 90, 100%)
11. When I think of my mother's house, I consider him and my mother to be parents to the same degree. (0%, 10, 20, 30, 40, 50%, 60, 70, 80, 90, 100%)
12. I consider him to be a father to me. (0%, 10, 20, 30, 40, 50%, 60, 70, 80, 90, 100%)
13. I address him by his first name. (0%, 10, 20, 30, 40, 50%, 60, 70, 80, 90, 100%)
14. If I were choosing a greeting card for him, the inclusion of the words *father* or *dad* in the inscription would prevent me from choosing the card. (0%, 10, 20, 30, 40, 50%, 60, 70, 80, 90, 100%)

Scoring

First, reverse the scores for items 10, 13, and 14. For example, if you circled a 90, change the number to 10; if you circled a 60, change the number to 40; and so on. Add the percentages and divide by 14. A 0 percent reflects that you do not regard your stepdad as your parent at all. A 100 percent reflects that you totally regard your stepdad as your Dad. The percentages between 0 percent and 100 percent show the gradations from no regard to total regard of your stepdad as your Dad.

Norms

Respondents in two studies (one in Canada and one in America) completed the scale. The numbers of respondents in the studies were 159 and 156, respectively, and the average score in the respective studies was 45.66 percent. Between 40 percent and 50 percent of both Canadians and Americans viewed their stepfather as their parent.

Source: Developed by Dr. Susan Gamache. 2000. Medical Centre, #119, 3195 Granville Street, Vancouver, B.C., Canada, V6H 3K2 sg@susangamache.com. Details on construction of the scale including validity and reliability are available from Dr. Gamache. The PSI Scale is used in this text with the permission of Dr. Gamache and may not be used otherwise (except as in class student exercises) without written permission.

Alzheimer's Quiz

	True	False	Don't Know
1. Alzheimer's disease can be contagious.	___	___	___
2. People will almost certainly get Alzheimer's disease if they live long enough.	___	___	___
3. Alzheimer's disease is a form of insanity.	___	___	___
4. Alzheimer's disease is a normal part of getting older, like gray hair or wrinkles.	___	___	___
5. There is no cure for Alzheimer's disease at present.	___	___	___
6. A person who has Alzheimer's disease will experience both mental and physical decline.	___	___	___
7. The primary symptom of Alzheimer's disease is memory loss.	___	___	___
8. Among people older than age 75, forgetfulness most likely indicates the beginning of Alzheimer's disease.	___	___	___
9. When the husband or wife of an older person dies, the surviving spouse may suffer from a kind of depression that looks like Alzheimer's disease.	___	___	___
10. Stuttering is an inevitable part of Alzheimer's disease.	___	___	___
11. An older man is more likely to develop Alzheimer's disease than an older woman.	___	___	___
12. Alzheimer's disease is usually fatal.	___	___	___
13. The vast majority of people suffering from Alzheimer's disease live in nursing homes.	___	___	___
14. Aluminum has been identified as a significant cause of Alzheimer's disease.	___	___	___
15. Alzheimer's disease can be diagnosed by a blood test.	___	___	___
16. Nursing-home expenses for Alzheimer's disease patients are covered by Medicare.	___	___	___
17. Medicine taken for high blood pressure can cause symptoms that look like Alzheimer's disease.	___	___	___

Answers: 1–4, 8, 10, 11, 13–16 False; remaining items True.

Source: Neal E. Cutler, Boettner/Gregg Professor of Financial Gerontology, Widener University. Originally published in 1987, in *Psychology Today*, 20th Anniversary Issue, "Life Flow: A Special Report—The Alzheimer's Quiz," *21*(5):89, 93. Reprinted with permission from *Psychology Today* magazine, © 1987 Sussex Publishers, LLC. The scale was completed by sixty-nine undergraduates at East Carolina University in 1998. Of the respondents, 40 percent identified less than 50 percent of the items correctly.